Cleaning with Solvents:

Methods and Machinery

John B Durkee II

ELSEVIER

Amsterdam • Boston • Heidelberg • London
New York • Oxford • Paris • San Diego
San Francisco • Singapore • Sydney • Tokyo

William Andrew is an imprint of Elsevier

William Andrew is an imprint of Elsevier
The Boulevard, Langford Lane, Kidlington, Oxford OX5 1GB, UK
225 Wyman Street, Waltham, MA 02451, USA

First edition 2014

British Library Cataloguing in Publication Data
A catalogue record for this book is available from the British Library

Library of Congress Cataloging-in-Publication Data
A catalog record for this book is availabe from the Library of Congress

ISBN: 978-0-323-22520-5

For information on all William Andrew publications
visit our web site at books.elsevier.com

Printed and bound by CPI Group (UK) Ltd, Croydon, CR0 4YY
14 15 16 17 18   10 9 8 7 6 5 4 3 2 1

Contents

Preface

In many shops into and through the early 1990s, solvent cleaning was done using 1,1,1-trichloroethane (TCA) or "Freon" (CFC-113), with little control of emissions. It was then discovered that emitted solvents catalyzed deterioration of the stratospheric ozone layer above Earth, and their manufacture was banned per the Montreal Protocol.

In that time coffee cans were often used as solvent storage tanks. Paint brushes acted as cleaning machines. Cosolvent machines weren't known. The emission control device was the open air.

Two decades later, solvent cleaning is done with and in vapor degreasers designed to replace the coffee can with a condensation system. Some machines recover condensed solvent at ambient pressure, some do so when operating under vacuum. Both liquid and vapor spray devices have replaced the paintbrush. And the open air has been replaced with a reusable bed of activated carbon adsorbent. These cleaning machines can be complex and very expensive.

This book was created to dissolve that complexity about cleaning equipment by means of straightforward explanations coordinated with simple illustrations and supplemented by informative footnotes and endnotes.

Despite advances in written and electronic communications, I believe that verbal communication is still how most technical information about cleaning technology – at any level, gross, precision, or critical – is transmitted. Too often, technology transfer through publication of scientific papers isn't intended to educate, but to impress.

In the 1980s, I was assigned to develop and manage new cleaning technology for a large chemical company. I learned about the equipment technology through conversations, with fellow professionals, suppliers, customers, operators and technicians, anyone. And, of course, I learned by using the equipment! There were no academic offerings, no books, no trade magazines or professional journals. My cohorts reported similar experiences.

As a consultant, I found my clients had primarily learned about equipment from suppliers[1]. As someone providing training in cleaning technology, I found the previous experience of my students originated from the same source and methods as had my clients: chiefly, verbal contact with suppliers. Essentially, this technology transfer was done by personal contact.

Is this how we are going to train future practitioners of industrial cleaning?

This book is an answer to that question. The answer is NO!

[1] Some feel there is something inherently wrong about learning from suppliers of cleaning equipment. I see the situation as a bargain. On the one hand, suppliers know their technology and the pitfalls of it use better than anyone, and should have the capability to communicate it users. On the other hand, their point of view isn't, and shouldn't be, balanced and open about alternatives. On the third hand, suppliers know applications involving their equipment; from where is one to learn of basic theory of how it works?

Acknowledgments

The genesis of this book came about in 2011 during completion of the broad-based and intensely detailed book *Cleaning with Solvents: Science and Technology* (ISBN-13: 978-1455731312). It was the realization that only a portion of the story was told in that book – that to clean with solvents one had to have and use equipment, and that story hadn't been told.

To this author, it was evident that the Science and Technology book was but a down payment on a charter to provide a full explanation of all aspects of solvent cleaning. It was also evident that global interest in safe, environmentally responsible, simple, and reliable solvent cleaning technology was growing. This book is an outcome of those beliefs.

A book may bear the name of a single author on its spine. But no published book – especially this one — is the product of just one person. The persons noted below have contributed to this book.

My chief critic, advisor, and reviewer of the this book was and is Steven Abbott, Ph.D., University of Leeds, U.K. In no small measure is the quality inherent in this book due to his help; the defects are my responsibility. His unique ability to review and suggest revisions to the chemistry science I am trying to communicate and simultaneously be rigorous on grammar has never been equaled.

Friends, advisors, teachers, and reviewers included Bob Benedetti, National Fire Protection Association, who as in 2006 reviewed my writing on management of flammability; Art Gillman, Unique Equipment; and Joe McChesney, Control System Designs LLC. Your support was needed and is appreciated.

Supporters, teachers, and advisors from whom I have benefitted in the preparation of this book include (in alphabetical order):

Jeff Beard, Honeywell Performance Products
Bill Breaut, Petroferm
Erik Finger, Ph.D., Serec Corporation
Peter Gebhard, IMPCO
Don Gray, Ph.D., Vacuum Processing Systems
Troy Hurley, Finish Thompson
Edward Kallis, Vapor Engineering
Ken Kikta, Carbtrol
Bill McCormick, Serec Corporation
Ken Schaeffer, Carbon Resources
Joe Schuttert, Vacuum Processing Systems.

The production of this book was complicated, so thanks to Matthew Deans, Senior Publisher, Elsevier Science & Technology Books, who exhibited the inordinate patience needed for me to complete this research and write about it, and then to enable its printing in color. Thanks also to David Jackson, Associate Acquisitions Editor, at Elsevier, who provided wise counsel every time there was a choice to be made.

Perhaps the two most important persons who contributed to the production of this book are Pauline Wilkinson, Book Production Project Manager at Elsevier, who made insoluble production problems disappear with dispatch, and John Hutchison, esteemed proofreader, who found and corrected a seemingly endless stream of errors that were formerly unknown to all.

The one to whom I owe the most is my wife, Dorothy, who gave me the support I needed to complete the long journey of experiences and learning which produced this book in a short amount of time. Thank you, Dorothy. Know you're needed, loved, and cherished.

Disclaimer

This book is replete with specific and general recommendations about the technology within. The author is a consultant; that is what consultants do – make recommendations. But one learns as a consultant to never make any recommendations to clients without knowing, from a first-hand evaluation, their local circumstances.

That's why all the recommendations made in this book (about solvent selection or replacement, hazard management, material selection, management approaches, regulatory compliance, purchasing methods, etc.) are made without any knowledge of local circumstances and certainly without first-hand evaluation.

ALL recommendations made in this book are based on the technology being described, and should in no case be adopted without evaluation of the local circumstances.

What You Can Do with This Book

This is a book about the engineering (and perhaps chemistry) of cleaning surfaces with solvents. It is a book based on the disciplines of engineering, and the experiences of this author and coworkers.

If the guidance found in this book is followed to a moderate degree, the user will have the information and basis to:

- *Make good decisions* about selection of equipment and facilities to be both effective in dissolving soils from surfaces, and protecting the personnel who use them as well as the environment into which they are used, and to
- *Have knowledge necessary to implement those decisions.*

Study within this book will lead to an understanding of:

- Why parts aren't effectively dried.
- Why vacuum drawdown takes too much time.
- Why vacuum capability isn't used for cleaning.
- Why rinsing in a soy/HFE cosolvent machine leaves residues.
- Which combinations of solvents shouldn't be combined in a cosolvent machine.
- Why costs of cleanup of "tailpipe emissions" of solvent by adsorption can be excessive.
- Which vacuum pumps offer the least potential for failure.
- Why one shouldn't use lip vents.
- Why open-top machines are normally sited in a pit.
- How parts should be organized in a cleaning basket.
- How air amplifiers may be the most valuable part of a solvent cleaning machine.

A Note on Organization

The content of this book is arranged in three layers. The top layer is an illustrated story. As a writer about technology and its use, I strive to communicate as a storyteller.

The story is always about the technology: the reasons for its use, or methods by which it was created, the science underlying it, the engineering by which it is applied, and the expected outcomes produced by it. A good story has a beginning (that's the reasons and methods), a middle (that's the science and engineering), and an end (that's the outcomes). A successful teller of stories uses images to explain, and perhaps augment. Most of the chapters in this book were prepared by first developing the story through technical research, then planning the sequence of images which would tell the story without text, and finally writing the text. *One may read the main text (the story) and understand all that was intended.*

The second layer is within the footnotes (and boxes), and is found on the page of the citation. This material contains information not essential to the story, but of interest to some readers. The second layer contains internet references for sources which are thought to be useful. These may be reliable monographs published by suppliers, reports published by government agencies, websites maintained by academic sources who provide either additional or background information or relevant opinions, the technical basis for cited calculations, a chronology of events, restatement in different language of points made in the story, background about persons who contributed to the development of the chemical or physical technology used in solvent cleaning, or recognition of sources for images not prepared by this author. Of course, the problem with citation of internet references is that their existence and URL identification, while current as of this writing, is not necessarily permanent.

The third layer is within the endnotes found at the end of each chapter. These are conventional literature references, such as citations of journal articles or books. The most complete citation available is provided; if page or volume or issue numbers aren't given, it's because they couldn't be found.

Substantial detail in the endnotes includes details of calculational methods, bases for cost estimations, basic engineering information needed to implement solvent cleaning technology, supplementary discussion about topics covered in the main chapters, background about environmental regulations, and results of specific evaluations conducted by the author.

This book is a collection of interrelated, intact, and intensely cross-referenced chapters. Together they tell a story about the use of solvents in cleaning operations. But they do not comprise a story where each chapter must be read in sequence to understand the story. One can start their understanding of the story anywhere in this book.

Units Used in This Book

It would be easy to write that the units used in this book are those commonly used in the country in which this book was written (the United States), or to declare adherence to a specific system of units (English, SI, International, natural, etc.) because that system is believed to be the one most globally recognized or most frequently used or more based in science. But not all readers will be located in the U.S., nor will all have declared allegiance to one specific system of units. Today the span of any activity is both global in reach and compartmentalized by discipline. So it is not so easy to specify what system of units should be used in a book to be used in any country.

Instead, this book uses the units most commonly associated with the parameter being noted. Said another way, *the units used in this book are those recognized and used by readers actually doing solvent cleaning.*

Those persons may be familiar with or educated in chemistry or physics; chemical, mechanical, or environmental engineering; industrial hygiene, industrial safety, or something else. In other words, those practicing solvent cleaning will have many and various preferred systems of units. Consequently, the system of units used in this book is mixed: The flammability classifications of the globally-recognized National Fire Protection Association (NFPA), for example, are temperatures scaled in degrees Fahrenheit – although they could be based on scales of degrees Celsius (Centigrade), Rankine, or Kelvin. So where flammability classifications based on temperature are given in this book they will be denominated in degrees Fahrenheit (°F). But the default unit used in this book for scaling of temperature is degrees Centigrade (°C). For convenience, a temperature may be noted in both scales, °C and °F.

The default unit used in this book for scaling of positive pressure is the pound force per square inch (psi). Likewise, the default unit used in this book for scaling of negative pressure is the Torricelli of vacuum (TorrV)[1].

The default unit used in this book for scaling of energy is the BTU and, the default unit used in this book for scaling of area is the square foot (SF), So the default unit used in this book for scaling of length is the foot. For convenience, smaller amounts of length may be noted in other scales, feet, inches, mils; metres, or cm; Angstroms or nm.

The unit most commonly associated with scaling of solvent density[2] is the gram per cubic centimeter (g/cc)[3]. But the default unit used in this book for scaling of mass is the pound (lb). For convenience, often smaller amounts of mass will be noted in other scales, grams (g), or ounces (oz).

The default unit used in this book for scaling of volume is the cubic foot (CF). For convenience, often smaller amounts of volume will be noted in other scales, cubic centimeters (cc).

The unit most commonly associated by chemists with molecular weight is the gram mole. So the default unit used in this book for molar volume is the cubic centimeter per gram mole.

[1] Often noted elsewhere as millimeters of Mercury or mm Hg, 1 Torr is equivalent in pressure to 1 mm Hg. When levels of pressure are denominated by either mm Hg or Torr, 0 values of each are ambient pressure (not absolute pressure). When levels of vacuum are denominated in mm Hg, the 0 value is total vacuum. When levels of vacuum are denominated in Torr, the 0 value is ambient pressure. In other words, total vacuum may be expressed as 760 TorrV or 0 mm Hg; ambient pressure may be expressed as 0 TorrV or 760 mm Hg.

[2] Most ignore the fact that values of density denominated in g/cc are actually values of specific gravity: density relative to that of water.

[3] Often noted elsewhere as kilograms per cubic metre.

Finally, the unit most commonly associated with a solubility parameter is the square root of a pressure denominated in megapascals, or $MPa^{-1/2}$[4,5]. The unit most commonly associated by suppliers and users with surface tension of solvents is surface energy per area[6] denominated in dyne/cm[7], or dynes.

To summarize, units used in this book are chosen for recognition by those involved with solvent cleaning, not by those inhabiting a particular country, subscribing to a certain system, or educated in a specific discipline. The reason for this deliberate, if unorthodox, choice is to make this book more useful to those involved with cleaning of surfaces with solvents.

Units used in these book include the following:

Absolute temperature values	°K
Density	g/cc
Dipole moment	Debye
Energy	BTU
Flash point temperature values	°F
General temperature values	°C
Length	foot (ft)
Mass	pound (lb)
Molar volume	cc/g mole
Negative pressure	torrV
Positive pressure	pound force per square inch (psi)
Power	kW[8]
Refractive index[9]	dimensionless
Solubility parameter	$MPa^{-1/2}$
Surface tension	dyne/cm
Universal gas constant	0.00788 BTU/g mole-°K (1.98588 cal/g mole-°K, or 8.31447 J/g mole-°K)
Volume	cc

Common Unit Conversions

1 BTU	= 0.25199 calories (cal)	= 1,055.05 Joules (J)	= 1.05505 kiloJoules (kJ)
1 cc	= 3.5315E−005 cubic feet (CF)		
1 dyne/cm	= 1 milliNewton/metre (mN/m)	= 0.001 kg/sec^{-2}	= 6.852×10^{-5} lb force/ft
1 centipoise	= 0.001 Pascal-sec	= 0.001 kg/metre-sec	= 2.0885×10^{-5} lb force-sec/SF
1 g/cc	= 1000 $kg/metre^3$		

[4] While often noted elsewhere as $mPa^{-1/2}$ for other purposes, the nomenclature $MPa^{-1/2}$ is most commonly used to denominate solubility parameters.

[5] Solubility parameters may also be scaled in units of square root of pressure scaled in calories per cc, or $(cal/cc)^{-1/2}$. To get values of $MPa^{-1/2}$, one multiplies values of $(cal/cc)^{-1/2}$ by 2.0455. Pressure is also energy divided by volume.

[6] An equivalent representation for surface tension is force per length.

[7] An equivalent unit for surface tension is the mN/m (milli-Newton per metre).

[8] Systems of units are further confused by the conventions about capitalization associated with the prefixes of units. The prefix mega (1,000,000) is noted as M, while the prefix kilo (1,000) is noted as k, the prefix milli (0.001) is noted as m, and the prefix nano (0.000000001) is noted as n.

[9] Measured at the yellow doublet sodium D-line, with a wavelength of 589 nm, as is conventionally done.

(continued)

Common Unit Conversions

1 foot	= 30.48 cm	
1 lb	= 453.6 gram mass (g)	
1 inch	= 2.54 centimeter (cm)	
1 mil	= 0.001 inch	= 0.00254 cm
1 KW	= 3,412 BTU/hr	= 238.84 cal/sec
Temp, °C	= (1.8 × Temp) + 32 (°F)	

For those who prefer consistency of units, there is a simple, elegant, and free solution. One may download and use the freeware program Master Converter, available from Savard Software (http://www.savardsoftware.com/masterconverter/). It is pre-configured to convert over 800 units in 42 categories, and can be configured to support any other unit in any other category.

External References Cited in This Book

This book is written to inform, to support ongoing or planned work with solvent cleaning systems, enable selection of the appropriate cleaning solvent, control safe and economical operations, and explain how desired results may be obtained if they are not originally produced. One hopes it may also be enjoyed.

This book is not intended to stand alone as a reference or management tool for all cleaning operations. This book is about solvent cleaning. It refers to and is supported by information available in these other sources.

Other corollary references by this author support other aspects of surface cleaning:

Management of Industrial Cleaning Technology and Processes, published by Elsevier (ISBN: 978-0080448886), 2006. Coverage includes:

How all modern cleaning technologies work and should be used
U.S. and global environmental regulations
Health and safety hazards associated with all cleaning agents
Control of industrial cleaning processes
Testing for cleanliness
Equipment used in cleaning, and their components
Statistical procedures for management of cleaning (or other) operations
Description of analytical procedures for cleanliness testing
Challenging situations in critical, precision, and industrial cleaning

In this book, the *Management* book is cited as needed to support discussions. It is always referred to as **Ref 1.**

Cleaning with Solvents: Science and Technology, published by Elsevier (ISBN: 978-1455731312), 2014. Coverage includes:

Design of solvents for specific purposes
Solubility parameters, including those by Hansen, Hoy, and many others
Use of solubility parameters to select solvents to clean specific soils
Solvent substitution to achieve a variety of goals
Toxicology of cleaning solvents
Technology and politics of VOC exemptions (in the U.S.)
Wipe cleaning with solvents
Economics of solvent use including the environmental regulations which affect cost of use
Cleaning with solvent aerosols Stabilization of solvents Q & A about use of solvents in cleaning operations

This book is supported by a coterie of appendices which describe the evolution and implementation of the various systems of solubility parameters. Also included as appendices in the book are: a database of properties about 210 cleaning solvents, a database of recommended glove and sealing materials useful for each of many cleaning solvents, description of blend rules for solvents and examples of their use, a treatise on the tropospheric reaction chemistry of VOC solvents, and a description and example usage of the Ohnesorge Number.

External References Cited in This Book

In this book, the *Science and Technology* book is cited as needed to support discussions. It is always referred to as **Ref 2.**

Handbook of Cleaning Solvents (in preparation). Coverage includes, for each type of solvent:

Its origin and synthesis
Key applications where they or similar materials might be found as components of soil materials
Analysis of any unique naming system
Properties compared to those of other solvents
Properties of specific solvents compared to those of others of the same type
Significant hazards such as flammability and how and when to avoid them
Significant environmental regulations which affect use
Values of solubility parameters
Comparison to specific soils for which use as a cleaning solvent is or is not recommended
And how their specific structure determines their specific properties.
A database of binary azeotropes is also included.

In this book, the *Handbook* is cited as needed to support discussions. It is always referred to as **Ref 3.**

Economics of Cleaning with Solvents – Bases, Outcomes, and Comparisons (in preparation). Coverage includes:

The basis for any analysis of cost – the book-kept cost sheet
Conventional, and unconventional assumptions made to support any analysis of cost
Costs for operating an open-top vapor degreaser, by solvent type
Costs for operating an enclosed vapor degreaser, by solvent type
Costs for operating a Type I cosolvent machine
Costs for operating a Type IIA or B cosolvent machine
Costs for operating a Type III cosolvent machine
Costs for operating a Type IV cosolvent machine
Costs for operating an adsorption system for scrubbing solvent vapor from air emissions
Total cost for operating all solvent cleaning machines including adsorption treatment of emitted air
Analysis of operating cost for specific solvent cleaning tasks

In this book, the *Economics* book is cited as needed to support discussions. It is always referred to as **Ref 5.**

The author is also planning a series of Digital documents, which will include:

- *The Complete Naming System for Fluorinated Solvents*
- *Cleaning Test Results with Plant-Based Solvents*

Chapter | 1 |

Open-Top Cleaning Equipment for Vapor Degreasing

In references[A] about solvent cleaning, the focus tends to be all about solvents and not about the cleaning machines in which they are used, and the processes in which they are implemented within those machines.

This book is that content. It's about cleaning machines[1] *and* processes so that the various cleaning solvents described in Ref. 2 and Ref. 3 can be successfully used to clean parts.

Figure 1.1[B] is an image of an unusually large vapor degreaser, and Figure 1.2[C] is an image of an abnormally small vapor degreaser.

Figure 1.2 Very Small Open-top Vapor Degreaser

Figure 1.1 Large Open-top Vapor Degreaser

Organization of this book starts with the identification of five common types of cleaning machine. These are differentiated less by the cleaning process being implemented in each but more by the way in which each machine controls emissions of solvent. Different cleaning machines may implement the same cleaning processes, but control emissions in different ways.

A cleaning process is a sequence of steps. Although cleaning processes may be lengthy of time or require repetitive steps under different operating conditions, a cleaning process is usually composed of only three different steps. Each of these steps is described and illustrated in detail.

And there is one common function which every cleaning machine must complete or its owner will become bankrupt, or incarcerated. That is recovery of solvent after use in cleaning work. The equipment necessary to complete that function is also described and illustrated in detail.

Then detail is provided about each of the five types of cleaning machine.

[1] This topic is covered in considerable detail in Ref. 1, Chapter 7, pages 339 to 390; with Section 7.14, pages 390 to 395 covering open-top vapor degreasers.

In that work the focus was on the components from which cleaning machines are assembled. In this chapter the focus is on the cleaning process being implemented by the cleaning machines.

http://dx.doi.org/10.1016/B978-0-323-22520-5.00001-0

1.1 THE CLEANING MACHINES

Solvent cleaning can be done in at least five types of equipment (machines):

- *"Open"*[2] *tanks* in which the solvent is usually not heated[3], and so the cleaning process is known as cold cleaning.
- *"Open-Top"*[4] vapor degreasers (OTVDs) in which the solvent is heated to its normal[D] boiling point, and so the cleaning process is known as vapor degreasing.
- *"Airless"* enclosed degreasers (machines) which operate at ambient pressure and at the solvent's normal boiling point, in which the first processing action is to remove the air and replace it with nitrogen, while the last processing action is to reverse the former action.
- *Vacuum* vapor degreasing (VVDG) in which the pressure in an enclosed degreaser is reduced by removal of vapor so that emissions may be best contained and the boiling point of the solvent may be reduced.
- *Cosolvent* cleaning in which two solvents are used (vs. only one in the above four machines). The purpose of the twin solvent aggregation is to reduce concerns about VOC emissions and flammability, while using a cleaning solvent which by itself does raise both concerns.

Each of these five types of solvent cleaning machine will be covered separately in this book. The focus for each type of machine will be on: (1) how they do what they do; (2) the type(s) of cleaning and drying processes each machine can complete, how those should be done, and why they might be selected; (3) the lowest exposure limit each machine can be expected to meet for several solvents of interest; and (4) an expectation for the future commercial vitality of each type of machine.

1.2 THE CLEANING PROCESSES/METHODS

Note is made in this book of both cleaning machines and cleaning processes. The two words are not synonyms of one another. Machines are the facilities in which cleaning work is done. Processes are the methods used within cleaning machines to do the cleaning work. Cleaning methods are selected steps of operation of which the cleaning work is comprised. The three steps in cleaning processes noted in this book are:

- *Immersion in liquid.* This process step can be practiced in any of the five types of machines noted above. Mechanically driven agitation may be involved.

Depending upon the type of machine selected, the liquid sump (pool or bath) into which the parts are immersed may be at ambient temperature, normal boiling temperature, or a superheated temperature; and also the liquid may be at ambient pressure, reduced pressure, or increased pressure.

Within a cleaning process, the immersion sump may be alternately heated and cooled, or the reverse. There are frequently more than one immersion sump within a cleaning machine. For example, in one immersion sump cleaning with boiling solvent is done, and the solvent becomes soiled. In the other, adjacent immersion sump, rinsing of part surfaces to remove soiled solvent is done with externally supplied purified solvent. A weir separates the two sumps[5]. This is shown in Figure 1.3[E].

- *Vapor contact.* This process step, which can be practiced in enclosed machines, and to an extent in OTVDs, is a multifaceted action. In an enclosed machine, the vapor is sprayed onto parts. In an OTVD, the vapor exists in a transparent layer above the two immersion sumps from which the solvent is evaporated.
- The vapor zone in an OTVD is maintained by the forces of buoyancy and the unit operation of heat transfer.
- Vapor produced by evaporation from the immersion sumps rises because of buoyancy[6].
- That vapor is cooled as it passes through a zone bounded by cold thermal coils around the walls of the degreaser. At and near the cooler walls, solvent condenses into droplets, some of which are collected in a trough (channel) beneath the cold heat transfer coils. External to the degreaser, liquid from the trough drains into a receiver, from which it is recycled to the rinse sump (perhaps through a water separator).
- Above the cold heat transfer coils, cooled solvent vapor is heavier than the ambient air above it, and no longer rises to penetrate the air.
- Hot rising vapor forms a cloud (layer), which should fill the volume between the cooling coils, and then is cooled by them. Within the cloud, some vapor is condensed. Those liquid drops "rain" down back to the sumps.

---

[2] The adjective "open" is a commonly used misnomer. Only if the solvent has a highly elevated boiling point, so that emission of VOCs is within acceptable limits, are these tanks truly open to the ambient atmosphere. VOC solvents (in the US) such as glymes, some other glycol ethers (Ref. 3, Chapter 14), or methyl esters of dibasic (Ref. 3, Chapter 19) or soy-derived acids (Ref. 3, Chapter 6) might be used in this manner. Otherwise, the "open" tank might well be an OTVD in which no heat is applied.

[3] Actually, the description cold cleaning refers to solvent cleaning done at any temperature short of the normal boiling point. So, cold cleaning may include situations where some heat is applied.

[4] The adjective "open" is certainly a misnomer in the context of an OTVD. These machines are always constructed with some type of movable cover which is only removed as parts baskets are inserted into or removed from the machine.

[5] The weir is sized and positioned so that cleaner rinse fluid overflows into the less clean cleaning sump. Else one would be adding soil to one's rinse fluid – defeating the point of rinsing the parts.

[6] The word buoyancy simply means that something is lighter than something else. In this case, vapor is less dense than liquid.

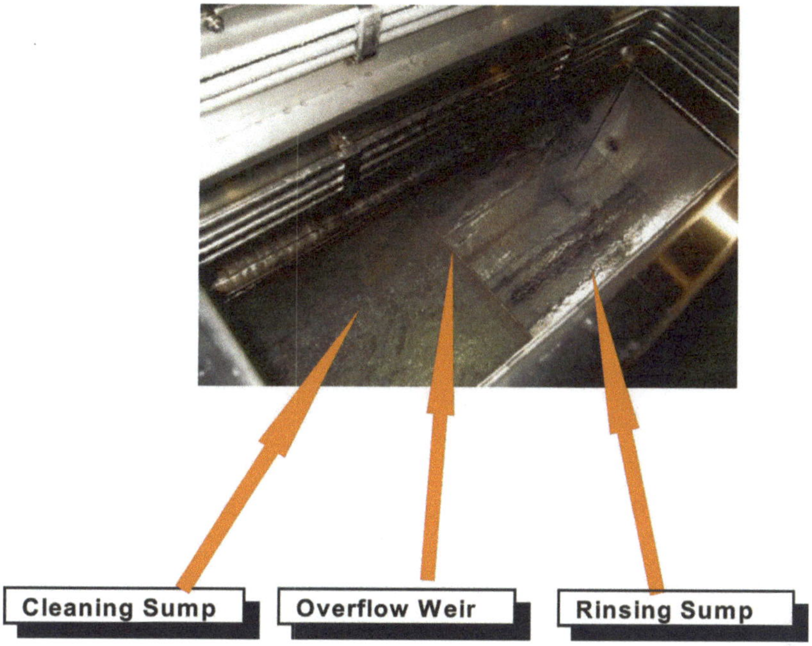

Figure 1.3 Internal View of Two-Sump Open-top Vapor Degreaser – Showing Arrangement of Sumps

- The outcome of this force balance is a metastable and movable zone (cloud) of vapor which exists above the liquid immersion sumps and below the top cover of the degreaser.
- At locations far from the walls where the cold heat transfer coils are located, the solvent vapor is less cooled. So, it can continue to rise further. The result is that the shape of the vapor cloud is bulged upward at its center – farthest from the walls. Figure 1.4 is a diagram of this situation.
- Naturally, this vapor cloud is generally[7] transparent, though its effects are not. See Figure 1.5[8] for a photograph of the equipment and conditions shown in the diagram of Figure 1.4.

[7] This is not always so. The presence of water vapor can affect the transparency of the vapor phase, making it appear cloudy. Some refer to this as "ghost vapor." The source of the water is not the normal intrusion of tramp impurities (Chapter 3.6.11, Footnote 57), rather it is water released as water-soluble coolants and lubricants are cleaned as soils from surfaces. In a short time, this water is vaporized, condensed, and removed in the water separator (Chapter 3.6.11, Figures 4.40 and 4.41), and the vapor phase returns to being transparent.

[8] The image of Figure 1.5 shares the same source as Figure 1.1. Boiling fluid can clearly be noted in the cleaning sump.

The unit pictured in this image has a second set of cooling coils, above the ones shown in the diagram of Figure 1.4. This upper set is of a different kind, and serves a different purpose than the lower set. The upper set serves the purpose of reducing solvent emissions in order to meet a local environmental or health regulation. The heat transfer fluid is not water, cooled from a cooling tower; rather it is a water glycol solution (called brine), cooled by refrigeration. Details are given in Chapter 1.19.

Hot solvent vapor can heat part surfaces, evaporating whatever liquid solvent was wetting those surfaces given sufficient exposure time. In this way, parts are rapidly or gradually dried.

Or if the part surfaces are cooler than the boiling point of the vapor, the vapor will condense – wetting/rinsing the surface with distilled (pristine) solvent.

And vapor can diffuse more readily through porous materials, thereby wetting them when the vapor condenses.

- *Liquid spray with force.* Instead of being sprayed with vapor, parts may be sprayed with cold liquid to rinse them of cleaning solvent and soil, or hot liquid to complete a pre-cleaning operation. This step is in addition to liquid sprays under immersion.

A cleaning process is a selection of steps from the above short list, a sequence (order) in which they are to be used, a schedule of durations that each will be implemented, and a specification of what operating conditions will be maintained in each step.

Some processes will have iterations of these steps, such as: immersion in hot solvent within the cleaning sump, immersion in cold(er) solvent within the rinse sump, vapor contact to act as a rinse above both sumps, followed by a repeat(s) of these same three steps (Figure 1.9).

If the wrong process steps are chosen, the parts won't be made clean even if the most expensive cleaning machine or longest cycle time is used. If the wrong cleaning machine is chosen, it may not be possible to implement the chosen process steps.

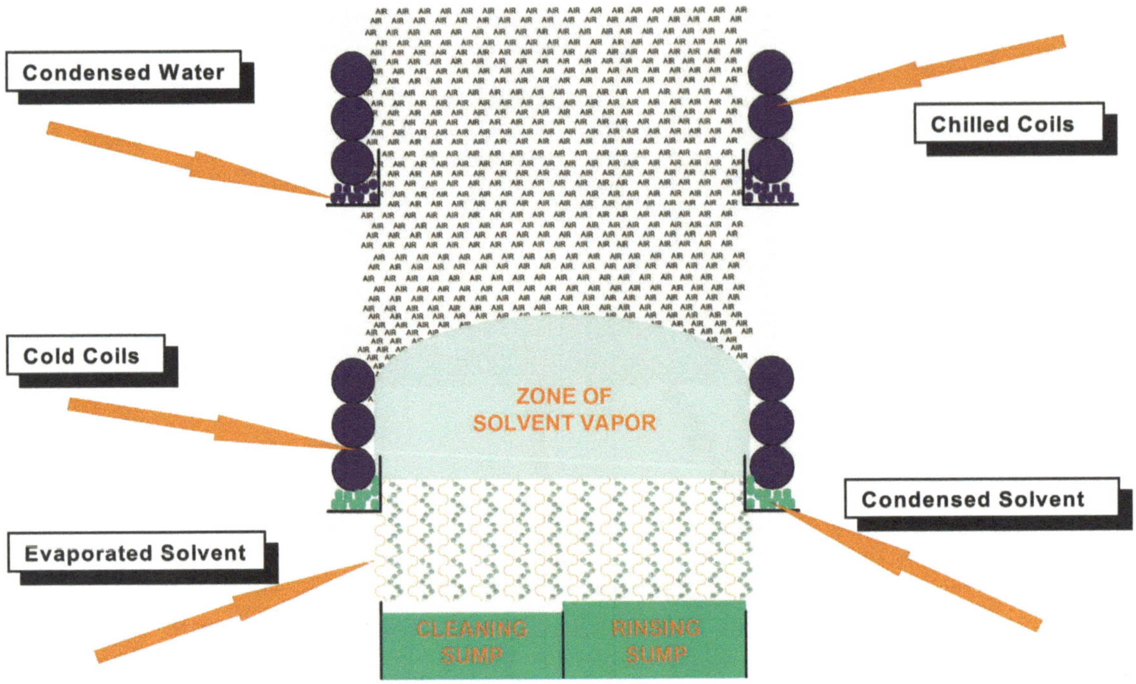

Figure 1.4 Idealized View of Arrangement of Typical Components of Vapor Degreaser

Figure 1.5 Internal View of Open-top Vapor Degreaser – Showing Cooling Coils

Not a process step, but movement of a basket of parts by a hoist is essential. This is the lynchpin of any cleaning process. It moves the parts in space and in time from one process step to another. It inserts and removes them into and from the equipment, and holds them within the equipment at the required position for the required time so the planned process step can be completed. This material handling action provides the most controlled and efficient operation.

1.3 A BASIC PRINCIPLE ABOUT CLEANING WITH SOLVENTS

The principle is simple. It is that *all* solvent gets dirty doing cleaning work.

This principle has two aspects. The most important aspect is that in a solvent cleaning process, all of the solvent has dissolved some of the soil.

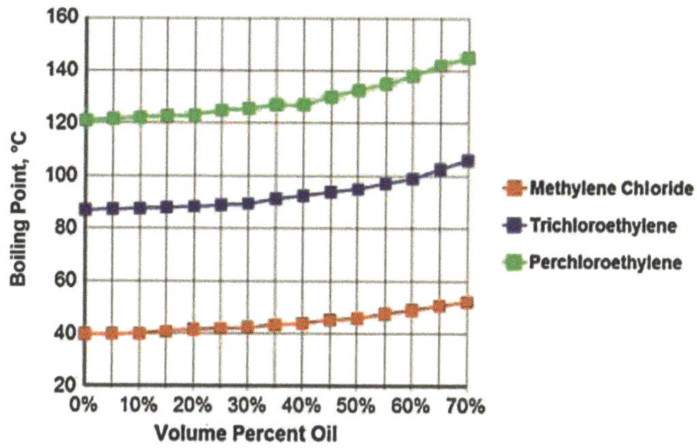

Figure 1.6 Effect of Oil Accumulation on Solution Boiling Point

- This means that in a solvent cleaning process the parts become clean(er) of soil, and *all* the solvent becomes more dirty with soil. There is no soil-free ("clean") solvent within a solvent cleaning process, unless it is produced externally to the cleaning sumps.

Every solvent cleaning process (being implemented in any of the five cleaning machines above) has three additional tasks to perform in addition to removing the soil from a surface by dissolving it into a solvent. They are to: (1) refine the soiled solvent for reuse as a flush or rinse fluid, (2) concentrate the soil removed from parts into a small amount of solvent, and (3) remove that concentrate of soil from the cleaning machine so the concentrate may be disposed of properly[F].

The second aspect of the principle is that the solvent is changed in the process of doing the cleaning work. It is made less pure.

- So the physical properties of the soiled solvent become different than those of the pure solvent.

This second aspect is absolutely important to management of solvent cleaning. It controls cleaning quality, because one can't produce clean parts from dirty solvent!

1.3.1 Effect on Properties

The solvent/soil solution has different properties (boiling point, density, viscosity, solubility parameters, etc.) than one would expect from knowing the values for the pure solvent[G]. This is the second aspect of the basic principle.

Data[H] for three chlorinated solvents show this effect, in Figure 1.6 (boiling point) and Figure 1.7 (density).

This outcome is critical in solvent cleaning technology[9]. Simply, it controls cleaning quality.

If soiled parts are contacted with soil-laden solvent, the wet parts can be no more free of soil (i.e. more clean) than the amount of soil which is dissolved in the cleaning solvent[I].

Suppose the cleaning solvent has been used to where it contains 20% mineral oil by volume, and will be used to "clean" the next part. The fluid remaining on that part after a single stage of cleaning will contain 20% oil. When the volatile solvent is evaporated from the part, the non-volatile mineral oil remains behind. So, in passing that part through the solvent cleaning process, did the part become more clean or more dirty?

As is shown in Figure 1.6, mineral oil has only a slight[10], and probably non-detectable, effect on the boiling temperature in the degreaser at the 20% level of mineral oil[11]. However, as Figure 1.7 shows, that outcome can be detected by measuring the density of the solvent active in the cleaning machine. Refractive index can also be used to the same effect.

---

9 The counterpart phenomenon is probably more critical in aqueous cleaning technology.

10 This is not always so. As Figure 1.6 shows, accumulation of oil into halogenated solvents only slightly elevates the normal boiling point until a substantial amount of oil has been accumulated. When a substantial amount of oil has been accumulated in the cleaning solvent, the parts may have been made cleaner, but they are certainly not clean.

But other soil materials dissolved into other solvents may elevate the mixture's boiling point to a greater degree. The extent of elevation of the boiling point of an ideal mixture containing only a non-volatile soil and a pure solvent can be calculated from the solvent's ebullioscopic constant and the molar concentration of the soil (and not the nature of the soil). Values of this constant may be found in the CRC Handbook of chemistry and Physics, Vol. 15. Larger values of this constant mean a larger elevation of boiling point. Solvents with large values of the ebullioscopic constant are decane (6.10), perchloroethylene (6.18), n-octanol (5.06), trichloroethylene (4.52), *o*-xylene (4.25), heptane (3.62), toluene (3.40), and methylene chloride (2.42). Highly polar compounds such as water (0.51), methanol (0.86), ethanol (1.23), and isopropanol (1.58) have lower values which mean that the same concentration of soil will less elevate the mixture boiling point.

11 This means that one can't manage the cleanliness of the cleaning bath, and the parts within it, by management of the bath's boiling temperature.

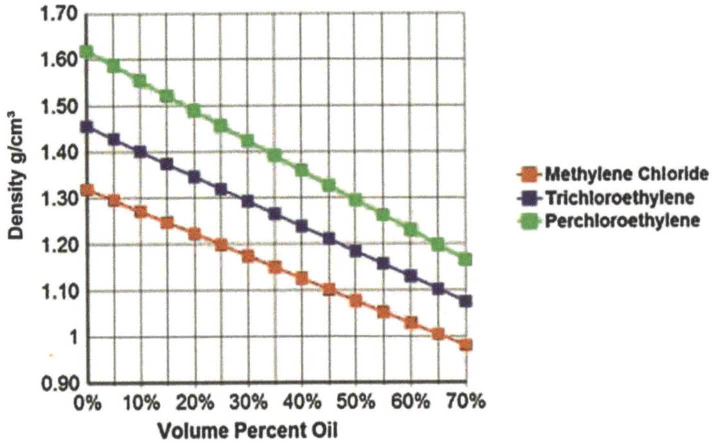

Figure 1.7 Effect of Oil Accumulation on Solution Density

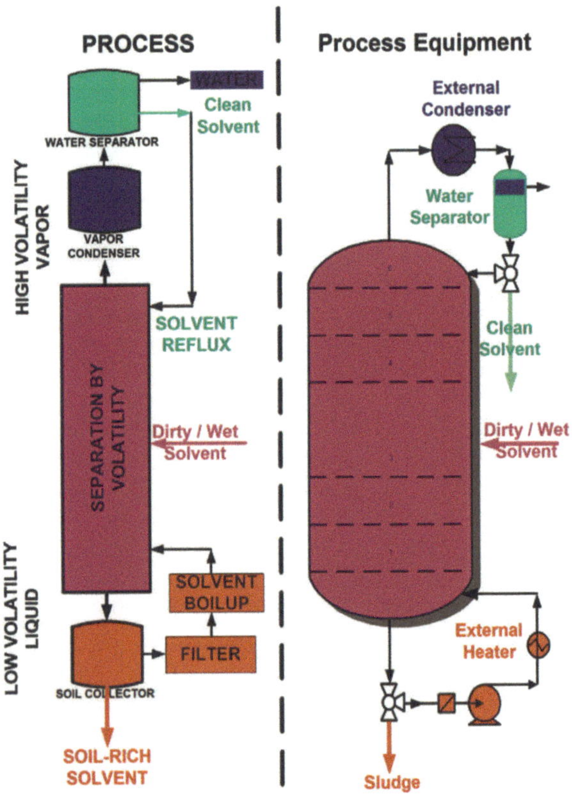

Figure 1.8 Process Elements, and the Equipment to Implement Them, for Solvent/Soil Management

In all solvent cleaning processes described below, cleanliness must be maintained by adequately removing soil via the process and equipment shown in Figure 1.8, and must be also managed in any other solvent cleaning processes. Simply, soil must be continuously removed at least at the rate it enters the cleaning machine on the parts, or it will accumulate over time, and so will dirty parts!

These are the two chief outcomes recognized as necessarily following the first principle of solvent cleaning processes.

1.4 SUITABLE PROCESS DESIGN

Not often said, but often true – it is generally not difficult to remove soil from parts. The necessary and more difficult task is to remove soil, first from the solvent, and then from the cleaning machine.

- That's what a cleaning machine does, it implements a designed process – a series of steps which together manage soil[J] concentration and location. That process must be designed to get the soil off the parts; and then concentrate it for efficient disposal, effect that disposal, and refine clean solvent so the process can be repeated[12]. This is the first aspect of the basic principle above.
- But it must do so while coping with the second aspect of the basic principle noted above – meaning that all the solvent around the parts is dirty, and any surface contacted by dirty solvent will become dirty.

So, any cleaning process (aqueous or solvent) which fails to account for the above principle will be unsuitable and fail for its users – no matter what equipment is involved.

[12] In general, the same actions are necessary for a useful process based on aqueous cleaning technology.

1.5 PROCESS DESIGN FOR SOLVENT CLEANUP (RECOVERY)

A generalized soil/solvent management process is shown in Figure 1.8. It may be (and should be) practiced with any (or all) of the five types of equipment above:

- As one of the modules integral to the solvent cleaning machine (Ref. 2, Figures 7.27 and 7.34), or
- Externally to the solvent cleaning machine but on site (Ref. 2, Figures 7.29 to 7.31), or
- Externally by shipping the soil/soil mixture to a toll processor (Ref. 2, Chapters 7.15 and 7.16).

Conventionally, this generalized process for soil management is embodied in equipment such as an internal or external distillation column, and is usually accompanied by a filter[K], a condenser, and a decanter.

Both the process for managing soil and solvent, and its required equipment, are shown in parallel in Figure 1.8 – the process modules in "blocks" on the left; the process equipment in conventional symbols for each equipment component on the right. The same color coding is used in between both presentations.

- The main process element (shown in purple) separates fluids (solvent from soils) on the basis of differential volatility.
- The secondary process element (shown in red) separates particulate solids from fluids on the basis of size-exclusion.
- The tertiary process element (shown in light blue) separates water from solvent on the basis of immiscibility (Ref. 2, Chapters 11.8.1 and 11.8.2).

When the soil-laden/wet solvent is boiled in the primary separator:

- Higher volatility materials (solvent and water) rise upward in the separator – because vapor is less dense than liquid,
- Lower volatility materials (soils) flow downward in the separator – because liquid is more dense than vapor.

In this straightforward way, the soils are separated from the solvent – producing soil-free solvent which contains water (at the top of the separator), and soil-rich solvent which may contain solids (at the bottom of the separator).

If the soils are relatively non-volatile[13] relative to the solvent, essentially all of the soil is discarded[14] from the bottom of the separator in a soil-rich sludge[15].

Some liquid solvent is filtered to remove particulate and then returned to the separator to increase its efficiency, and to supply heat to enable the vaporization (called boilup).

The solvent vapor is condensed, and is often sub-cooled below its normal boiling point. It should be quite free of soil, but laden with water.

The water is immiscible with and more buoyant than halogenated solvents (though less buoyant than most oxygenated solvents) and so the water floats as a top layer in a decanter (Ref. 2, Figure 11.7).

In summary, these process elements of soil and solvent management are much more straightforward than their counterparts in aqueous cleaning, but:

- Are essential to any solvent cleaning process,
- Can be implemented with common process equipment (the right side of Figure 1.8, and Ref. 2, Figures 7.29 to 7.31).

In all five general solvent cleaning processes described below, these distillation-based process elements will be found.

1.6 OPEN-TOP VAPOR DEGREASERS (OTVDs)

Long the standard solvent cleaning machine because of its effectiveness, simplicity, and low required investment, the open-top vapor degreaser is being/will be gradually replaced with cleaning machines which release less solvent to the environment. Open-top machines are still the dominant machine for solvent cleaning, but during the useful lifetime of this book that is likely to change.

1.6.1 Background about Vapor Degreasing

The manufacture and maintenance of numerous products, particularly those made from metal or glass[16], have contaminants including chips, grease, lubricants, oil emulsions, dust and dirt, as well as grinding and polishing pastes. For the first half of the last century, hydrocarbon solvents[L] were used to remove these materials from contaminated surfaces.

In general, the equipment then used was known as a wiper (Ref. 2, Chapter 9 and Ref. 3, Chapter 7.6.1) or a cold cleaner (Ref. 3, Chapter 7.6.2), as those solvents weren't boiled. In other words, vapor degreasing was not done.

In the second half of the last century, concerns about flammability of hydrocarbons led to their replacement with non-flammable (Box 1.1) (supposedly[M]) "safer" alternatives;

[13] If the soils are volatile relative to the solvent, this means either: (1) a more volatile (and therefore likely to be flammable) solvent should be chosen, or (2) a cleaning technology other than solvent cleaning should be chosen.

[14] In aqueous cleaning, this process step is considerably more problematic. If, as is common, the soil is "coordinated" in water as an emulsion with a surfactant, that emulsion must be broken by a change in temperature or surfactant composition.

[15] In aqueous cleaning, this process step involves the water and the immiscible soil being separated in a decanter. In this process step, the surfactant often partitions with both the water and the soil phases – the latter being a loss of cleaning agent. Consequently, this process step is also problematic.

[16] Products made of plastic may have the same (and other) soil materials as surface contamination. But, solvent cleaning may not be a useful option for removal of this contamination as it is likely/possible that the solvent which dissolves that contamination will also swell, soften, craze/crack, or dissolve the plastic surface. Specific guidance can be found in Ref. 2, Appendix A4.

It's not the same with glass surfaces, which should not be soluble in the solvents used for parts cleaning.

Box 1.1 **History of the Flash Point Test**

The flash point test is frequently noted in this and other books, because it is the metric most commonly and globally used (or misused) to classify some hazards of flammability. Concern about its misuse arises because the flash point test is used to measure the flammability of materials for which it was never designed.

The flash point test was needed in the UK in the 1860s for the purpose of characterizing the amount of low-boiling (lights) material in the kerosene which was used for heating homes. Cheaper low-boiling material adulterated good quality kerosene by making it more easily ignitable, and less safe for home heating. In 1879, the UK government established the Abel closed cup test to measure the flammability of kerosene, and set 73°F as the lower limit for non-flammability.

In 1869, a New Orleans ordinance stated that liquids for home heating were flammable, and not useable, if their flash points measured by a similar procedure were below 110°F. In 1871, the US Coast Guard used a similar procedure to categorize and regulate transportation of hydrocarbon liquids by ships. By 1890, this work was done globally – for volatile hydrocarbons.

In 1918, the American Society for Testing and Materials (ASTM) published method D56–18T as a closed cup flash point tester developed by a man named Tagliabue. It was known as the Tag closed cup tester, and applied to hydrocarbons found in closed containers.

In 1921, ASTM, responding to requests for a flammability measurement which represented use in open tanks, published method D1310 as the Cleveland open cup tester. The same year ASTM published method D93 for use with high-boiling hydrocarbons.

Around that time, ASTM published D93 which is the Pensky-Martens closed cup method useful for highly viscous hydrocarbons as well as paints and coatings.

Closed cup testers normally give lower values (typically by 5–15°F) for the flash point than open cup and are a better approximation to the temperature at which the vapor pressure reaches the lower explosive limit (LEL).

In 1972, the US Occupational Safety and Health Administration (OSHA) adopted the closed cup method (D56-18T) for regulation by the Federal Hazardous Substances Act. In 1976, the new US Department of Transportation (DOT) adopted similar regulations based on D56-18T.

Recognition that hydrocarbon blends used as jet fuels required a different flash point test led to the British-developed Setaflash tester in 1973, whose ASTM method became D3828. The difference is that the conventional closed cup and open cup methods involved an instantaneous ignition (or not) when a spark is applied, whereas the Setaflash method is one where the vapors and the liquid are at equilibrium.

Around 1980, the International Standards Organization (ISO) developed two additional closed cup methods for general petroleum products (ISO TC 28) and paints and varnishes (ISO TC 35).

This history clearly shows how the development of flash point test methods and testers was matched to the character of the fluid being tested. Today, D56-18T (or its current embodiment) is used to characterize not volatile hydrocarbons, but mixtures of water and alcohol, HCFCs/HFCs/HFEs (Ref. 3, Chapter 22) and alcohols/hydrocarbons, multiple HCFCs/HFCs/HFEs, chlorinated and brominated hydrocarbons (Ref. 3, Chapters 20 and 21), mixtures of HCFCs/HFCs/HFEs and chlorinated hydrocarbons, and high-boiling alcohols/glycol ethers (Ref. 3, Chapter 14). All of these materials are separately used to clean parts.

Given that all halogen atoms are fire suppressants and that water is a heat sink as it doesn't participate in combustion reactions, the characterization of cleaning solvents by D56-18T and the like must be taken to be at best misused and at worst wrong-headed because none of these fluids resemble the hydrocarbon mixtures for which D56-18T was developed.

particularly halogenated solvents such as methylene chloride, 1,1,1-trichloroethane, trichloroethylene, and perchloroethylene[17], and later n-propyl bromide.

In general, the equipment used with these halogenated (fluorinated, chlorinated, brominated) solvents was known as an open-top vapor degreaser (OTVD). As differentiated from cold cleaning, the process of vapor degreasing obviously involves continuous boiling of the cleaning solvent(s) to produce vapor. The equipment involved was created to economically contain solvent emissions while compensating for the two aspects of the basic cleaning principle identified in Chapter 1.3.

Concerns about toxicity, depletion of the Earth's ozone layer, and groundwater contamination, have led many in Europe to a return to the use of hydrocarbon solvents for cleaning agents[N]. And, there has been a broad-based interest by chemical manufacturers with a global reach to synthesize fluorinated solvents which are neither flammable or toxic (Ref. 3, Chapter 22) – or especially useful in cleaning work (Ref. 2, Chapter 3). These solvents are/will be quite expensive because they: (1) are based on fluorine chemistry and so are difficult to synthesize, (2) will carry a significant cost of development, and (3) are new to the marketplace.

So the need for equipment in which vapor degreasing can be done exists today, and will not disappear tomorrow. However, the equipment as an OTVD is probably not adequate to fulfill that latter need for emission control.

[17] In the first half of the last century, carbon tetrachloride and chloroform were also used as cleaning solvents until their toxicity profiles became better recognized.

Box 1.2 Economic Life of the OTVD

In every situation, there are always two options: (1) the low-cost option, and (2) the high-quality option. In the past the OTVD was the option which contained solvent emissions to the extent that the investment for improvements in emission control could be offset by the value of solvent saved. That was the low-cost option.

The SHE regulations of today and tomorrow demand better emission control. Reduction of risks to employee health is/will be the chief basis for those improvements. The amount of solvent saved when the area exposure is reduced, for example, from 25 to 10 ppm, is puny and so are the cost savings. But the investment to do so is not! This is the high-quality option – emission control to protect the health of workers in which cost savings play little or no role.

The OTVD was the perfect low-cost option. Its useful life was measured in generations. The 1994 NESHAP, which significantly curtailed emissions of solvent, did so in ways which offered adequate reductions in the cost of solvent use to justify the investment required for compliance. That NESHAP didn't kill solvent cleaning, it validated it. More detail is in Chapter 1.18 about cycle times.

Until the level of solvent exposure limits in working areas must be reduced below ~5 to 15 ppm, the OTVD will have an economically useful life. When area exposure limits for n-propyl bromide, for example, become 1 ppm (as is contemplated in California), only enclosed machines will exist for use with most halogenated solvents.

This is because the OTVD does not collect and recycle an amount of: (1) expensive solvent to make it a cost-effective device, and (2) toxic solvent to make it compliant with safety/health/environmental (SHE) regulations (Box 1.2).

The three process steps (Chapter 1.2) incorporated in vapor degreasing are/will be done, not in so-called open-top machines, but in what are known as enclosed machines known as airless and vacuum respectively. They will be covered in Chapter 2.

1.7 THE PROCESS OF VAPOR DEGREASING

The appellation "vapor degreasing" is/was technically incorrect, and it has fostered confusion. While the cleaning process does degreasing work, and the solvent is present in the vapor phase, the soils aren't mainly removed by solvent in the vapor phase. They are removed, mainly, by immersion and solution in liquid solvent at its normal boiling point.

The vapor degreasing process includes two of the three simple steps noted in Chapter 1.2, which collectively (or individually) may be repeated as often as necessary:

Box 1.3 Don't Forget the Parts Basket

One must remember that nothing happens to the parts alone. These cleaning steps are applied to the parts and the basket in which they are contained. If the basket blocks fluid flow because it has too small an open area, the parts won't be well contacted with solvent. If it has a thermal mass approaching that of the parts, hot solvent vapor will condense on its surfaces instead of those of the parts. This is why it is important to choose a parts basket which has these two characteristics: (1) small (thin) support elements so the open area is as large as possible to not block fluid flow, and (2) as little metal as possible to minimize thermal mass so that heat transfer potential is not wasted (in Box 1.4 is described a situation where the opposite is desired). These characteristics minimize loss of the treatment effect by not "wasting" it on the parts basket.

- Immersion of parts (and basket (Box 1.3) into boiling solvent (the boil sump), which is where most of the soil is removed. Naturally, vapor is produced in this step; and naturally it rises above the boiling solvent (Figure 1.3). There may well be turbulation in the bath, produced by fluid pumped through nozzles. Contact time of immersion may be less than one minute to several minutes.
- *Immersion* of hot parts (and basket) into cooled liquid solvent (the rinse sump, Figure 1.3). There may well be turbulation in the bath produced by ultrasonic transducers or spray under immersion. Contact time is whatever is necessary to cool the parts to a value well above the temperature of the unheated bath (Table 1.1 and Chapter 1.6.1). Alternately, this

Table 1.1 Sizing of Cooling Load

Solvent	Amount of Needed Cooling, °C/°F
Trichloroethylene	1.17/2.1
Methylene Chloride	1.5/2.7
Perchloroethylene	1.11/2.0
n-Propyl Bromide	1.06/1.9
Acetone	1.06/1.9
Methyl Acetate	0.44/0.8
t-Butyl Acetate	0.72/1.3
Dimethyl Carbonate	0.89/1.6
Methyl Formate	1.06/1.9
Parachlorobenzotrifluoride	1.06/1.9

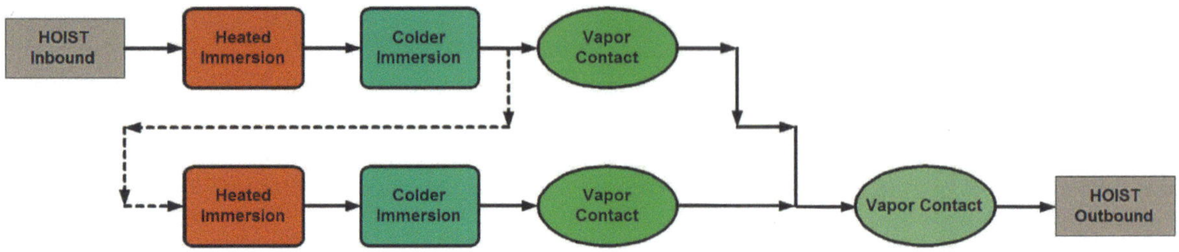

Figure 1.9 Three Step Vapor Degreasing Cleaning Process

process step may be followed by a return of the parts to the boiling sump, or

- *Lifting* of the slightly/somewhat cooler parts (and basket) upward from the colder immersion sump and *contacting* them with solvent vapor. Upon contact, the hot vapor condenses on the cooler surfaces – producing liquid solvent (Figures 1.4 and 1.5). Contact time is controlled by the time to reach the temperature in the cooled sump and the thermal mass of the parts (and basket). Rinsing with fresh condensed liquid stops when the parts (and basket) have been reheated by the hot vapor to the solvent's normal boiling point – because both temperatures are the same. This is the major and final rinse of the cleaning cycle. *At its completion, the parts will be surrounded by solvent vapor, and be wet with clean*[18] *liquid!*
- *Lifting* of the wet rinsed parts (and basket) above the zone where solvent vapor exists. Here the cooling coils have reduced the temperature of any vapor (air or solvent) present, the solvent vapor has mostly condensed, and the atmosphere is mostly cooled air. Under this contact with vapor, the parts (and basket) dry rapidly – because the heat of vaporization of the solvent is so low (Table 1.2), and the temperature difference between the hot parts and cool vapor is so high. The solvent vapor, evaporated from the parts, is more dense than air, sinks back into the zone adjacent to the cooling coils, is condensed to become liquid, and drains into the liquid sumps.

The outcome is clean dry parts, and dirty solvent.

A flow diagram of the multiple process steps of a vapor degreaser is shown in Figure 1.9[19]. Equipment arrangement diagrams, shown in Figures 1.10 to 1.15, describe the progress of the process steps of Figure 1.9[20].

The fictitious vapor degreasing machine shown in Figures 1.10 to 1.15 has two chambers – one for washing (in red) and one for rinsing (in light blue). Shown for

Table 1.2 Calculated Energy Supply Rates to Heat Parts (and Basket) to Boiling Point

Solvent	Boiling Point °C	Power Kilowatts	Current Amps @ 230 VAC
Trichloroethylene	86.7	23.4	58.8
Methylene Chloride	39.8	5.6	14.1
Perchloroethylene	121.1	36.5	91.6
n-Propyl Bromide	71.0	17.5	43.9
Acetone	56.2	11.8	29.7
Methyl Acetate	56.9	12.1	30.4
t-Butyl Acetate	98.0	27.7	69.9
Dimethyl Carbonate	90.0	24.7	62.0
Methyl Formate	32.0	2.7	6.7
Parachlorobenzo-trifluoride	139.2	43.3	108.7

[18] This presumes that the soil materials in the two sumps have a very low volatility relative to the cleaning solvent. If so, the solvent vapor evaporated from the sumps should be nearly pure (relative to soil content). Thus, so would be the liquid solvent condensed from that vapor, and then used for a rinse liquid. Fortunately, it is almost always true that this presumption is correct (Footnote 13).

[19] Note in Figure 1.9 that if the parts (and basket) have been rinsed by the condensation of hot vapor, they are not returned to the boiling immersion sump for additional cleaning. That would not be sensible because one would be immersing parts just rinsed in pristine solvent back into dirty solvent (Chapter 2.5.1)!

[20] The dotted line in Figures 1.9 and 1.17 indicates the use of ultrasonic transducers, assumed to be present in the rinse sump. That's the main reason why one would return rinsed parts to the dirty boil sump for exposure to a higher concentration of soil. The idea is to initially treat parts in the boil sump, rinse them in liquid solvent while exposing them to agitation provided by ultrasonic transducers in order to loosen remaining soil, then remove that soil when the parts are returned to the boil sump, and finally apply treatment with another rinse. The drawback of this idea is that the ultrasonic transducers or spray under immersion may remove (not loosen) soil in the rinse sump and so compromise the quality of the rinse fluid.

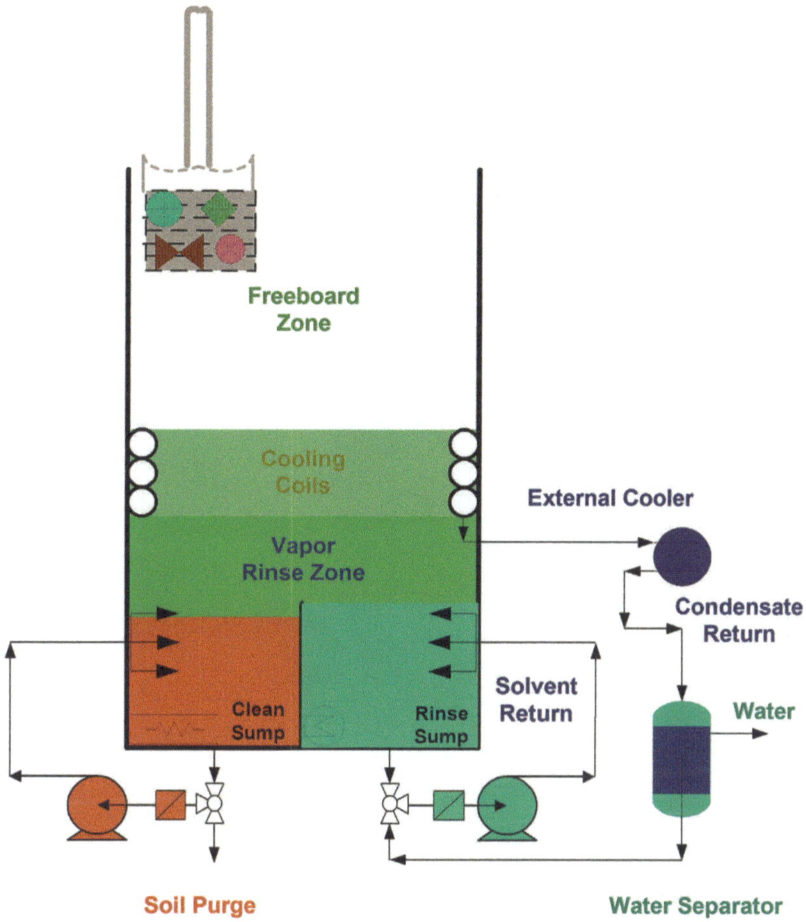

Figure 1.10 Parts Being Loaded into OTVD

completeness is the external hoist (in grey) which lowers a basket of parts into the machine.

Note the differences in rinsing method, environment, and material between:

- Figure 1.12 (immersion into cooled liquid containing soil), and
- Figure 1.13 (flushing in vapor zone with distilled solvent).

Note also that the three process steps practiced in Figure 1.9 (and Figures 1.11 to 1.13) can be collectively practiced in sequence more than once, and in more than one way.

Yet there is a "law of diminishing returns."

- The more soil removed from parts, the more the soil content in the boiling sump is increased.

As noted in Chapter 1.3.1 and Footnote 10, increased soil dissolved in solvent causes the solvent properties to change; but more importantly, more soil is left on the parts before they are rinsed.

1.8 HOW PARTS ARE DRIED

In any immersion sump, the parts (and basket) are wet. And they remain wet while in the vapor zone (green oval in the process diagrams such as Figure 1.9) because: (1) their temperature and the vapor temperature are identical (the solvent's normal boiling point), and (2) the concentrations of solvent in the vapor phase and in the liquid film on the parts are also identical (100%).

- Drying is an energy transfer process. In the vapor zone there is no driving force to transfer the latent heat of vaporization (condensation) – because the driving force is a temperature difference, and there isn't any.
- Drying is also a mass transfer process. In the vapor zone there is no driving force to evaporate solvent – because the driving force is a concentration difference, and there isn't any either.

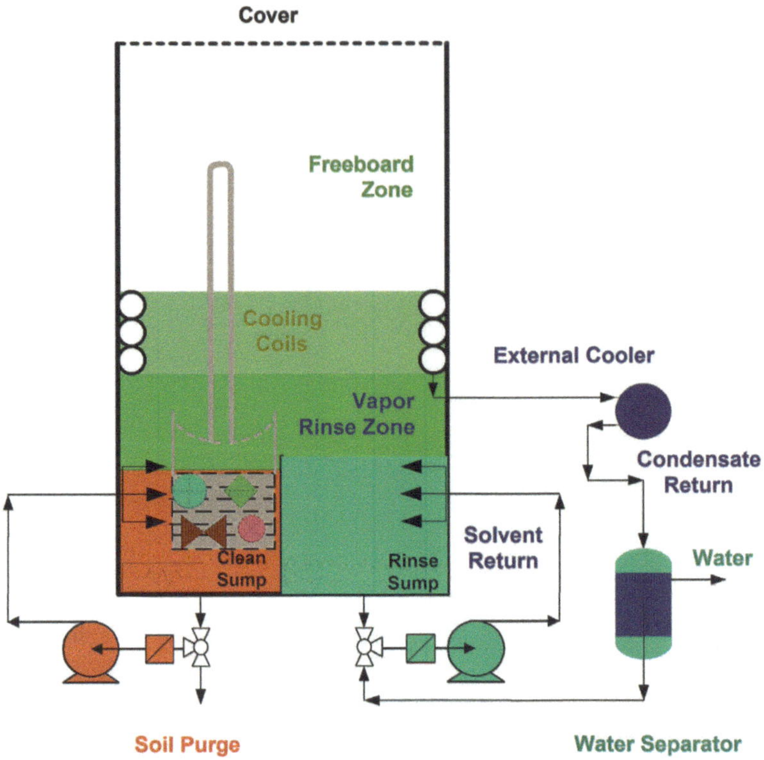

Figure 1.11 Immersed Parts Being Cleaned

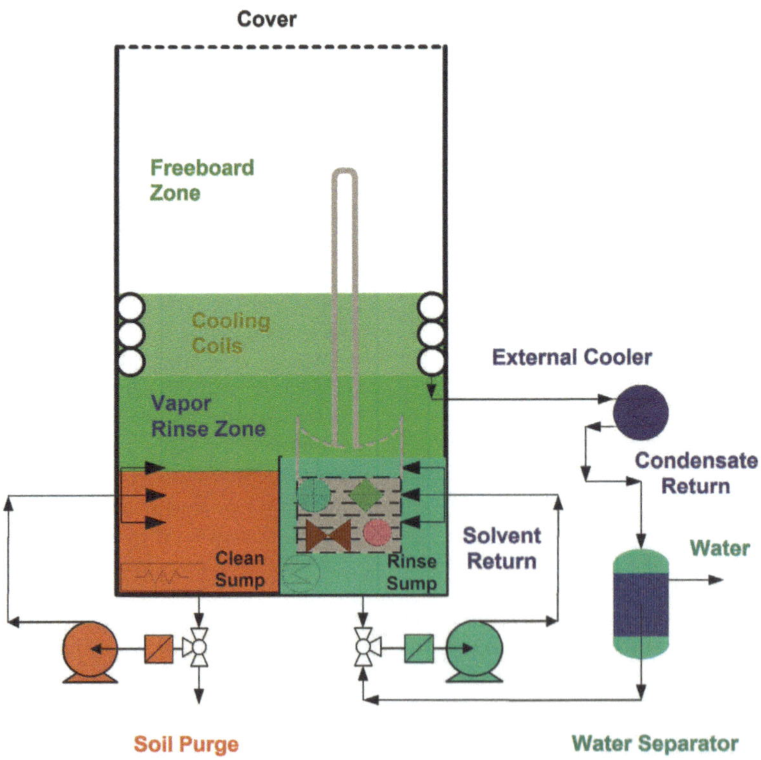

Figure 1.12 Parts Being Rinsed with Solvent

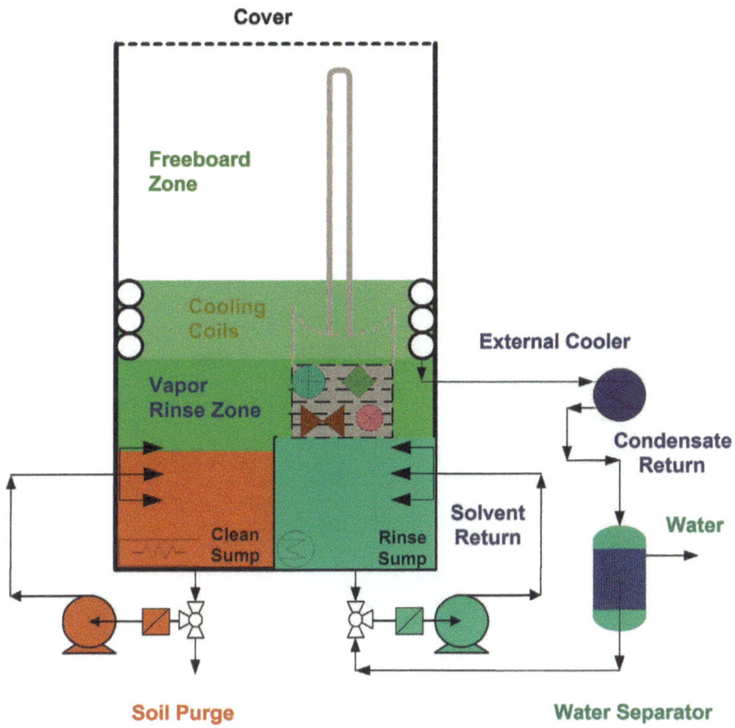

Figure 1.13 Parts Being Rinsed with Condensate

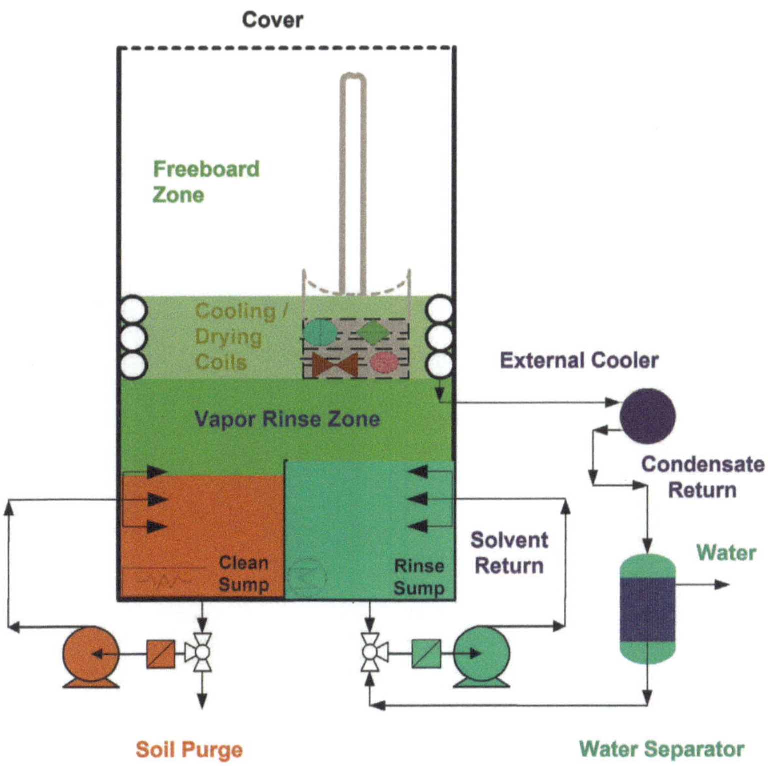

Figure 1.14 Parts Being Dried

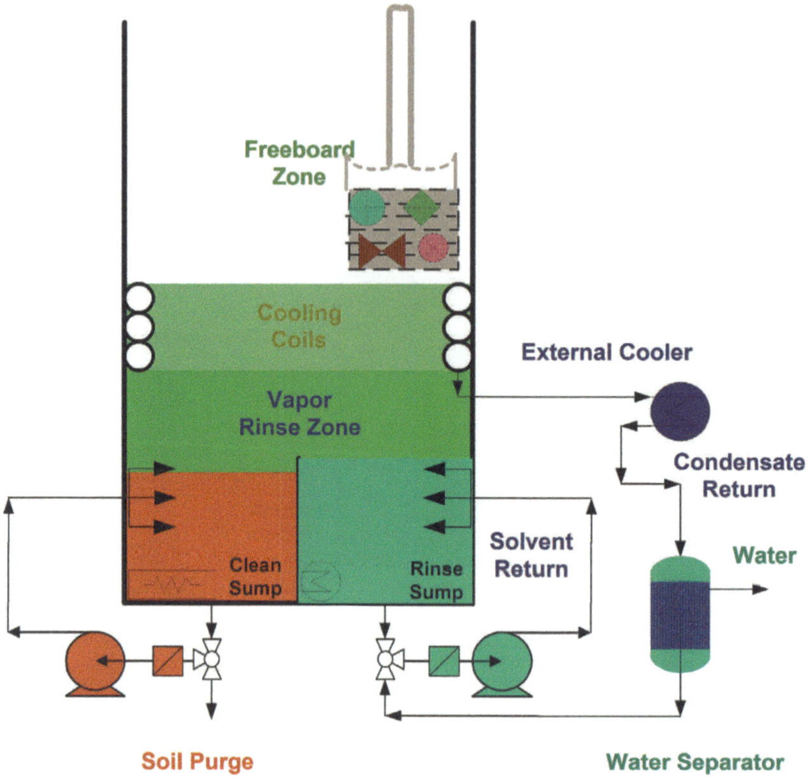

Figure 1.15 Parts Being Unloaded

Only when the parts are lifted upward above the vapor zone, into a cooler environment where there is a temperature and solvent concentration difference between the parts (and the basket) can drying commence.

So drying is done in the zone of a vapor degreaser where the cooling coils have their effect (Figure 1.16O and Chapter 1.19.1).

In the process diagrams, such as Figures 1.9 and 1.17, the zone where drying is done is shown in light green.

When the hot parts (and basket) are lifted into the cool vapor volume around the cooling coils (mostly air), the hot solvent is easily and rapidly vaporized because there exists a temperature difference for heat transfer and a concentration difference for mass transfer. The heat necessary to vaporize the solvent is provided by the thermal mass of the parts which are cooled in the drying process.

Heat and mass transfer are rapid. Sometimes it is not necessary to stop the hoist and leave the parts (and basket) stationary in the cool zone for a period of time. Depending upon the complexity and thermal mass of the parts, such a dwell time may be necessary.

Nearly as quickly, the solvent just evaporated from the parts is condensed in the cool vapor volume and "rains" back into the liquid sumps.

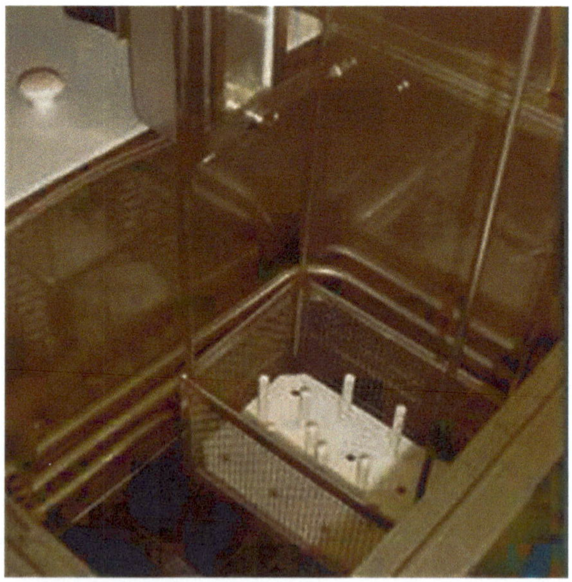

Figure 1.16

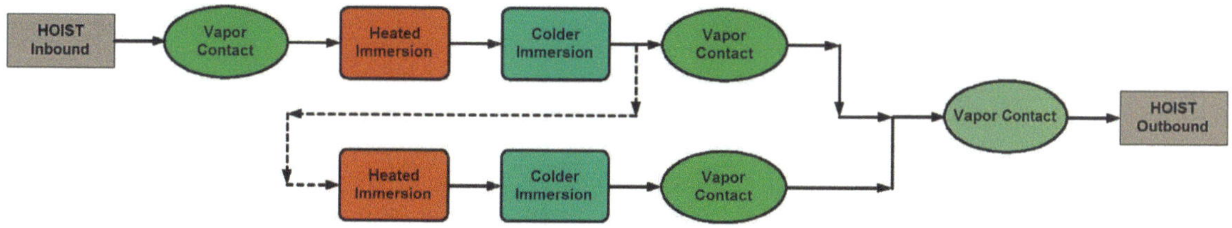

Figure 1.17 Four Step Vapor Degreasing Cleaning Process Diagram

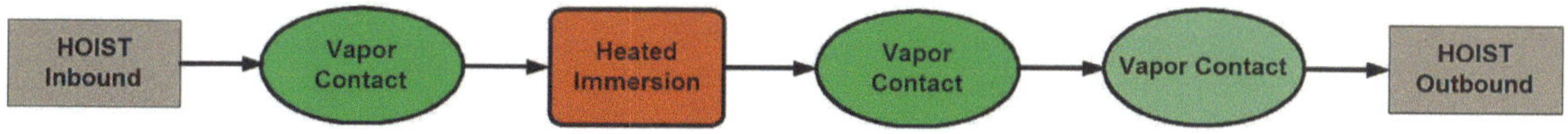

Figure 1.18 Single-Sump Vapor Degreasing Process

1.9 THE MEANING OF "FREE" RINSING[21]

Actually, the process shown in Figure 1.9 is an idealized one. The process steps shown are those one would design to complete one (or two, or more) cycles of the steps of vapor degreasing.

But there is a "free" step in this process. It is of vapor contact. This occurs because the parts fixtured in the basket are (probably) at ambient temperature *prior to entry* into the vapor degreaser. That means they are colder than the solvent in the vapor zone above the immersion sumps (see Figure 1.4). So solvent vapor condenses upon these part surfaces upon entry to this vapor zone – providing a pre-cleaning rinse[22]. This is shown in Figure 1.17 and described in Chapter 1.6.

While one can (and some have done so) use the initial contact of cold parts (and basket) with hot solvent vapor as the only cleaning step (omitting all immersion steps), that process would be very limited in its capability to provide clean parts. It would be only useful where the soil loadings were exceptionally low - a polishing or finishing step.

1.10 A SINGLE-SUMP VAPOR DEGREASER

Some vapor degreasers have only one sump – for cleaning. There is no rinsing sump. The "free rinse" noted above is available as the parts and basket are at ambient temperature when *inserted* into the machine. However, on *exiting* the cleaning machine, when the parts wet with boiling solvent pass through the vapor zone, there is no condensation of vapor on the parts; there is no final rinsing step! This is because the parts are at the temperature of the normal boiling point of the solvent, so there can be no change of phase.

This process is shown in Figure 1.18, the equipment arrangement diagram is shown in Figure 1.19, and a photograph of a single-sump vapor degreaser is shown in Figure 1.20[23].

Obviously, without a post-cleaning rinse, this is not a high-quality cleaning process – but it may be all that's needed, and the cycle time can be very short.

Without an efficient post-cleaning step of immersion rinsing, there is little reason for a second stage of immersion cleaning – because the solvent quality in the boiling sump hasn't improved.

While not efficient, or effective in removing the last traces of soil contamination, the process of Figures 1.18, 1.19[24], and 1.20 may be an excellent pre-cleaning step. It would be useful for highly soiled parts – to remove two thirds (or more) of the total soil loading. In this way, the next processing step, or the vapor degreaser in which the parts receive their final cleaning treatment, is spared contamination with a substantial amount of the large soil loading originally found on the parts. The outcome is cleaner parts, because there is less contamination of the

[21] This is a pun based on the description of aqueous cleaning agents with low surface tension. It is that they easily wet surfaces and leave little residue when rinsing is completed.

[22] One may argue, and be correct in the opinion of this author, that this rinsing step can be nearly useless because it precedes the first (and major) cleaning step. Nevertheless, it is normally present and can't be avoided – unless the parts are preheated!

Note that this "free" step of rinsing is only available once in a single-sump vapor degreaser – on entry of the parts to the vapor degreaser (Figures 1.10 and 1.11).

[23] The image of Figure 1.20 is courtesy of Parts Cleaning Technology, http://www.partscleaningbgky.com. Notice the two heating elements in the bottom of the sump, and the wide mesh screen which prevents one from placing the parts basket upon the heating elements. Note also, on the right, a hand-held wand (flexible hose) for directed rinsing of specific sections of the parts.

[24] Note that in Figure 1.19, no cooler is shown in the cleaning sump. This is because there is already a working heater in the same volume of solvent. So, the parts will not receive a final rinse with pristine solvent after the cleaning step.

Admittedly, one could install both a heater and a cooler in the single sump of a vapor degreaser. One could then cycle between heating and cooling the contents of that sump. The gain in doing so would be that the footprint (floor space requirement) of the vapor degreaser would shrink substantially.

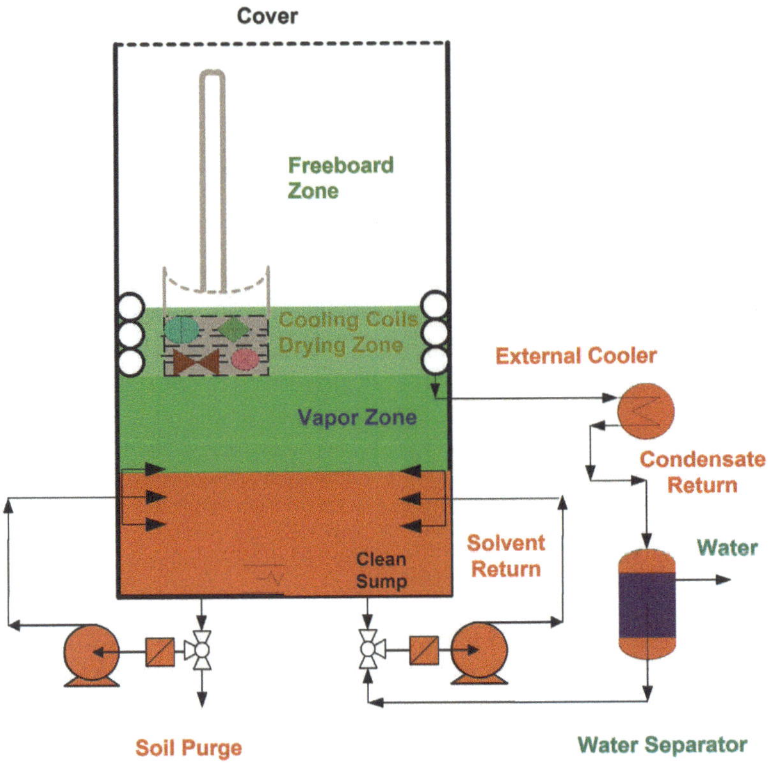

Figure 1.19 The Single Sump Vapor Degreaser

Figure 1.20 The Sump of a Single-Sump Vapor Degreaser

next processing step or the degreaser in which the final cleaning is completed (see Chapter 3).

1.11 ABOUT COOLING IN THE RINSE SUMP

Many two-sump vapor degreasers do not have a cooler located in the lower left hand corner of rinse sump as is shown in Figures 1.10 to 1.15. It is also not shown in the machine of Figure 1.19.

That cooler would have enabled the rinsed parts (under immersion) to be colder than the normal boiling point of the solvent. So when the those parts are lifted from the cooler solvent and reach the vapor zone solvent would have condensed upon their surfaces.

The absence of that cooler means that the parts processed in these machines, and the machine of Figure 1.19, will not be treated with a final rinse with pristine solvent. The solvent to be dried from the parts in the zone around the cooling coils will be solvent laden with soil. Simply, the parts won't be as clean as they could be (Chapter 1.3).

That's the reason the cooler has been added, to a second sump, as an enhanced capability to modern vapor degreasers – to enable a final rinse with pristine solvent.

One can use an energy balance[25] to estimate the degree to which the rinse sump needs to be cooled by the heat

[25] In this energy balance, an assumed 20 gallons of solvent in the rinsing sump must have an amount of heat removed from it by the sump's internal cooler equivalent to that necessary to condense the volume of solvent film attached to 10 square feet of part area at a thickness of 0.05 inches (50 mils).

exchanger shown in Figures 1.10 to 1.15. The results, shown in Table 1.1[26], are the necessary[27] reduction within the rinse sump of temperature below the solvent's normal boiling point.

- That these reductions are so small suggests what experienced operators already know. An internal cooler *may* not be necessary because normal heat losses from the unit will suffice to remove an adequate amount of heat from the rinse sump.

However, this author believes a small cooler[28] in the rinse sump is always desirable because (1) the energy balance only speaks to the amount of energy needed to be transferred, not the rate at which it is transferred, and (2) additional rinsing with pristine solvent is almost always a good idea.

1.12 KEEPING SOLVENT IN THE TANK (OR WHY VAPOR DEGREASERS ARE TALL)

That's one purpose of the basic design of all vapor degreasers – to keep the solvent in the cleaning and rinsing tanks (baths). If this purpose is unfulfilled, the owner of the vapor degreaser may well (as noted above) become bankrupt or incarcerated.

One may think the purpose of vapor degreasers is to clean parts – especially those which are difficult to dry such as those which are thin-walled (low thermal mass) or have capillary sections.

But, the ultimate purpose of a vapor degreaser is to contain solvent emissions while doing whatever the equipment owner wants to be done.

- If this weren't so, vapor degreasers would be short and stubby in height instead of being very tall and relatively narrow.

Solvent loss from vapor degreasers is not (or should not be) through leaks from piping or spills as solvent is transferred from storage. Losses are by diffusion of solvent through air from the two hot sumps, through the cooling zone(s), and leaking through the cover.

- The reason vapor degreasers are tall is to increase the length of the diffusion path so as to reduce emissive losses of solvent.

But, length of the diffusion path from the solvent sumps to the local environment does not adequately limit solvent emissions from OTVDs – either on the basis of mass amount of solvent emitted over some time period (a month), mass rate of solvent emissions, or local area concentrations in air.

1.12.1 Freeboard Is Not a "Ticket to Ride"

The parameter freeboard height is the length of the diffusion path. It is calculated as the distance from the top of the liquid solvent/vapor interface to the top of the degreaser.

Higher freeboard heights reduce diffusion losses by lengthening the diffusion path, and can reduce convection losses by minimizing the effects of external air currents.

The potential for diffusion losses of solvent is not only a function of the height of the degreaser, but also of its width. As the width of the degreaser increases, its susceptibility to losing solvent due to air drafts also increases. The way to compensate for increased susceptibility to air drafts is to lengthen the diffusion path – the freeboard height.

Consequently, it is common to refer to a degreaser's freeboard ratio, which is determined by dividing the freeboard height by the interior "width" of the equipment (the two sumps). The "width" is arbitrarily taken as the smaller of the two dimensions circumscribing the opening into the top of the vapor degreaser (Figure 1.21).

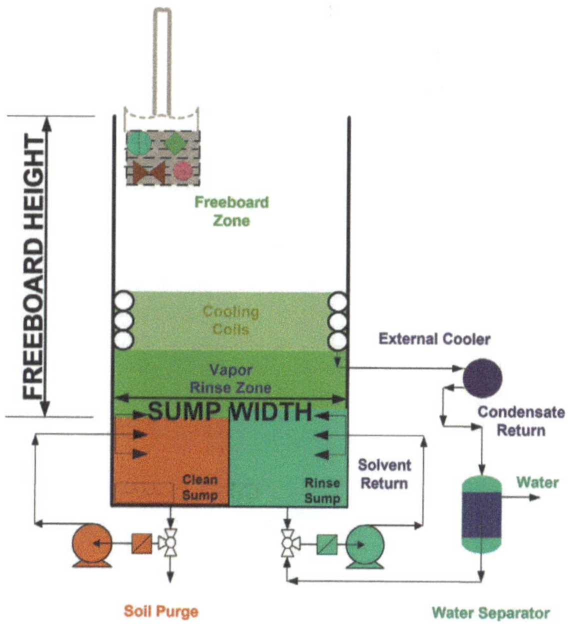

Figure 1.21 Basic Dimensions of Open-Top Vapor Degreaser

[26] The reason why values of necessary cooling are different in Table 1.1 is that the physical properties of each solvent are different.
[27] The temperature reduction below the solvent's boiling point is proportional to both the film area and thickness. This means that the needed temperature reduction is doubled if there are 20 square feet of part area, or is halved if the film thickness is only 2.5 mils. These film thicknesses are necessarily large (huge) because the part and basket surfaces won't have time to completely drain before reaching the zone where solvent is condensed.
[28] The least costly way to add such a cooler is to transfer the heat from the rinse sump to the ambient air via a small external finned-tube heat exchanger.

Figure 1.22 Open-Top Vapor Degreaser with Tall Freeboard Height, Located in Pit

Figure 1.23 Sliding Cover for Open-Top Vapor Degreaser Used with Highly-Volatile "Designer" Solvents

Older degreasers had freeboard ratios of 0.5 to 0.75, or so. The 1994 NESHAP required a minimum freeboard ratio of 1, while recommending a value of 1.25. The EPA estimated that increase of freeboard ratio from 0.75 to 1.25 would reduce diffusion-based solvent losses by about 25 percent.

Freeboard height appears cheap. It's simply higher walls made of stainless steel.

But, there is another cost involved. It's the cost of location of a very tall device in sites where tall ceilings may not be available. This cost usually appears not as the expense of raising a ceiling of a work area (Figure 1.1), but as the expense of producing a hole in it (Figure 1.22[P])! Simply, most vapor degreasers are found located in fabricated holes or, if not, are found surrounded by steps and railings.

1.13 ABOUT THE COVER

In the last decade of the last century, many suppliers of solvent and equipment sought to develop alternate designs for open-top vapor degreasers – which would allow their offerings to continue to compete for business.

There were three approaches:

- Increase the freeboard dimension, as above, and as claimed by Roger Cormack in USP 4,246,116.
- Add additional refrigeration capacity to the freeboard, as claimed by Bob Ramsey in USPs 5,142,873 and 5,048,548. This was done with a very high freeboard ratio (2.3) and three independent refrigeration systems (Chapter 1.19.5 and Figure 1.36).
- Strengthen the seal between the degreaser's freeboard area and the cover, as also claimed by Roger Cormack in USP 5,945,016 and Thomas Sisbarro in USP 5,216,736. Cormack's claims are unusual in that their embodiment has two covers – one in the middle of the machine in between refrigeration coils and one at the top of the degreaser. Sisbarro claims a movable cover which rides (slides) on two fixed tracks with a seal between the rail and the cover.

A conventional cover sliding for a small machine is shown in Figure 1.23 and Figure 1.24[Q] (see also Chapter 1.21). The small parts basket is shown in the upper left corner.

The most important point to make about covers for vapor degreasers is simple – *use them.* When the vapor degreaser is in idling mode (hot but not cleaning parts), the cover should always be in the sealed position. The only time the cover should be in the open position is when a parts basket is being inserted or removed. If maintenance

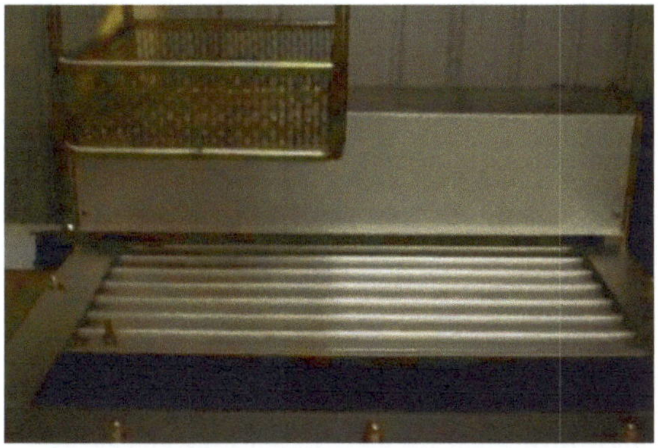

Figure 1.24 Sliding Cover for Open-Top Vapor Degreaser

is to be done on the vapor degreaser, the cover may be opened only if the unit is cold and the solvent drained to sealed storage.

1.14 STABILITY OF THE VAPOR CLOUD

The image of Figure 1.25 shows the portion of the vapor degreaser for which the equipment is named – the cloud of solvent vapor (also known as the vapor blanket). It's bounded by four equipment sidewalls, supported by the buoyant force of hot solvent vapor rising above a pool of boiling liquid solvent, and vertically limited by the downward buoyant force of cooled air.

Figure 1.25 The Vapor Cloud Exists Because of a Balance of Forces

Figure 1.26 The Vapor Cloud Collapses When Vapor is Condensed

Simply, the vapor cloud is the result of thermal effects. Figure 1.25 shows an idealized vision of these buoyancy forces and their net effect.

Consider the effect of inserting a colder (ambient or any sub-boiling temperature) basket of parts into that maelstrom of thermal effects. The hot vapor is condensed onto the colder surfaces, and the vapor cloud collapses. Figure 1.26 shows an idealized vision of that outcome.

Destruction and rebirth of the vapor cloud happens (at least once) every cycle of cleaning for which vapor degreasers are used. And it happens in a few seconds! Every dirty basket of parts has this effect on entry to the vapor degreaser. Every cooled and cleaned basket of parts has this effect as it is rinsed. Every one!

The reason this is so is that the thermal masses of the vapor cloud and the loaded parts basket are highly unequal. The vapor cloud is made of solvent vapor; the parts basket and parts are made of metal, glass, or plastic, and have several decades more thermal mass than does the vapor cloud.

- This means that the vapor cloud is easily destroyed by entry of parts (and basket[29]), but
- It is also easily reformed by solvent vapor generated by the heater in the boil sump – once the parts have passed through it and entered the boil sump.

Said as clearly as possible, parts entering an open-top vapor degreaser are rinsed with condensed solvent before they are washed in soil-laden solvent.

[29] Consider that the parts basket may have more thermal mass than the parts within it.

1.14.1 A Maelstrom of Energy and Mass Transfers

What happens in a typical open-top vapor degreaser is not static. It's dynamic, as:

- The cold parts (and basket) enter around the open cover, pass through the zone adjacent to the top cooling coils, and enter the vapor zone.
- In the vapor zone, the parts (and basket) are heated by contact with the hot solvent vapor cloud, which is cooled and collapses (Figure 1.26) as the parts (and basket) descend into the boil sump.
- In the hot boil sump, the still cooler parts (and basket) rapidly absorb heat from the hot solvent, and boiling stops as the temperature in the sump declines below the solvent's normal boiling point (based on the entering thermal mass).
- The temperature controller in the boil sump should immediately activate the boilup heater, and it should strive to regain the situation which was present when the parts entered the vapor degreaser by rapidly adding heat.
- Additional heat first elevates the solvent to its normal boiling point while simultaneously heating the parts (and basket) to that temperature, then provides the latent heat of vaporization to produce solvent vapor, and finally re-establishes the vapor zone.

The time span for all these events to happen must be measured in seconds because the entire cleaning cycle is programmed and required to last only a few minutes (or less). So the rate of heat transfer must be high.

It isn't that it is necessary to have all the cycle time spent by the parts (and basket) in the boil sump be at the normal boiling point of the solvent, it's that when the parts (and basket) are lifted from the rinse sump and into the vapor zone, the vapor zone must exist.

1.15 SIZING THE BOILUP HEATER

The boilup heater must have a high power capability or rating[30]. Its duty (job) is nothing like that necessary to compensate for convective and radiative heat losses to the environment so as to maintain the vapor degreaser unit at working temperature.

The heater rating can be as much as 3 to 10 or 20 (with a large thermal load) times that amount, so that: the vapor zone is re-established and stable, there is boiling solvent in the boil sump, and the temperature of the parts has nearly or actually reached the normal boiling point of the solvent.

- And the time for all this to be completed is perhaps 30 seconds – each and every cleaning cycle!

[30] Power rating is a rate of energy supply. The units are BTU/hr, calories/second, kilowatts, or tons of refrigeration (1 ton = 12,000 BTU/hr).

The heater may be powered by condensate, waste steam[31], natural gas, externally heated hot oil, but is more likely to be powered by electricity[32].

The rate of energy supply depends upon at least the thermodynamic properties of the solvent, and the size of the vapor zone as called for in the design of the degreaser. Some typical design[33] values for the needed rate of energy supply to: (1) heat the solvent in the boil sump to its boiling point, and to then (2) rebuild the vapor zone are given in Tables 1.2 and 1.3 respectively.

The first basis, noted above, for rating the boilup heater is to heat the parts (and metal basket) and the solvent to the normal boiling point of the cleaning solvent (which is different for each solvent). Requirements[34] of the boilup heater to fulfill this task are shown in Table 1.2[35].

[31] Condensate (hot water produced by the use of steam) or waste steam (steam at a temperature or pressure too low to be useful normally) could be used to provide boilup heat in a way which would not produce sparks when a flammable solvent is used (Ref. 1, Chapter 3.22). Natural gas could be used as a source of boilup heat if there were concerns about hot spots in the heating element(s). To avoid hot spots, electric heaters should be rated for no more than 20 to 25 watts/square inch. In this book, electricity is assumed to be the normal source of energy for heating because it is most commonly used, and it allows the degreaser to be mobile.

However, steam or hot oil may be preferred sources of heat energy for vapor degreasers and distillation systems because one can easily control the temperature to which the solvent (and parts) will be exposed. And there are no spark sources in these sources of heat energy! Heaters usually operate only in the ON/OFF mode, so that proportional control is impossible.

[32] If ultrasonic transducers are used in the rinse sump, it is certain the source of boilup heat would be electricity – because ultrasonic transducers are not powered by steam or natural gas.

[33] The design basis for this calculation is a ficticious vapor degreaser with sump dimensions of 22 inches width by 13 inches length for each of two sumps (4 SF total area), a vapor zone 12 inches high between the top of the rinse sump and the bottom of the cooling coils, and the requirement that the vapor zone be regenerated in 30 seconds. To compute the AC current requirement, a supply of 3-phase 230 VAC was assumed. The reason for calculation of the specific AC current load is that sites are often limited by their ability to supply and distribute electrical current through their breaker boxes (power panels).

[34] The basis for design included the items noted in Footnote 33 and that the combined weight of parts (and basket) is 50 lbs, and they are made of stainless steel. Heat duty is proportional to the mass of metal, and inversely proportional to the required recovery time from when the cold parts and basket disrupt the vapor zone until when the vapor zone has been restored by re-vaporization of solvent. More rapid recovery will require a boilup heater with proportionally more capacity, and the reverse.

This is an exceptionally large mass of metal to be heated in a small degreaser, but the value was chosen to make a point. The point is that not only is there a volumetric size limitation on parts which can dimensionally fit into a vapor degreaser, but there is a mass limit as well, which can require the boilup heater to have an exceptionally large capacity.

Even worse, the required current values at 230 VAC are, for some solvents, beyond the practical limits of available current supply in most shops, and many shops don't have a 460 VAC capability which can reduce the required current draws. So the workload can be limited by the capabilities of the supply of electricity.

Should the mass of parts and parts basket be reduced by a factor of 10 to only 5 lbs, the total current requirements in Table 1.3 would be reduced by about 80%.

[35] The values in Table 1.2 are proportional to the thermal mass (actual weight multiplied by specific heat) of the parts and parts basket. That mass, per Footnote 34 is 50 lbs. The base temperature of the parts and parts basket is assumed to be 25°C.

Table 1.3 Calculated Energy Supply Rates to Regenerate Vapor Cloud

Solvent	ΔH Vaporization, BTU/lb	Power, Kilowatts	Current, Amps @ 230 VAC
Trichloroethylene	103	4.4	11.0
Methylene Chloride	142	4.5	11.3
Perchloroethylene	90	4.4	11.1
n-Propyl Bromide	104	4.4	10.9
Acetone	216	4.5	11.2
Methyl Acetate	176	4.6	11.6
t-Butyl Acetate	142	5.2	13.0
Dimethyl Carbonate	180	5.2	13.1
Methyl Formate	205	4.7	11.8
Parachlorobenzotrifluoride	89	4.6	11.5

The second basis, also noted above for rating the boilup heater, is to vaporize solvent to form the vapor zone.

Note in Tables 1.2[36] and 1.3[37] that the required energy supply rate is somewhat a function of solvent properties, such as normal boiling point and heat of vaporization.

The boilup heater must independently fulfill both tasks: (1) heat the cold parts (and basket) from ambient temperature to the normal boiling point of the solvent (Table 1.2), and (2) replace the vapor zone by providing the latent heat of vaporization (Table 1.3).

That overall heat duty needed is shown in Table 1.4 – whose values are the sum of those in Tables 1.2 and 1.3.

One characteristic which can be used to differentiate among vapor degreasers from different suppliers is the rating of the provided boilup heater. Suppliers may provide a heater with a low rating so they can offer a machine with a lower price (or higher profit), and appear to offer a corollary benefit for the site – a reduced demand on the type[38] and supply of electricity.

This can be an unwanted bargain. A too-small boilup heater extends the production cycle time because the time to heat the parts (and basket) and establish the vapor zone are lengthened.

Yet a too-large boilup heater will generate hot solvent vapor at a rate that is faster than the cooling coils can condense it. This outcome will thrust hot solvent vapor up along the freeboard and out through the cover – potentially polluting the environment and exposing employees to high levels of undiluted solvent vapor.

Table 1.4 Calculated Energy Supply Rates to Heat Parts (and Basket) and Replace Vapor Zone

Solvent	Power Kilowatts	Current Amps @ 230 VAC
Trichloroethylene	27.8	69.8
Methylene Chloride	10.1	25.4
Perchloroethylene	40.9	102.7
n-Propyl Bromide	21.9	54.8
Acetone	16.3	40.9
Methyl Acetate	16.7	42.0
t-Butyl Acetate	32.9	82.9
Dimethyl Carbonate	29.9	75.1
Methyl Formate	7.4	18.5
Parachlorobenzotrifluoride	47.9	120.2

[36] One reason why parachlorobenzotrifluoride (VOC exempt in the US) is seldom used as a vapor degreasing solvent is its high boiling point (which might damage parts). But a second, and unexpected, reason is that its heat of vaporization is quite high (Table 1.3) – approaching that of water. The latter substantially increases heater duty.

[37] The values in Table 1.3 are a function only of the thermodynamic and physical properties of the solvent. It is assumed that when the vapor blanket collapses due to the entry of the cold parts basket that heat must only be supplied to re-vaporize the solvent. In other words, it is assumed that entry of the cold parts basket does not cause the solvent in the rinsing bath to cool below the normal boiling point of the cleaning solvent.

[38] Type means AC voltage. Suppliers might present their offering by proposing that higher voltage (230 or 460 VAC) or three-phase AC is not needed. This might be seen as an advantage by a user, as these utilities may not be part of the battery limits of the working site. But it is not an advantage if it limits the capability to provide a high current flow to meet local requirements for cycle time.

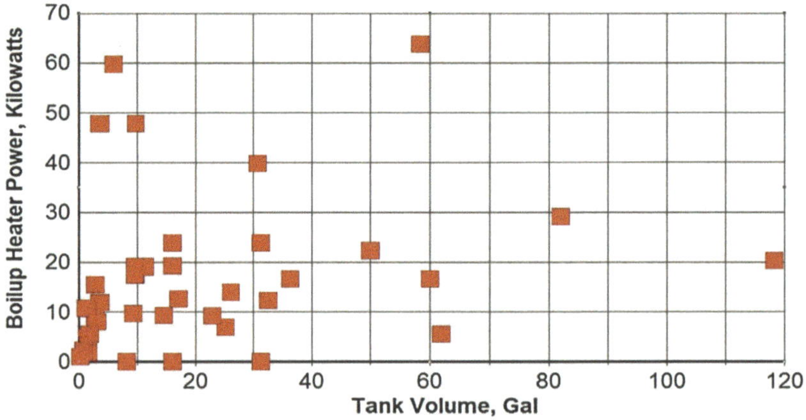

Figure 1.27 Heat Transfer Capability of Heaters Supplied in Commercial OTVDGs

A survey of more than 40 currently produced vapor degreaser models showed a great disparity in the capacity of the provided boilup heater. These details are plotted in Figure 1.27. Only 9 of these 40 machines met the substantial requirement for capacity of their boilup heater for every solvent as noted in Table 1.4.

A purchaser of a vapor degreaser should either specify the size of the boilup heater using the information above, or specify the required recovery time and solvent, and allow the supplier to provide the specification for the heater.

In summary, the boilup heater may appear to be a minor appliance in the overall package of components which is a vapor degreaser; but it's not. The boilup heater directly controls the cycle time of the process completed in the degreaser. It controls the quality of cleanliness achieved because of the rinsing provided within the vapor zone. And if it's oversized, it can supply hot vapor at a rate faster than the cooling coils can condense it and return it to the liquid sumps.

1.16 COOLING OFF

It's conventional to oversize the cooling coils in a vapor degreaser. This is because they are the main line of defense which protects the user from letting solvent vapor escape from the cleaning machine – causing at worst injury, and at best pollution.

This author agrees with Durney[R] that the cooling coils should be at least 100% oversized (on a rate basis) vs. the *instantaneous* capability of the boilup heater to produce vapor. Spatially, this places the top of the vapor zone in the middle of the cooling coils – leaving the portion of the cooling coils above the vapor zone as an "insurance policy," and a location where the drying process can be completed (Chapter 1.17).

One regulates the capability to remove heat within the vapor zone by regulating the flow of coolant passing through the cooling coils (Chapter 1.19[39]).

1.17 A TIME FOR DRYING

Liquid solvent just "flies off" (Figure 1.28) simple surfaces in response to a large difference in temperature between the solvent's normal boiling point and the air temperature between the cooling coils.

The operative word is "simple." Surfaces which are flat, and don't have crevices, pockets, cracks, blind or tapped holes, cupped structures, or knurled surfaces are those identified as "simple." Simple surfaces don't trap droplets of liquid. Simple surfaces are wet with films which evaporate rapidly.

- Complex surfaces will require extension of the cycle time to allow trapped liquid to be evaporated.

Note the position of the parts basket in Figure 1.28 – it is above the vapor cloud (zone) but within the volume affected by the cooling coils.

That location might not continue if the parts basket continues on an upward trajectory attached to the hoist before the liquid collected in complex surfaces is vaporized. In that case, three negative consequences will eventuate:

- The parts won't be fully dried,
- Liquid evaporated when the parts are outside the zone affected by the cooling coils will become pollution (possibly harming workers), and
- Operational cost will be increased.

[39] The temperature of cooling tower water (CTW) is normally only adjustable to a minor extent. Alternately, a mechanical refrigeration system offers considerable more capability to adjust the temperature of the coolant, albeit at a considerably higher cost than CTW (Table 1.6).

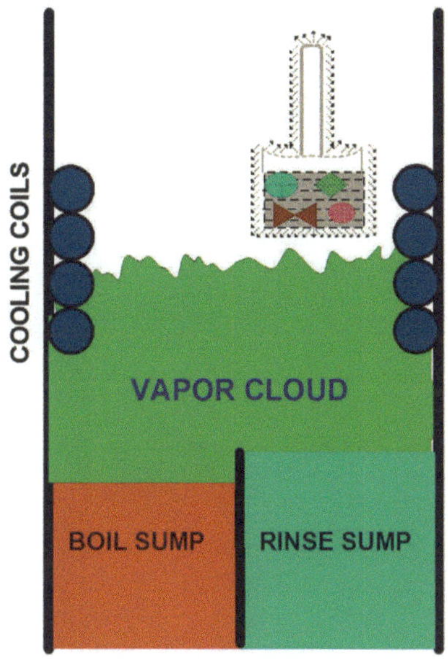

Figure 1.28 Hyropthetical Arrangements Within an OTVDG – Showing How Drying is Accomplished

1.18 CYCLE TIMES FOR VAPOR DEGREASING

Yet, cleaning with solvents is usually (Box 1.4) done very quickly.

While cycle times with aqueous technology last 10 to 30 minutes, those with solvent technology often last less than 5 minutes. The reason for this is that solutioning of soil takes little time if the solvent is well matched to the soil materials; and a very long time if it is not. And the drying process described in Chapter 1.17 should require little time!

> **Box 1.4 The Pre-Soak Step**
>
> It is often possible to operate a cleaning machine so that there is a "pre-soak" step before the step of immersion in boiling solvent is initiated. One does this by imposing a delay before the boilup heater (Chapter 1.15) is activated. The length of the time delay is the duration of the "pre-soak" step. What makes this possible is that insertion of the cold parts and their basket into a sump full of boiling solvent has a chilling effect. It stops boiling causing the vapor cloud to collapse as no vapor is generated (Figure 1.26). And the sump temperature is reduced below the normal boiling point. Only when the boilup heater is activated does boiling resume and the vapor cloud regenerated.
>
> One enables the "pre-soak" step to soften or swell soil materials, making their later removal more efficient. Obviously, cycle time is lengthened by choice.
>
> Should the thermal mass of parts be low, the thermal shock to the boiling solvent may not be adequate to stop boiling. In this case, the "pre-soak" delay can still be implemented by using a parts basket with increased thermal mass.

As noted above, the process of vapor degreasing typically involves four steps[40], and two major movements of the parts basket. These are listed in Table 1.5.

A graphical accounting of the time spent in each of the steps noted for the same fictional cleaning process as in Table 1.5 is shown in Figure 1.29[41].

The time spent in the boil sump (red fill in Figure 1.29) is the longest interval because (1) regeneration of the vapor zone and completion of the heating of the parts (and basket) is done in this interval, and (2) all the soil must be loosened or removed because the remaining process steps are rinsing and not cleaning.

1.19 COOLANTS ARE COOL

Refrigeration technology is unseen in the spectrum of chemical, mechanical, physical, and mathematical sciences necessary to practice solvent cleaning technology. But, without refrigeration technology, solvent isn't easily contained within an open-top vapor degreaser, some EPA environmental emission limits and OSHA health exposure limits can't be met, parts won't be well dried of solvent, and an enclosed degreaser will exhaust itself of its supply of solvent.

Two types of coolants are used: (1) water[S] from a cooling tower {CTW), and (2) chilled mixtures of glycols and water, or of soluble salts and water. The mixtures are called "brines" (Box 1.5), and are produced by mechanical refrigeration.

A cooling tower operates by evaporating some water into air and cooling the water not evaporated by extracting the needed heat of vaporization from it.

- Use of CTW[T] has long been common with cleaning solvents having high boiling points – especially perchloroethylene and some hydrocarbons. But, water at the lower temperatures which can be produced by a cooling tower is practically useless to condense solvents which have lower boiling points such as HFCs, HCFCs, HFEs, HFOs (Ref. 3, Chapter 22), and methylene chloride (Ref. 3, Chapter 20).
- Use of mechanical refrigeration in a vapor degreaser is a two-step process. First, heat is transferred from hot

[40] However, using the descriptions of Chapter 1.2, only two different steps are involved – immersion in liquid and vapor contact.

[41] The image of Figure 1.29 suggests an incorrect conclusion – that drying of solvent from parts (in a basket) is done outside the vapor degreaser. It's the absence in Figure 1.29 (and Table 1.5) of a noted time interval for drying which prompts that misconception.

The correct conclusion is that drying of parts is always done within the parameters of the cleaning cycle shown in Figure 1.29. Said another way, evaporation of liquid from simple part surfaces happens almost instantaneously once the part (and basket) is lifted above the vapor zone and into the zone opposite the cooling coils.

Table 1.5 Process Steps in a Cycle of Vapor Degreasing

Step	Description	Elapsed Time
0	Loading of parts into parts basket (Figure 1.10)	
1	Lowering of filled parts basket into vapor degreaser	Movement @ 11 ft/minute over a distance of ~3+ feet
2	Holding of cold parts basket in vapor zone, while solvent condenses on them as pristine rinse (Figure 1.26)	:15 to :45
3	Holding of parts basket immersed in boiling solvent, with turbulation produced by flow from jet nozzles (Figure 1.11). This is the cleaning step.	:30 to 3:00
4	After relocation, holding parts basket immersed in colder solvent, possibly with ultrasonic agitation (Figure 1.12). This is the rinsing step.	:30 to 3:00
5	Movement of cooler parts basket upward to vapor zone, and holding them while solvent condenses on them as pristine rinse (Figure 1.13)	:15 to :45
6	Lifting and holding basket filled with cleaned parts within freeboard zone to dry, and removing cleaned/dried parts from vapor degreaser (Figure 1.15)	As step 1
7	Unloading of parts from parts basket (Figure 1.10)	

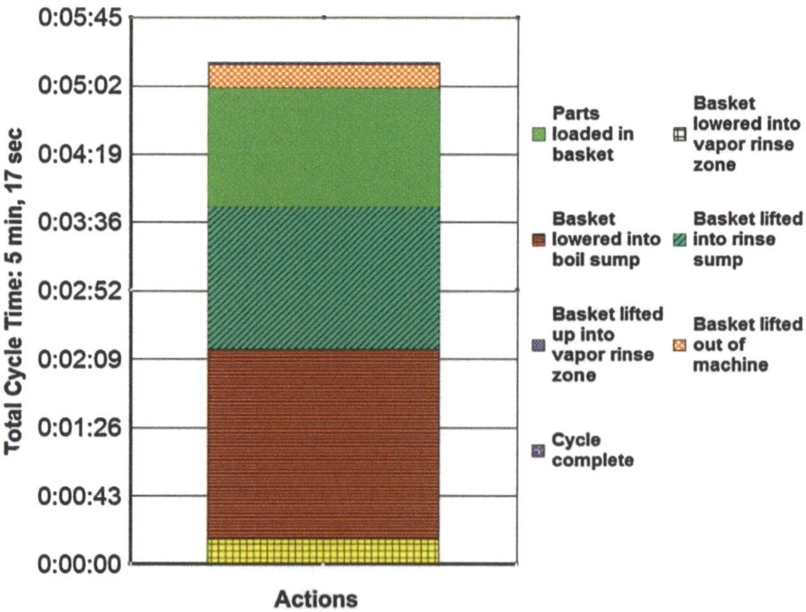

Figure 1.29 Graphical Arrangement of Sequence of Events in Operation of an OPVDG

Box 1.5 **Characteristics of Brines**

The name "brine" refers to a liquid used in a refrigeration system, which is usually an aqueous solution of calcium chloride or sodium chloride, that is externally cooled by an evaporating refrigerant (Figure 1.31). The addition of salt lowers the freezing point of the solution. The lowest temperature which can be reached by a solution of sodium chloride is about −21°C (−7°F). That for calcium chloride is −52°C (−62°F).

Water solutions of ethylene and propylene glycol are also commonly used as heat transfer agents in refrigerated service, and also share the name "brine." Their freezing points are −40.0°C (−40°F) and −28.9°C (−20°F) respectively.

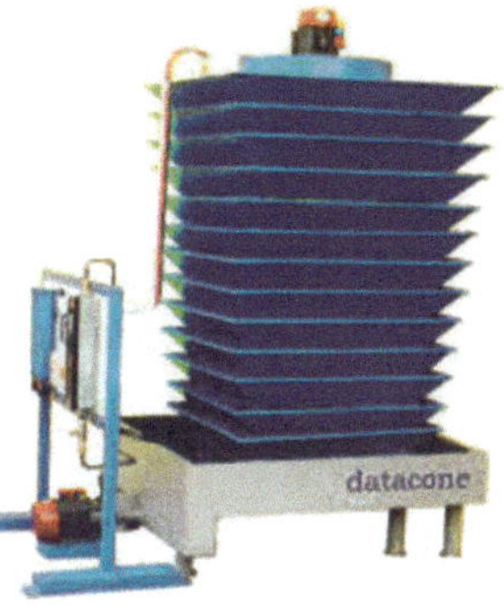

Figure 1.30 A Small Cooling Tower, as Would be Used to Support any VDG

solvent vapor inside the vapor degreaser to a coolant circulated through cooling coils located (usually) on the external walls of the vapor degreaser.
- Then, the heat within the coolant is transferred from the cooling coils to the ambient atmosphere through a mechanical refrigeration unit (Box 1.6) that employs a second fluid known as a working fluid (refrigerant).

The coolant is the glycol/water or chloride salt/water mixture (brine) noted above, and it is never in contact with the refrigerant. Figure 1.31 illustrates the integration of the multiple heat transfer systems.

Box 1.6 **The Basics of Mechanical Refrigeration**

Mechanical refrigeration is a process in which work is done to remove (or move) heat from one location to another. This is traditionally accomplished by the mechanical work of a compressor which is powered by electricity. Heat is removed from a high-temperature source (the hot parts and basket) and rejected to a low-temperature sink (the ambient air adjacent to the refrigeration machine) with the help of externally supplied work (Figure 1.31). Some refrigeration machines reject the heat to water from a utility.

Mechanically driven refrigeration processes are cyclical — a refrigerant (such as HFC-134a or its replacement HFO 1234yf (Ref. 3, Chapter 22 and Appendix 3) is compressed to be a vapor at a higher temperature, but still below the vapor pressure at that temperature. Then the pressurized vapor flows to a condenser that transfers heat to the ambient air, which is circulated by a blower or fan. The condenser cools the hot refrigerant vapor (not solvent vapor) until it starts condensing, and then condenses the refrigerant vapor into a liquid by removing additional heat (the heat of vaporization). The next stage in the refrigeration cycle is to expand the still-pressurized refrigerant liquid by passing it through a specially designed valve. Expansion produces a mixture of liquid and vapor at a lower temperature and pressure. The cold liquid-vapor refrigerant mixture then flows to the evaporator unit (Figure 1.31) where the refrigerant fluid is converted to a vapor by absorbing heat from the heated working fluid passing from cooling coil tubes within the vapor degreaser. Then, the cycle of use of the refrigerant fluid starts anew.

The net effect of this vapor compression refrigeration cycle, and its incorporation into a vapor degreaser, is to transfer heat from the hot solvent vapor above the boiling and rinsing sumps to ambient air adjacent to the cleaning machine.

The units of refrigeration are always units of power. Commercial refrigerators may be rated in kJ/s or BTU/hr of cooling, watts or kilowatts; or more commonly and simply in tons of refrigeration where one ton is not a unit of weight but is equivalent to 12,000 BTU/hr.

Nearly always the refrigeration equipment is integrated into the vapor degreaser to produce a single package. This is shown in the image above (courtesy of Branson Ultrasonic, http://www.bransonultrasonics.com/). And it's common for the refrigeration equipment to comprise 25 to 50% of the volume of that package.

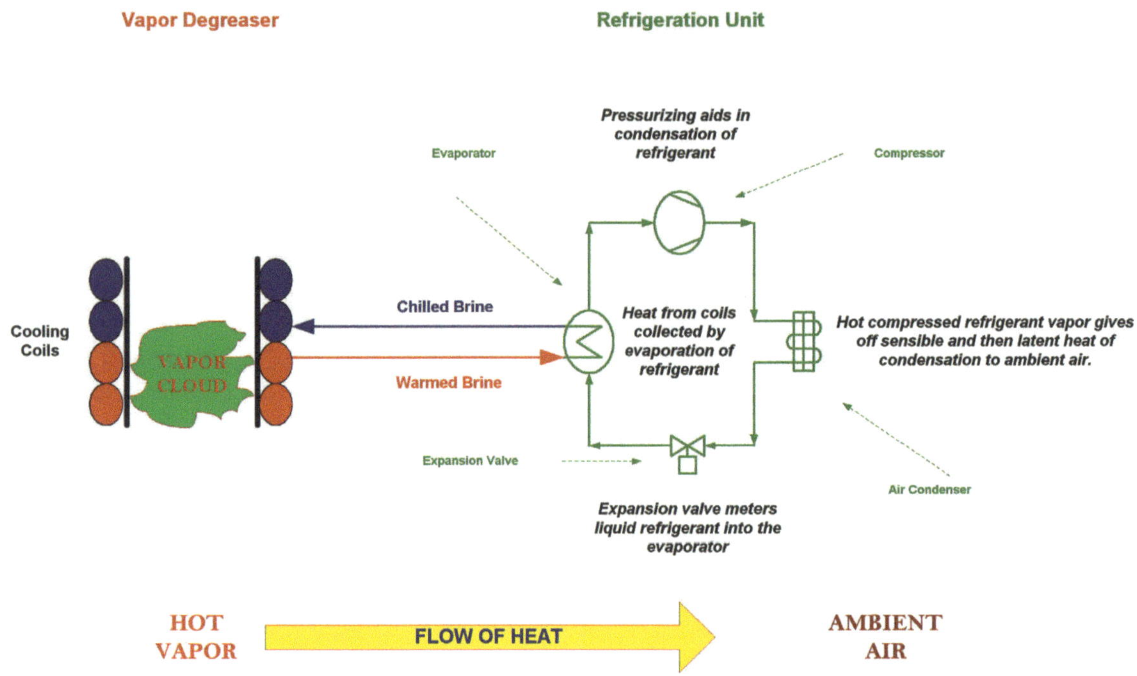

Figure 1.31 The Multiple Heat Transfer Processes Operating Within an OTVDG

There are fundamental differences between the use of cooling towers and mechanical refrigeration units. Cooling towers are considerably more efficient in the use of electricity than mechanical refrigeration units, and would always be chosen for that reason – if their use were practical. Generally, with vapor degreasers, such use is not practical because the needed low temperatures can't be produced by a cooling tower.

Some differences between cooling tower and mechanical refrigeration are shown in Table 1.6.

1.19.1 Demands on Coolants

Vapor degreasers have two distinctly different sets of demands for coolants:

- One demand is to quickly rebuild and maintain the vapor zone above the sumps, keep boilup vapor from escaping the freeboard volume, and provide an environment for rapid drying of parts. This primary cooling is implemented along the freeboard just above the vapor zone.
- The other demand is to reduce the concentration of solvent in the air[42] entrained with the outbound parts (and basket). This secondary cooling is implemented somewhere below the top of the freeboard area.

While these two demands are complementary, they are not fulfilled at the same location in the vapor degreaser. Although a user may use the same source of cooling to fulfill both demands, separate streams of the same coolant (refrigerant) are used; each at the same temperature (probably[43]), and almost certainly at a different flow rate.

That's why a conventional vapor degreaser has two (at least) sets of cooling coils – one to meet each of the two demands for cooling.

- The primary coils are located just above where it is desired to maintain the vapor zone[44] as is shown in Figures 1.10 through 1.15, 1.19, and 1.21.
- The secondary coils are normally located just below the top of the degreaser as shown in Figure 1.32.

[42] Solvent exiting a vapor degreaser is contamination, uncontrolled pollution, wasted materials (money), and a potential threat to the health of persons adjacent to the degreaser.

[43] It is always desirable to incorporate a secondary cooling system that operates at a "low" temperature. The definition of "low" is so that the solvent concentration in the vapor exhaust from the top of the degreaser, diluted by normal area ventilation, allows a comfortable margin for the work area to always (1) meet all environmental and safety regulations, and (2) avoid a feeling of discomfort among workers. It is likely that a temperature this low in the primary coils is not necessary to constrain solvent emissions upstream of the freeboard zone.

[44] The height of the vapor rinse zone is generally chosen to be a minimum value, suitable to comfortably contain the parts basket, and little higher. After all, an increase of this height reduces the freeboard height for the same overall unit height.

Table 1.6 Comparison between Cooling Tower and Mechanical Refrigeration

Item	Cooling Tower	Mechanical Refrigeration
Temperature of Coolant (°C/°F)	+ 24 to 32/+75 to 90	−15.5 to +10/−60 to +50
Useful with Solvents	Perchloroethylene, Hydrocarbons, Glycols & Glycol Ethers, any high-boiling solvent	Any
NOT Useful with Solvents	HFCs, HCFCs, HFEs, HFOs, Methylene Chloride, Trichloroethylene, n-Propyl Bromide	None
Working Fluid (participates in refrigeration cycle)	Air	A refrigerant
Coolant (passes through cooling coils in vapor degreaser)	Water	A brine
Heat Sink (to which heat from vapor degreaser is exhausted)	Ambient Air	
Refrigeration Duty, tons	10+	5+
Electricity Requirement, kW	<1	~15 to 20+
Coolant Flow, gpm	~25+	<5 to 15
Procurement	Separately Purchased Purchased with Site	Provided with Vapor Degreaser
Physical Size (Volume)	Huge (located outside)	Small (located in same package as vapor degreaser)
Number of Compressors	None	One[U]
Materials of Construction	Fiberglass	Stainless Steel

However, another approach is possible (Box 1.7)[V] where the cleaning solvent is presumed to be perchloroethylene[45].

1.19.2 Selection of Coolants

In designing a cooling system for a vapor degreaser one has to be conscious of the two sets of demands, and select a coolant (and the refrigeration system to produce it) that simultaneously meets both sets of demands. Purchasing two different refrigeration systems with two different coolants would require additional investment, operating cost, maintenance attention; and certainly engender curiosity – unless CTW can be used as the primary coolant because of its exceptionally low cost.

The selection of coolant (not the working fluid) can determine compliance with environmental and safety regulations, as well as determining overall cycle time, operating cost, and the quality of the parts drying operation.

- Users may select from among CTW, or one of the three brines. Normally, CTW is a desirable but not practical choice.

Selection of a coolant is based on repeatedly producing the low temperature(s) needed in the air opposite (above) both sets of cooling coils with the refrigerant solution that is useful to produce the desired low temperature.

- The lower the temperature of the chosen coolant, the greater will be the rate of transfer of heat between the coil walls and the solvent-rich air adjacent to them.
- Higher internal heat transfer (cooling) rates from the coil walls mean greater rates of condensation of vapor in the vapor zone. When condensation is enhanced, parts dry

[45] The boiling point of perchloroethylene is quite high (121°C, 250°F). This allows CTW to be used as the primary coolant because the exit temperature of the CTW is around 95°F – allowing 155°F for a temperature difference to provide condensation of the perchloroethylene solvent. Certainly, CTW is the least expensive type of cooling for fulfilling this primary demand for cooling.

But the secondary demand for cooling cannot be fulfilled with CTW. The reason is that it is not cold enough to condense any solvent (including perchloroethylene) such that air emitted at the top of the degreaser is adequately free of contamination. Refrigerated brine is necessary for the secondary cooling coil.

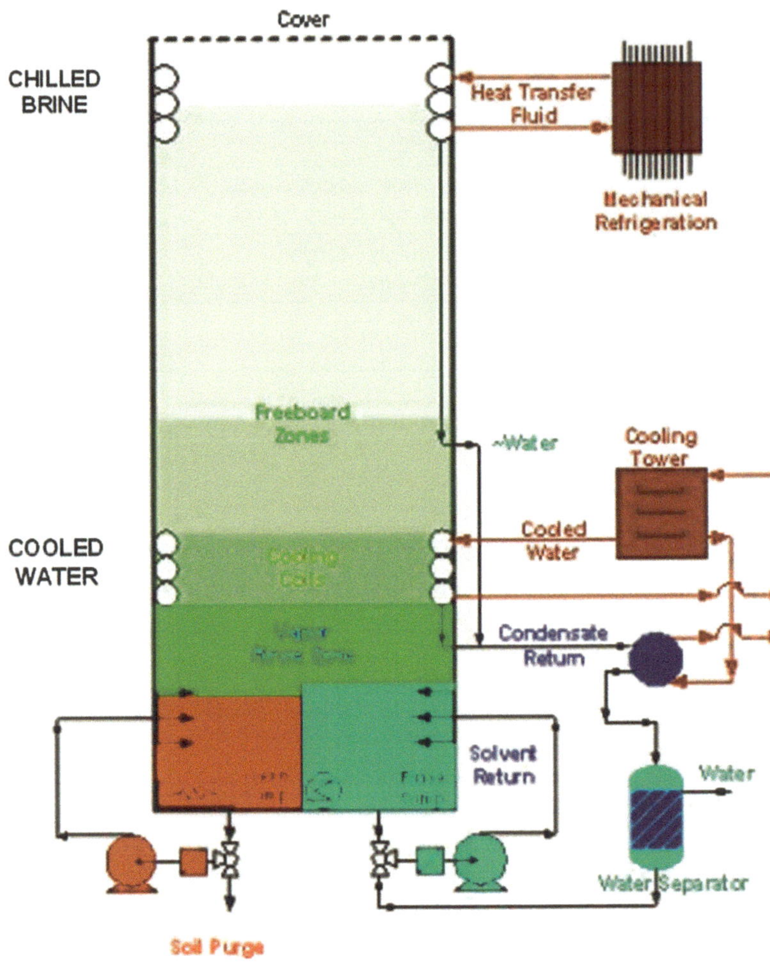

Figure 1.32 Arrangements of Cooling Apparatuses in an OTVDG

faster and likely more completely (Chapter 1.8), and when the cover is opened, the escaping air contains less solvent[46] with less need for lip vents to control emissions.

So selection of the coolant for an open-top vapor degreasing system is normally a choice between the three types of brines, or possibly CTW.

Based on experience, the designer of a vapor degreasing system should strive for a temperature *difference* between the cooling coil walls and the solvent normal boiling point of at least +23.9°F (75°C), and preferably as much as +51.7°F (125°C).

The selections of preferred coolant temperature in Table 1.7 were made by subtracting 51.7°F (125°C) from the solvent's normal boiling point to get a preferred coolant temperature. Then, the information about the freezing points of various possible coolants in Box 1.5 was used to make a selection of a coolant. One can note that if +23.9°F (75°C) were taken as the acceptable temperature difference between coil walls and solvent boiling point, the selection of preferred coolants would not change – however, the refrigeration duty, operating cost, cycle time, and solvent emission rate would change.

1.19.3 Selection of Coolants – Keeping Solvent In and Water Out of the Degreaser

The cooling coils at the top of the freeboard zone are nearly always kept at a lower temperature than the cooling coils above the vapor zone. This is because the former have the more rigorous and significant requirement – to keep solvent emissions low enough so that workplace limits are not exceeded.

[46] Recall that good practice requires the top cover to always be closed except when the parts basket enters and leaves the "open-top" degreaser.

Box 1.7 Use of Freeboard as the Final Control Element

The normal location for the secondary cooling coils is adjacent to the top of the degreaser freeboard – at or near the cover.

But some designers prefer to use an extra-tall freeboard length as the ("final" or topmost) limiting control element to retain solvent emissions within a degreaser – rather than a mechanical cooling unit. An increase in freeboard height extends the diffusion path which solvent molecules must take to exit from the top of the degreaser (Chapter 1.12.1). These designers believe that the use of freeboard height is a more economically efficient way of controlling emissions than is electrically powered cooling. So they locate the secondary cooling coils in their designs lower in the freeboard height – not adjacent to the cover – as shown in the image below. Both approaches are examined in more detail in Endnote V.

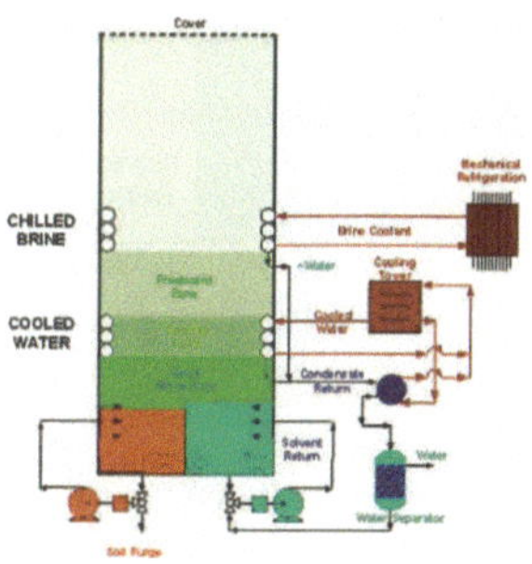

While the lower locations in a vapor degreaser contain pure solvent vapor, the material at the top of the freeboard section is air much less than saturated with solvent at some low temperature. It may also be contaminated with suspended droplets of solvent. As the parts (and basket) are lifted upward by the hoist, this air is entrained with them and pushed (or pulled) to the top of the freeboard, and so out into the work area.

A useful visualization may be to think of the parts basket as a very loose-fitting piston and the freeboard zone as a cylinder. The downward-moving basket displaces air when inserted into the freeboard area of the vapor degreaser. That air, escaping into the work area, is nearly saturated with solvent at whatever is the external temperature at the top of the freeboard zone. Correspondingly, when the parts basket is lifted upward from the freeboard area, air is entrained with it. That air is also presumably less than saturated with solvent at whatever is the internal temperature at the top of the freeboard zone, and it too escapes into the work area. This can be visualized using Figure 1.33 (basket removal from degreaser) and Figure 1.34 (oversized basket entry into degreaser), especially around the parts basket.

Table 1.7 Selection of Refrigeration Coolants for Producing Rapid and Effective Drying

Solvent	Boiling Point °C	Preferred Coolant Temp., °C	Preferred Coolant (in Water)
Trichloroethylene	86.7	17.3	Sodium Chloride
Methylene Chloride	39.8	−29.7	Propylene Glycol
Perchloroethylene	121.1	51.7	Sodium Chloride
n-Propyl Bromide	71.0	1.6	Sodium Chloride
Acetone	56.2	−13.2	Sodium Chloride
Methyl Acetate	56.9	−12.6	Sodium Chloride
t-Butyl Acetate	98.0	28.6	Sodium Chloride
Dimethyl Carbonate	90.0	20.6	Sodium Chloride
Methyl Formate	32.0	−37.4	Ethylene Glycol
Parachlorobenzo-trifluoride	139.2	69.6	Cooling Tower Water
Isopropanol	82.0	12.6	Sodium Chloride

Guidance about the relationship of the basket size to the size of the degreaser opening is that the basket should never (ever!) occupy more than one half of the opening into the degreaser[W]. Furthermore, the speed of basket entry or removal should never exceed 11 ft/min (0.055 m/s).

This emission of solvent can't be avoided unless the freeboard zone is flushed with nitrogen (as is done with enclosed machines). However, solvent emissions can be minimized by reducing the temperature of the air at the top of the freeboard zone so that more of the solvent is condensed, and retained within the degreaser. Liquid solvent can be removed from the entrained air because it is heavier than air, but vaporized solvent cannot be removed and escapes with air from the degreaser into the work area.

A second reason for reducing the temperature of the air at the top of the freeboard zone is to remove humidity (water) from the air entrained with the parts (and basket) incoming to the vapor degreaser.

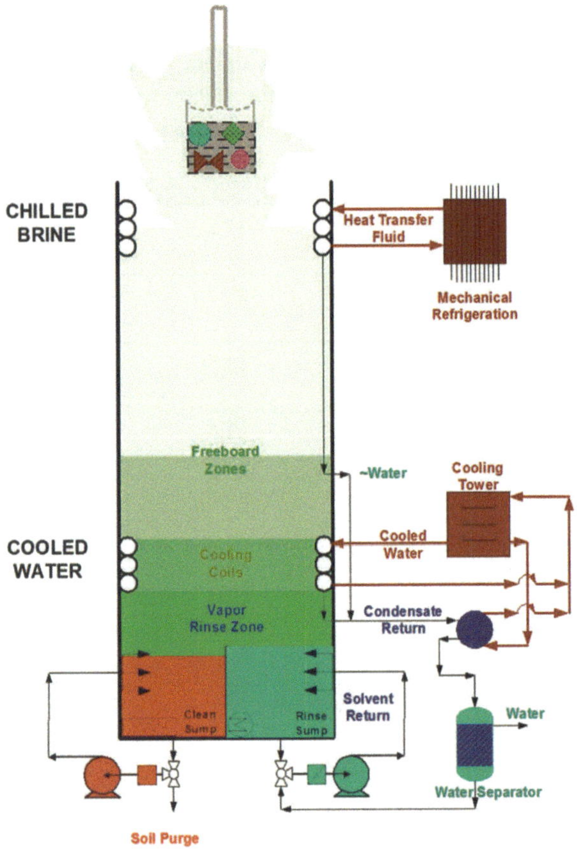

Figure 1.33 Basket Exit

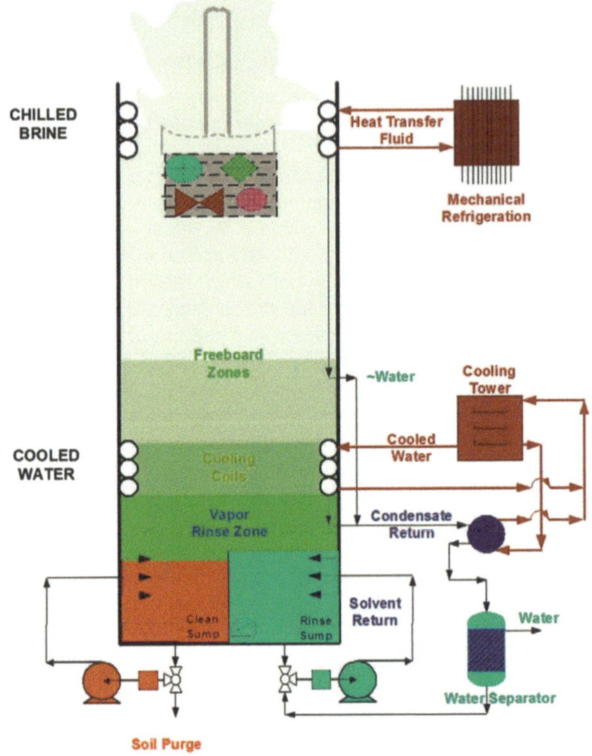

Figure 1.34 Basket Entry (Oversized Basket)

The top left side of Figure 1.5 shows material collected as condensate. This is a mixture of solvent and water. It is likely that the water concentrations will be higher than those of solvent among various machines and solvents.

Selection of a coolant for the top level of coils is based on a calculation[X] of the temperature at the top of the freeboard section which will enable the air/solvent emissions to meet the required exposure limit.

When one compares the refrigeration necessary to dry parts effectively with the primary cooling coil (Table 1.7) with the requirement that the secondary cooling coil is to keep emissions from exceeding the exposure limit in the work room (Table 1.8), the latter requires a lower coolant temperature for nearly every[47] solvent. This is the normal outcome – coolant temperature at the top freeboard coils is nearly always lower (sometimes substantially lower) than that just above the vapor zone.

[47] Note the exception to the "normal" outcome – acetone: a solvent with a high and easily met exposure limit, and a low boiling point.

Designers of vapor degreasers normally use only a single coolant fluid for each machine, with two different coolant loads (flows at the same temperature) directed to each set of coils to achieve the intended different results. This outcome is shown in Table 1.9.

Note that cooling tower water, useful with high-boiling solvents to dry parts (Table 1.7), is not useful with any solvents to control emissions at the top edge of the freeboard zone.

1.19.4 Degreasers with Two Sets of Refrigeration Coils

This arrangement is the one most commonly found in commercial vapor degreasers, as shown in Figures 1.32 to 1.35:

- One set of cooling coils ("primary condensing coils") operating at a relatively high temperature located just above the vapor zone, whose chief purpose is to:
  - Rapidly and effectively dry the parts, while keeping the boilup heater from filling the work room with solvent.

Table 1.8 Selection of Refrigeration Coolants for Controlling Emissions from the Top of the Freeboard

Solvent	Exposure Limit ppm	Calculated Top Temp., °C	Preferred Coolant (in Water)
Trichloroethylene	10	−44	Calcium Chloride
Methylene Chloride	50	−47	Calcium Chloride
Perchloroethylene	25	1	Sodium Chloride
n-Propyl Bromide	5	−65	Too Volatile
Acetone	750	4	Sodium Chloride
Methyl Acetate	200	−23	Propylene Glycol
t-Butyl Acetate	200	9	Sodium Chloride
Dimethyl Carbonate	200	17	Sodium Chloride
Methyl Formate	100	46	Calcium Chloride
Parachlorobenzotrifluoride	25	10	Sodium Chloride
Isopropanol	400	26	Cooling Tower Water

Table 1.9 Summary of Required Coolant Temperatures for Drying (Table 1.7) and Exposure (Table 1.8)

Solvent	For Drying	For Exposure	Preferred Coolant (in Water)
	Temp., °C		
Trichloroethylene	17.3	−44	Calcium Chloride
Methylene Chloride	−29.7	−47	Calcium Chloride
Perchloroethylene	51.7	1	Sodium Chloride
n-Propyl Bromide	1.6	−65	Can't Meet Exposure Limit
Acetone	−13.2	4	Sodium Chloride
Methyl Acetate	−12.6	−23	Propylene Glycol
t-Butyl Acetate	28.6	9	Sodium Chloride
Dimethyl Carbonate	20.6	17	Sodium Chloride
Methyl Formate	−37.4	−46	Calcium Chloride
Parachlorobenzotrifluoride	69.6	10	Sodium Chloride
Isopropanol	12.6	26	Sodium Chloride

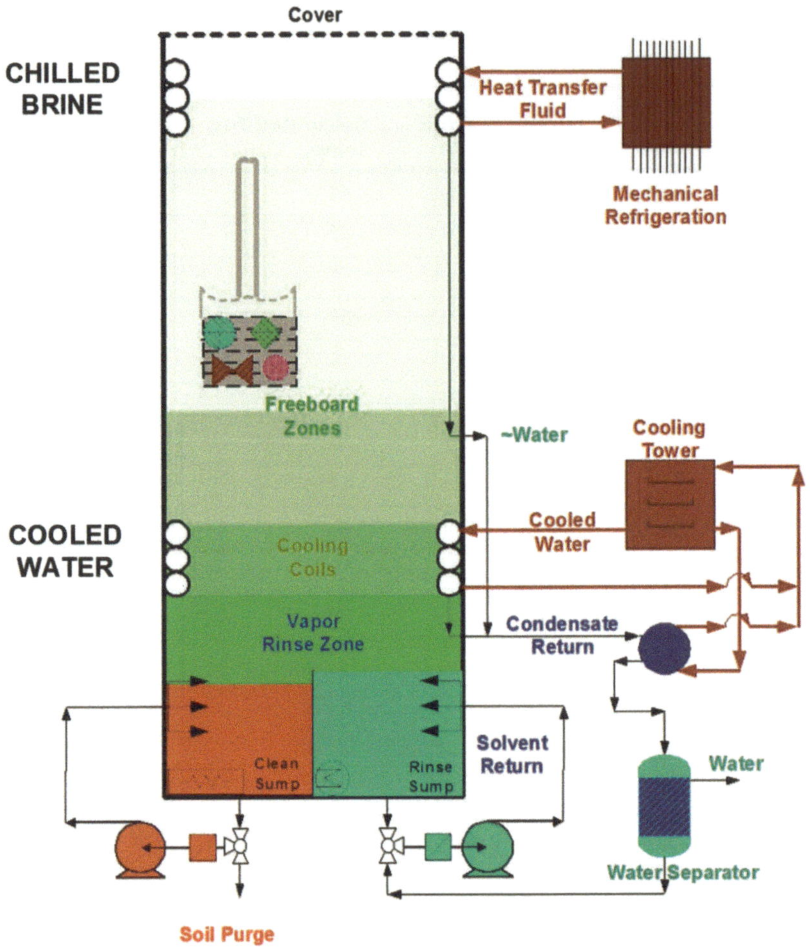

Figure 1.35

- A second set of cooling coils ("freeboard chiller") operating at a relatively low temperature located just below the top of the freeboard zone, whose chief purpose is to:
  - Contain solvent within the degreaser to meet emission requirements, while condensing as much water as is possible from the inlet air.

Water from a cooling tower (CTW) may be used in the primary refrigeration coils, but it is normally not adequately cold enough to constrain emissions to be used in the secondary refrigeration coils (Box 1.6).

A common refrigeration package feeds coolant to both sets of refrigeration coils, even though the operating temperatures of each are quite different. It is the flow of coolant to each which is varied.

The equipment arrangement of Figure 1.35 shows how it is recommended to feed two different coolants to each set of coils in a vapor degreaser.

- The coldest coolant, that emanating from the bottom of the cooling tower or the refrigeration machine, should flow to the uppermost coil of the set.

In this way, the vapor leaving the vapor degreaser is contacted with the coldest coolant, thereby effecting the greatest possible reduction of solvent content in the emitted air at that upper location. This is shown in Figures 1.32 to 1.35.

- This second (upper) refrigeration coil is the third in a series of barriers which vaporized solvent must overcome to be emitted into the work area.

  Obviously, the initial and most relied-upon is the primary chiller just above the vapor zone. The second is the length of freeboard between the primary chiller and the upper refrigeration coil – the longer this diffusion path is, the less amount of solvent will diffuse to the top of the degreaser (exceptionally short in Figure 1.5). Also obviously, the third barrier is the upper refrigeration coil. The effects of the second and third barriers are compared in Endnote V.

Note also in Figure 1.35 that the material condensed in the top coils is presumed to be mostly water with a small amount of solvent (because of the three barriers noted above) as it is fed to the water separator.

1.19.5 Degreasers with Three Sets of Refrigeration Coils

Such an arrangement of refrigeration coils as shown in Figure 1.36, is claimed in USPs 5,142,873 and 5,048,548, and noted in Chapter 1.13. It is not in common use today, but this technology was a staple for use with CFC-113 before that solvent was banned by the Montreal Protocol.

The two upper refrigeration coils can be fed with mixtures of salts or glycols in water (brines) – mixtures which have freezing points lower than that of pure water.

The lower refrigeration coil can be fed with any cold fluid, CTW or brine – depending upon the boiling point of the solvent used. CTW is shown used in Figure 1.36.

The temperature of the air cooled by the middle refrigeration system must be no more than 0°C (+32°F). Propylene glycol in water can be used as that refrigeration fluid.

- Liquid in the air contacted by the middle refrigeration system, presumed to be pure solvent, is condensed, and then mixed with the solvent condensed by the lower refrigeration system.

The temperature of the air cooled by the uppermost refrigeration system must be no more than 13.3°C (8°F) lower than the temperature of the middle refrigeration system (when the patented three-coil configuration is

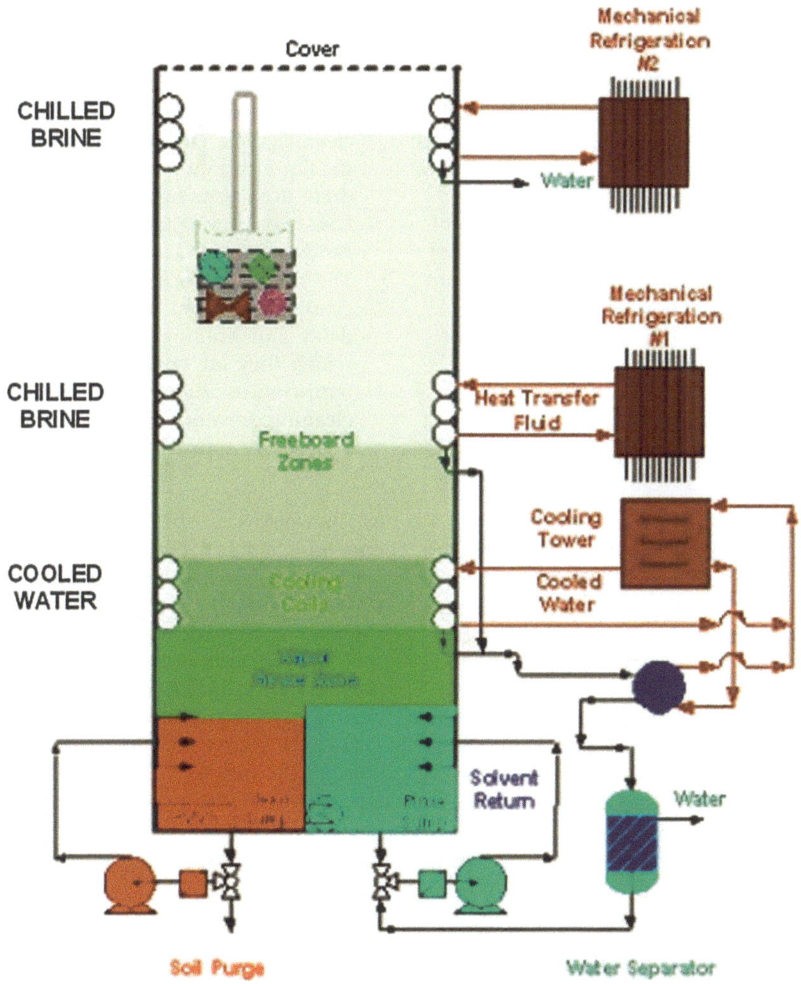

Figure 1.36

used). Calcium chloride in water can be used as that refrigeration fluid.

- The uppermost refrigeration system is tasked with condensing water from whatever air enters the top of the degreaser when the cover is removed. That condensate is discarded, so as to not overload the water separator.

While three different refrigeration systems could be used as shown in Figure 1.36, it is highly likely that only one (or perhaps two) would be used in the same degreaser.

1.19.6 Summary of Criteria for Choice of Refrigeration Systems

The two basic requirements for refrigeration are to (1) dry parts well and quickly, and (2) keep solvent within the degreaser. The selected coolant is one which efficiently allows attainment of the desired internal temperature and allows the degreaser to fulfill those requirements.

Specific requirements are:

- *Drying* – The selected coolant in the primary cooling coil will provide a temperature difference between the normal boiling point of the solvent and the coolant temperature of at least 23.9°C (75°F), and possibly as much as 51.7°C (125°F). Examples are given in Table 1.7.
- *Solvent Containment* – The selected coolant in the secondary cooling coil will provide a concentration of solvent in the work area no more (and hopefully less) than the posted exposure limit when the emission from the degreaser is diluted with normal ventilation. Examples are given in Table 1.8. The combination of both requirements are given in Table 1.9

1.20 SUPERHEAT IS HOT!

A core problem with aqueous cleaning systems is *dragout* – films of water/detergent/soil which attach to parts, clinging to crevices or lips or threads or wires, filling holes (blind or open), or being trapped within the internal conformation of multiport valves or cupped sections or annular sections. So, dragout is the solvent film remaining on parts when they exit the degreaser.

The property of fluids which engenders dragout is surface tension. Even with surfactants added to water to reduce that property from 71 to ~50 to 40 dynes/cm, the propensity to form dragout is still a significant problem with aqueous cleaning technology.

Dragout is also a problem with those organic cleaning solvents whose surface tension may range from 31 (perchloroethylene) to 29 (trichloroethylene) dynes/cm, though somewhat less so with values around 24 dynes/cm (acetone and methyl acetate), and significantly less so with values around 14 to 15 dynes/cm (HFEs/HFCs/OSs).

Dragout is therefore a negative outcome of parts cleaning because of both the surface character of the parts and the surface tension of the cleaning solvent.

- The specific negative outcome is still-soiled wet parts. Dragout fluid, whether water or solvent based, contains soluble or suspended soil, and unwanted liquid.

Parts that are still-soiled and wet after removal from the degreaser are of concern to two organizations: (1) those who manage the cleaning work because they are concerned about quality of cleaning, and (2) those who manage environmental affairs because they are concerned about solvent emissions. Dragout is generally the largest source of degreaser emissions. Solvent losses through dragout are three to eight times greater than diffusion losses (Chapter 1.14).

Designers of solvent cleaning systems employ several tactics to minimize the propensity of solvent to form dragout films and/or their effects.

- The most profitable approach to avoid formation of dragout films is to design parts, and the baskets into which they are fixtured[48], so that fluid is not retained; and dragout is not created.
- The most profitable approach to managing dragout, if it is formed, is to impose a delay (called dwell time) in the cleaning cycle. During this time the parts are held above the liquid sumps and given time for dragout fluid to drain from parts under the influence of gravity[49]. Low-frequency agitation or rotation of the parts basket can help as well.
- The final approach to managing dragout fluid is to evaporate it from part surfaces by imposing a time delay and holding the parts (and the basket in which they are contained) in a zone where the temperature is above the boiling point of the cleaning solvent.
  - This last approach is superheat technology (Figure 1.42).

As indicated by the title of Chapter 1.14.1, vapor degreasers are a maelstrom of energy and material flows. Superheat is yet another source of both – happening in a small volume!

Energy is supplied to the solvent vapor by convection and radiation from heated tubes. The tubes are often finned (Figure 1.37[Y]) to increase the area of heat transfer from the heating tube to the solvent vapor.

And the source of the heat is often hot oil passing through the inside of the finned tubes. The oil is heated externally from the vapor degreaser to eliminate a source of ignition energy.

---

[48] It is quite common that the parts basket becomes the chief source of dragout – more so than the parts! Choices for basketing are covered in Ref. 1, Chapter 7, especially Figures 7.25 and 7.26.
[49] This time delay, to be effective, can be as lengthy as one half of the total cycle time. The reason for this is explained in detail in Ref. 1, Chapter 1, Section 12.5, especially Figure 1.5.

Figure 1.37 Interior View of Open-Top Vapor Degreaser Showing Superheat Coils

Note the entire assembly of superheating and adjacent cooling coils within a vapor degreaser in the diagram of Figure 1.37 – the superheat coils are below the normal cooling coils.

An internal thermocouple is shown immediately above the superheat coils in Figure 1.37. Its purpose is to verify by continuing measurement that the vapor temperature is above, by the required amount, the solvent boiling point.

A minimum value for the increase is 10°F (5.6°C)[50]. A maximum value is about 50°F (27.8°C) (Box 1.8).

One must notice in Figure 1.37 the relative proximity of the superheat coils to the cooling coils. And so note that some of the heat added by the superheat coils must be removed by the cooling coils before the parts (and their basket) can be returned to the work area – clean, cool, and dry.

Delays in a cleaning cycle are never wanted – except when they bring a positive return for the time expended. The delay imposed by superheat technology is typically determined by the manufacturer based on their experience. But typical values are between 15 and 45 seconds. A programmed hoist (Chapter 1.22) is essential to assure that this delay (called dwell) occurs at the right time for the right duration.

[50] The 1994 NESHAP required a minimum of 10°F (5.6°C) increase for superheat to be considered acceptable as a control device to reduce solvent emissions (the reference of Endnote Y).

Box 1.8 **Technology of Parts Drying with Superheat**

Per USP 6,026,588

USP 6,026,588 (to Forward Technology Industries) describes a vapor drying system for use with isopropanol which reveals considerable detail about alternate designs of superheat technology. Automated continuous machines are also described which can clean small baskets of parts with a short cycle time. An example is shown in the figure above.

USP 5,085,238 (to Branson Ultrasonics) claims a more conventional system as described in Figure 1.42.

Per USP 5,799,677

USP 5,799,677 (to Finishing Equipment Corporation) cleverly claims treatment in an automated mode with a solvent in a hermetically sealed enclosure. Drying of surfaces by raising of the environment to a temperature above the normal boiling point isn't claimed. What's claimed is an apparatus, divorced from the main degreaser, in which surfaces and volumes of parts can be exposed to temperatures above the normal boiling point. This apparatus was most commonly applied in the 1990s to vapor degreasers in which methylene chloride or isopropanol was the solvent. The nature of this apparatus is that it adds significant height to an already tall vapor degreaser. Opposite to the arrangement shown in Figure 1.37, the superheat coils are above the cooling coils, as shown in the image above.

1.21 LIP VENTS – NOT A REALLY GOOD IDEA

Lip vents are narrow slits in the degreaser freeboard zone. They are almost always located at the top of the freeboard zone – adjacent to the top cover. And they almost always cover nearly the entire periphery of the degreaser.

Vapor degreasers, filled with boiling solvent, have long had lip vents to reduce workplace area concentrations of solvent. Cold cleaning tanks commonly (and unfortunately) use lip vents for this purpose as well.

Lip vents represent a bargain. The bargain specifics are to provide reduced solvent content in the air adjacent to the OTCDG in return for increased solvent loss in air emissions from the apparatus. Almost always, it's a poor bargain. However, lip vents installed on a vapor degreaser do provide a benefit – just not at a fair price.

- Lip vents do/can reduce local concentrations of cleaning solvent in the air around the work area in which the vessel with lip vents is located.

In operation, an exhaust blower, mounted external to the degreaser, pulls vapor from the top of the freeboard zone, through the lip vents, and out of the work area[51]. Presumably, the removed vapor is rich in solvent fumes.

A diagram of the top of a hypothetical vapor degreaser with lip vents and top cover is shown in Figure 1.38.

- When the cover is sealed, there is no need (and little ability[52]) for the external blower (or air amplifier) to try and remove vapor.
- When the cover is loosened, the seal at the top is broken, the air amplifier can remove solvent-laden vapor from the vapor degreaser.

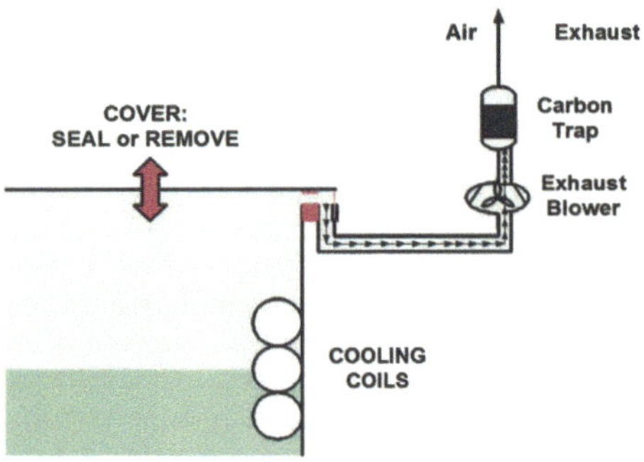

Figure 1.38 Arrangement of Components of Lip Vent System

[51] The flow rate of air exhaust removed by the blower should not exceed about 60 CF/min-SF of degreaser opening. However, air flow rates through lip vents are commonly 5 to 10 times that flow rate.

[52] Open-top vapor degreasers are not capable of supporting vacuum, nor are exhaust blowers capable of producing it.

The vapor removed, which is mostly air ripe with solvent, is a moving stream of air pollution. Traditionally, the stream was directed to an exhaust vent from the work area. As with many aspects of solvent cleaning, the 1994 NESHAP for halogenated solvents mandated that, under the equipment control options, the exhaust from lip vents be directed to a bed of activated carbon[Z] for removal of solvent from the moving stream by adsorption.

That's the problem with lip vents. Use of them trades reduction of solvent concentrations in the work area (Box 1.9) for two costs – purchase of lost solvent, and the very inefficient (Ref. 5 use of activated carbon adsorbent to capture it – a costly and losing tradeoff [53,54]!

But there is an even worse outcome of misusing lip vents:

When the cover is removed, the exhaust blower can (and will) remove copious quantities of solvent-laden vapor from the vapor degreaser. Even worse, solvent vapor can diffuse into the work area, raising the concentration of solvent in ambient air.

This disaster is shown in Figure 1.39. Note how the flow streams created by the blower entrain vapor from the freeboard area toward the opening made by removal of the cover. This outcome is properly described by the second word of this paragraph.

In the event that it is decided/necessary to use lip vents, direct their output to a bed of adsorbent and not to a vent to the external environment (Chapter 4.6 and Figure 4.11).

> **Box 1.9 How to control workplace Emissions**
>
> The presence of noticeable (by measurement or by smell) concentrations of solvent vapor in the work area does not mean that lip vents should be engaged. The culprit causing the presence of solvent vapor is likely solvent dragout on parts (longer dwell time or superheat needed), splash out by sprays (improved cover or operator training needed), or leaks in external fittings (plumber needed). Lip vents will not capture or restrict these sources of emissions.

[53] *"...Using a degreaser without lip vents, or with the vents turned off, will release 15% less solvent emissions than a degreaser with a vent. If lip vents are required to meet workplace health rules, use covers on the degreaser and shut off the vent when the cover is closed...."*

Source: The Minnesota Technical Assistance Program (MinTap), http://www.mntap.umn.edu/mach/resources/70-VaporDegreasers.htm.

[54] *"...Although the lip vent lowered workplace concentrations, it also greatly increased TCA emission rates. Eliminating the lip vent with the simultaneous addition of a combined freeboard extension/add-on chiller provided about the same workplace concentration of TCA, but with a decreased discharge rate of TCA. The rate of consumption of TCA was reduced by about 75% with this configuration while maintaining the same degree of worker safety as provided by the lip vent."*

Source: Szabo, M.F. and Nutter, M.T., EPA report #600S2–90052, March 1992.

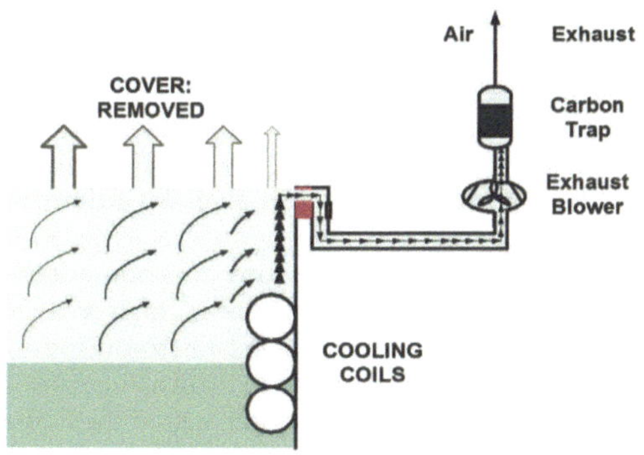

Figure 1.39 Operation of a Lip Vent System

In summary, if lip vents are provided with a vapor degreaser, don't use them. If the supplier recommends lip vents, consider another supplier.

1.22 PROGRAMMABLE HOISTS – A REALLY GOOD IDEA

Like a kit for measuring cleanliness sold with aqueous cleaning agents, one ought to find a programmable two-axis hoist sold with every vapor degreaser. Alas, neither situation prevails in the commerce of cleaning technology.

Such a device allows for simple, safe, repeatable, and precise use of a vapor degreaser, by:

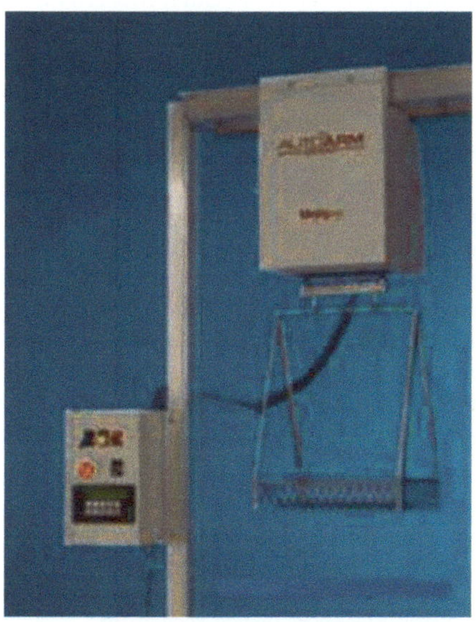

Figure 1.40 A Modern Programmable Hoist System for a Small OTVDG

- Assisting the operator so they may attend to other events during the cleaning cycle and aren't physically burdened by trying to hold a heavy load of parts at a certain position for a certain time within the degreaser.
- Assisting the organization, and the operator, by minimizing or eliminating their contact with solvent vapor.
- Minimizing levels of solvent emission by:
  - Maintaining an entry and exit speed for the basket to not exceed 11 ft/min.
  - Holding the parts basket (and parts) in the vapor, superheat, or freeboard zones for the appropriate time periods.
  - Jogging the basket to enhance drainage of liquid from the parts basket (and parts).
- Conducting the preferred cleaning cycle of holdup times in each sump and positions every time.
- Maintaining an electronic record of compliance (or not) with NESHAP or local requirements.

In addition, modern degreasers have the capability to link their temperature, other control systems, and ultrasonic transducers to a programmable hoist. One such device is shown in Figure 1.40[55].

1.23 AVOID THE DRAFT

Inside the perfect vapor degreaser there should be no movement, since one doesn't want to mix solvent vapors with air. Unfortunately, perfection is not possible as every vapor degreaser was constructed, purchased, and used to wash baskets of parts which do move into, within, and out

55

The image of Figure 1.40 is courtesy of Unique Equipment Corporation, http://uniqueequip.com. In machines with larger solvent volume, where parts are larger and heavier, hoists are commonly used.

Use of programmable hoists is an opportunity to both save labor and improve reproducibility of cleaning operations. A commercial example is pictured above from the same source. In use, the human operator fills the parts basket at left, and activates the process with the "Start" button: the basket is lifted, the cover opens, the basket is lowered, all steps of the cleaning cycle are completed, the basket is raised, the cover opens, the basket is placed on the open shelf at right, and the cover is closed. The operator is only needed when dirty parts constrain operation or when the clean parts are needed.

of the degreaser. It can't be helped; air and solvent vapor will mix. So the closest approach to perfection, then, is to minimize the mixing of solvent vapor from within the vapor degreaser with air from the work area. That means minimizing room drafts.

Room drafts cause turbulence within the freeboard volume of the vapor degreaser, which increases the mixing of solvent vapors and room air. Room drafts then sweep the solvent vapor from the solvent cleaning machine and emit the vapors to the work environment. The higher the draft velocities, the greater the emissions of solvent.

The most effective method of isolating solvent vapor from the room draft within the vapor degreaser is to use a cover[56] (Chapter 1.13). There is no alternative. A proper cover should slide easily over the mouth of the freeboard section (see Figure 1.25). A cover which is lifted or hinged should not be used. Each time a hinged or lifted cover is opened, drafts and turbulence carry out vapor. Covers must slide to be a control measure under the 1994 Halogenated Solvents NESHAP (Endnote W).

At some work sites, room draft can't be reduced because of other ongoing activities or the need to relieve heat stress of workers. In that case, one can enclose the open-top degreasers – the second most effective method of isolating solvent vapor from room air.

Full enclosures (Figure 1.41) eliminate external drafts and greatly reduce the opening for diffusion losses. They can reduce volumetric solvent loss by up to 40%. Enclosures can reduce exposure of some workers by creating separation between the solvent emissions and operators in the work area outside the enclosure. Yet, as shown in Box 1.8 (overly tall degreaser) and Figure 1.22 (degreaser in a pit), enclosures can be difficult to fit to a large vapor degreaser. The crucial drawback of enclosures is that they can concentrate whatever solvent vapor is emitted into the small volume of an enclosure – producing elevated exposures to the workers within the enclosure. In some work areas, a partial enclosure may protect all workers.

Figure 1.41 Enclosure for Operating Vapor Degreaser

The 1994 Halogenated Solvents NESHAP limits air speeds above a degreaser to less than 50 ft/min unless idling covers are used.

Generally, substantial reductions of room drafts are difficult to obtain. This is because a high velocity of ventilation, if present, was installed for some continuing reason. To the extent that reason isn't still valid, substantial reductions are possible. Said another way, if the work area is not air conditioned, and workers require movable high-velocity fans for comfort, reduction of room air draft to reduce solvent emissions from a vapor degreaser will be difficult for workers to accept.

An ideal outcome would be to provide a ventilation experience that would: not suspend particles whose size is 10 microns or larger (to keep clean parts clean), have a relative humidity not exceeding 50% (to minimize water intrusion into the degreaser with the parts), provide an adequate volumetric turnover of fresh air of 5 to 10 changes per hour (for employee comfort), and not exceed a laminar face velocity from a unidirectional diffuser greater than 30 ft/min (0.15 m/s)[57] (to minimize solvent diffusion from the degreaser).

1.24 SUMMARY ABOUT EMISSION CONTROL TECHNOLOGIES

Call it cost reduction through solvent conservation, or call it environmental protection, or call it concern for employee health and comfort, reduction of emissions was a good idea in the 1990s to comply with the NESHAP; and if it hasn't been fully implemented through publication of this book, it is then long overdue!

Significant items of concern are listed throughout this chapter. Their quantitative effects on emissions have been estimated by the US EPA, suppliers of equipment, and local regulatory agencies.

Naturally, degreasers differ in their design and features, installation requirements, and use parameters. So, specific percentage reductions in emission will be different for every degreaser.

If one is considering reducing the risk of either non-compliance with local or federal regulations, or the threat

[56] The desired device is called a working mode cover, because it is easily movable when a basket of parts is to be inserted or removed. Covers which are only in use when the vapor degreaser is idling between workloads are not working mode covers, and are next to useless.

[57] As to the method of introduction of clean air and removal of room air, the 2010 ASRAE handbook recommends:

> *"...the delivery of air from the ceiling, with a downward movement to several exhaust inlets located on opposite walls, is probably the most effective air movement pattern for maintaining the concentration at an acceptable level...."*

Table 1.10 Impact of Options to Reduce Emissions

Option to Reduce Emissions	Potential for Reduction
Install working mode cover	40%
Increase freeboard ratio from 0.75 to 1	20%
Increase freeboard ratio from 1 to 1.25	10%
Increase freeboard ratio from 0.75 to 1.25	25%
Add above- or below-freezing freeboard chillers	40%
Reduce room drafts	50%
Add lip exhausts to a carbon adsorber	40%
Reduce hoist speed from 60 to 11 ft/min	35%
Increase dwell time	30%
Superheated vapor	50%
Operate closed-loop equipment (Chapter 2.1)	95%

to employees of overexposure to solvent vapor, the question of where to start is appropriate.

This author's answer is to start anywhere. Specific predictions are provided in Table 1.10[AA]. This author's opinion is that given the differences among operating machines, all the items listed in Table 1.10 may be considered as good places to start, or finish[58], as they all have "similar" expected percentage reductions in emissions.

The important point is not the method (it's not the journey), but the outcome (it's the destination). Every open-top vapor degreaser should be capable of performing vapor degreasing cleaning with a solvent loss rate not exceeding 0.04 lb/hr-SF; no exceptions, and hopefully less.

1.25 A VAPOR DEGREASER – AN ENERGY TRANSFER UNIT

The image of Figure 1.42 (open-top vapor degreaser with superheat coils) illustrates the essence of solvent cleaning technology – heating and cooling of the right solvent at the right time and for the right duration.

- Heating of parts thins oils and greases, making them more mobile when immersed in the right solvent in the cleaning sump (red color).
- Cooling of parts (light blue color) makes it possible to first rinse with more clean liquid and then for solvent vapor to condense upon the surface of the parts – providing further rinsing with pristine solvent (lighter green color).
  - This pair of operations may be repeated as often as necessary.
- Reheating (superheating) the parts to rid them of entrained liquid films (dark green color).
- Cooling, and rapidly drying the parts by evaporation of solvent vapor (light green color).
- Lengthening the diffusion path for solvent vapor to keep it within the cleaning machine (very lightest green color).
- Condensing some water from the humid air entrained into the degreaser with the parts and basket, in addition to acting as the "goalie" in saving solvent from exiting to the work area (white color).

It's interesting to note that aqueous cleaning technology consumes substantially more energy (more total BTU, kilocalories, or joules) than does solvent cleaning technology, but energy transfers are more significant in solvent cleaning technology. This higher energy consumption occurs because water has a much higher specific heat and a much higher latent heat of vaporization than do vapor degreasing solvents.

Yes, selection of the right solvent is crucial for successful solvent cleaning operations (Ref. 2, Chapters 2 and 3), but so is transfer of energy to the solvent and by the solvent in the right amount at the right time for the right duration.

1.26 LOSS OF VAPOR CONTAINMENT IN LARGE VAPOR DEGREASERS

The image of Figure 1.43 suggests a potential problem – that the cooling effect produced at the sidewalls of a vapor degreaser may not extend far enough into the maw of the freeboard zone above the cooling coils.

That suggestion has come to fruition in the idealized image of a large vapor degreaser shown in Figure 1.44, and in the outcome of a loss of containment of solvent vapor.

- For it is the cooling coils that maintain a zone of solvent vapor, which is the primary barrier that keeps nearly all the cleaning solvent inside the open-top degreaser. The barrier rejects passage of additional solvent vapor upward through it by condensing the vapor and returning the condensed solvent to the sumps as liquid droplets (light blue circles in Figures 1.44 and 1.45).

What if the flow of coolant to the coils is deliberately stopped, unintentionally or intentionally blocked or

[58] Don't make the mistake of treating the estimates of Table 1.10 as additive. They are not. At best they are multiplicative. A total reduction of 70 or 80% of prior solvent use would be an excellent outcome.

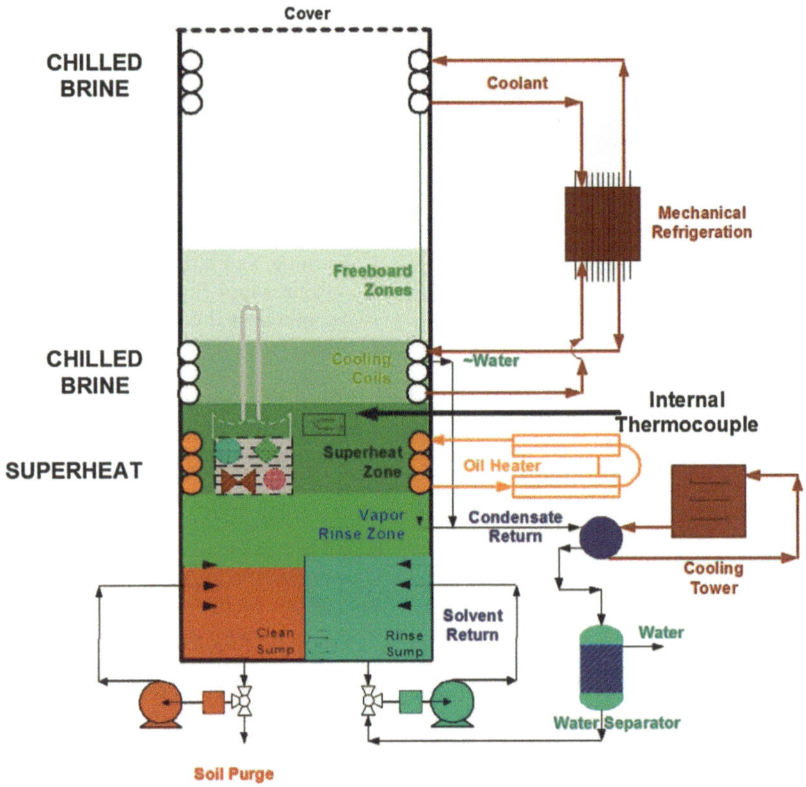

Figure 1.42 Open-Top Vapor Degreaser with Integral Superheat Coils

limited, while the boilup heater maintains its supply of heat? Why then, the vapor zone (the vapor barrier or blanket) starts to disappear, because the cooling effect which supports it is diminished or depleted.

Another way that can happen is to increase the size of the vapor degreaser relative to the capability of the cooling coils to remove heat. This is shown in Figure 1.44. Note in this figure that the vapor barrier (blanket) has collapsed in the center portion of the vapor degreaser. This is because there was not adequate cooling capacity supplied from adjacent to the degreaser walls to maintain it.

Figure 1.43 Cooling Overpowers Solvent Vapor

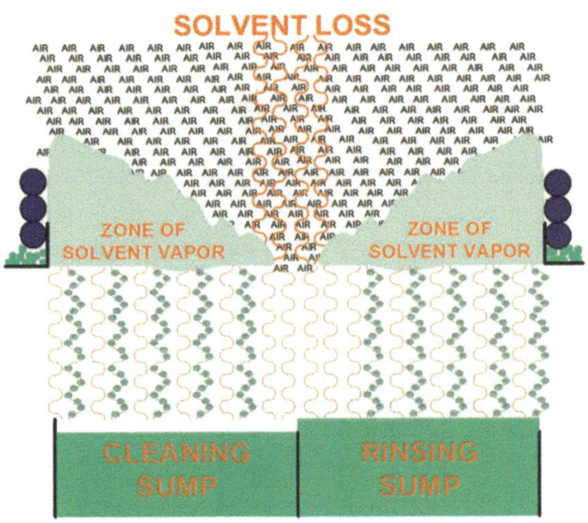

Figure 1.44 Solvent Vapor Overpowers Cooling

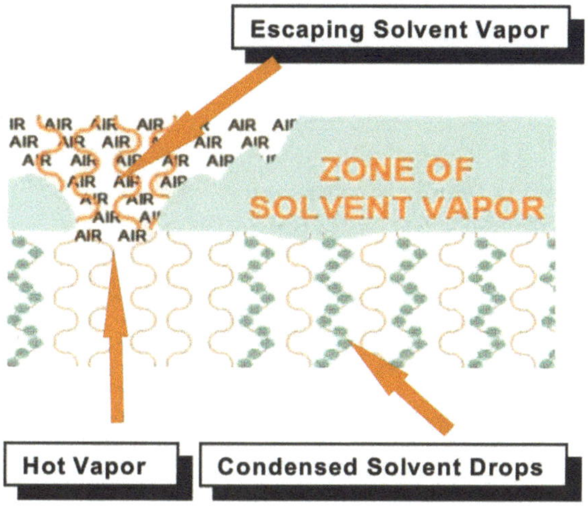

Figure 1.45 Closeup View of Figure 1.44 Showing How Inadequate Cooling Allows Hot Vapor to Penetrate Solvent Vapor Cloud

Figure 1.45 is a cutaway section from Figure 1.44. In the former one can see that without the cooling capability supplied from adjacent to the walls, there is nothing to stop vaporized solvent from rising to the top of the freeboard zone, and almost certainly escaping into the work area.

Note also in Figure 1.44 that where the vapor barrier has collapsed, there is no return of liquid droplets (light blue circles) to the degreaser sumps through the condensate trays[59].

There are two causes for the inadequacy of the cooling effect: (1) reduction of coolant supply for whatever reason, or (2) overlong separation of the center area of the vapor degreaser from the cooling coils adjacent to the degreaser walls. The former cause is an operational affair; the latter cause is that the cooling unit is undersized for the combined width of the two sumps of liquid solvent.

The latter cause is a real limitation on the implementation of vapor degreasing technology. Simply, one generally can't clean in a vapor degreaser parts of some large width dimension – such as airplane wings. Transfer of heat through vapor by convection and radiation is inefficient and ineffective over distances measured in feet.

The design *width* of the vapor degreaser is the key parameter, as shown in Figure 1.46. It's the shorter distance between opposite cooling coils. Cooling must be effectively implemented over a maximum of one half of that distance.

[59] Recall Figures 1.4 and 1.5 which show and illustrate, respectively, how condensed solvent normally returns to the sumps after condensation *upon* the primary cooling coils. It drips from the coils and then falls into a narrow channel or tray (noted at left in Figure 1.5). From the tray, condensed solvent drains into the water separator where it is decanted (separated by gravity) and returned to either the solvent storage tank or the rinse sump.

Solvent condensed within the vapor zone drains immediately to whatever sump is beneath it – bypassing the water separator.

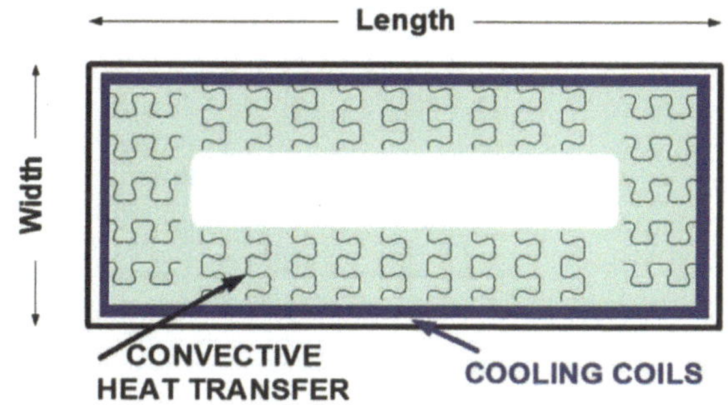

Figure 1.46

Widths of degreasers seldom exceed 30 inches; a degreaser width of 48 inches would probably exceed that of most commercial units.

The white center area of Figure 1.46 represents inadequate cooling relative to the width of the degreaser; it represents an area through which solvent vapor can escape upward from the vapor degreaser. The light blue peripheral area of Figure 1.46 represents adequate cooling relative to the width of the degreaser; it represents the area where the vapor barrier is established.

One can test the integrity of the vapor barrier by measuring the temperature in the center of the degreaser at the level of the freeboard area[BB] (above the vapor barrier). It should be significantly lower than the normal boiling point of the solvent – with the understanding that it will not be as low as the coolant temperature. Should it approach the former high value, the vapor barrier has been destroyed; should it approach the latter low value, the vapor barrier is sufficiently intact.

- A metric recommended by the US EPA[CC] is that the measured temperature in the centerline of the degreaser at the level of the freeboard zone (above the cooling coils) should not exceed 30% of the solvent normal boiling point.

Values of the maximum allowable (per EPA NESHAP) temperature at the centerline of the degreaser (Figure 1.45) are given in the second column from the right (column head in pink) in Table 1.11. The cooling coils at the wall must keep the vapor temperature in the center of the degreaser below this value (26°C for trichloroethylene).

- This goal will be more achievable the more narrow is the degreaser and the colder is the cooling coil.

So the width of the hypothetical degreaser can be no more than whatever width causes the vapor temperature at the centerline to rise from the freezing point of the coolant to the maximum allowable temperature (–21°C and 26°C respectively for trichloroethylene – a 47°C rise).

With no exceptions in Table 1.11, the width of the degreaser can be expanded for every solvent until the

Table 1.11 Limitation on Temperature at Center of Vapor Barrier (Blanket)

Solvent	Coolant Chosen	Coolant Freezing Point, °C (Footnote 41 & Box 1.5)	Available Temp. Rise as Width of Degreaser Expands °C	Maximum Vapor Barrier Temp., °C (BP x 30%)	Boiling Point, °C
	Chosen to Meet Exposure Limit (Table 1.9)				
Trichloroethylene	Sodium Chloride	−21	26 − (−21) = 47	26.0 = 86.7 x 30%	86.7
Methylene Chloride	Propylene Glycol	−28.9	11.9 − (−28.9) = 40.8	11.9	39.8
Perchloroethylene	Sodium Chloride	−21	36.3 − (−21) = 57.3	36.3	121.1
n-Propyl Bromide	Sodium Chloride	−21	21.3 − (−21) = 42.3	21.3	71.0
Acetone	Sodium Chloride	−21	16.9 − (−21) = 37.9	16.9	56.2
Methyl Acetate	Sodium Chloride	−21	17 − (−21) = 38	17.0	56.9
t-Butyl Acetate	Sodium Chloride	−21	29.4 − (−21) = 50.4	29.4	98.0
Dimethyl Carbonate	Sodium Chloride	−21	27 − (−21) = 48	27.0	90.0
Methyl Formate	Ethylene Glycol	−40	(9.6 − (−40) = 49.6	9.6	32.0
Parachlorobenzo-trifluoride	CTW	0.0	41.8 − (0) = 41.8	41.8	139.2

temperature in the center of the vapor barrier has increased at the degreaser wall by *about* 45°C (column head in yellow) from the freezing point of the coolant. It is impossible to know what this width is in units of length because the character of the parts and their basket are not known.

- The point of this calculation is that a large degreaser used with one degreasing solvent can probably be used with another solvent and another coolant with the same loading of parts without violating the EPA guidance about centerline temperature. Obviously, the two solvents may impose other concerns.

If a coolant with a lower freezing point had been chosen, such as calcium chloride (FP = − 52°C), the width of the degreaser could be increased without violation of the EPA guideline about emission control so as to accommodate larger baskets of parts. Obviously, that would substantially increase the energy costs for providing chilled coolant.

One can see from this discussion, and the associated images of Figures 1.44 to 1.46, and Table 1.11, that low temperature refrigeration is critical to successful vapor degreasing in open-top machines. Operating cost spent on electricity to provide refrigeration is well-justified (usually). Low temperature refrigeration: keeps the solvent in the tank to save cost, protect employees, and the environment; enables rapid drying of parts; and more than occasionally allows lower-price open-top machines to perform well enough to avoid consideration of the more expensive enclosed machines (Chapter 2.9).

The important point in Table 1.11 isn't just that there are preferred values for the coolant and freeboard air temperatures. The important point is that these values affect the temperature at the centerline distance from the degreaser

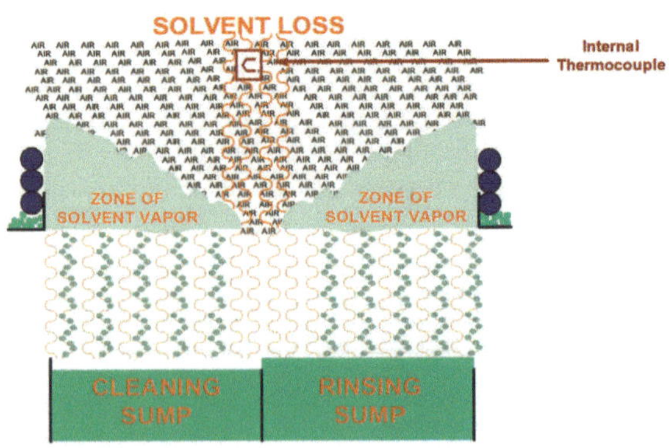

Figure 1.47

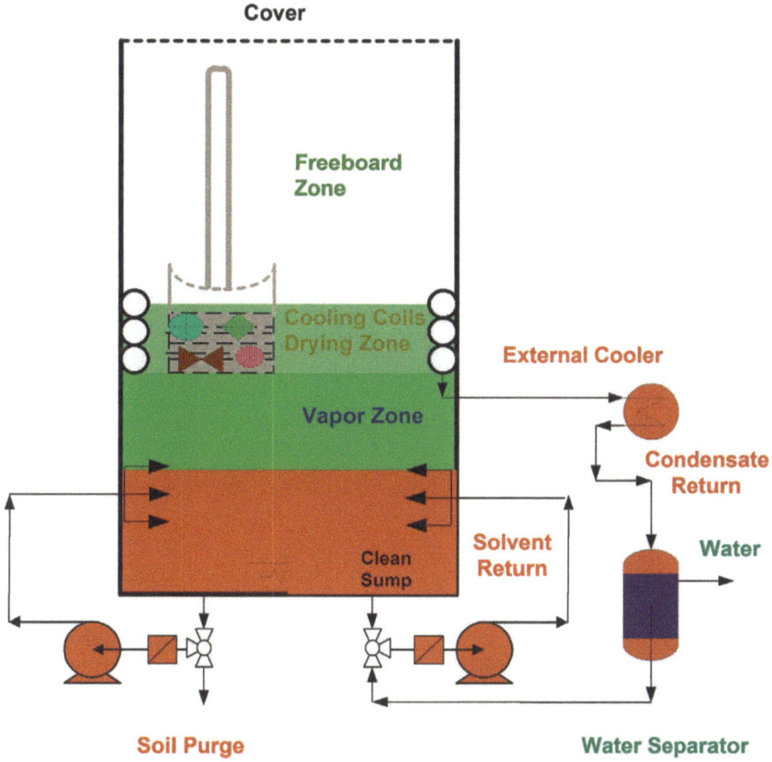

Figure 1.48 The Single Sump Vapor Degreaser (and Figure 1.19)

walls and this value must be managed. Where this is to be done, and what should be detected, is identified in Figure 1.47.

The word managed, above, means that the internal thermocouple should be interlocked to the boilup heater, so that if the measured temperature within the freeboard area exceeds a selected value (such as that in the column second from the right in Table 1.11), then the boilup heater is immediately disabled. This automatically invoked management decision can stop the unchecked flow of solvent vapor to the work area. Such a feature is an essential one for every open-top vapor degreaser of any size, large or small[DD]. And there should be a manual reset so the degreaser may be restarted after a cleaning load with a smaller thermal mass has been selected for processing.

1.27 CONTINUOUSLY OPERATING BATTALION DEGREASERS (INLINES)

Not all vapor degreasers operate in batch mode. Were that so, some sites would need phalanxes of operators and battalions of batch cleaning machines to deal with an ongoing flow of manufactured or repaired parts produced by other machines operating in continuous mode[60].

Yet the population of inline (or in-line) vapor degreasers is sparse. One company reports building eight inline machines over the past two years, while in the same time they built several dozen batch vapor degreasers.

Every inline vapor degreaser is different – because of the different natures of the parts being cleaned, and the work flows being managed.

Yet, every inline vapor degreaser has the capability to implement the same functions found in batch vapor degreasers, such as the single-sump vapor degreaser shown in Figure 1.48.

Figure 1.49 is the image of a hypothetical inline single-sump vapor degreaser. It has a zone of low volume for vapor rinsing, because there is no facility to cool the parts after cleaning by liquid immersion.

Some functions found in both the batch and inline machines are, using the common color code:

[60] In the context of solvent cleaning machines, continuous operation means that soiled parts or baskets of parts are fed to the machine before the previously fed parts, or baskets of parts, have completed their cleaning cycle.

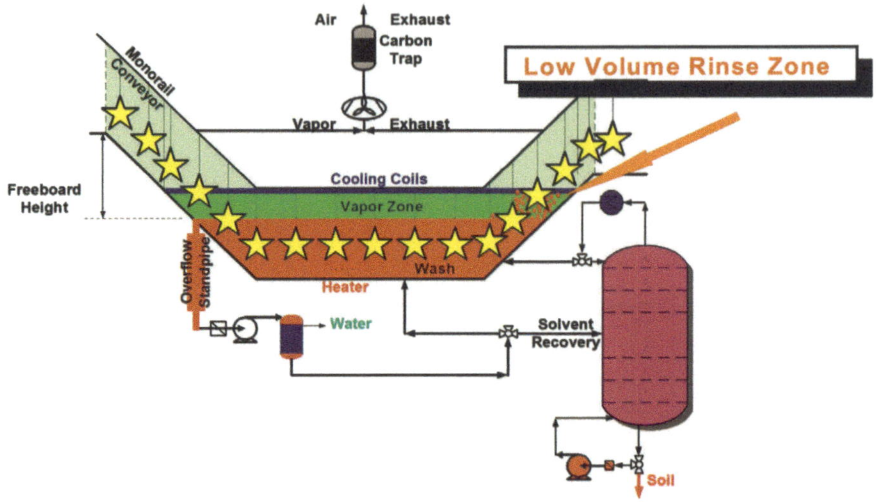

Figure 1.49 The Continuous Open-Top Vapor Degreaser

- One liquid sump for immersion cleaning (color code red),
- Cooling coils above the sump for restraining the volatility of the cleaning solvent (color code blue),
- A zone above the vapor zone where parts can be dried of liquid (color code light green),
- A freeboard zone above the vapor zone which can limit diffusion of solvent from the vapor zone into the work area (color code very light green),
- A water separator to decant water content in the solvent above its solubility limit (color code blue),
- A soil removal system so that soil can be discarded and solvent can be recovered for reuse (color code purple[EE]),
- A weir (overflow standpipe) to allow containment of liquid at a designated level.

In addition, inline machines have other capabilities not found in batch vapor degreasers, which are necessary with this type of machine:

- A vapor exhaust system to keep solvent vapor from being released to the room. This capability is not found in batch machines that have covers which can be positively sealed to the degreaser freeboard (Figure 1.24). Inline machines have, instead, movable curtains[FF] or other facilities for restricting emissions as parts (or baskets) pass *continuously* into or out of the degreaser.
- However, one can provide in an inline machine the function provided by a sealable cover in a batch machine by maintaining a slight negative pressure within the inline degreaser. This assures all vapor flow is in the inward direction. A low-pressure blower as an exhaust system can maintain a slight negative pressure[GG] relative to the room environment.
- An exhaust treatment system, as a bed of activated carbon (shown in Figure 1.49), to dispose of (not treat[61]) solvent fumes contained in the air exhaust.
- Dilution of emitted solvent does not adequately protect workers. Using Table 1.8 as a reference, a dilution ratio[62] of at least one thousand to one would be necessary to dilute solvent emissions from a degreaser to be within their 8-hr exposure limits. That level of room ventilation is not normally practical. Furthermore, ventilation does nothing positive to reduce the overall mass emission rate of solvent vapor.

Inline machines also can be designed to have other capabilities found in batch vapor degreasers, such as the unit shown in Figure 1.50. This includes two sumps, each having different solvent quality (as shown in Figure 1.51), allowing:

- Separate wash and rinse capabilities (color codes red and light blue),
- A rinse sump which can be cooled (color code light blue),
- A large volume zone above the rinse sump allowing for rinsing with freshly condensed solvent (color code green),
- A zone of larger volume in the freeboard area where drying of parts (color code light green), if incomplete, can be augmented, and

[61] The difference between disposal and treatment of waste components lies in the fate of those components. When one uses a bed of activated carbon to capture waste components by absorption (Chapter 4.2), the saturated bed is usually discarded and either put to landfill or incineration. That's disposal. Treatment would be if those components were recovered from the bed and recycled or converted for another type of reuse (Chapter 4.16).

[62] This value reflects the amount of room ventilation which would be necessary to dilute solvent-saturated air as effluent from a degreaser to the 8-hr exposure limit.

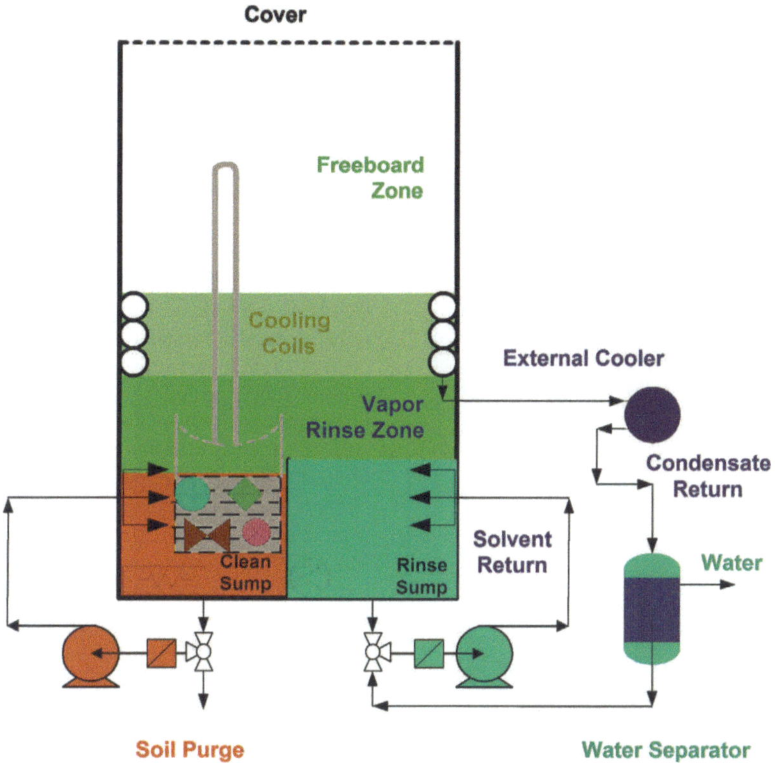

Figure 1.50 Immersed Parts Being Cleaned (and 1.11)

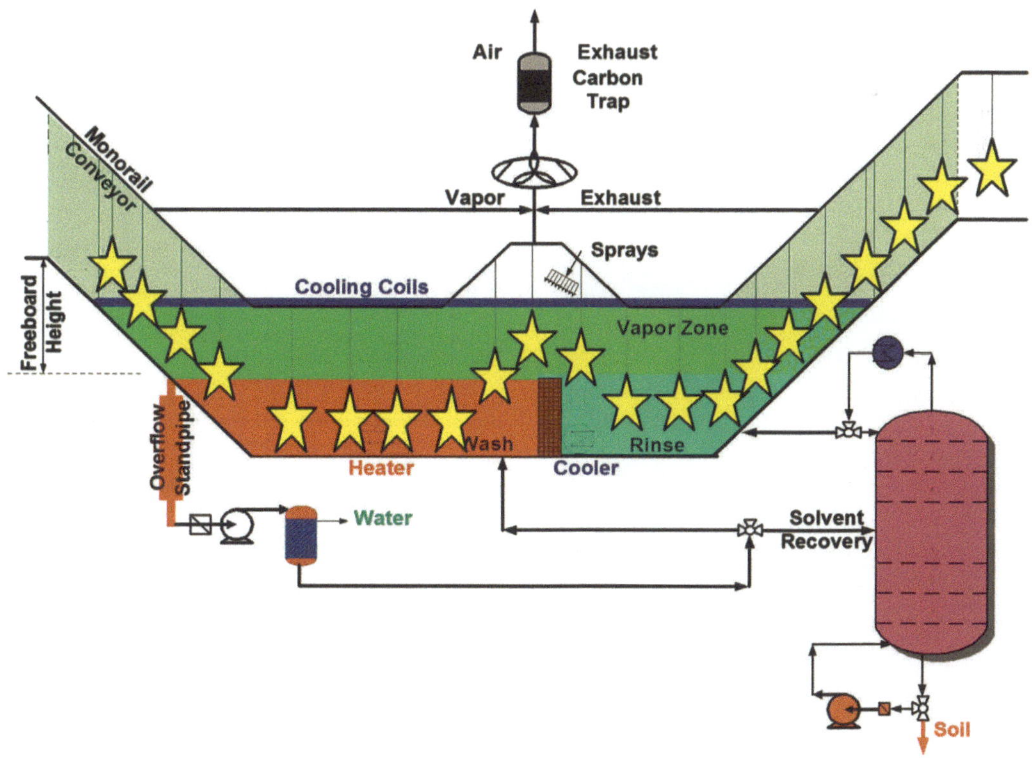

Figure 1.51 A Continuous Open-Top Vapor Two-Sump Degreaser with Emission Control Facilities

Figure 1.52 Parts Conveyor for Continuous Open-Top Vapor Degreaser

- A zone, located above and between the two liquid sumps (color code white), where parts can be sprayed[HH] with pressurized hot or cold solvent.

Part conveyance shown in Figures 1.49 and 1.51 is being done with an overhead transport. Here, parts (or baskets) are supported from arms attached to a moving chain.

Other types of transport are moving conveyor belts (Figure 1.52[II] and Figure 1.53) which may be loaded with parts or baskets by human operators or conveyors connected to/from other processing centers.

Inline vapor degreasers are not inexpensive, relative to their batch-only siblings. Investment requirements are highly variable because of the need for customization, but are probably four to eight times the investment required for a single batch vapor degreaser – but they save the purchase and operating cost of more than that number of single batch machines.

Figure 1.53 Construction of an Continuous Multi-Stage Open-Top Vapor Degreaser

Part transport systems, and the automation of same, are a most significant portion of the needed investment.

1.28 EMISSION CONTROL – THE HALOGENATED SOLVENTS NESHAP

From December 1994 to around the millennium, little else commanded the attention of firms and users involved with solvent cleaning. The aim of the Halogenated Solvents NESHAP was to limit exposure of workers doing solvent cleaning to six solvents which had been previously classified by the US EPA as Hazardous Air Pollutants (HAPs).

Written with considerable input from industry, this NESHAP was based on two approaches. First was to change the engineering design of all solvent cleaning machines in which these six solvents were used to improve their containment efficiency to a common standard[63,64]. Second was to change the workplace methods by how the modified machines were used. Considerable latitude was allowed by the EPA in meeting both goals of the NESHAP.

Yet, this NESHAP doesn't apply to a cleaning solvent in common use, despite significant concern about its effects on human reproductive systems and internal organs. The reason is that while the six affected[65] solvents (carbon tetrachloride, chloroform, methylene chloride, trichloroethylene,

[63] It is important to note that the "middle initial" of a NESHAP is "standard." The US EPA did not, and generally does not, mandate specifically how emissions of HAPs from equipment should be curtailed. Rather, in 1994, the EPA mandated that users must achieve one of two outcomes: (1) selection of engineering design parameters from a diverse menu of choices, or (2) a mass emission rate when the machine was in use (30.7 lb/month-SF) or while it was ready for use (0.045 lb/hr-SF = 32.85 lb/month-SF).

A rare example of the EPA mandating specific action concerns ozone-depleting chemicals (ODCs). Their manufacture was banned!

[64] That standard was known as MACT – Maximum Achievable Control Technology. MACT standards are designed to reduce HAP emissions to a maximum *achievable* degree, *taking into consideration the cost of emission reductions and other factors.*

When developing a MACT standard for a particular source category, the EPA looks at the current level of emissions achieved by best-performing similar sources through clean processes, control devices, work practices, or other methods. These emissions levels set a baseline, often referred to as the "MACT floor" for the new standard. At a minimum, a MACT standard must achieve, throughout the industry, a level of emissions control that is at least equivalent to the MACT floor. The EPA can establish a more stringent standard when it makes economic, environmental, and public health sense to do so – which was its aim with the update proposed in 2007.

[65] Actually, because of concerns about toxicity and ozone depletion potential, the 1994 NESHAP practically applied to users of only three of the six cleaning solvents. Carbon tetrachloride and chloroform raise concerns about human toxicity and their use as cleaning solvents is and has been practically nil for decades; 1,1,1-trichloroethane has been classified as a Class II Ozone Depletion Compound, and its manufacture was banned in the US by 1996.

perchloroethylene, and 1,1,1-trichloroethane) are HAPs, n-propyl bromide is not so classified[66,67,JJ].

More than two hateful words were expressed to the US EPA about this regulation. Yet more than a decade after its deadline for compliance, it's clear that interest in cleaning dirty parts with solvents has not declined but has increased. Essentially, the Halogenated Solvents NESHAP did not destroy solvent cleaning technology, rather it validated its use.

As of publication of this book, essentially all open-top vapor degreasers are compliant with the design or emission requirements of the 1994 NESHAP. There are three reasons for this: (1) it is and has been the law (in the US), (2) the concomitant reduction of solvent emissions has saved the cost of solvent purchase, and (3) it has been effective in meeting current workplace exposure limits. The 1994 NESHAP remains in force today.

1.29 THE AMENDED 1994 HALOGENATED SOLVENTS NESHAP

Imagine playing a game and someone unexpectedly notes a rule about which you hadn't known, and the new rule affects your play. Confusion! Sabotage! That's what happened in 2007 with the 1994 Halogenated Solvents NESHAP.

The key word in the above paragraph is unexpectedly. Confusion about the rules of the NESHAP should not have been unexpected; but it was. The rule change was mandated under the 1990 US Clean Air Act[KK] (CAA).

Not only were few users (and consultants) aware that regulation by NESHAP is a multi-stage process over time; even fewer persons were aware that a subsequent stage of regulation could be done on a basis completely different than was done with the 1994 NESHAP.

Contrast these two situations:

- The 1994 NESHAP limited the mass of solvent emissions via implementation of engineering technology into the design of new and used cleaning machines, incorporated specific work practices into site work rules, and required reporting of exceedances of emission limits.
- The second-generation Halogenated Solvents NESHAP used the outcome(s) of the 1994 NESHAP as a baseline. Then, the 1990 CAA required the EPA to review whether or not at that baseline there is residual risk to the US population caused by users of halogenated solvents; and required in the event residual risk is found by the EPA, that they take action to reduce that risk to acceptable levels[68].

In 2007, the EPA announced they had completed their review. Their stated purpose was to identify the remaining risks to public health and the environment following implementation of the 1994 technology-based rule; and then – via additional regulation – "...provide an ample margin of safety to protect public health...."

But what was the residual risk remaining after the 1994 NESHAP was fully adopted? And what is an ample margin of safety?

- For these three cleaning chemicals, the risk[69] is of cancer – because they are carcinogens. In 1999 the EPA reported to the US Congress that if that risk was not less than a "1-in-a-million" value, the EPA must promulgate residual risk standards. The EPA must also "consider costs, energy, safety, and other relevant factors in doing so."

The EPA's review in 2007 concluded that 93% of the reporting sites emitting solvent nationwide were already in compliance (with the proposed limits of the *amended* NESHAP).

Second, the EPA's review exempted some use sectors covered by the 1994 NESHAP because of compliance feasibility and cost. Those use sectors were: aerospace, web cleaning, and narrow tube manufacturing (Chapter 2, Footnote 4). Within each sector there were significant technical challenges in providing clean parts. And the EPA determined that those challenges could not be met after proposed changes to equipment and operations within those sectors. So there were no proposed changes for those three manufacturing sectors.

Thirdly, the EPA's review revised the basis for compliance. Compliance with the 1994 NESHAP required that the user choose between installation of proven engineering features or meeting a set emission level. Further, it was *each cleaning machine* that was regulated.

In the revised NESHAP, *the overall emissions from all machines on a site* were regulated.

[66] In October 2010, the US EPA announced that it had received from the Halogenated Solvents Association (http://www.hsia.org) a petition for the consideration of n-propyl bromide to be a HAP (it can be downloaded from http://www.hsia.org/news/nPBPetitionsigned.pdf). The basis was: emissive use by humans, past multiple findings of reproductive damage in exposed laboratory animals, and a recent finding about carcinogenicity. Any US citizen can file similar petitions.

[67] Neither is *trans*-1,2-dichloroethylene classified as a HAP (Ref. 3, Chapter 20.5). In other words, not all chlorinated or halogenated cleaning solvents are classified as Hazardous Air Pollutants.

[68] Specifically, the 1990 CAA requires the EPA to revise all NESHAPs "as necessary, taking into account developments in practices, processes, and control technologies, no less frequently than every 8 years." Obviously, this statement means that there can be additional NESHAPs in sequence.

[69] The risk is of cancer. Methylene chloride (MC), trichloroethylene (TCE), and perchloroethylene (PCE, or "perc"), which produce acute and/or chronic, non-cancerous health effects at sufficient concentrations, have also been classified as probable or possible human carcinogens by either the EPA or other governmental or international agencies (Ref. 3, Table 20.1).

Table 1.12 Analysis of Emission Levels Based on Risk

Site	US Population		Industry	
Blended Emission Limit, kg/yr	**Risk Level per Million**	**Annual Cancer Incidence**	**Additional Emissions Limit tons/yr**	**Annual Financial Impact, Dollars**
Current Levels	100	0.55 Cases		
60,000	**20 to 50**	**0.36 Cases**	**1594**	**+1.3 MM**
40,000	20	0.21 Cases	1759	−0.13 MM
25,000	10	0.24 Cases	2351	−1.2 MM

1.29.1 The EPA's Determination of Residual Risk

The descriptive phrase is "blended risk."

Quantitative risk is derived from documented epidemiological experience. MC, TCE, and PCE beget different cancer Unit Risk Estimates (UREs[LL]). The EPA recognized that these three solvents have quite different boiling points and solubility parameters, and that choice of a solvent should be based on the nature of the job – not the URE value of the solvent being used.

To recognize the difference in the level of hazard presented, and to not mandate how cleaning work should be done, the EPA chose to homogenize[70], or blend, risk on the basis of URE values for the three solvents. MC, the least hazardous, was assigned the value of 1. On that basis, TCE was assigned 4.25 and PCE was assigned 12.5.

The EPA proposed on May 3, 2007 to amend the 1994 NESHAP by regulating the annual emission of the *mass* quantity of emissions from all cleaning machines on a site[71]. This allowed users to choose their solvent and their technology approach for compliance.

But what level of risk to the entire US population was "adequate to provide an ample margin of safety"? And what did that mean to measurables at a site – such as emission rates or concentrations?

$$\sum\{[MC \times 1] + [TCE \times 4.25] + [PCE \times 12.5]\} \quad (1.1)$$

The calculated risk (using Equation 1.1) to the overall US population was that used by the EPA to determine the total emission quantity in a calendar year.

Three cases were considered, each with a different blended mass limit of solvent emissions. The outcomes which the EPA considered are shown in Table 1.12[72], with the highlighted case being chosen by the EPA[MM,NN].

Readers can make their own judgments about risk level. This author believes that all three cases above offer similar risk and expected cancerous harm, and impose negligible financial burdens[73] on industry.

1.29.2 Challenges to the EPA's Ruling

This author's opinion was not held by all! Strongly worded petitions[74] were filed requesting the EPA to reconsider its emission standards under the May 3, 2007, final rule[OO]. On October 20, 2008, the EPA issued a proposed

[70] In terms of the risk presented of cancer.

[71] MC, TCE, and PCE are the mass quantities emitted of methylene chloride, trichloroethylene, and perchloroethylene respectively.

[72] The EPA's review allowed different treatment of military maintenance facilities. They must meet a facility-wide emission limit of 100,000 kg/yr of MC-equivalent emissions.

[73] The annual financial impact noted in Table 1.12 is the difference between the financial value of the solvent not needed to be purchased if emissions are reduced, less the financial penalty of making the reduction of solvent use. In Table 1.12, the reduction from current total emissions at sites to 60,000 kg/year is estimated to produce a net positive financial impact for sites (solvent savings exceed conversion costs). But further proposed reductions of annual emissions save less solvent and cost more to accomplish, making their financial impact not be positive for sites.

[74] As an example of the fervor inherent in the written petitions consider the following:

> *"...EPA has adopted the same, flawed logic in concluding that it has adequately discharged its responsibility to consider environmental justice implications with respect to regulations on halogenated solvent cleaning ... EPA's willingness to ignore the distributional impacts – including significant distributional benefits – of regulations addressing air pollution from refineries, power plants, and industrial sources that use halogenated solvents is particularly egregious because it is common knowledge that operations of the aforementioned industries pose disproportionately high and adverse human health risks to minority populations and low-income populations...."*

Source: letter Wesley Warren, National Resources Defense Council, to Kevin Neyland, Office of Information and Regulatory Affairs in the Office of Management and Budget, March 30, 2009.

reconsideration of the "final rule." In November 2009, the EPA was granted a remand[75] of this rule.

As of publication of this book, the EPA has not proposed[PP] a rule addressing the concerns expressed by those who filed petitions to point out that (1) some facilities should not have been exempted, and (2) the blended risk emission limit chosen should have been the lowest one noted in Table 1.12, or at worst the 40,000 kg/yr emission limit[QQ].

Whatever the ultimate outcomes of these court challenges are, readers should expect changes in the basis for environmental regulation from mass, volume, or concentration limits to regulation of calculated outcomes. It turns out that the "low-hanging fruit" has been picked, and the fruit higher up on the tree tastes different.

1.30 COLD CLEANING WITH VOC-EXEMPT SOLVENTS

Before local, regional, and federal VOC regulations became stringent, cold cleaning in open tanks was commonly done. It is much less so today. Solvents with lower volatility were used to avoid concerns about flammability, and to reduce the cost of replacing evaporated solvent. Tanks were truly open – with either no covers or loose-fitting covers. And so these tanks were very inexpensive!

Paint stripping was a common cold cleaning application where high-boiling solvents such as n-methyl pyrrolidone (boiling point (BP = 202.2°C (396°F)), propylene carbonate (BP = 242.2°C (468°F)), esters of dibasic acids (BP = 228.9°C (444°F)), or Texanol (BP = 243.8°C (471°F)) were used. This cleaning was not thermally "cold," as the bath was heated to temperatures below these high normal boiling points. But no solvent was heated to its boiling point – hence use of the descriptive term cold cleaning.

That cleaning process was as simple as could be imagined – immersion of fairly large parts usually having a fairly large thermal mass, allowing them to soak for a considerable time (hours), and then removal of them for drainage/drying. Generally there was little or no agitation in the cleaning bath, and no machine-based capability for drying of treated parts (because the solvent was extremely non-volatile). Certainly, this was not vapor degreasing!

1.30.1 Another Type of Cold Cleaning

Going forward from publication of this book, the process of cold cleaning deserves fresh consideration about use in other applications. Significant reasons for reconsideration of cold cleaning are simply, simplicity. This is a desire to enjoy: (1) freedom from VOC regulations; (2) avoidance of solvents about which there is safety, health or environmental (SHE) concern; (3) reduced cost for purchase of cleaning solvents; and (4) reduced energy costs.

- Recall that cold cleaning is any solvent cleaning process in which the solvent is heated but not boiled. For the cold cleaning process under discussion in this chapter, the solvent is not to be heated, at all. Cleaning work in this process is to be done at ambient temperature – thus saving energy.

The solvent cleaning process being discussed is to be conducted in the enclosure that is a vapor degreaser, but it will be a cold cleaning process – with no heat added, and all known spark sources eliminated.

Unfortunately, cold cleaning cannot be a process in which zero energy is added. Although the cleaning solvent is not to be heated, it may be necessary to provide cooling in order to control its airborne concentration in the workplace to meet two requirements: (1) less than 25 percent of the lower explosive limit (LEL), and (2) as close as possible to the 8-hour exposure limit.

In this cold cleaning work, one would use very different solvents, in a different process, implemented in familiar but different equipment, with different parts, to remove different soils.

1.30.2 VOC-Exempt Cold Cleaning Solvents

The different solvents are those which are VOC exempt (in the US), highly volatile, highly flammable[RR], and not generally expensive as they are commodity chemicals.

Some VOC-exempt solvents are listed in Table 1.13 in order of estimated preference – most favored to least favored – to replace n-propyl bromide by matching its Hansen Solubility Parameters[76].

Any solvent listed in Table 1.13 offers a major advantage – one need not be concerned about how much

[75] A remand is an opportunity to *reconsider*. In this case, the remand was granted by federal courts to whom petitioners had appealed for the EPA to stay its 2007 proposed regulation.

The reason that the EPA needed a remand is because the CAA requires review and action by them by a certain time. The remand absolves the EPA of their requirement to review and act – so that they can re-review and re-act. In other words, a remand to the EPA is a delay of the requirements of the CAA to allow them to respond to the petitions.

[76] Technically this means that the R_A value between n-propyl bromide and any specific solvent is less than 8 $MPa^{½}$. This parameter and the limiting value are described in Ref. 2, Chapters 2.10.4 and 3.29 respectively.

Note that it is only the Hansen Solubility Parameters which are being matched; the boiling point of n-propyl bromide is not being matched. A lower boiling point may not provide adequate thinning of grease and oils to allow the level of soil removal by physical means to be as done with n-propyl bromide.

Table 1.13 Properties of "Usable" VOC-Exempt Solvents (in US)

Name	Exposure Limit ppm	Relative Evaporation Rate (n-BuAc=1)	Vapor Pressure @ 20°C mm Hg	Boiling Point, °C	Flash Point °F	NFPA Ratings	LEL Vol %	δ disperse (MPa$^{1/2}$)	δ polar (MPa$^{1/2}$)	δ H-Bonding (MPa$^{1/2}$)	R_A vs. n-Propyl Bromide
Methyl Acetate	200 (ACGIH)	5.5	57.0	56.9	14.0	Flammable Class IB	3.1	15.5	7.2	7.6	3.40
Acetone	750 (ACGIH)	5.8	56.0	55.5	1.0	Flammable Class IB	2.6	15.5	10.4	7.0	3.79
Tertiary Butyl Acetate	200 (OSHA)	2.8	98.0	98.0	34.0	Flammable Class IB	1.3	15.0	3.7	6.0	5.19
Dimethyl Carbonate	200 (CEL)	3.2	90.0	90.0	64.0	Flammable Class IB	4.2	15.5	8.6	9.7	5.27
Methyl Formate	100 (OSHA)	100.0	31.7	31.7	−2.0	Flammable Class IA	4.5	15.3	8.4	10.2	5.85
Parachlorobenzo-trifluoride	25 (PEL)	0.9	139.0	139.0	109.0	Combustible Class II	0.9	18.5	5.2	1.7	5.88
HFC-365mfc	1000 (CEL)	(Rapid)	40.2	40.2	−13.0	Flammable Class IB	7.0	14.0	4.8	2.0	5.94
HFC-245fa	300 (PEL)	104.0	14.9	14.9	NONE	NOS	NA	14.3	5.6	1.1	6.05
HFE 8200	200 (PEL)	33.0	76.0	76.0	NONE	NOS	1.8	13.0	4.0	1.0	8.71
HFE-7100	750 (AIHA)	49.0	60.0	60.0	NONE	NOS	NA	13.7	2.2	1.0	8.72
HFE-7200	20 (CEL)	33.0	76.1	76.1	NONE	NOS	2.4	13.3	2.0	1.0	9.36
HFC43-10mee	200 (CEL)	29.0	54.4	54.4	NONE	NOS	NA	14.2	1.1	0.0	9.41
OS-10 Hexamethyl di-siloxane	200 (CEL)	3.8	100.6	100.6	30.0	Flammable Class IB	1.2	14.0	1.0	0.0	9.67

Table 1.14 Volatility of Commodity Chemical VOC-Exempt Solvents

Name	Boiling Point, °C	Calculated Vapor Conc. @ 27°C (~80°F), ppm	Calculated Vapor Conc. At Cooling Coils (Table 1.7), ppm	25% of Lower Explosive Limit (LEL) ppm	8-hr Exposure Limit, ppm	Relative Evaporation Rate (n-BuAc=1)[SS]
Methyl Acetate	56.9	342,000	7,150	7,750	200	5.5
Acetone	55.5	334,000	6,550	6,500	750	5.8
Tertiary Butyl Acetate	98.0	52,800	2,390	3,250	200	2.8
Dimethyl Carbonate	90.0	53,200	7,870	10,550	200	3.2
Parachlorobenzo-trifluoride	139.0	7,345	789	2,250	25	0.9

of it is emitted relative to a VOC-limiting environmental regulation. Any of these solvents can be used in the US without a VOC emission permit.

The first eight listed solvents (green colored background) should be expected to perform similarly to a conventional vapor degreasing solvent – n-propyl bromide. Each is also prized as being priced as a commodity chemical. And the first four listed are characterized as being relatively non-hazardous because compliance with their exposure limits should be easily attained by emission and ventilation control.

But the first four share another distinction as well – all are classified by the NFPA, based on their flashpoint, as Flammable Class IB (nearly the least safe) because their flash points do not exceed 23 °C. The meaning of this outcome is that all equipment in which these solvents are used cannot have any places in which a spark could be generated[77].

Pertinent information about the volatility of the first five solvents is given in Table 1.14[78,79]. Note the high Relative Evaporation Rates (RER values) for each of the first four solvents listed.

Significant operating issues can be addressed with the information in Tables 1.13 and 1.14.

- Evaporation of *all five* cold cleaning solvents in Table 1.14 can be restrained as needed by being condensed through contact with coils in which brine from a refrigeration unit acts as the working fluid[80,TT]. The refrigeration coils should be located immediately above the two liquid sumps.
- In a normal scenario with all of the five solvents[81] listed in Table 1.14, concentrations of solvent normally emitted to the work area from the cleaning machine are at or within safety limits of 25% of the lower explosive limit (LEL)[82].
- Under the same scenario, these concentrations are substantially above the posted 8-hour exposure limits for all five VOC-exempt solvents. Substantial local ventilation is necessary to dilute solvent concentrations in the work area.

[77] It isn't simply that this equipment must be classified as Class I (ignitable fluids), Division 1 (fluid normally present), Group D (containing gasoline, hexane, naphtha, acetone, etc.). It must be so classified – per the NFPA! But, simply, this equipment used for cold cleaning must not contain any sources of ignition.

[78] The temperature of 27°C (~80°F) is taken as representative of what might be expected as ambient temperature.

[79] Actually, the low boiling point of methyl formate renders it as not being preferred for this application. This is because the coolant with the lowest freezing point, calcium chloride, cannot adequately cool methyl formate to reduce its concentration in the freeboard zone to less than 25% of the lower explosive limit (LEL).

[80] As noted in (Box 1.5), the lowest temperature at which calcium chloride brine can be used is −52°C (-62°F). Assuming a +6.1°C (+11°F) heat exchange approach (temperature difference), the value of −45.9°C (−50.6°F) is the lowest temperature to which solvent vapor can be chilled (condensed).

[81] This is a most conservative case, assuming no reduction of concentration of solvent which has risen through the freeboard zone (Endnote S).

In Table 1.14, calcium chloride brine must be used with an 11°F approach to contain the two most volatile solvents (methyl acetate and acetone); ethylene glycol brine may be used with an 18°F approach for tertiary butyl acetate; sodium chloride brine may be used with a 23°F approach for dimethyl carbonate and parachlorobenzotrifluoride. In each case, the concentration of solvent in the undiluted effluent from the cooling coils will not exceed 25% of the LEL.

[82] The comparison of calculated temperature at the calcium chloride brine cooling coils and 25% of the LEL is identified by both values for all five solvents being surrounded by wide black lines.

While reduction of the effluent solvent vapor concentration to less than 25% of the LEL can be done by condensing the vapor, it can also be done by maintaining a static pressure in the freeboard zone of inert purge gas (nitrogen) slightly above that in the work area. This retards entry of air (oxygen). Without a cooling coil, this modification will meet the requirement to control solvent concentration vs. 25% of the LEL. It will also present two drawbacks: (1) the nitrogen flow outward will increase the mass loss of solvent by entraining solvent vapor and (2) the freeboard volume may be a safety hazard because it won't support human respiration.

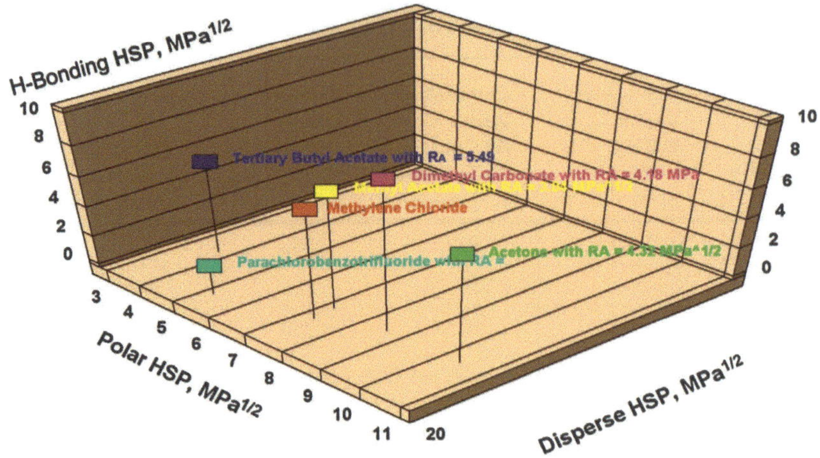

Figure 1.54 Three-Dimensional Plot of HSP for VOC-Exempt Cold Cleaning Solvents vs n-Propyl Bromide

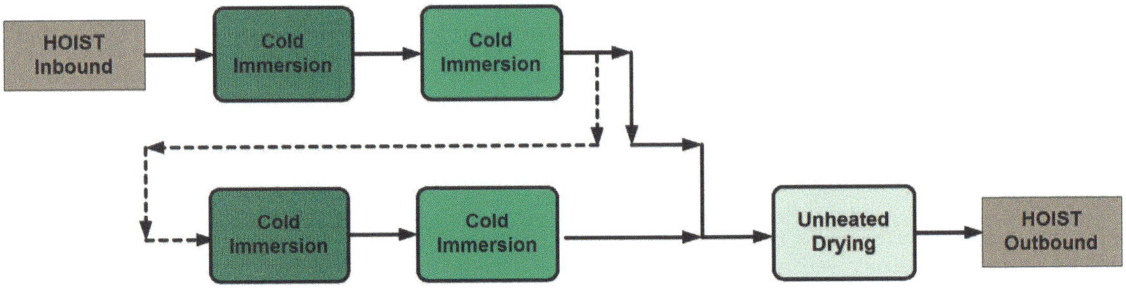

Figure 1.55 Process Diagram for Two-Stage Cold Cleaning Apparatus

- Replacement of n-propyl bromide as a vapor degreasing solvent with another solvent as a cold cleaning solvent should be a credible option to clean parts, given the low values of R_A tabulated in Table 1.13. In addition, plotted in Figure 1.54[UU], the values of R_A vs. those of methylene chloride are also well within the threshold value for good solvent cleaning of 8 $MPa^{1/2}$.

1.31 A UNIQUE COLD CLEANING PROCESS

This different process is one in which the hazards of flammability are doubly[83] avoided – a process in which *no heat is applied*[VV]. That's true cold cleaning.

A flow diagram of the multiple process items in an unheated cold cleaning degreaser is shown in Figure 1.55.

This cold cleaning machine has two liquid sumps – one for washing (color code blue-grey) and one for rinsing (color code light blue), and a freeboard zone for drying (color code very light blue) – the same color code used above. Unheated drying is done by holding parts above both sumps. Shown for completeness is the external hoist which lowers a basket of parts into the machine, and returns the basket from the machine. No heat is added in either sump – to eliminate the possibility of spark sources in those locations.

This process is not conventional vapor degreasing (Chapters 1.6 and 1.7[WW]). There is no generation of vapor; parts are not dried by application of heat. But there are significant similarities.

- *Degreasing* is done by solvation[84] of the soils in the liquid phase, although the temperature is ambient (Figure 1.56).

[83] Coincident existence of three factors is necessary to produce a fire: (1) oxygen, (2) fuel in the proper ratio to oxygen, and (3) a spark with sufficient energy. Removal of any one will prevent a fire. But, fail-safe design (in which this author believes) removes two factors: (1) ignition sources by facility design, and (2) condensation of fuel vapor to below one quarter of the lower explosive limit (LEL) by refrigerated coils.

[84] HSPs are not a strong function of temperature. In other words, for a non-viscous soil which does not have to be thinned by temperature, one would not expect the performance of the cold cleaning process to be substantially different than with vapor degreasing.

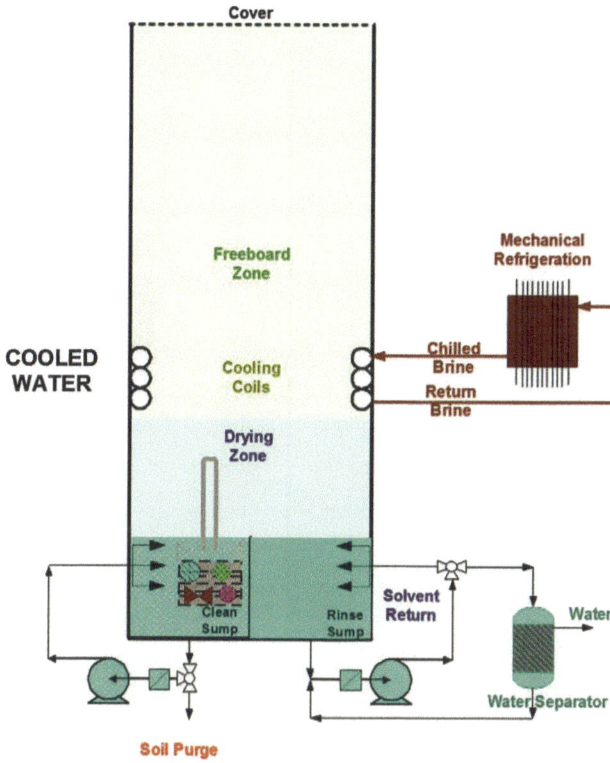

Figure 1.56 Cleaning of Parts in Cold Cleaning Apparatus

- *Rinsing* is done by immersion in a second sump (color code very light blue) containing liquid solvent that is cleaner.

The major differences between this cold cleaning process scheme and conventional vapor degreasing are:

- *Drying* of parts is done by evaporation of solvent, as with vapor degreasing; but no heat is applied to speed, or
- enhance, the drying process. Solvent released from surfaces in the drying process is contained within the cold cleaning machine by the two methods already discussed for vapor degreasing – *condensation* as the parts (and basket) pass through a cold zone produced at the machine walls by a single coil with refrigerated fluids (Figure 1.57), supplemented with a lengthy freeboard zone (Figure 1.58) to act as a *diffusion barrier* (Endnote S).
- *Soil management.* The soil-laden solvent can only be distilled within the enclosure that is the vapor degreaser when *heat from non-sparking sources*[85] *is added* to the degreaser.

However, the solvent-soil mixture can be: (1) distilled on site but away from the cleaning machine using any heat source, (2) transported in sealed containers to any off-site distillation system[86], or (3) distilled adjacent to the cleaning machine with non-sparking heat sources.

- It is done at a *low temperature* (ambient) so viscous soils (such as greases) cannot be thinned by an increase of temperature (Ref. 2, Figure 3.11) – because *no heat is added.*
- *Ultrasonic transducers* can't be used to provide surface-based agitation to remove particles and other debris.

[85] One may safely use, if available, non-sparking heat sources such as steam, steam condensate, or hot oil.

[86] This situation is an example of cleaning being soil management – energy is normally required to remove the soil from the parts, *and* later to remove the soil from the solvent. It is difficult to imagine another method for removal of soluble oil and grease soils from solvent without evaporation of the solvent from the relatively non-volatile soils – and evaporation takes energy.

The basic premise of cold cleaning without energy (heat) addition is that it is restricted to only the removal of soil from the parts, and not to the necessary removal of soil from the solvent – unless one uses the non-sparking energy sources of Footnote 85.

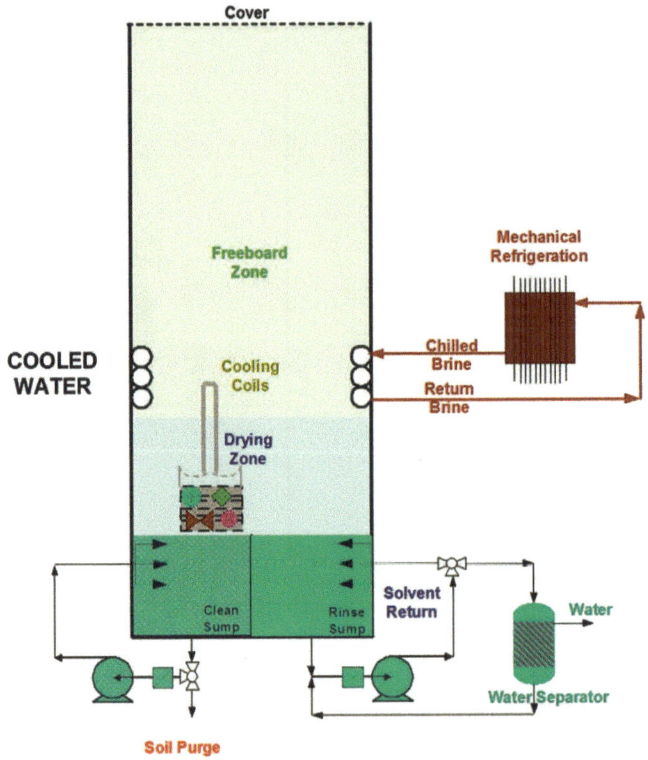

Figure 1.57 Drying of Parts in Cold Cleaning Apparatus

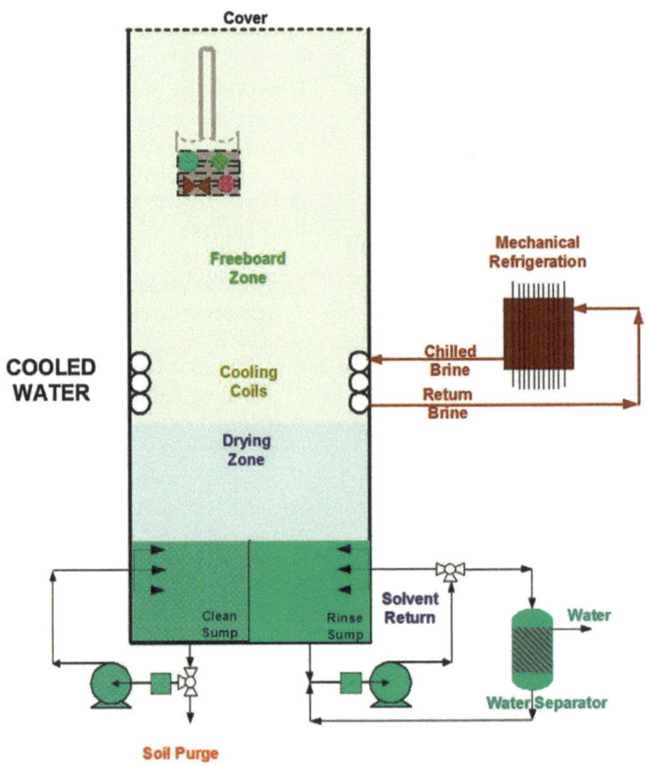

Figure 1.58 Removal of Parts from Hi-Freeboard Cold Cleaning Apparatus

Table 1.15 Aspects about the Presence of Water with Commodity Chemical VOC-Exempt Solvents

Solvent Name	Solubility of Water Vol %	Reactivity with Water	Type of Water Separator
Methyl Acetate	8.2	Negligible level of hydrolysis to produce acetic acid	Adsorber (a bed of activated alumina or silica, which is regenerated by a flow of hot nitrogen)
Acetone	100	None expected	
Tertiary Butyl Acetate	Low	Negligible level of hydrolysis to produce acetic acid	
Dimethyl Carbonate	Modest	None expected	
Parachlorobenzotrifluoride	0.024	Hydrolysis not expected	Decanter or adsorber

The electrical signals which power ultrasonic transducers can be sources of ignition[87].

- A concern is the need for and nature of *removal of water* (Ref. 2, Chapter 11.8). The cold cleaning solvents of Table 1.14 both share and don't share the solubility and reactivity characteristics of halogenated cleaning solvents. The situation is summarized in Table 1.15.

  So the concerns about water intrusion with cold cleaning solvents are different and significantly less than the same with halogenated cleaning solvents.
- Finally, per the specifications noted in Footnote 78 and their outcomes as noted in Table 1.14, the undiluted solvent fumes emanating above the cooling coils contain *no more solvent than 25% of the LEL* and so can't be ignited.

  So the reason no heat is added – to avoid spark sources which might ignite the flammable solvent; and no solvent concentration is permitted above 25% of the lower explosive limit, so as to have no ignitable mixtures – is to be *doubly* protective against the serious hazard of ignition of flammable solvents.
- That's the compromise one makes to receive the benefits of: (1) freedom from concern about emissions – VOC or flammability limits, (2) commodity pricing, and (3) dissimilarity to other solvents about which there are many credible SHE concerns.

This author believes that is not a compromise, but a rational choice. Diagrams of one facility designed to implement this choice are shown in Figures 1.56, 1.57, and 1.58[88].

1.32 INTRINSICALLY SAFE EQUIPMENT FOR COLD CLEANING

Almost any vapor degreaser can be used – as long as it has no sources of ignition. This constraint is not quite as simple as disconnecting the electrical connection.

Actually, electrically powered instruments and controls can be used, as long as they are intrinsically safe[89,XX].

These facilities are commonly found in chemical plants, refineries, laboratories; and cleaning equipment designed to use flammable solvents such as isopropanol or acetone (or gases such as ethylene and propylene).

Any experienced maintenance organization can coordinate the replacement of conventional electrical equipment with intrinsically safe equipment.

[87] The generator provides high voltages of 500 to 1000 volts to the transducer. Current levels are low, measured in milliamps. So, each one of a set of transducers in an assembly may receive 30 to 50 watts of power. A powerful spark, adequate to ignite flammable solvents, can be generated with these transducers.

But many cleaning tanks containing flammable solvents have been safely operated with ultrasonic transducers because two design modifications were made: (1) the immersed transducers were insulated with a very tight skirt to keep fluid out and a purge stream of nitrogen was continuously fed around them when they were in use, and (2) the generators were either isolated from the flammable solvents by a significant distance – dozens of feet; or were also purged with nitrogen.

[88] One might consider adding a second set of refrigeration coils at the top of the freeboard zone. However, they would not be at a lower temperature than those coils above the solvent sumps as calcium chloride has the lowest operating temperature among commercial refrigerants. The only purpose of the second set of refrigeration coils is to provide additional *time* for solvent condensation – which may be desirable.

[89] Of the three general ways by which a fire or explosion can be prevented when flammable materials are present in air (Oxygen), one involves management of ignition sources. One can: (A) contain the explosion or fire via enclosures [so-called explosion proof]; (B) pressurize or purge the affected volume with an inert gas such as Nitrogen or occasionally carbon dioxide, and manage the concentration of flammables vs. 25% of the lower explosive limit [LEL]; or (C) use intrinsically safe facilities (Endnote MM).

The design favored by this author, and described above, requires *both* (B) and (C) above.

1.33 A METHODOLOGY FOR COLD CLEANING

The different parts being cleaned via this cold cleaning process are not large. They are not those parts noted above being stripped of paint.

- The parts being cleaned in this process have low thermal mass, and are fittings, fixtures, and small housings; screws, bolts, and nuts; electrical and optical components; strip, wire, and cable; and cutlery, eyeglasses, toys, and jewelry.
- The soils being cleaned in this process are not intractably cured paints or coatings.
- Rather they are common light oils (either pure or oxygenated hydrocarbons, or vegetable-based), fingerprints and some other human-based fluids, biological or microbial materials, some food waste, and low-viscosity fluorinated or silicone greases.

Two of the three process steps (immersion and liquid spray with force) from Chapter 1.2 are practiced in the process of Figure 1.55 and the equipment diagrams of Figures 1.56 to 1.58. But Figure 1.55 shows the opportunity to practice each step more than once.

Consider this process sequence of items 1 through 5, with items 1 and 5 (Figure 1.59) being use of the hoist to insert and remove (respectively) a basket of parts into the cold cleaning machine:

- **Item 1** – A Basket is filled with parts to be cleaned. The parts are arranged to allow full and efficient contact of the parts with the cleaning solvent.
- **Item 2** – A basket of parts is lowered (Figure 1.60) into an immersion sump (color code cross-hatched light blue in Figure 1.55), where the contact time may be several minutes, but probably not less than one minute.
- Certainly, the contact in this sump can be augmented with turbulation and the forceful impact provided by pressurized fluid spray flow from nozzles.
- **Item 3** – Then the hoist lifts the basket of parts above the cleaning sump and allows them to drain prior to immersion into the rinse sump (Figure 1.58). Liquid drainage into the cleaning sump is the soil-laden solvent clinging to the parts (dragout). The volume of retained dragout should be minimized by holding the parts for the longest time that can be spared in

Figure 1.59 Use of Parts Basket with Hoist to Insert and Remove Parts from Open-Top Vapor Degreaser (Items 1 and 5)

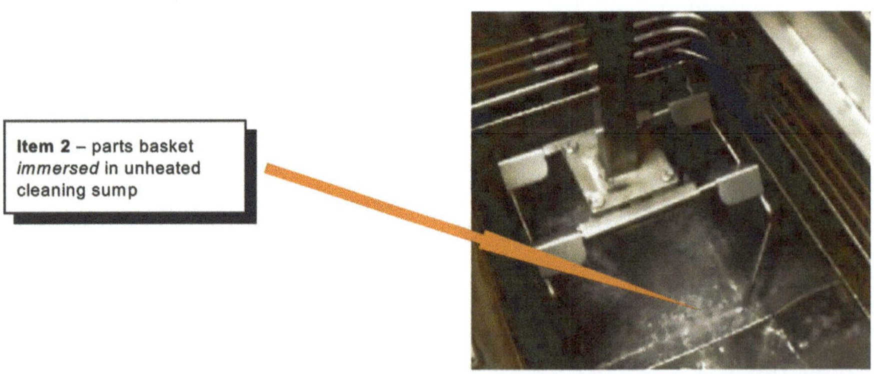

Figure 1.60 Immersion of Parts Basket into Cleaning Sump (Item 2)

the cleaning cycle (Ref. 1, Figure 1.5). Then the hoist lowers the basket of parts into an unheated immersion sump (color code light blue in Figure 1.55), where the shorter contact time is whatever is necessary to dilute the retained dragout. Again, fluid-based agitation and impact may be involved in item 3.

These two items may be repeated as desired, recognizing that re-immersion in the cleaning sump will not expose the parts to solvent that is cleaner than was in the rinse sump from which they were just removed.

- **Item 4** – Next, the hoist lifts the basket of parts into the vapor zone (color code very light green) where they are held so that free solvent droplets can drain into the rinse sump (Figure 1.58). The remainder of the solvent on the parts (and basket) flash evaporates. Contact time is short – whatever is necessary to adequately dry the parts by evaporating liquid solvent to vapor.

Finally, the basket of parts is slowly lifted upward by the hoist along the cooling coils and the freeboard to clear the top of the cleaning machine (Figure 1.58).

Solvent vapor, in equilibrium with the liquid in the two immersion sumps and also that which had been liquid dragout films on the parts (and basket), is constrained from leaving the machine because it is condensed by the cooling coils in the freeboard.

Remember that whatever solvent is evaporated from the parts when outside the enclosure that is the vapor degreaser is not subject to VOC-control regulations and not at a concentration which exceeds 25% of the LEL. It is, however, at a concentration which exceeds the 8-hour exposure limit (Table 1.14), and so normal ventilation will be required.

1.34 NOTES ABOUT THE DESIGN AND OPERATION OF THE COLD CLEANING PROCESS

All five items, the three operational items noted above plus the two items involving the hoist, are described and located not in an equipment arrangement diagram, but adjacent to photos of a small vapor degreaser[YY] (Figures 1.59 to 1.62).

- **Item 1** – To reduce the opportunity for a spark to be created by friction between moving metal parts, the parts basket and all moving parts of the hoist must be

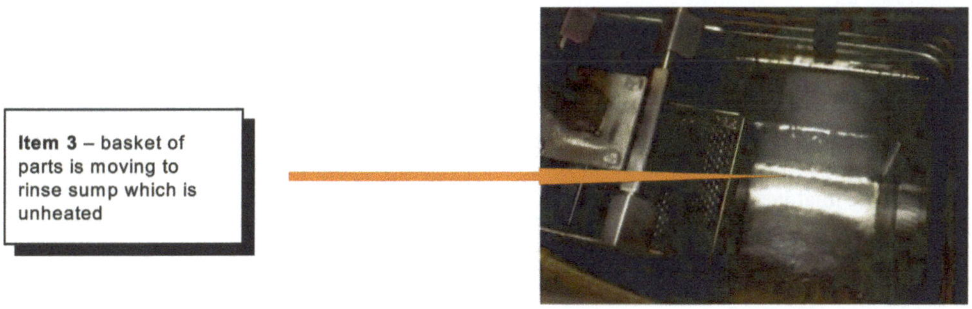

Figure 1.61 Movement of Parts Basket from Cleaning to Rinsing Sump (Item 3)

Figure 1.62 Removal of Parts Basket from Immersion so they can be Dried (Item 4)

made of plastic[ZZ] or some non-sparking metal[90]. To minimize[91] displacement of the solvent vapor by the downward-moving parts basket, two design features should be implemented: (1) the hoist should be restricted from moving upward or downward faster than 11 ft/min, and (2) the open area of the machine should be at least 50% larger than that of the parts in the basket so that solvent vapor is not "pushed" or "pulled" from the liquid or vapor zones as the load is inserted or removed[92]. After item 1 is completed, the parts basket is fully immersed in the liquid contained in the cleaning sump.

- **Item 5** – The "speed limit" of 11 ft/min is still enforced to avoid pulling vapor up through the freeboard zone of the degreaser.

Item 2 – The noted five VOC-exempt commodity solvents are highly flammable and volatile. Although it is possible to heat the contents of the cleaning sump to less than the normal boiling point (and comply with the academic definition of cold cleaning), there is little reason to do so and the (negligible) risk of an ignition is one reason not to do so.

Using the same design approach, electrically powered ultrasonic transducers are not present (Footnote 87). Needed agitation will come with flow from nozzles aimed at the parts basket. The flow will be produced by pumps driven by compressed air – to eliminate spark sources.

- **Item 3** – The rinse sump (Figure 1.60) remains at the same temperature (ambient) as does the cleaning sump. The main purpose of item 3 is to enable a rinse with a cleaner solvent.
- Items 2 (wash) and 3 (liquid rinse) may be repeated as necessary.
- **Item 4** – This is the final item in the process of cold parts cleaning. The parts are allowed to flash dry within the confines of the freeboard of the vapor degreaser, but beneath the cooling coils. This is the same procedure used in heated vapor degreasing.

1.35 WARM COLD CLEANING

The design approach of Chapters 1.31 to 1.34 is deliberately very conservative[AAA] – no heat is applied, temperatures are not raised, vapor pressures aren't increased, solvent concentrations in air exiting the machine don't exceed 25% of the lower explosive limits (LELs), and there are no ignition sources!

Yet, one may without undue risk use the cleaning solvents of Tables 1.13 and 1.14 in a cold cleaning operation where these solvents are heated, and cooled.

This is not vapor degreasing, as the solvents are heated to sub-boiling temperatures.

There are two reasons for doing this. First, to allow some grease and oil soils to be more easily cleaned by reducing their viscosity. Viscosity is plotted for SAE 30 motor oil in the right vertical axis (red square symbols) of Figure 1.63.

The second reason is to enhance the rate of drying of parts by increasing the vapor pressure of five VOC-exempt solvents as plotted on the left vertical axis of Figure 1.63.

But there is a package of basic (and necessary) modifications that must be made to the apparatus noted above.

- Implementation of a safe supply of heat to the cleaning sump is the fundamental modification. It must be produced by non-sparking sources. Three examples of non-sparking heat sources which are commonly found in industry are: steam, condensate (hot water from steam), and externally heated hot oil (heat transfer agent). These sources of heat are relatively slow to respond to a reduction in sump temperature (via the boilup heater, Chapter 1.15), vs. electricity. Another non-sparking source of energy is natural gas, which is more quickly responsive to changes in sump temperature – but, of course it is flammable.
- An associated and necessary modification is that the supply of heat must have a thermal cutout switch which is installed in the cleaning sump. The purpose of the switch is to stop the supply of heat to the cleaning sump when the temperature in it exceeds a preset value. That preset temperature value[93] is one which doesn't exceed the normal boiling point (otherwise the work would not be cold cleaning; and more importantly, the freeboard zone would be expelling solvent vapor).
- A complementary and necessary modification is that the refrigerating capacity (heat transfer rate) of the cooling coils must be upgraded to include the capability of removing from the solvent vapor at least the heat added in the cleaning sump[94]. This is likely done by increasing the surface area of contact (the number or

[90] Metals such as brass, bronze, Monel metal (copper-nickel alloy), copper-aluminum alloys (aluminum bronze), copper-beryllium alloys (beryllium bronze), and titanium.

[91] The reason for these actions is simply good practice, concern for operators, and cost conservation. The solvents of Table 1.14 were chosen to not threaten LEL limits when used with a brine with a low freezing point.

[92] This second feature is necessary to avoid the "piston effect," where the area of the parts in the basket can act as a piston and force solvent vapor away from the parts and upward through the open cover and out into the work area (Figure 1.34).

[93] Another value of temperature which could deactivate the heater is the temperature above the cooling coils, as is shown in Figures 1.38 and 1.47.

[94] Recall from Chapter 1.16 and the reference of Endnote R, that the cooling coils should be sized to remove no less than twice the amount of heat added in the cleaning sump.

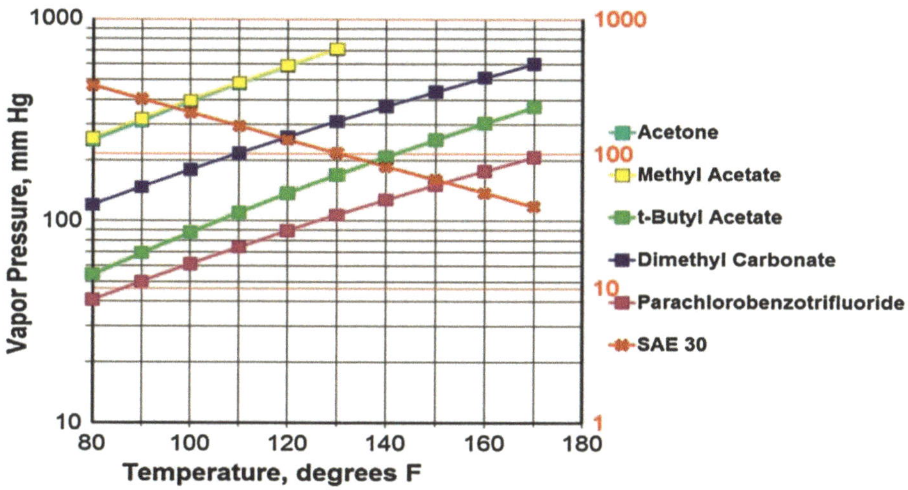

Figure 1.63 Thinning of Soils (Reduction of Viscosity) with Increasing Temperature

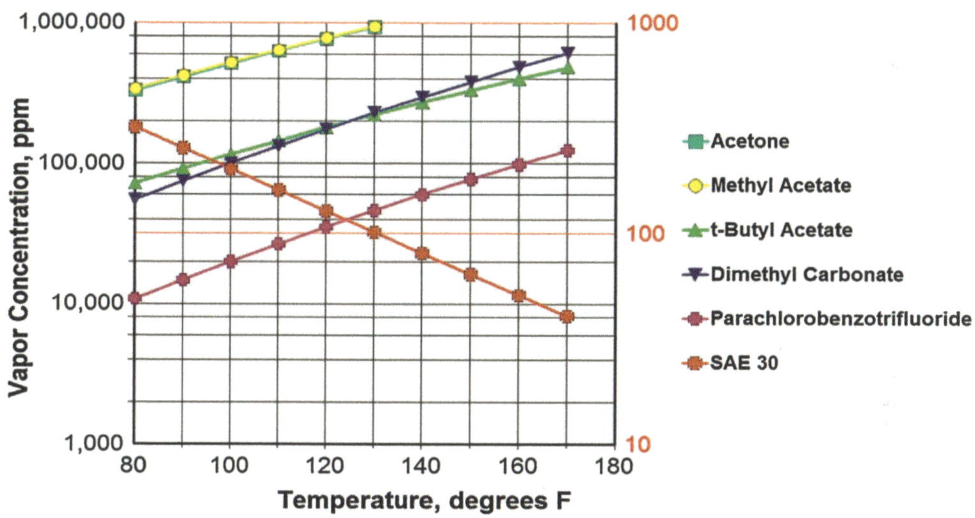

Figure 1.64 Volatility, and Drying Rate, Increase as Soils are Thinned with Increasing Temperature

diameter of coils) and likely not done by reducing the coolant temperature[95].

- Two non-essential, but highly recommended modifications should be implemented external to the vapor degreaser. They are "insurance"[96] against the cooling coils not being able to maintain the same temperature in the freeboard zone with heat applied to the solvent as was present in the freeboard zone without application of heat (Figure 1.64 shows how vapor concentrations of solvent change with temperature). These modifications are:
  - To use an explosimeter (Figure 3.22) to measure the solvent concentration being emitted from the cleaning machine and compare it to the safety limit of 25% of the LEL (Ref. 1, Chapter 3, Footnote 29).
  - To use personal monitoring (sampling) facilities to assure that adequate ventilation is present so employees are not exposed to concentrations of solvent above the 8-hr exposure limit (Table 1.14).

[95] The reason is that it is more expensive in terms of capital and operating cost (energy) to use a refrigerant with a lower freezing point than it is to add heat transfer area. An outcome which would reverse this recommendation is if the needed incremental amount of heat transfer area distorted operation within the cleaning sump by taking too much space.

[96] An "enlightened" user would adopt these two modifications as normal practice when using flammable solvents in *any* open apparatus.

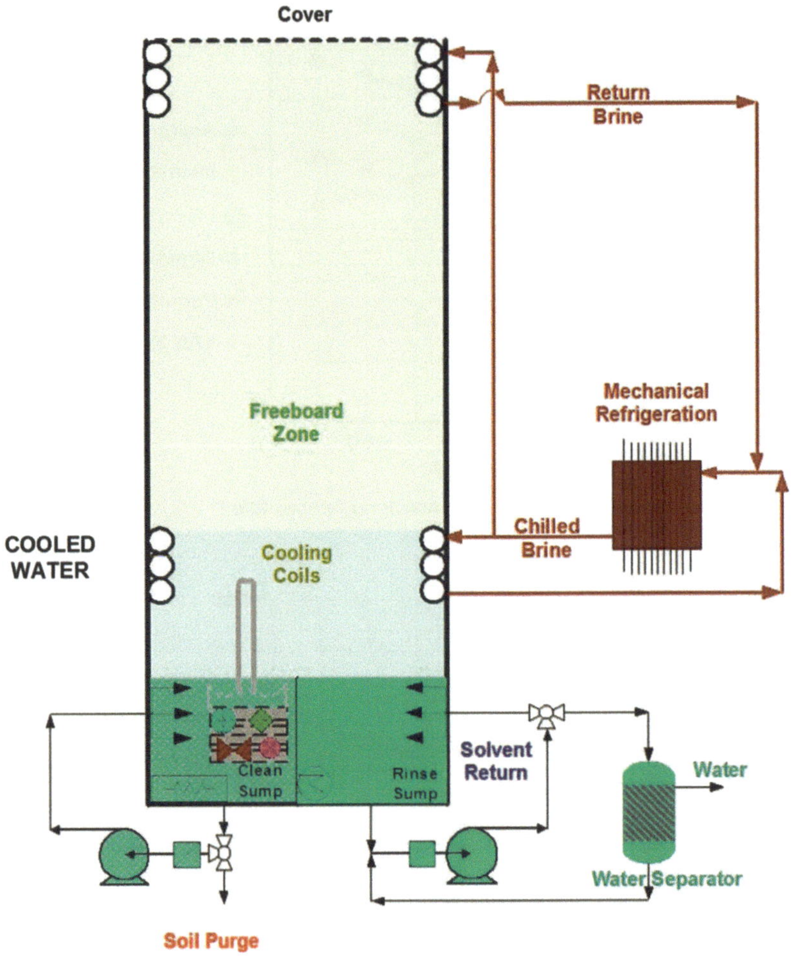

Figure 1.65 Cold Cleaning Degreaser with Two Cooling Coils

- An additional "insurance policy" is to install a second set of cooling coils at the top of the freeboard zone. They would use the same low-freezing refrigerant as used in the coils above the liquid sumps – to avoid the additional cost of installing and maintaining two different refrigeration systems.

The purpose is simple – to increase the contact time of vapor with cooling coils. This is not the normal purpose of a set of cooling coils installed at the top of the freeboard section. The dual coil apparatus is shown in Figure 1.65.

It's not clear whether the value of sub-boiling heated cleaning supports the extra cost of its practice, but as with a belt and suspenders, a second coat of paint, or mouse and a keyboard touch-pad, some circumstances may justify it.

1.36 SUMMARY ABOUT COLD CLEANING

Cold cleaning, as described above, is founded on four factors:

- Five VOC-exempt (in the US) solvents have HSPs similar to those of a solvent about which there are significant SHE concerns (n-propyl bromide),
- Values of HSPs are only weak functions of temperatures, so for some soils it is not essential to heat solvents to insure good cleaning,
- Two of the three factors necessary to ignite flammable solvents can be easily controlled (avoidance of ignition sources and of solvent concentrations above 25% of the LEL),

- Not all cleaning work requires a rinse with pristine solvent or thinning of soils via increased temperature.

Could the cold cleaning process above be used with n-propyl bromide or TCE? Absolutely! But, why would one do so when VOC-exempt solvents are available which don't raise concerns about carcinogenicity.

Endnotes

A. Available references, all by this author, are:

Ref. 1 Management of Industrial Cleaning Technology and Processes, 2006, Elsevier Science, ISBN-13: 978–0080448886.

Ref. 2 Cleaning with Solvents: Science and Technology, 2013, Elsevier Science, ISBN-13: 978–1455731312.

Ref. 3, Handbook of Cleaning Solvents, 2014, Elsevier Science, ISBN-13: 978–1455731442.

B. The image of Figure 1.1 is courtesy of ChemLogic, http://www.vapordegreasing.info.

C. The image of Figure 1.2 is courtesy of Unique Equipment, http://www.uniqueequip.com.

D. There is no abnormal boiling point. The word normal refers to the system pressure at the solvent boiling point being 1 atmosphere.

E. The image of Figure 1.3 shares the same source as Figure 1.1. Figure 1.3 is an image of a working (boiling) vapor degreaser.

F. Much more details are available about this important topic in Ref. 2, Chapter 7.8.

G. A database of properties of 210 cleaning solvents is available as Appendix A1 of Ref. 2 and Appendix I of Ref. 3. The properties included are general, physical, environmental, safety, and solubility. Properties of mixtures can be calculated using the blend rules in Ref. 2, Appendix A2 – if the properties of the soil component(s) are known.

H. The data source is Dow Chemical, see http://www.dow.com/gco/steward/manual.htm, Figures 4, 5, and 6. See also Chapter 35.7.2.1 and Figure 1.35.21 for the general effect on cleaning quality.

I Ref. 3, Chapter 1.3, Footnote 1, pages 5 to 6.

J. For a more detailed discussion of cleaning and solvent cleaning in general, see Ref. 1, Chapter 1, Sections 1.1 to 1.4, pages 2 to 7.

K. Ref. 2, Chapter 7.14.1 and Footnote 67 therein.

L. Ref. 3, Chapters 1 to 5.

M. **Flammability and Solvent Structure:**
The issue of supposed non-flammability of most halogenated solvents is explained with peer-reviewed data in Ref. 1, Chapter 3, Section 3.4, pages 100 to 115, especially Table 3.9.

For example, using the database prepared by the Chemistry Department at the University of Akron (http://ull.chemistry.uakron.edu/erd/) which is considered to an academic work, the flash point of perchloroethylene is reported as 113°F. In addition the Royal Society of Chemistry's data portal Chemspider (http://www.chemspider.com/Chemical-Structure.29106.html) reports a value of 83.4°F. However, various internet sources, such as Wikipedia (http://en.wikipedia.org/wiki/Tetrachloroethylene), the Center for Disease Control's Pocket Guide to Chemical Hazards (http://www.cdc.gov/niosh/npg/npgd0599.html), and most MSDSs report that the flash point value is "NA" which is presumed to mean "none."

The discontinuity is explained by noting that the molecule (C_2Cl_4) contains both fuel (two carbon atoms) and fire suppressant (four chlorine atoms), and that the ASTM closed cup test (D3278) requires only a momentary flash (1 second ignition) for a flash point to be "recognized." In other words, the conventional flash point test is "fooled" by not being reliably able to distinguish whether the atoms of fuel or ignition suppressor dominate in producing a visual outcome.

N. **The Return of Propane:**
A major discovery of the past decade (or so) in Europe and Australia is that fear of flammability can be overemphasized. Since 1995, all the major companies in Europe have adopted so called "Greenfreeze" technology – use of propane as a refrigerant in both fixed (commercial, home) and movable (cars, boats, aircraft) refrigeration systems. The reason for this adoption was to reduce the threat of global warming due to release of HFCs. Supporters of this evolution report that since this adoption there have been few or no reports of safety or flammability incidents. Ironically, or humorously, the few incidents of fires that have been reported have been with fluorocarbon refrigerants contaminated with the oil vapor of gases discharged from these systems!

The general outcome is the same in the US Aerosol propellants have been converted from HFC-134a to propane, isobutane, and n-butane without a plague of flammability incidents.

In both cases the reason for these positive outcomes is that the conventional safety guidance offered for use of cleaning solvents in Ref 1., Chapters 3.10 to 3.10 and Ref. 2, Chapters 10.10 to 10.14 has been followed. Briefly, that guidance notes that three factors (spark, oxygen, and fuel in the right ratios) are necessary for ignition, and very safe practice is to remove two of them from being able to occur by both required operating procedures and equipment design.

O. The image of Figure 1.16 is courtesy of Ekimloaders whose web site is http://www.ekimloaders.com.

P. The image of Figure 1.23 was produced by this author from images courtesy of Reliance Specialty Product, Inc. (www.solventequipment.com).

Q. The image of Figure 1.24 was created by this author from images courtesy of FraserTech, http://www.vmaxfct.co.uk/equipment.html.

R. Durney, L.J., Electroplating Engineering Handbook, Springer, 4th Edition, 1984, ISBN 0412741105, page 133.

S. Open-top vapor degreasers constructed in the past used drinking water from city utilities for the purpose of cooling. Some of those machines survive, consuming water and discarding it after one use.

T. In a wet or open circuit cooling tower, a fan draws ambient air past (usually countercurrent to) a flow of water. Depending upon the inlet humidity (water content) of the ambient air, more (low humidity) or less (high humidity) water can be evaporated into this air. When the water evaporates, the heat of vaporization required by that portion of the water which is evaporated is taken from the remaining mass of water, thereby reducing its temperature.

Because of the extremely high heat of vaporization of water (about 970 BTU/lb, 2256 kJ/kg), only a modest reduction in temperature can be achieved in the best circumstances. Typical performance is a reduction of water temperature by 3.9°C (7°F) to 6.7°C (12°F) – from 35.0°C (95°F) to 28.3°C (83°F). The limit on achieving a higher level of cooling of water in a cooling tower is the saturation limit of water in air, which is very low. For example, the just-described performance is an increase from 0.010 to 0.020 lb water/lb dry air.

The best operating conditions for a cooling tower are those where the air is hot and dry – where more water can be evaporated into the air. In other words, cooling towers aren't useful in all site locations. Cooling towers are not normally found in tropical (too humid) or northern latitudes (too cold). And cooling towers don't provide the same performance in all seasons, although they may be used in winter. During that season, some sites continuously operate cooling towers with 40°F (4°C) water leaving the tower. Basin heaters, tower drain-down, and other freeze protection methods are often employed in cold climates.

Because of their nature, cooling towers aren't incorporated into the overall enclosures that are packaged commercial vapor degreasers. These are large, emit thousands of cubic feet of wet air per minute, and older models can make significant noise. See Figure 1.30 which is courtesy of http://www.ditrade.com.

U. Some very large open-top vapor degreasers will have two and occasionally three small compressors with the refrigeration units not centrally located.

V. Freeboard is a low-cost emission control device. No moving parts are involved. No energy cost is imposed. Additional freeboard can be retrofitted, assuming there is adequate vertical space in the work area for the newly elongated degreaser.

Freeboard functions as an emission control device by being a barrier which escaping solvent molecules must surmount by diffusing through it. Freeboard acts as "flypaper" – acting to keep all solvent molecules "sticking" within the degreaser. A taller freeboard is a more effective emission control.

Diffusion is a continuous process. Freeboard is a lengthy tunnel through which the solvent vapor molecules can escape to the work area only by diffusing along the entire length of the freeboard to above the cooling coil. Fortunately diffusion is very, very slow. Diffusion of solvent molecules upward only ends when the solvent concentration at the top of the freeboard volume is the same as that above the cooling coil. That duration, in an undisturbed volume, is nearly forever.

But, the freeboard volume isn't undisturbed. In every cycle of cleaning the volume is disturbed – twice. When parts are inserted, the parts basket acts as a piston – pushing the vapor contents within the freeboard downward (Figure 1.34). When the parts basket is removed, the vapor volume within the freeboard is entrained upward with the movement of the basket. So the contents of the freeboard volume are reset every cycle of cleaning (Chapter 1.19.3).

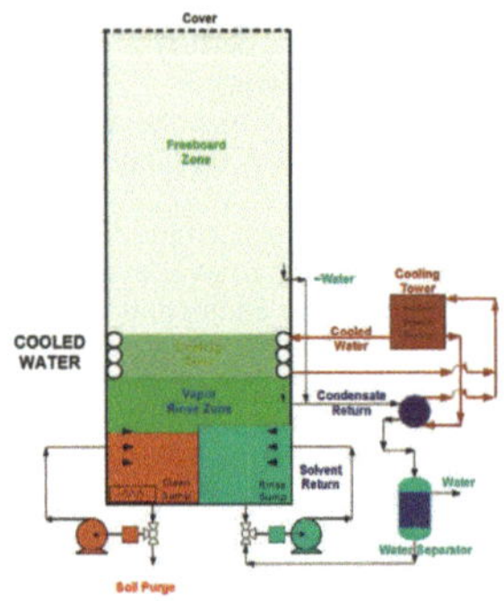

This freeboard emission control device is used in series with another – a single cooling coil. This is shown in the diagram of the vapor degreaser above. This degreaser has only a single cooling coil – there is no coil at the top of freeboard as is shown in the degreaser of Figure 1.32.

As noted in Box 1.7, the author has modeled the heat and mass transfer processes associated with controlling emission of solvents from degreasers through the use of freeboard. The complicated, non-steady state, three-dimensional situation was simplified by using the Chapman-Enskog approximation, which produces a one-dimensional steady state equation with a diffusion coefficient that is a function of both composition and temperature. Two useful references are: (1) Marrerro, T.R., and Mason, E.A., "Gaseous Diffusion Constants," Journal of Physical Chemical Reference Data, Vol. 1, No.2, 1971; which can be downloaded from http://www.nist.gov/data/PDFfiles/jpcrd1.pdf., and (2) Ray, H.S., and Ghosh, A., Principles of Extractive Metallurgy, John Wiley and Sons, 1992, ISBN-13:978–0470218112, pages 174 to 178, Equation 7.34 and Tables 7.1 and 7.2.

The degreaser under consideration is the one described in Footnote 33; width (depth) of 22 inches, and length of 26 (2 X 13) inches.

The freeboard ratios considered in this study are chosen to be extra-long – beyond the freeboard ratios considered necessarily effective by US EPA (Table 1.10). The maximum value of freeboard ratio considered in that table is 1.25. At freeboard heights of 30 and 36 inches, the freeboard ratios considered in this study are 1.36 (30/22) and 1.63 (36/30) respectively.

Two types of cooling fluid in the single cooling coil were included in the models, CTW and various brines. The solvent used in the models was perchloroethylene – chosen because of its high boiling point (121°C), which makes it more easily condensed by CTW.

The issues to be investigated in the modeling work were: (1) an overall evaluation of the two emission control devices located in series with both CTW and brine, (2) evaluation of various heights of freeboard (freeboard ratios as in Table 1.10), and (3) evaluation of the location of a secondary cooling coil. The two locations considered were: (1) at the top of the freeboard, and (2) above the primary cooling coil.

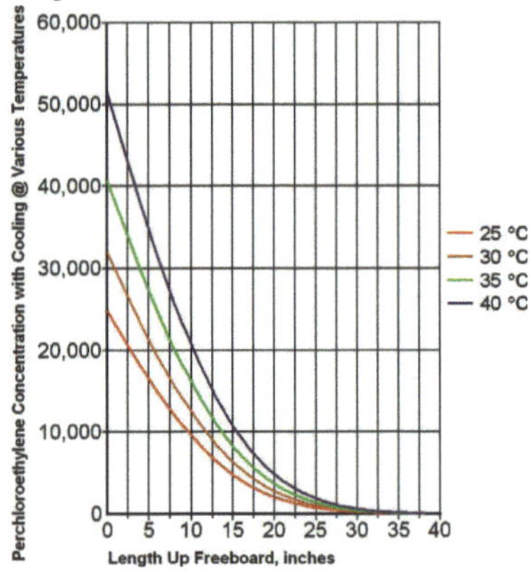

The figure above depicts use of CTW at four different temperatures to control emission of perchloroethylene from the degreaser described in Endnote X. One can see how solvent diffusion is inefficient in transfer of solvent mass along the freeboard length.

One also can't avoid notice of the substantial effect of variations which appear small in the temperature of CTW. Yet, a 5°C change in temperature of CTW is not practically small. It may not even be achievable in some local circumstances of humidity and temperature!

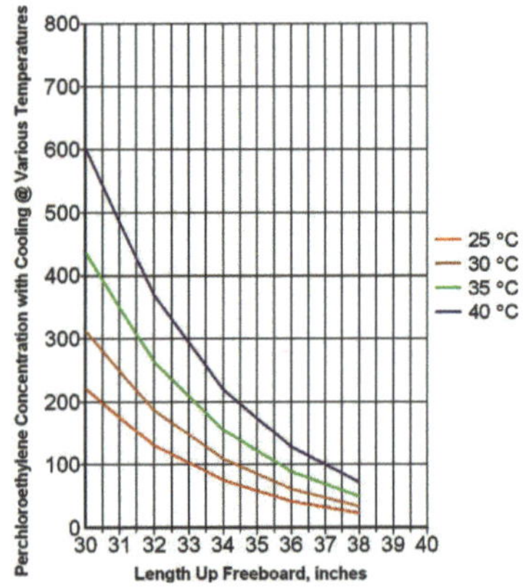

The figure above shows the calculated effect of freeboard upon emission control after the cooling coil using CTW has produced its effect. That effect was to reduce perchloroethylene concentration from 1,000,000 ppm (pure solvent in the vapor rinse zone) to ~25,000 to ~50,000 ppm depending upon the assumed temperature of the CTW. The effect of freeboard is to reduce the perchloroethylene concentration from those levels to a few hundred ppm. The latter level of emission is easily diluted with only a modest level of room ventilation to produce concentrations of perchloroethylene in the work area well below the exposure limit (25 ppm per ACGIH and 100 ppm per OSHA).

One can learn which method is more powerful if one compares these two methods of emission control as if the reduction in perchloroethylene concentration was achieved by a first order chemical reaction.

The relevant metric is

$$\ln\left\{\frac{C}{C_o}\right\},$$

Comparison of Methods of Emission Reduction $\left[\ln\left\{\frac{C}{C_o}\right\}\right]$

Method	CTW Temperature			
	25°C	30°C	35°C	40°C
Condensation by Cooling Coil	$-3.7 = \ln\frac{24,857}{1,000,000}$	−3.4	−3.2	$-3.0 = \ln\frac{51,500}{1,000,000}$
Restriction by Freeboard Diffusion Barrier	$-7.7 = \ln\frac{22.8}{24,857}$	−6.8	−6.7	$-6.5 = \ln\frac{72.3}{51,500}$

where C is the perchloroethylene concentration in ppm after either control method and C_o is the same before the control method. These metrics will always be negative, as C is always less than C_o. More relative concentration reduction will have been achieved *when the absolute values of* $\ln\{\frac{C}{C_o}\}$ *are larger.*

The comparison is shown in the table below, where it can be seen that freeboard is considerably more restrictive of passage of perchloroethylene vapor in air than is condensation by CTW. On the left and right ends of the result line are given the values of solvent concentration which are used to make the logarithmic calculation.

The analysis may be repeated by replacing CTW with brine at various temperatures. The outcomes are shown in the two figures below and at the top of page 66.

The values plotted with red lines are those calculated for use with CTW (the same as in the above two graphs). The values plotted with blue lines are those with brines at various temperatures, from +10°C to −20°C.

The effect of expending resources to provide refrigerated cooling is striking in both figures. It's the difference between the values plotted with the blue vs. the red lines.

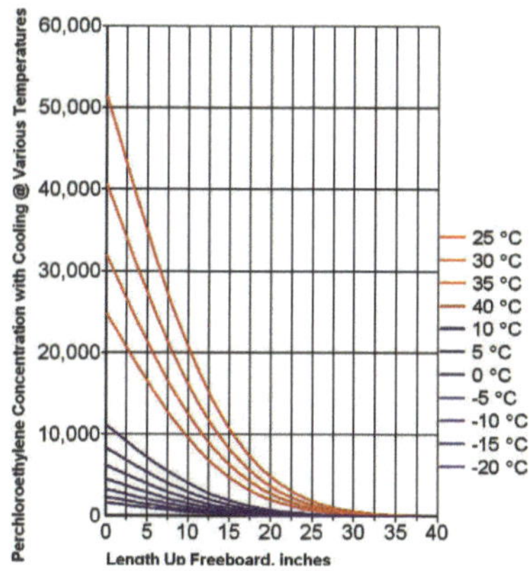

One can compare the two emission control methods, brine cooling vs. freeboard, to examine as above which has a more powerful effect (it will have the larger negative value of the logarithmic term). The comparison can be made with information in the table on page 66. On the left and right ends of the result line are given the values of solvent concentration which are used to make the logarithmic calculation.

As with the use of CTW, the relative effect of extra-long freeboard extension on reduction of perchloroethylene concentration exceeds the relative effect of condensation with brine in cooling coils. It should be clear from the two tables that freeboard extension is a credible complement to, though not a substitute for, a second cooling coil.

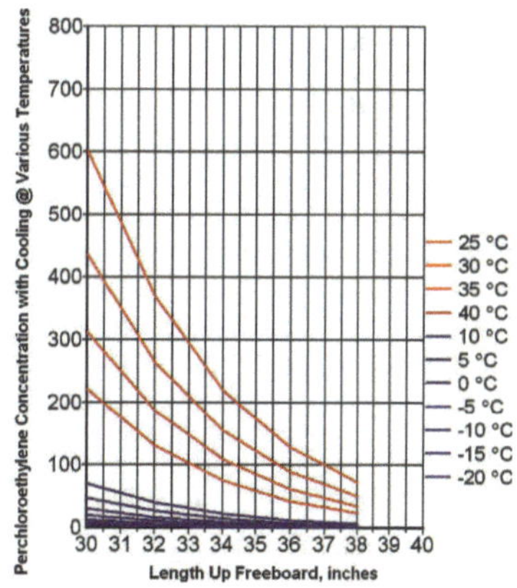

These idealized, simplistic, though not simple, calculations overpredict the extent to which emissions of solvent vapor are reduced by both cooling coils and freeboard extension. EPA recommendations, presumably based on actual operating experience, are listed in Table 1.10. They show both the value of a cooling coil using brine and freeboard extensions to be considerably less than the value noted for both above.

Comparison of Methods of Emission Reduction $\left[\ln\left\{\frac{C}{C_o}\right\}\right]$

Method	Brine Temperature						
	+10°C	5°C	0°C	−5°C	−10°C	−15°C	−20°C
Condensation by Cooling Coil	$-4.5 = \ln\frac{11,106}{1,000,000}$	−4.7	−5.0	−5.4	−5.7	−6.0	$-6.4 = \ln\frac{1,665}{1,000,000}$
Restriction by Freeboard Diffusion Barrier	$-7.5 = \ln\frac{22.8}{11,106}$	−7.7	−7.9	−8.1	−8.3	−8.5	$-8.7 = \ln\frac{0.26}{1,665}$

- A major reason for this discrepancy is that one of the assumptions necessary to make the calculations relatively simple – that the freeboard column of vapor is static and undisturbed – is certainly not true.

After all, insertion and removal of the parts basket disturbs the composition and temperature within the volume contained within the freeboard. The same effect occurs upon every occasion in the cleaning cycle when the basket is lifted from the boiling sump to the rinsing sump and thence to the vapor zone, and then returned to the boiling sump. In other words, every time the vapor barrier (Chapter 1.14) collapses and must be re-established, that event destabilizes the equilibrium assumed to be continually present in the calculations above.

Nevertheless, this author believes the modeling work above makes a valid relative if not absolute point.

- It is that extra-long freeboard extensions on vapor degreasers offer to users a valuable, low-cost, energy efficient, and underused option for control of solvent emissions.

W. For more details, see the EPA's Guidance Document for the 1994 NESHAP for Halogenated Solvents, available at http://www.epa.gov/ttnatw01/degrea/solv-cln4-30-04.pdf.

X. These calculations are based on a hypothetical but representative vapor degreaser. The basis was the following degreaser with a(n):

- Top opening of 38" wide by 30" in length (dimensions of one commercial degreaser),
- Freeboard ratio of 1.25 (meaning that the height of the freeboard is 1.25 times the degreaser length) so that the cover height above the boiling fluid level is 30" x 1.25 or 37.5", so that the
- Internal volume is 37.5" x 38" x 30" = 24.7 CF.

Furthermore, we also assume that:

- Twenty percent of that internal volume of air in the degreaser is entrained, pushed, or pulled out into the work area every time a basket of parts is removed from the top of the freeboard area, and
- The volume of air is (at worst case) saturated with solvent, and
- The working area is a room 10′ x 10′ x 10′ or 1000 CF

The calculation can be completed for any solvent whose Antoine coefficients are known (Appendix 1, Table A 2) so its vapor pressure can be estimated as a function of temperature.

An intermediate calculated result is the ratio of the volume of solvent vapor (removed from the vapor degreaser with the parts basket) divided by the volume of the room. This is the vapor phase concentration of solvent. For simplicity, it is also assumed that the room ventilation rate maintains this concentration.

Output of the calculation is the temperature at which the vapor phase concentration is equal to the solvent's exposure limit (which is given for each solvent in Ref.2, Appendix A1, Table A3. This value of temperature is listed in Table 1.7. The coolant which will meet this temperature limit is selected from the information in Box 1.5.

The calculation is validated by noting that the lowest achievable vapor concentration of n-propyl bromide in the work room is between about 8 and16 ppm for most selections of different hypothetical open-top vapor degreasers. This is consistent with industrial experience. The calculation also is consistent with industrial experience, in that it shows that lower coolant temperatures (more refrigeration) are needed to maintain exposure at less than the stated exposure limit when the vapor degreaser is dimensionally larger.

The phrase "too volatile" in Table 1.8 means that the refrigeration available as noted in Footnote 40 can't adequately condense n-propyl bromide at its lowest operating temperature in order to reduce calculated exposure to below the stated limit.

Y. The image of Figure 1.37 is courtesy of the same source noted in Endnote O.

Z. There is more information in the reference of Footnote 48, pages 4–2 and 6–3, and Figure 1.124.

AA. The source of information in Table 1.8 is the California Air Resources Board (CARB), in their "Risk Reduction Plan Worksheet for Halogenated Solvents Degreasing Operations." It can be downloaded from http://www.arb.ca.gov/AB2588/rrap-iwra/dgdoc.pdf. The same information from the US EPA or state regulatory or technology transfer organizations is slightly different. This author believes those differences are insubstantial considering the difference among equipment and installations.

BB. One can easily hang a thermometer probe using a wire from the cover of the vapor degreaser.

CC. The 1994 NESHAP for halogenated solvents can be downloaded as www.epa.gov/ttnatw01/degrea/solv-cln4-30-04.pdf. This document, written in 2004, covers the involved emission sources, the compliance options which included specific equipment requirements OR emission performance requirements, the record keeping requirements, *de minimis* limits, and guidance for preparing for a compliance inspection.

DD. It is also required by the EPA per the 1994 Halogenated Solvents NESHAP (Endnote V).

EE. This item is not shown in Figure 1.19. but details can be found in Figure 1.8.

FF. Shown in the upper right corner of Figure 1.49.

GG. Small deviations in pressure from atmospheric pressure, or some other standard pressure, are frequently given in the units of inches of water (less often cm of water). An inch of water pressure is only 0.00249 bar or atmospheres, or 0.036 psi, but 1.866 torr or mm Hg. The negative pressure within some inline vapor degreasers is typically around 10 inches of water, which would be the height of water in a tube of a manometer arranged to measure negative pressures.

HH. Only partially shown in Figure 1.51, to minimize complexity.

II. Figures 1.52 and 1.53 are images of the same inline vapor degreaser under construction. These images are courtesy of Parts Cleaning Technology, http://www.partcleaningbgky.com. The image of Figure 1.53 shows the same arrangement of cleaning facilities as is shown in Figure 1.51, and the same part transport facility as shown in Figure 1.52.

JJ. The reference about carcinogenicity can be found in Ref. 3, Chapter 21, Footnote 22.

KK. The reference is Sections 112(f) and 112(d)(6), codified at 40 CFR Part 63, Subpart T.

LL. Available current URE data can be found and downloaded from http://www.epa.gov/iris/standal.html.

MM. The rule can be downloaded from http://www.epa.gov/ttn/oarpg/t3pfpr2000s.html. The authority for a risk-based review is from the 1990 Clean Air Act Section 112(f)(2).

NN. The US EPA has stated

> "...*It was determined that a 97 percent reduction in emissions would result from switching from an existing solvent cleaning machine to a vacuum-to-vacuum cleaning machine. This determination came from data taken from Vacuum to Vacuum systems operating in the field for over 10 years. Units have now been in the field for over 15 years...*"
>
> Federal Register, *Vol. 71, No. 159, Thursday, August 17, 2006,* Proposed Rules, *page 47682.*

Vacuum-to-vacuum enclosed cleaning machines are covered in Chapter 2.1.

OO. The final rule was published in the Federal Register on October 20, 2008, Vol. 73, No. 203, pages 62383–62408. The docket ID was EPA-HQ-OAR

PP. As of December 16, 2012, status of this regulation can be found at http://www.reginfo.gov/public/do/eAgendaViewRule?pubId=201004&RIN=2060-AP07, and / or http://yosemite.epa.gov/opei/rulegate.nsf/byRIN/2060-AP07#4.

QQ. The US EPA has stated

> "...*To achieve the emission limit of 40,000 kg/yr MC-equivalent, nearly 31 percent of the affected facilities may need to select vacuum to vacuum cleaning machines to achieve necessary emission reductions. We estimate the annualized capital costs plus the operation and maintenance (O&M) costs at nearly $4.4 million for these machines, yet with a solvent savings of nearly $8.9 million,*

the total annualized control costs would ultimately save the industry nearly $4.5 million for this emission control...".

Op. Cit. Endnote JJ, page 47684.

Vacuum-to-vacuum enclosed cleaning machines are covered in Chapter 2.1.

RR. There is more information from another point of view in Ref. 3, Appendix 2, in the answer to question 20.

SS. Relative Evaporation Rate (RER) is a parameter used to compare the evaporation of liquids in a simple standardized situation. Numerically, it is not used to calculate absolute evaporation rate. It is used in the important task of comparing solvents so a selection can be made. Values of RER (with n-BuAc rated as 1) above 1 represent solvents whose evaporation at ambient conditions may be considered as rapid. Values of RER on this basis are inversely proportional to the time required to evaporate 90% of a mass of solvent. Values of RER are tabulated for 210 cleaning solvents in Appendix 1, Table 2.

TT. The temperature of the refrigerant in these cooling coils, −52.2°C (−62°F), is significantly higher than the freezing point of three VOC-exempt solvents: methyl acetate −97.8°C (−144°F), acetone −95°C (−139°F), and tertiary butyl acetate −96.7°C (−142°F); so there should not be a concern about freeze-up of the cooling coils with these solvents.

UU. Plots showing the relationships between the five VOC-exempt solvents and n-propyl bromide (plot at top) and trichloroethylene (plot at bottom).

Proximity in HSP Space with n-Propyl Bromide & 6 VOC-Exempt Solvents

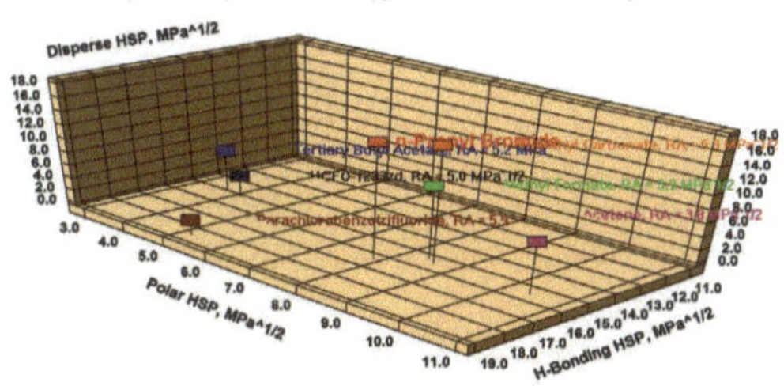

Proximity in HSP Space with Trichloroethylene & 6 VOC-Exempt Solvents

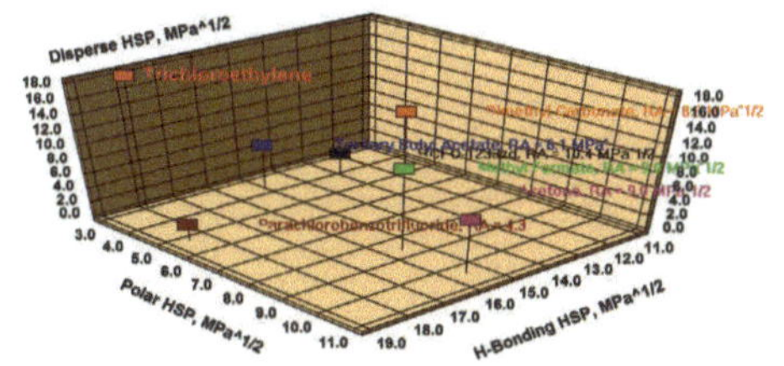

VV. A heated cold cleaning process is described and evaluated in Ref. 2, Chapter 3.32.

WW. See Endnote SS.

XX. Intrinsically safe (IS) is a practice (or condition or characteristic) in which the energy level of available sources is restricted. This prevents an electrical short or fault or a hot surface or contact of metals which produces more than a certain amount of energy – an amount tested to be less than that required for ignition. The International Electrical Code (IEC) states "...Equipment must not store or generate more than 1.2 volts, 0.1 amperes, 20 microjoules, or 25 milliwatts."

The practice was developed for the safe operation of process control instrumentation in hazardous areas, particularly around petrochemicals. The certification process for IS practices in North America is different from that in Europe, but the two certifications produce nearly identical results.

While specific design procedures are outside the scope of this book, they are knowable. The "bible" about IS is Application Note 9003 (AN9003). An excellent resource, provided by a commercial firm manufacturing IS devices, can be downloaded from www.mtl-inst.com/images/uploads/datasheets/App.../AN9003.pdf.

The requirement for use of IS devices does not restrict a designer. Rather it directs them to certain choices of equipment which have been designed for and are certified by IEC as available for use in hazardous areas with intrinsically safe barriers including: 4–20 milliamp (mA) DC two-wire transmitters, thermocouples, resistance temperature detectors (RTDs), strain gages, pressure, flow, & level switches, current/pneumatic (I/P) converters, solenoid valves, proximity switches, infrared temperature sensors, potentiometers, LED indicating lights, and flowmeters with magnetic pickups.

YY. The image of Figure 1.59 is courtesy of Petroferm, http://petroferm.com/.

ZZ. The images of Figure 1.61 and 1.60/1.61 are of different vapor degreasers.

AAA. Throughout this book, specific recommendations are offered. Many are general guidance. Some are directional – approaches to improvement. Others, as above, propose the outcome of an optimization. Some recommendations are firmly made – usually related to SHE issues where cost is not the critical factor. Most recommendations are aimed at cost or quality improvements.

In many specific situations, the choice recommended in this book is shut-ended – because capital is not available to allow investments to reduce operating costs, or local situations/opinions are not as assumed by this author. In all cases there is no commercial bias supporting any recommendation found in this book; the bias is that of the author's experiences about what should be done.

Chapter 2

Enclosed Vapor Degreasers

Enclosed cleaning machines are the "next new thing," but they have been that for at least a generation.

This simple, basic idea is true – whatever cleaning process can be conducted in an open-top machine can be conducted in an enclosed vapor degreaser with the benefit of reduced solvent emissions. That's the usual reason enclosed machines are built – to house some cleaning process and limit the emissions of solvent from it.

A second reason enclosed machines are built is to conduct unique processes which could not otherwise be practiced under ambient atmospheric conditions. Not all cleaning processes are vapor degreasing with solvents at their normal boiling points, or immersion cold cleaning.

Enclosed vapor degreasers have become popular and necessary in the last two decades (Box 2.1). The reason is that non-solvent cleaning methods (particularly aqueous cleaning technology) don't provide adequate performance, and some users desire or need to use solvents about which there are SHE (safety, health, and environmental) concerns – particularly perchloroethylene, methylene chloride, trichloroethylene, and n-propyl bromide (in that order).

2.1 TYPES OF ENCLOSED VAPOR DEGREASERS

Enclosed vapor degreasers are differentiated both by the type of process enclosed within them and also by the degree and method by which they are sealed from the ambient environment.

There are three basic types of enclosed vapor degreasers: (1) externally sealed, (2) airtight, and (3) airless.

- **Externally sealed** enclosed vapor degreasers are not sealed for pressure or vacuum; they can't be, because they are externally sealed. This type of Vapor degreaser is sealed to restrict interaction of the internal environment (within the enclosure) with the ambient environment[A,B,1]. A perfect, but overly small, example of an externally sealed enclosed vapor degreaser is a glove box. Described another way, externally sealed vapor degreasers are open-top vapor degreasers in an isolation chamber with an airlock and a means of flushing its contaminated volume for maintenance (Figure 2.1).
- The cycle time of a cleaning process practiced in an externally sealed enclosed machine can't be less than the cycle time for the same process practiced in an open-top vapor degreaser. This is because additional time is needed to operate the airlock on both the entry and exit process steps. Furthermore, there is an additional expense beyond that of an open-top vapor degreaser: operating cost for labor to operate the airlock, or investment for automated equipment to do the same.
- An unobvious advantage of externally sealed enclosed vapor degreasers is that solvents with low flash points or exposure limits can be used because they are contained.
- **Airtight** enclosed machines are those which seal against internal pressure, as does an autoclave. They can be basically tanks with spray nozzles (Figure 2.2), or more useful arrangements.
  - In an airtight enclosed machine one could use methylene chloride[C] to clean at a much higher temperature than the temperature at which

[1] In the mid-1990s these externally sealed machines were exhibited at various trade shows. They drew some interest, but few sales. The reason was that other technologies, such as airless enclosed machines, provided better value. Externally sealed enclosed machines do constrain emissions at the price of increased initial investment and somewhat lengthened cycle time. Airless and airtight enclosed machines do also constrain emissions with similar drawbacks; but they enable cleaning processes not possible (pressurized and vacuum operation) with externally sealed enclosed machines, which are basically open-top machines with an airlock as a cover.

http://dx.doi.org/10.1016/B978-0-323-22520-5.00002-2

Box 2.1 Some History

Though enclosed vapor degreasers of various types have been known for generations, they were not commercially attractive because the use of perchloroethylene, methylene chloride, and trichloroethylene in open-top vapor degreasers was not significantly threatened by regulatory fiat or managerial fear.

It was the general need to clean carbonaceous residues from heat treated metals by spraying/soaking them in hot oil which led to cleaning machines in which the pressure was elevated above ambient. Other soil materials about which there was the same need were asphaltine tars, coke residue, aged grease, oxidized heat treat oils, waxy carriers, etc.

It was the specific need to clean greases rapidly and efficiently from parts by soaking them in boiling perchloroethylene without exposing the parts to an associated high temperature which led to vapor degreasers in which pressure was reduced below ambient. The normal boiling point of perchloroethylene is 121°C (250°F).

In a sense in these applications, the enclosed machine was an autoclave, a bariatric chamber, a chemical reactor, or possibly a storage vessel. These were not cleaning machines designed to just conduct conventional cleaning operations, such as to apply mechanical force to parts, apply solvent to complex sections, remove dragout to make rinsing more efficient, flush (rinse) remaining dragout, or evaporate all solvent from all part sections.

perchloroethylene is used in a vapor degreaser. A user would do this to obtain use of a solvent with the HSP (Hansen Solubility Parameters) values of methylene chloride[D] but employ it at a high temperature in order to thin a viscous chlorinated paraffin or grease soil[E,F]. Naturally, the internal pressure within that airtight enclosed machine would be higher than ambient.

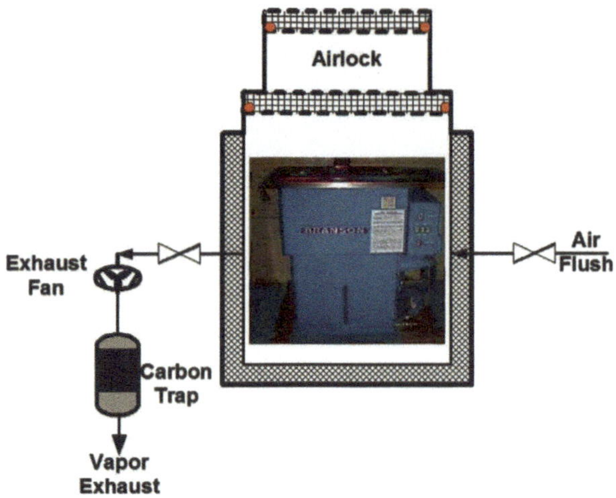

Figure 2.1 Equipment Diagram of Externally-Sealed Enclosed Vapor Degreaser

- In another (or the same) airtight enclosed vapor degreaser, one could find the opposite situation. A user could want to use perchloroethylene as the cleaning solvent in boiling service (to match HSP values of a specific soil), but not expose the parts to the high normal boiling temperature (121 °C) of that solvent. One could easily reduce the internal pressure within the airtight enclosed vapor degreaser so that the boiling temperature of perchloroethylene would be half that of the normal boiling point, or less; or more.
- In still another (or the same) airtight enclosed vapor degreaser, one could use an ultra-expensive light-unstable perfluorinated butyl iodide solvent Ref. 3, without exposure to either UV light (Ref. 2, Figure A III-1) or the threat of unwanted exposure to openings for emissions. Practically, PFBI must be used in an airtight enclosed vapor degreaser.

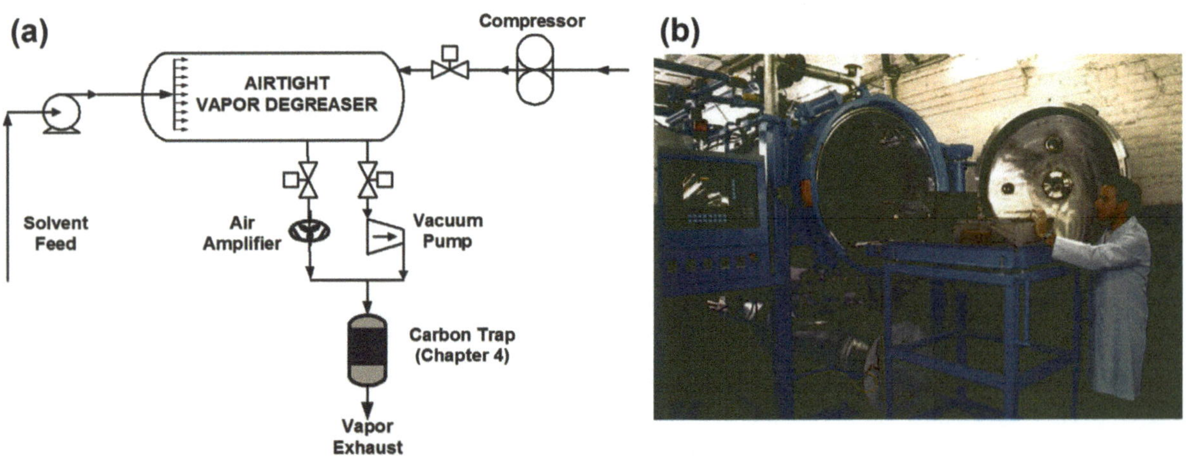

Figure 2.2 (a) Equipment Diagram of Airtight Enclosed Vapor Degreaser. (b) Example of Airtight Vapor Degreasers, this image is courtesy of Serec (http://www.serec-corp.com)

Box 2.2 Cyclic Operations in an Enclosed Machine

The information in Figure 2.3 represents a general extreme of practical capability. Another example of a capability limit of a vacuum pump is one that can remove air (or nitrogen) from a chamber whose capacity is a few cubic feet so that the pressure is reduced from one atmosphere to around one Torr absolute in about 30 seconds. There is more information in Endnote Q.

A still further example is intermediate to the situation portrayed in Figure 3, and that of conventional ultrasonic transducers. For the former the interval of one cycle of pressure is 160 million microseconds. For the latter the interval is around 25 microseconds.

With the process of Vacuum Cavitational Streaming (VCS) aka Vacuum Cycle Nucleation (VCN) the interval can be on the order of 1 to 30 million microseconds. VCS produces pulsations in the overall fluid which repeatedly produce bubbles on soiled surfaces. Collapse of the bubbles generates mechanical force to liberate soils.

VCS is technology developed (Endnote PP) for cleaning surfaces in an enclosed chamber with pulsations in a solvent such as methylene chloride or perchloroethylene. Water may also be used, and may contain a chemical such as hydrogen peroxide.

There is less importance on the rapid implosion (physical mechanism) of the bubble, and more focus on the controlled formation and collapse (as opposed to implosion) of the vapor bubble.

A liquid ring vacuum pump or a mechanical dry pump is used. Operating pressures are in the range of 1 to several hundred Torr over the time interval of one cycle. Operating pressures and temperatures are just below the saturation temperature for the operating fluid.

A valuable and technically difficult application for VCS is cleaning of long thin tubes to be made into medical syringes.

- In a final use of another (or the same) airtight enclosed vapor degreaser, one could perform cyclic operations – an increase of pressure (hence temperature) followed in some time by a reduction of pressure (hence temperature). The purpose of the pressure/temperature change would be to subject soil structures to thermal stresses, thereby weakening them, or their bonds to surfaces. An example is plotted in Figure 2.3 where the pressure is cycled about (2/3) of an atmosphere around atmospheric pressure over a time period of about two and one-half minutes (Box 2.2).
- **Airless** enclosed vapor degreasers are just that – vapor degreasers in which the internal volume is not air. The word airless has been used generically, and as a trademark. Some airless machines are also known as vacuum-to-vacuum vapor degreasers[G].
  - In an airless vapor degreaser, a person familiar with the issues, problems, and solutions described in Ref. 2, Chapter 11 (Stabilization of Solvents) can find relief. An environment where solvent degradation can't occur is one in which the air (and its contained humidity) has been removed. This is usually achieved by flushing the system volume with nitrogen, thereby removing the oxygen and water; conducting the cleaning operations; and then replacing the atmosphere with clean dry air at ambient pressure (Figure 2.4).
  - In a second type of airless vapor degreaser, the air is removed to produce a vacuum. This type of airless machine is the most common – a vacuum-to-vacuum machine, a vacuum vapor degreaser. This is done to expose solvent soiled surfaces which are complex and difficult to thoroughly wet, flush, and dry. Examples are multiport injectors, aluminum/alloy honeycomb structures used as panels in construction of aircraft, metal or glass filtration media, long tubing[2], and completed assemblies of components.

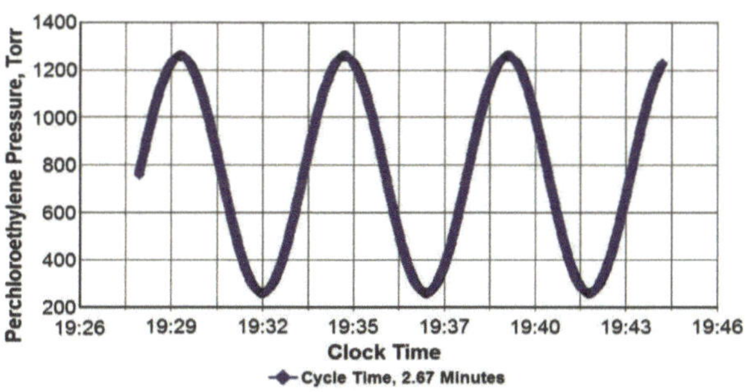

Figure 2.3 Cycling of Pressure in an Airtight Vapor Degreaser

[2] An application which has long frustrated end-users, equipment and solvent developers, and environmental regulators is cleaning of long sections of thin drawn tubing. The problem is that to clean drawing fluids from the internal tube surfaces, flush the tubing, and dry the inside of these pieces requires a volatile solvent, with low surface tension, and the solubility parameters associated with halogen atoms.

The available solvents for open-top vapor degreasing have been trichloroethylene (Ref. 3, Chapter 20) and n-propyl bromide (Ref. 3, Chapter 21) – both of which raise concerns about their effects on human health. Few credible options have been developed (Ref. 2, Chapter 2.12.3.1) to clean these parts which don't involve emission of these hazardous materials. That is why this application was exempted from further regulation in the updated NESHAP for Halogenated Solvents (Chapter 1.28).

Recently, enclosed vacuum cleaning machines have been constructed in which narrow tubes up to 85 feet in length have been cleaned with these two solvents (Box ?).

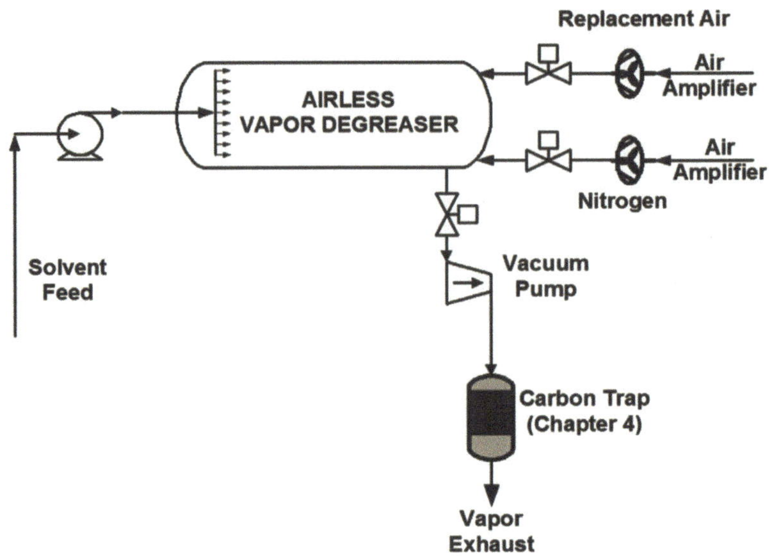

Figure 2.4 Equipment Diagram of Airless Enclosed Vapor Degreaser

Table 2.1 Comparison of Enclosed Machines

Type	Externally Sealed	Airtight	Airless
Recognizable Feature	An airlock	A sealed vessel	No contact between solvent and air
Purpose	Contain and manage normal emissions of solvent from open-top vapor degreaser	Allow control of environment in which solvent contacts parts	Avoid contact between solvent and air (oxygen and water)
Has an air-solvent interface[I]	**YES**	**YES**	**NO**
Pressure capable[3]	**NO**	**YES**	**YES**
Vacuum capable	**NO**	**YES**	**YES**

- In a third type of airless vapor degreaser, the air is replaced with nitrogen (or another inert gas), and then pressurized[H]; the desired cleaning operations are conducted; and then the inert pressure is relieved and the atmosphere replaced with clean dry air at ambient pressure.

2.1.1 Summary of Types of Enclosed Machines

The three types from above are compared in Table 2.1.

[3] Assuming that the vessel has been designed with adequate thickness to contain the applied stress (Equations 2.3 and 2.4).

Note that these distinctions are not mutually exclusive. Both airtight and airless enclosed cleaning machines are sealed vessels. They differ in how they are used.

2.2 VACUUM PUMPS

These devices are found in nearly all airless machines, because most are vacuum vapor degreasers.

A vacuum[J] is a volume of space which is to a degree void of matter. It's technically impossible to produce a vacuum by suction, because fluids can't be pulled. Rather, the vacuum-producing device expands the volume of space to

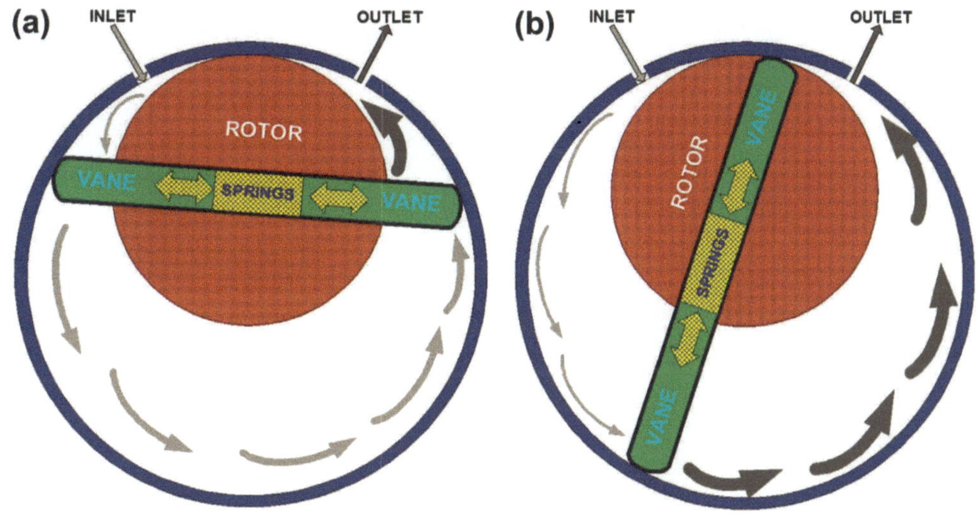

Figure 2.5 Operational Sequence in a Rotary Vane Vacuum Pump

be evacuated into some larger volume, thereby reducing the pressure[4] in the volume being evacuated.

There are many types of vacuum pumps, designed to produce various levels of vacuum for specific types of research, development, and production work, and for use in enclosed cleaning machines. The four types most commonly used with enclosed cleaning machines are described below.

2.2.1 Rotary Vane Pumps

These are positive displacement devices – also called fixed displacement pumps[5]. That means that the volume of movement within the pump is the volume of fluid which will be removed from the work chamber during every revolution of the pump.

Of course, the mass of fluid being removed within the volume of the work chamber during every revolution depends chiefly upon the pressure and temperature within the work chamber, and so varies within the cycle time of evacuation.

The most common type is a single vane rotating inside a circular cavity. The rotor and the cavity are not concentric – their centers are offset (eccentric).

There is a movable vane attached to the rotor, which can slide on springs toward and away from the rotor – always contacting the cavity wall.

The volume between the rotor vane, the rotor, and the circular cavity is a chamber which traps fluid. As the rotor moves, the chamber rotates, moving a volume of fluid from the work chamber to somewhere else. Two functional diagrams of the interior of the same rotary vane vacuum pump are given in Figure 2.5a and b. The rotor is in a different position in each of the diagrams. Oil is used as a lubricant, coolant, and gas sealant for the vanes.

Purchase of a vacuum vapor degreaser is likely to include a rotary vane vacuum pump among the installed components. Commercial rotary vane vacuum pumps also may have more than two rotating vanes as are shown in Figures 2.5a and b. Any even number of vanes is possible. Figure 2.6[K] shows an image of a rotor with six vanes.

2.2.2 Jet Eductor[6] Pumps

These pumps have no parts which rotate, or move. The eductor effect of a converging-diverging nozzle is to convert the pressure energy of a motive fluid to kinetic energy.

---

[4] Representations of levels of vacuum can be confusing. First, they are denoted with the same units as are levels of pressure – Torr, inches of mercury, millibar, psi, atmosphere, etc. This is because pressure and vacuum are the same entity – the force outward (pressure) or inward (vacuum) exerted by contained matter over an area.

Second, they may be referenced to atmospheric pressure at some standard condition of temperature (often 15.6°C (60°F)), sea level elevation, and humidity (often 50% relative). Unfortunately there is no standard standard condition.

Third, the distinction between gauge pressure and absolute pressure is not always made, or made clearly. A reading of zero pressure on a gauge means 760 Torr (corrected for standard conditions), or one atmosphere absolute pressure.

Fourth, the distinction between pressure and vacuum is not always made clear. Zero Torr (0 TorrV) of vacuum is standard atmospheric pressure, or 760 Torr of pressure. 760 Torr of vacuum is total evacuation (full vacuum), written as 760 TorrV, or –760 Torr.

[5] Other fixed displacement pumps for liquids are inline piston and diaphragm pumps. Piston pumps can be made variable-displacement by inserting springs in line with the pistons. That displacement is not positively controlled, but decreases as back-pressure increases. However, not all vane pumps have springs. Some rely on centrifugal force to seal along the pump's internal housing.

[6] The eductor or eductor-jet pump is also called an injector, a vacuum ejector, a water eductor, a steam-jet ejector, or an aspirator. Every chemistry laboratory has many.

Figure 2.6 Rotor from a Rotary Vane Vacuum Pump

That creates a low pressure zone that draws in and entrains a suction fluid (some of the contents of the work chamber). This is the mechanism by which a vacuum is created in the work chamber.

After the combined fluid (suction and motive fluids) passes through the throat (diffuser) of the jet eductor, it expands through the diverging (or exit) nozzle. In that nozzle, kinetic energy (basically velocity) is converted back to pressure energy. So, the pressure is increased at the exit as velocity is reduced.

The motive fluid can be any pressurized fluid compatible with the contents of the work chamber – water, steam, or air – whose availability has a low price, and whose disposal is convenient. A functional diagram of an eductor is shown in Figure 2.7, and a commercial model is shown in Figure 2.8.

An eductor would be an excellent choice for the vacuum pump in an enclosed machine where the parts are badly soiled with large or abrasive particulate matter, or mobile gelatinous or viscous material. This is because there are no clearances to be fouled in an eductor and the liquid

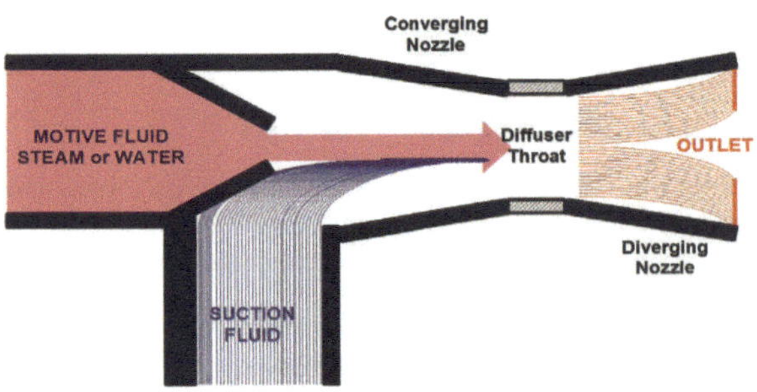

Figure 2.7 A Jet Eductor Vacuum Pump

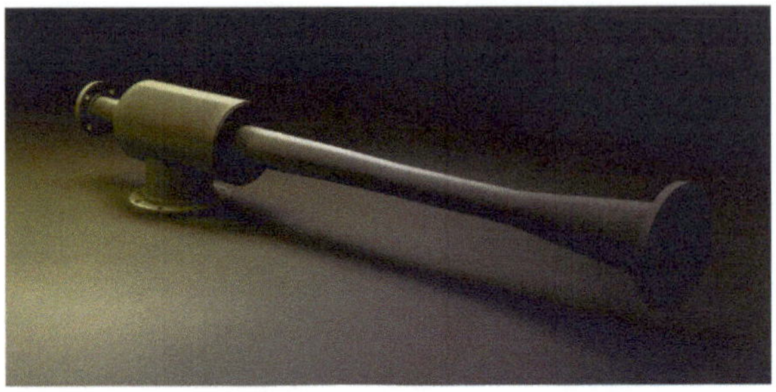

Figure 2.8 Commercial Jet Eductor Vacuum Pump

discharged from an eductor can't be harmed by contamination with the fluid soil material.

An eductor is not an excellent choice to achieve low absolute pressures. A single stage can produce pressures around 600 TorrV; two stages can produce pressures < 730 TorrV; and four stages can produces pressures < 750 TorrV.

2.2.3 Liquid Ring[L] Pumps

This popular type of rotating impeller positive displacement vacuum pump is often used in vacuum enclosed cleaning machines.

The liquid ring vacuum pump shares but one trait with the rotary vane vacuum pump – an eccentrically mounted rotor, driven by an electric motor.

But the liquid ring pump has: (1) multiple vanes (fins) vs. a single one (often) for the rotary vane pump; (2) no vanes which contact the perimeter of the cavity as it's essential for all to do so in a rotary vane pump; (3) a point of entry for the vapor within the rotor and not at the circumference of the casing as is typical of other vacuum pumps; and (4) a "pool" of liquid which acts a "collector" to remove vapor from the work chamber and also as a piston to compress the vapor not a good descriptive diagram of a liquid ring vacuum pump is shown in Figure 2.9.

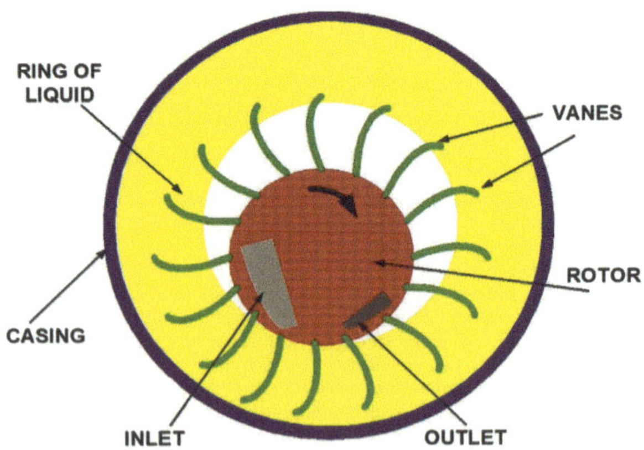

Figure 2.9 Diagram of Liquid Ring Pump Being Fed with Inlet Gas

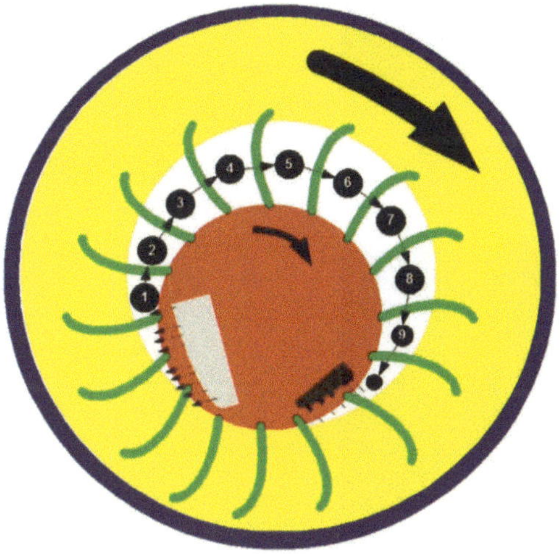

Figure 2.10 Diagram of Liquid Ring Pump Discharging Gas Causing a Vacuum

Figure 2.11 Rotor from a Commercial Liquid Ring Vacuum Pump

When this pump is not operating, it is partially full of a liquid – usually water. When the eccentrically mounted rotor is activated, the vanes lift water from the bottom of the casing and fling it to the circumference of the casing. This is centrifugal force in action – forcing the liquid to form a stable ring. While the rotor is not mounted coaxially with the casing, the axis of the ring of liquid is coaxial with the casing.

The volume of liquid within this pump is just sufficient for all vanes to be wetted with liquid at all times. This is ensured by the continuous flow of liquid to the pump. Usually liquid is removed from the pump, externally cooled, and returned to the pump.

The point of entry of vapor from the work chamber into the pump is within the rotor – not the casing – the light grey zone shown in Figure 2.9.

A liquid ring pump draws vapor into its interior by increasing the volume available to it where it exists. A clever arrangement this, as the vapor is pulled into the available volume of the pump and, like the liquid, is "flung" by centrifugal force outward toward the liquid ring.

Once into the liquid ring pump a charge of vapor is confined within a volume bounded by the rotor, two adjacent rotor fins, and the liquid ring. In a well-designed and maintained pump that confinement is maintained and vapor doesn't leak to an adjacent confinement volume of lower pressure.

Note in Figure 2.10, which is a diagram of the same idealized unit shown in Figure 2.9, how when the rotor moves clockwise the confinement volume for each charge of vapor initially increases. So vapor is continually pulled into the pump from the work chamber. In Figure 2.10, the volume of confinement increases through volumes numbered 1 through about 6.

But as rotation continues to volumes numbered about 7 to 9 and beyond, the volume of confinement decreases. This increases the pressure of the vapor just before it's discharged from the pump through the dark grey zone of Figure 2.10.

A liquid ring vacuum pump is a three-dimensional device – the rotor shown in Figures 2.9 and 2.10 actually has substantial length. Its shape is that of a cage, as shown in Figure 2.11[M].

2.2.4 Rotary Piston Pumps

These are the fourth type of positive-displacement vacuum pumps used as the "engines" of enclosed vacuum cleaning machines.

The various components of a rotary piston pump are shown in Figure 2.12.

Rotary piston vacuum pumps are based on the rotational action of an eccentric cam – not a piston producing a reciprocating action in a linear direction. Like linear piston pumps, it is the piston in this pump which is exposed to the material being removed from the work chamber.

A key part of a rotary piston pump is the slide valve, and especially the inlet sleeve within it. One end of the slide valve (a cylinder or tube) is continually completely open to the work chamber; the other end continually contacts (rides upon) the moving rotary piston.

The slide valve is a thin-walled cylinder through which gas flows from the work chamber, through the open inlet valve, through the cylindrical slide valve, and exiting into the pump cavity through the inlet sleeve. The inlet sleeve is a slit in the sidewall of the slide valve. This flow path for the fill stroke is shown in Figure 2.13.

The open end of the inlet sleeve is mounted in what is basically a bearing, but which is identified by manufacturers as a hinge pin. The hinge pin is housed into and rotates within the pump casing.

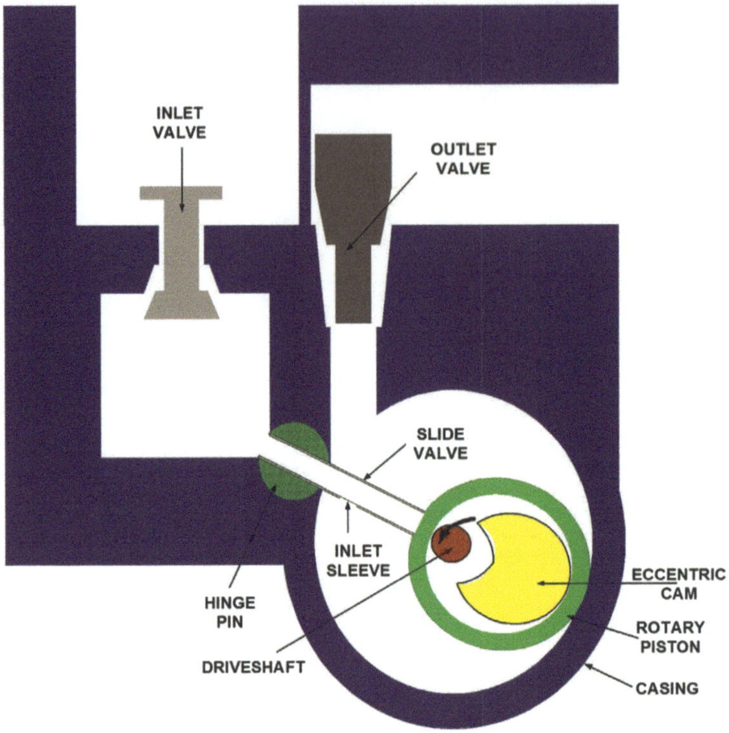

Figure 2.12 Arrangement of Components in a Rotary Piston Vacuum Pump

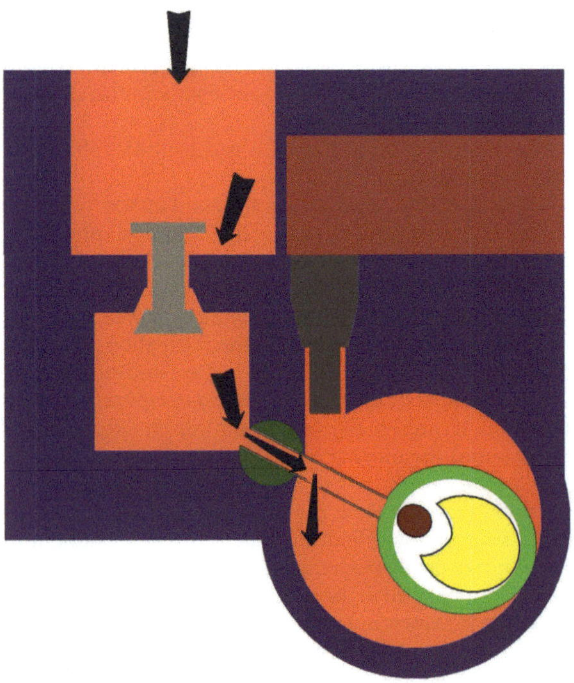

Figure 2.13 Intake Stroke (Volume Expansion) of Rotary Piston Vacuum Pump

Inside the circular piston, an eccentric cam moves relative to a driveshaft. As the driveshaft turns the cam, the rotary piston undergoes a rocking motion within the casing (or stator). This rocking motion produces a reciprocating action with the slide valve riding on the rotary piston and moving within the hinge pin. Each up/down revolution of the valve completes one cycle: intake, compression, and exhaust.

As with rotary vane vacuum pumps and liquid ring vacuum pumps, the vapor contents of the work chamber expand into some of the volume of the pump – thus lowering the pressure in the work chamber. The expanded volume within the pump volume is shown in Figure 2.14.

Further rotation of the rotary piston pushes the slide valve into the hinge pin so that the inlet sleeve lies within that bearing. This blocks flow of vapor into the pump, and is so noted by the position of the red arrow within Figure 2.14. Note also that the inlet valve has closed, and that the volume of vapor within the pump cavity has been partially compressed as shown by the somewhat darker red color.

When the vapor contents of the pump volume have been compressed by further rotation, the exit valve is opened and the further-compressed contents of the pump cavity are expelled. This, and the still darker color of the vapor volume, is shown in Figure 2.15.

Figure 2.14 Compression Stroke of Rotary Piston Vacuum Pump

As an oil-sealed pump, the rotary piston pump uses a thin film of oil between the rotating piston and the pump casing (stator). When the rotor moves along the stator, the oil prevents the migration of gas from the compression stage to the low-pressure input.

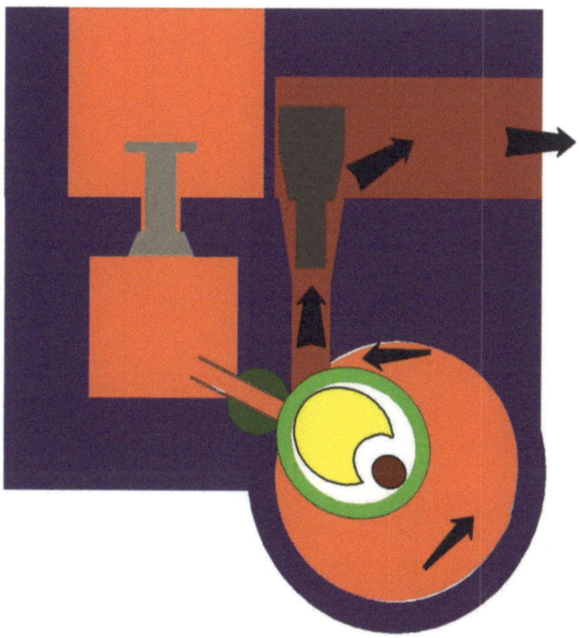

Figure 2.15 Discharge Stroke of Rotary Piston Vacuum Pump

A minimum gap of about 50 microns (.002 to .003 inches, 2 to 3 mils) exists between the rotor and the stator. This gap is larger than the gap in rotary vane pumps. The size of the gap makes the rotary piston the pump of choice when the vapor loads are heavy with particulates loosened from parts.

2.2.5 Summary of Types of Vacuum Pumps

It is important to note the requirements for the vacuum pumps used in enclosed vacuum cleaning machines. They do not have to:

- Produce lower vacuum levels than about 1 Torr absolute[N], which is not really considered "high vacuum"[7].
- Produce that level of vacuum in an extremely short time, such as less than 30 to 45 seconds (although typical pumpdown times can be several minutes in length).
- Operate continuously during their useful life, although the pressure-reduction portion of a vacuum vapor degreasing cycle is periodic. Many vacuum pumps are designed for continuous duty.

All four of these types of pumps act to both extract vapor from a work chamber and then to increase its pressure – although there is usually little value of the latter outcome in an enclosed cleaning machine.

Features, capability, and requirements of these four types of vacuum pumps are listed in Table 2.2[O].

2.3 TEMPERATURE REDUCTION AS A RESULT OF VACUUM

There is usually a significant energy component to creation of vacuum in a work chamber. But it is not expansion of vapor from the work chamber which produces the noticeable energy effect.

- The energy effect is noticed when liquids are expanded (pulled out of the vacuum chamber). For when a liquid

[7] Some who specialize in vacuum technology use the following gradation system for levels of vacuum: coarse vacuum (760 to 1 Torr absolute), rough vacuum (760 to 0.001 Torr absolute), high vacuum (10^{-4} to 10^{-8} Torr absolute), and ultra high vacuum (10^{-9} to 10^{-12} Torr absolute). To get to any pressure below 10^{-4} Torr absolute requires more than one vacuum pump.

At a vacuum level of 1 Torr absolute, 99.87 volume percent of the contents of the work chamber has been removed. In other words, the oxygen content of air in the work chamber would be reduced from the normal 21 volume percent to 276 ppm oxygen. And the water content, from inserting parts from a warm humid ambient environment (equal to 100% relative humidity in air at 90°F) can be reduced to around 50 ppm H_2O.

Table 2.2 Comparison of Vacuum Pumps[P,Q]

Issue	Rotary Vane	Jet Eductor	Liquid Ring	Rotary Piston
Sealing method against leakage	Oil sealed and air cooled	Not needed	Water sealed and cooled, or oil sealed and air cooled (preferred to avoid contamination)	Oil sealed and water cooled
Lubrication	Internal lubrication of bearings with oil	Not needed	Externally greased bearings	Internal lubrication of bearings with oil
Electrical efficiency	10 to 15% better	Not pertinent	Standard	~10% Poorer
Overall cost efficiency	~1.3X	~6X	X	~1.5X
Required investment (pump normally supplied with enclosed cleaning machine)	~$4000 to $8000	~$1500	~$4000 to $8000	~$4000 to $8,000
Ability to absorb unintentional input of liquid	Must be taken out of service to remove liquid	No concern	No concern	Must be taken out of service to remove liquid
Noise level	~80 dBa	~>80 dBA	~ 60 dBA	~> 80 dBA
Level of vacuum produced	≤ 1 Torr absolute	~10 Torr absolute (with a two-to-four stage unit)	≤1 Torr absolute (oil sealed) 30 Torr absolute (water sealed)	≤ 1 Torr absolute
Waste produced	Negligible	Torr Absolute — lb steam/lb dry air (w steam @ 100 psi) 252 — 1.8 150 — 3.3 100 — 6.6 49 — 5 (two-stage unit)	Negligible	Negligible
Coolant needed	Air	All steam must (probably) be condensed, for recovery of energy	Air (preferred)	Water

is expanded it is first vaporized. Recall that vaporization of liquid requires a supply of energy (the heat of vaporization). Then the produced vapor is further expanded.

The heat involved in vaporizing liquid contamination, such as moisture, within the work chamber[8] can only be drawn from its surroundings, namely the water (or other liquid contamination) inside the system, the parts, or the cleaning machine itself.

"Ice" crystals (of water or other liquid) so formed can remain in the work chamber as an impediment to good cleaning. They can also stay in the flow circuit throughout the evacuation process and be a threat to the proper functioning of the pumping system.

Without process modification, the parts become chilled (possibly frosted) and the remaining water/other liquid becomes frozen. This situation is much more critical with water than with organic contamination because the heat of vaporization of water is ~1000 BTU/lb and that of organic materials is ~200 BTU/lb[9].

The standard process modification is to add heat externally, so the heats of vaporization are not supplied by the contents of the work chamber. This supply of heat may come from hot air[10], radiant heat, initially heated parts, solvent added for cleaning, or hot water[11].

2.4 SIZING OF VACUUM PUMPS

There is a simple equation which guides sizing of vacuum pumps. It is derived from a mass balance and is given in Equation 2.1[12]. The evacuation rate is the size of the vacuum pump.

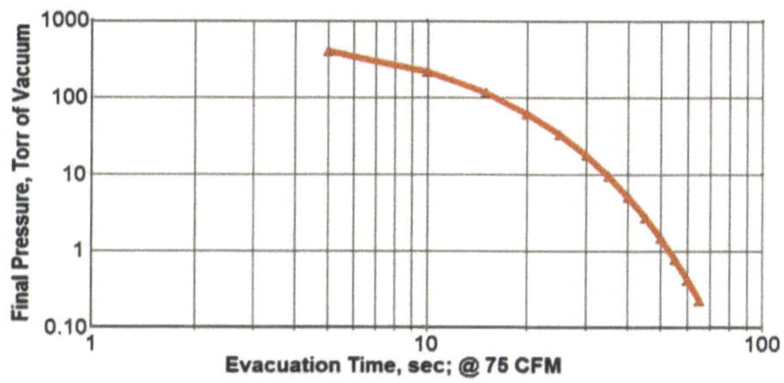

Figure 2.16 Calculated (Equation 2.1) Pressure Response to Vacuum Evacuation of a Work Chamber

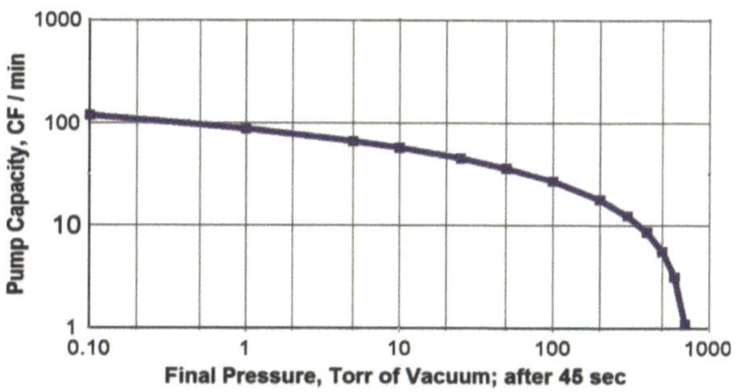

Figure 2.17 Sizing (Equation 2.2) of a Generalized Vacuum Pump to Produce a Given Level of Vacuum

$$P\ final = P\ initial \times Exp\left[\frac{(Time) \times (Evacuation\ Rate)}{Volume}\right] \quad (2.1)$$

Figure 2.16 shows the relationship involved using Equation 2.1, with the initial pressure being ambient, a modest system volume of 10 cubic feet, and a small vacuum pump rated at 75 CFM. To reach 1 Torr, a time of just less than one minute is required.

Usually, the vacuum pump is sized (Q) based on the time required (t) to evacuate the work chamber from one pressure (760 Torr, or P_1 in the general case) to another (P_2), and the volume of the chamber (V). So Equation 2.1 becomes Equation 2.2.

$$Q = \frac{V}{t} \times ln\frac{760}{P_2} \quad (2.2)$$

Figure 2.17 shows the relationship involved with Equation 2.2 for the same system volume and initial pressure as plotted in Figure 2.16.

Equations 2.1 and 2.2 are general representations and are not dependent upon any specifics about the vacuum pump used. In specific situations, other outcomes will prevail.

---

[8] The operation of evacuation is presumed to take place immediately after parts are inserted into the work chamber, but before solvent is added to initiate the cleaning operation. So what is being removed is air, and the water (humidity) and oxygen it contains

Removal of solvent vapor from a work chamber isn't normally done by production of a vacuum. It's done by purging with an inert gas such as nitrogen. So there generally isn't a concern about providing the heat of vaporization for the cleaning solvent.

[9] In other words, vaporization of one pound of water has the same energetic effect as vaporization of five pounds of solvent. Fortunately, water is a minor (and unintended) component of the contents of a vapor degreaser afterward of the parts being inserted.

[10] Hot air is less useful as a heat sink because of its low heat capacity value. A large volume of air provides or adsorbs little heat energy.

[11] Hot water is more useful as a heat sink because of its high heat capacity value. But, of course, it must be removed by evaporation so the parts are dry.

[12] The units involved in Equation 2.1 are straightforward: both initial pressure (P_1, most often 760 Torr) and final pressure (P_2) must have the same units, usually TorrV or Torr absolute; time (t) is given in minutes; evacuation rate or pump speed (Q) is given in volume per minute, usually cubic feet (CF)/min; and the volume of the work chamber (V) is in the same units as the evacuation rate, usually CF.

2.5 VESSELS FOR ENCLOSED MACHINES

A prisoner of the size and character of the parts is the size and shape of the work chamber. There are no decisions to make here – the parts have to fit into the work chamber, *and* the work chamber has to be able to house the cleaning process.

There are also no decisions to make about the shape of the work chamber; it must have the shape of the parts – or of the collection of parts if they are small (Figure 2.19). Enclosed vacuum machines constructed in the early 1990s had work chambers with a cylindrical shape – for all parts. This is because the walls of a cylindrical pressure vessel can be thinner[13] (and so cheaper and more available) to constrain a given level of internal stress (vacuum or pressure) if they are round[14].

But that situation has changed rapidly. Motivated by a developing interest in enclosed vacuum machines, and recognition that only a relatively thin level of metal was needed to retain full vacuum (Equation 2.3), a variety of shapes and sizes of work chambers were commercialized.

Various examples of commercial shapes and sizes of enclosed vapor degreasers are shown in Figures 2.18[R] through 2.28[S].

Figure 2.18 (Front View) Small (10 gal) Cylindrical Chamber

2.5.1 Sizing of Work Chambers for Enclosed Machines[T]

There are two key characteristics of the work chamber given its shape. They are its dimensions (diameter, or straight sides), and its wall thickness. The latter is substantially dependent upon the former. Vessels containing substantially larger volumes require substantially thicker walls.

The dimensions (diameter, or straight sides) of a cylindrical vessel to be used as the working chamber of an enclosed cleaning machine will be a compromise. Vessel volume is increased by the need to: (1) connect all process flow inputs and outputs to the vessel and make them capable of performing their functions, (2) provide adequate volume in which to immerse the parts in liquid solvent or vapor spray and disperse both over all of their external surface area, (3) maneuver the parts within the work chamber to make them accessible for fluid contact, and (4) allow them to be inserted and removed from the chamber easily and without damage. Vessel volume will be decreased by realization of the dramatic increases of investment required for purchase of machines of increased size (vessel volume) such as that in Figure 2.22.

- The issue of investment cost usually dominates. So the choice of vessel size is often to choose the smallest volume into which the parts can be placed – not always the best choice because it may not provide for the best fluid contact[U].

2.5.2 Sizing of Chamber Wall Thickness

The wall thickness of a cylindrical (or any shape) vessel is sized to be safely able to contain the pressure or vacuum loads imposed upon it, even if some wall material has been lost due to corrosion.

Calculation of the vessel wall thickness for cylinders (and all other shapes) is done according to an internationally-recognized procedure[15,V,W].

[13] A spherical vessel is the shape which will contain a given volume of material at a given pressure and temperature with the least required thickness of metal walls. But that is not usually the vessel shape which best contains the parts to be cleaned, and best implements the desired cleaning process.

[14] The end caps (heads) of the cylindrical work chamber may be made of sections of spheres, cones, or ellipses. The latter shape is used quite commonly with cylindrical work chambers (Endnote BB).

[15] The procedure, while internationally recognized, is not internationally consistent. The procedure is based on an analysis of the stresses applied to a "body" of certain shape and dimensions; the materials of which the vessel is to be made; and an understanding of how the vessel is to be fabricated. The procedures are collected into a code. A vessel designed and fabricated according to procedures is given a "code stamp."

Box 2.3 Stressful Activities

Ksi (or Kpsi) is an acronym for thousands of pounds per square inch of stress. It is the maximum allowable working stress for materials at the maximum design temperature. The maximum design temperature is the maximum anticipated temperature of use plus a significant safety allowance of about 50 to 100°F. Values of Ksi are in Figure 3.2 of the reference of Endnote V.

Values given for SA 516–70 Carbon Steel are 20.0 Kpsi at 149°C (300°F), and for Type 316L Stainless Steel are 16.7 Kpsi at the same temperature. Whatever safety factor ["S" (safety factor)] is selected or required is included in these values of Ksi.

Values of Ksi are calculated from measuring the response (deformation or strain) of metals to increased loads of stress, as that's how a vessel can fail. Basically all metals deform an amount proportional to the level of applied stress – up to their elastic limit, according to Hooke's Law. The elastic limit is the limit beyond which the material will no longer go back to its original shape when the load is removed. Since one doesn't want to use pressure vessels under stress loads which produce permanent bulges in the vessel's shape, the elastic limit is a certain amount of stress which should not be exceeded. That limit is divided by a safety factor to define the maximum allowable levels of working stress used in design of pressure vessels. Safety factors range between three and four. "S" is essentially the elastic limit of the metal at operating conditions, divided by a safety factor.

The equation from which a vessel wall thickness of a straight-sided cylindrical vessel (Figures 2.21 and 2.22 for example) is to be calculated, using the ASME code, is Equation 2.3[X].

$$t = \frac{[0.01934 \times P \times (\frac{D}{2})]}{[(S \times E) + (0.0116 \times P)]} + CA \qquad \text{2.3}$$

The following nomenclature is used:

t – Vessel wall thickness, inches;
P – Working pressure[16], Torr;
D – Diameter of vessel, inches;
S – Maximum allowable stress, for the chosen material at the *maximum expected use temperature*, Ksi (Box 2.3);
E – Efficiency of welds[Y], dimensionless;
CA – Total lifetime corrosion allowance[17,Z], inches.

The similar equation for calculation of wall thickness of an ellipsoidal end cap (head Box 2.4) is Equation 2.4, which uses the same variables and units as in Equation 2.3.

$$t = \frac{[0.01934 \times P \times D]}{[(2 \times S \times E) - (0.003867 \times P)]} + CA \qquad \text{2.4}$$

These equations, and others like them, are used for large (Figures 2.21 and 2.22), small (Figure 2.19), and non-cylindrical[AA] (Figure 2.20) vessels to be used in enclosed vacuum machines.

Figure 2.19 Example of Work Chamber Shaped to fit the Work to be Cleaned

Vessels for vacuum service in enclosed machines are normally made from an austenitic stainless steel, to achieve excellent corrosion resistance and unusually fine mechanical properties. The most common austenitic alloys are iron-chromium-nickel steels and are widely known as the 300 series – chiefly 304[BB] and 316[CC] stainless steels. While carbon steel is certainly less costly than either grade of stainless steel, there is not enough savings to warrant its use in purchase of an enclosed vacuum machine.

Pipe, not tubing[DD], is normally the raw stock from which vacuum work chambers are fabricated. This is because a work chamber can simply be a length of pipe with two end caps – one welded and one flanged. The flanged end cap can be sealed and unsealed to the vessel as parts are inserted and need to be removed.

[16] For full or partial vacuum conditions, the design pressure is applied externally and is the maximum pressure difference that can occur between the atmosphere and the inside of the pressure vessel. This applied external pressure is normally 760 Torr for full vacuum conditions.

[17] Corrosion allowance (CA) for stainless steel vessels is normally taken to be zero.

Box 2.4 **Capping it Off**

End caps (heads) of pressure vessels have a variety of shapes which are chosen based on cost or availability for a given chamber diameter and working pressure. Factors affecting that choice are: size of flat blank needed, ease and cost of fabrication, need for welded nozzles in the cap, retained fluid volume, etc. The choice of type of end cap should be left to the supplier of the enclosed cleaning machine as it has no bearing on cleaning operations. For completeness, various types of end caps are briefly described below.

A common type is ellipsoidal. Vessels made from this design are more compact and so take less space. For an ellipsoidal end cap (head), the height is just one quarter of the diameter. So its radius is not constant – it varies between the major and minor axis.

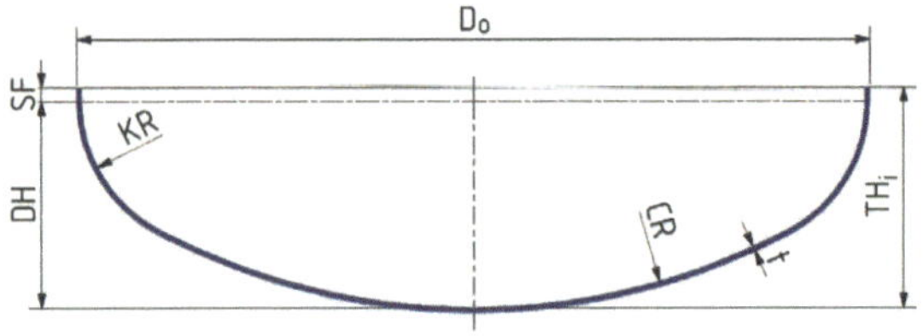

Another common type of end cap is torispherical which has some of the same virtues as does the ellipsoidal type. An image of a torispherical end cap is shown below. Torispherical end caps have a dish with a fixed radius (CR = crown radius), the size of which depends on the type of torispherical end cap. The transition between the cylinder and the dish is called the knuckle (KR = knuckle radius) which has a toroidal shape.

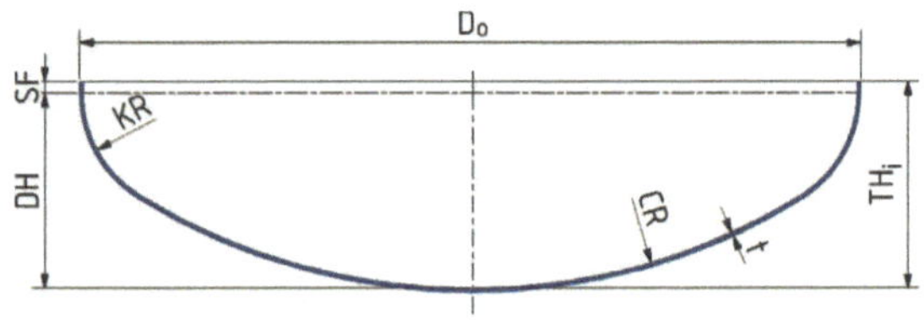

A less commonly-used type is hemispherical. This end cap is a halfsphere, with the height being the same as the diameter. The radius of the end cap is constant – it equals the radius of the cylindrical part of the vessel. Hemispherical end caps require the least wall thickness (weight of metal, difficulty of welding, fabrication cost, etc.) for any shape of end cap. But, this difference is small for vessels with diameters less than about 24 inches.

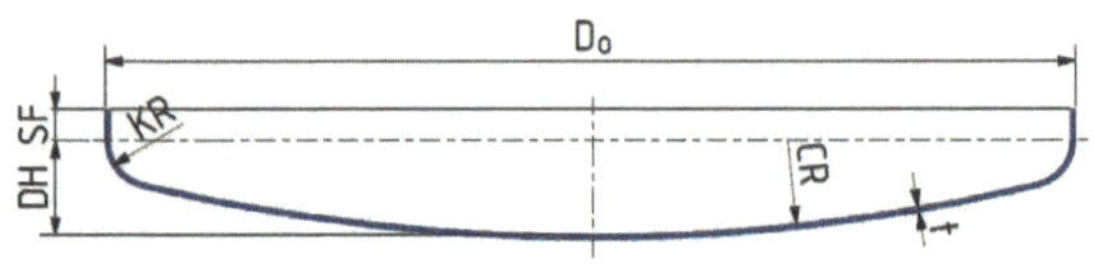

Dimensions for the ellipsoidal, torispherical, and hemispherical end caps are compared in the table at right.

A fourth type of end cap commonly-used for vessels in low pressure service is a flat head (Figure 2.23).

Torispherical End Caps (DIN 28011)	**Parameter**	**Ellipsoidal End Caps (DIN 28013)**	**Hemispherical End Caps**
DH = 0.1935Do – 0.455t	DH = depth of dishing	DH = 0.25Di	Not Specified
THi = SF + DH	THi = total internal head height	THi = SF + DH	Not Specified
CR = Do	CR = crown radius	CR = 0.9Di	CR > Do
KR = 0.1Do	KR = knuckle radius	KR = 0.17Di	KR per request
Not specified	Di = internal head diameter	Di = Do – 2t	Not Specified
Not specified	Do = external head diameter	Not specified	Not Specified
Not specified	t = wall thickness	Not specified	Not Specified
SF = 3.5t or per request	SF = straight flange height	SF = per request	SF = 20-30 mm

Figure 2.20 "Pizza Oven" Shape for Aerospace Honeycomb Panels

The surfaces of both end caps should join smoothly to the straight pipe so that there is the least possible volume of crevices where particles, soiled solvent, and rinse fluid can be trapped. In open-top vapor degreasers, these junctions are called cove corners.

Surface finish of the work chamber need not be rated #3 and above, or "mirror," for applications which are not considered to be critical cleaning. Such finishes are produced by grinding. The so-called "2B" finish, produced by cold rolling, is a cost-effective choice for most solvent cleaning operations.

It is not necessary for the work chamber to be integral with the solvent management process. This is shown for a small system in Figure 2.23, and in Figure 2.24 for a large system. Note that the work chamber shape is not always the same shape as are the parts, as shown in Figure 2.24.

2.6 ENCLOSED VAPOR DEGREASERS FOR CLEANING TUBING

As noted in Footnote 4 and Chapter 1.29, cleaning of the *inside* bore of tubing (especially of small diameter) with

Figure 2.22 Very Long Work Chambers for Cleaning Tubes

Figure 2.23 (1 ft Diameter) Isolated Work Chamber for 35 ft Long Tubes

chlorinated (Ref. 3, Chapter 20) or brominated (Ref. 3, Chapter 21) solvents is both desirable and environmentally difficult. Some coils of tubing can be immersed, flushed, and dried in the sumps of liquid maintained in open-top vapor degreasers. But, there are limits to what can practically be done:

- Long (50 ft more or less) straight tubes won't fit within such a sump,

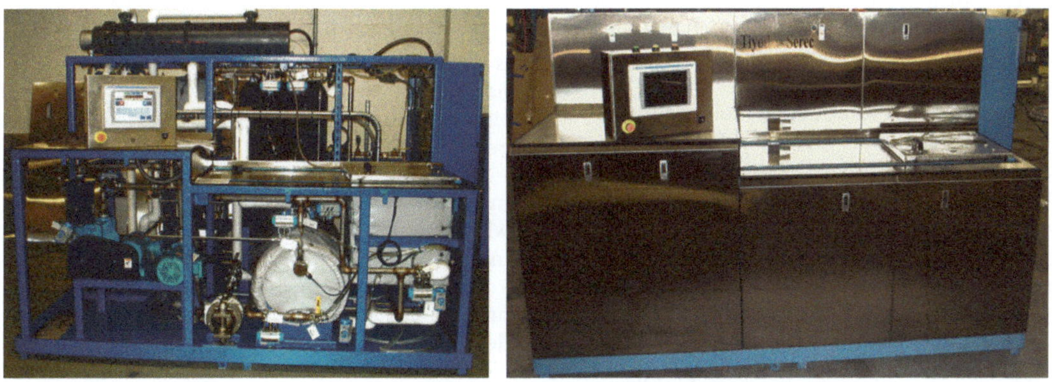

Figure 2.21 Inside and Outside of Enclosed Vapor Degreaser Rated for Service with Flammable Solvents.

Figure 2.24 (5 ft Diameter) Isolated Work Chamber

Figure 2.25 Dual Chamber Unit for Long Tubes

- Tubes with narrow (1/16 to 3/8 inch but not the 1 inch diameter shown in Figure 2.19) diameters can't be flushed by flow in such a sump, or dried at ambient pressure by exposure in a vapor zone, even if their lengths are short, and
- Levels of emission of some of these solvents from open-top degreasers are tightly regulated at federal (NESHAP), and more local (by the Los Angeles basin's South Coast Air Quality Management District [SCAQMD]) levels of government.

An enclosed vacuum vapor degreaser was conceived, designed, and constructed for just this situation. But it wasn't until the latter part of the last decade that such machines became commonly available. Their existence would appear to contradict the NESHAP (Chapter 1.24) finding in 2007 that tube cleaning machines should be exempt from its requirements (Ref. 2, Chapter 4.15).

Figure 2.26 High-Throughput Dual Chamber Unit

Various commercial vapor degreasers, constructed with dual work chambers for high productivity are shown in Figures 2.25 and 2.26.

A collage of views of a vapor degreaser constructed for long (35 ft) narrow diameter tubes to be used in medical applications is shown in Figure 2.27.

And in the image of Figure 2.28, one can plainly see that handling of parts in an important component (and cost) of the cleaning process.

During the useful life of this book, the size and shape of parts or required productivity are not likely to be impediments to the use of enclosed vacuum machines. There should be only a single impediment – as discussed in the next section.

2.7 THE CHILLING EFFECT OF PURCHASE PRICE

One might think that the popularity of enclosed machines is ever-expanding. After all, their use circumvents the hazardous potential of chemicals from being realized to either human users or the environment. This is an outcome everyone would favor.

But not so. Widespread adoption of the enclosed machine awaits. It's the reduction of selling price which limits commercial acceptance. And that isn't imminent.

Which came first – the chicken or the egg? Which came first – production of 100 standard enclosed vapor degreasers in a calendar year by a single firm, or a price reduction from one firm which enabled sale of 100 enclosed vapor degreasers in a calendar year? Answers: to the former, I don't know; to the latter, neither.

Figure 2.27 Cleaning of Small Long Tubes for Medical Service

Figure 2.28 Handling of Tubing

- All (certainly those shown in the above images) enclosed cleaning machines are custom-made for specific applications, and
- A major European firm reports that they sell a half a dozen highly automated[18] enclosed cleaning vapor degreasers and other enclosed cleaning machines within the US in a year. Total revenue from these sales is a few million dollars US.
- A major US firm reports in 2013 that all they lack to produce 25 small (Box 2.5) vacuum machines of a common size and design at a commodity selling price is several million dollars of venture capital.

Perception of economics-in-use for enclosed machines is generally incorrect, both financially and politically. The savings enabled by an enclosed machine are of: reduced costs to treat emissions of solvent and air (Ref. 5) and therefore less cost for purchase of solvent, reduced use of labor because enclosed vapor degreasers tend to be more automated than open-top vapor degreasers, and very often of elapsed time in the cleaning cycle. Only the former might find prominence on an actual cost sheet for the operating area that does parts cleaning.

Box 2.5 **The Diminishing Effect of Capacity**

A major, and in some cases dominant, component of cost is the control package. It costs almost the same to provide control facilities for a small machine with a chamber volume of 2 cubic feet as it does to provide the same capability for a machine operating the same process with a chamber volume of 20 cubic feet. Cleaning machines with a larger capacity always require less amortized investment per part. A small "commodity" machine will have a disproportionate share of its purchase price consumed by the necessary control package. This aspect, more than any other, is the limiting factor in commercializing a small commodity enclosed vapor degreaser.

[18] As with managing the purchase price of a new car by being selective about what options are ordered, purchase of a cleaning machine, or a surface treatment system with automated parts handling, is an opportunity to stress one's management skills. This author's advice as a professional consultant to clients is to fully automate the cleaning process *or* to use human labor everywhere possible. The reason is that one seldom knows the value of automation of any specific task – with one exception which is the simplest to automate. The exception is imposing a hold time between washing and rinsing to minimize dragout.

As to the value of consistency or repeatability of cleaning work done via an automated process, that's in the eye of the beholder (or user).

And as to the eagerness of US-based management to pay purchase prices of three to five times that of an open-top vapor degreaser in advance of accumulating those savings and other benefits, that zeal has yet to be realized in the US[19] by financial commitments.

2.8 USEFUL LIVES

The anticipated useful life of a carbon steel[EE] pressure vessel in an enclosed vapor degreaser is considerably longer than that of any non-enclosed aqueous cleaning vapor degreaser. That for a stainless steel pressure vessel should be considered as nearly infinite because they are not susceptible to general corrosion[20].

Useful life for mechanical vacuum pumps is considerably shorter than for the vessel which it evacuates. This is simply because it has parts moving at a high frequency and making contact with one another. When evaluating the purchase of an enclosed vapor degreaser with a vacuum pump, one ought to insist on a warranty life of the vacuum pump of at least 2 years, and consider premium value as one which has a 3-year warranty.

Useful life for an aqueous cleaning machine is only 3 to 5 years (*at best*) because of the generally corrosive character of aqueous cleaning agents.

That of open-top vapor degreasers is generally similar (*or less*) because of the same character of halogenated cleaning solvents exposed to water and oxygen.

So, if the question is whether there is an advantage in useful life for enclosed cleaning vapor degreasers vs. open-top vapor degreasers or aqueous cleaning machines, the answer is that it depends upon type. An enclosed vapor degreaser becomes a useless storage tank after around 2 years if it implements vacuum technology because of the lifetime of the vacuum pump. But an enclosed vapor degreaser operated at ambient pressure may well last a generation.

2.9 SOLVENT MANAGEMENT IN ENCLOSED MACHINES

An enclosed cleaning machine is not simply a sealed container. The container exists only to allow or enable execution of the processing of cleaning solvent, and application of that processed solvent to soiled parts.

It's the processing *of solvent* which enables the cleaning apparatus to wash parts. And it's the enclosure which is designed and constructed to enable the used solvent to be processed.

There are five actions, or functions, which must be incorporated into whatever work is done in an effective enclosed solvent cleaning vapor degreaser[21]:

- *APPLICATION* of clean solvent to dirty parts (which produces dirty solvent),
- *SEPARATION* of dirty solvent from cleaned parts (and replacement with clean solvent),
- *EVAPORATION* of clean solvent from cleaned parts,
- *PURIFICATION* of dirty solvent by removal of soil and other materials, and
- *REMEDIATION* of soil materials discharged from the enclosed cleaning vapor degreaser.

There is nothing especially illuminating about the analysis above, as it was covered in a previous reference[FF]. But it simply notes that one has to wash (clean), rinse, and dry parts as well as not get them dirty by contacting them with soil-laden solvent and not violate environmental constraints about disposal of soil-enriched solvent. But what this analysis does is to note how solvent is managed in an enclosed vapor degreaser (not how parts are cleaned[GG]).

2.9.1 Solvent Management in Externally Sealed Enclosed Machines

Figure 2.29 shows the simplest equipment arrangement (not to scale) in which these five actions or functions are incorporated into the facilities of a basic *externally sealed* enclosed vapor degreaser. The parts are washed (APPLICATION of solvent), rinsed (SEPARATION of solvent), and dried (EVAPORATION of solvent) in the work chamber. PURIFICATION of solvent occurs in a distillation column. Similarly located is where REMEDIATION of solvent occurs so that soils can be collected, concentrated, and disposed (reused).

But more (and less) can be seen in the idealized Figure 2.29, which is not to scale. Note these facilities and capabilities:

- That all these essential facilities are enclosed (within the dashed line) from the ambient environment.

[19] Though that zeal is commonly found in Europe, particularly in Germany.

[20] This statement is based on the belief that if halogenated solvents are used within enclosed vapor degreasers, the degreasers will be operated in an airless manner by first purging the oxygen and water which enter with the parts. Without that action the solvent, possibly unstabilized, will react as it does in an open-top vapor degreaser.

[21] All five, and essentially no others, are the necessary functions which must be incorporated into any successfully operated cleaning apparatus – whether an open-top solvent vapor degreaser, enclosed solvent vapor degreaser or a multistage machine practicing aqueous technology.

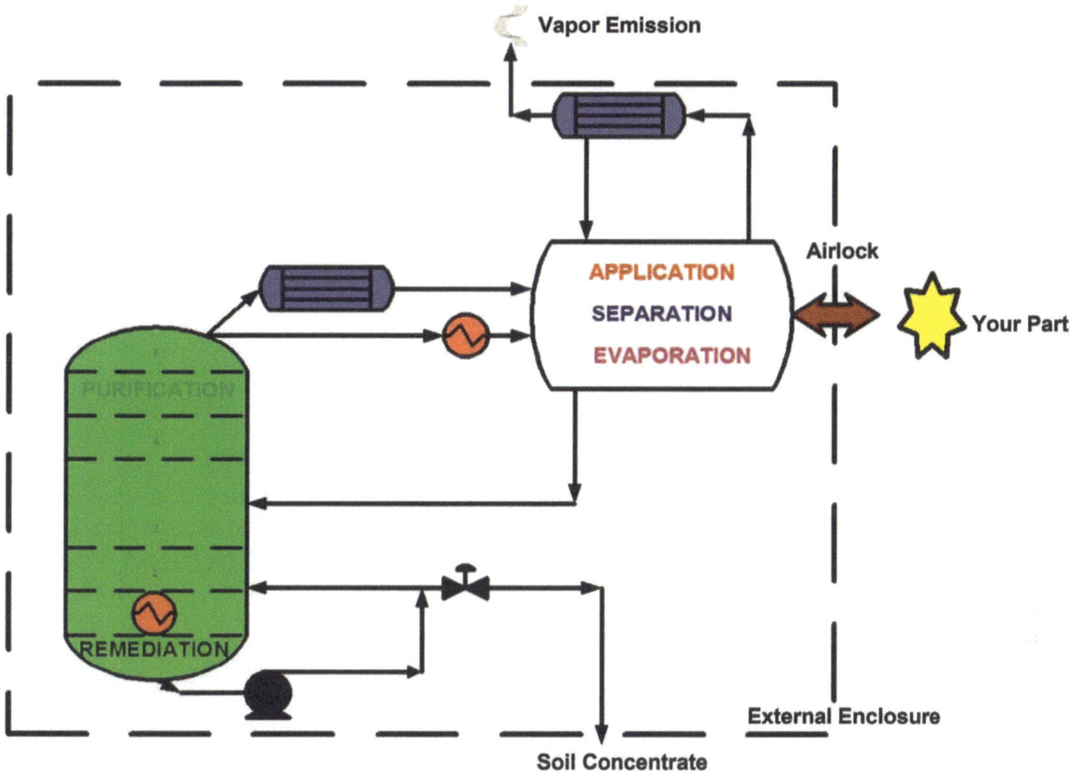

Figure 2.29 Externally Sealed Enclosed Vapor Degreaser

- The airlock where dirty parts enter and clean parts leave (the dual arrow).
- The work chamber (the white tank).
- The multi-plate[22] distillation column (the light green vessel [not to scale][23]) with an internal heater (the red heat exchanger).
- A cooler to condense and return hot or cooled liquid solvent to the work chamber and a second cooler to condense and return hot or cooled liquid solvent from the distillation column to the work chamber (the two blue heat exchangers).
- A heater to provide hot vapor as an environment or a spray to the work chamber (the second red heat exchanger).
- A pump (black) to be used with the distillation column, and other facilities not shown, to remove a concentrate of soil from the enclosure.
- Two emissions from the enclosure: (1) the soil/solvent concentrate, and (2) non-condensables (mostly air) which enter with the parts. Note the latter (wisps in light grey) at the top of Figure 2.29.

Not shown are both essentials (feed and waste material storage or holdup, instrumentation, controls, supports and frame, insulation and heat-tracing, filters, valves, baskets, manuals, MSDS, etc.) and non-essentials (observation or sampling ports, stainless steel skins with a 2B finish, and automated materials handling facilities).

Note that the facilities shown in Figure 2.29 are neither those of an airtight or airless enclosed vapor degreaser. They are the facilities of a conventional open-top vapor degreaser enclosed within a suitable housing.

[22] It is not always necessary to incorporate a multi-plate distillation column into the design of facilities. The need for it depends upon the difference in volatility between the solvent and the soil materials. When the volatilities are more similar, the need is greater for a multi-plate distillation column. Alternately, where the soils are of low volatility (greases, high-boiling oils, etc.), and the solvent has a boiling point less than that of trichloroethylene, pressure control in a solvent storage tank may allow the tank to act as a single-plate (single-stage) "flash drum," and effect the necessary separation.

This author believes that there must be at least 15°C difference in normal boiling points for a useful separation to be made in a single stage of distillation.

A multi-plate distillation column is noted in Figures 2.29 to 2.34 to support the general point that a nearly complete separation between solvent and soil components must be made.

[23] It is certainly possible that the distillation column will not be located within the enclosure. But it is both cheaper and usually simpler to do so.

2.9.2 Solvent Management Airless Enclosed Vapor Degreasers

An airless non-vacuum vapor degreaser would be described by the image of Figure 2.30. It is diagrammatically different from the image of Figure 2.29 by removal of the external sealing chamber, and spreading of the light grey lines at the top.

- Figure 2.30 diagrammatically, does not have an airlock. Parts enter the work chamber and leave through any sealable opening.
- Figure 2.30 is nearly silent on a significant requirement – the need to treat the substantially greater[24] volume of exhaust air which is presumably saturated with solvent vapors (note the top of Figure 2.30). The near silence is because this requirement must be met, but is done so outside of the battery limits of the enclosed vapor degreaser.
- APPLICATION, SEPARATION, EVAPORATION, PURIFICATION and REMEDIATION are conducted identically in both Figures 2.29 (externally sealed) and 2.30 (airless non-vacuum).

The power of the airless enclosed cleaning vapor degreaser of Figure 2.30 is mighty – it is that solvent cleaning can be done in a totally inert environment (no water, no oxygen), or in a solvent-only environment, or mixed environment of inert (likely nitrogen gas) and solvent – at ambient pressure.

In one sense, the airless enclosed vapor degreaser is the perfect open-top vapor degreaser because: (1) no tall freeboard is required to contain solvent vapors, allowing the apparatus be compact and easily located in a work area; (2) a fresh supply of pristine distilled solvent is always available for cleaning and rinsing so there is no space consumed by a second liquid sump, and no outcome where the parts are rinsed before they are washed (Chapter 1.14); and (3) no programmable hoist is required as the solvent is "moved" rather than the parts in a basket being moved. Furthermore, no (or only a tiny quantity) of stabilizer is necessary because intrusion of oxygen and water is diluted by more than 99.8% with inert gas before solvent is introduced (unless the unusual operation noted in Footnote 13 is practiced).

Yet, in another sense, this airless enclosed vapor degreaser is the "anti open-top" vapor degreaser because it doesn't contain solvent emissions within its confines; it emits them to be treated in a downstream process Chapter 4, or contains them within some vessel external to the enclosed vapor degreaser.

But in summary, the power of the airless non-vacuum vapor degreaser is not paramount, because it doesn't enable solvent cleaning in a solvent-only environment – *at pressures more or less than ambient, and temperatures more or less than normal*[HH] *boiling points.*

2.9.3 Solvent Management in Airtight Vacuum Enclosed Vapor Degreasers

An airtight enclosed vapor degreaser which allows that is shown in Figure 2.31, as a vacuum vapor degreaser.

The addition of a vacuum pump (Chapter 2.2) to an airtight vessel rated to contain full vacuum (Chapter 2.4), produces a parts cleaning system known as a vacuum vapor degreaser. All else is diagrammatically the same between Figures 2.31 and 2.30, including the emission of solvent-rich vapor.

But what's not the same is the capability of the vacuum-enabled apparatus. One embodiment is that solvents may be used at temperatures lower than their normal boiling point. Another is parts may be dried to a greater degree of completeness. Still another is that one can achieve solvent contact upon all surfaces of the most convoluted part sections. Fitted with a pressure-capable pump instead of a vacuum pump, one can treat part surfaces in the work chamber with solvent at temperatures higher than their normal boiling point. All else is diagrammatically the same between Figures 2.31 and 2.30, including the emission of pressurized solvent-rich vapor.

Simply, the five actions noted above may be completed under vacuum or pressure – depending upon the chosen solvent and its temperature of use.

2.10 STORAGE OF SOLVENT IN ENCLOSED MACHINES

If one constructed a commercial cleaning machine using just the components shown in Figure 2.31, it would be a disaster. By analogy, it would be similar to preparing

[24] The volume of solvent in the air vented from the externally sealed apparatus of Figure 2.29 is small. This is because the cleaning chamber (the volume within an open-top vapor degreaser) never releases its full contents of vapor. Only air saturated with solvent is released when the parts basket displaces it into the airlock on entry and entrains it on exit. The released volume is probably measured in cubic inches per cycle. However, relative to the issues associated with solvent stability (Ref. 2, Chapter 11), it may be crucial to trouble-free operation.

But the volume of solvent in the air vented from the apparatus of Figure 2.30 can represent a substantial volumetric emission of a solvent which may well be a HAP (methylene chloride, trichloroethylene, perchloroethylene, and possibly in the future, n-propyl bromide). The volume of solvent-contaminated air is at least one and probably more volumes of the work chamber – probably one or more dozen cubic feet. This is because the entire volume of the airless enclosed vapor degreaser, nitrogen saturated with solvent at atmospheric pressure, is released for condensation and emission every cleaning cycle.

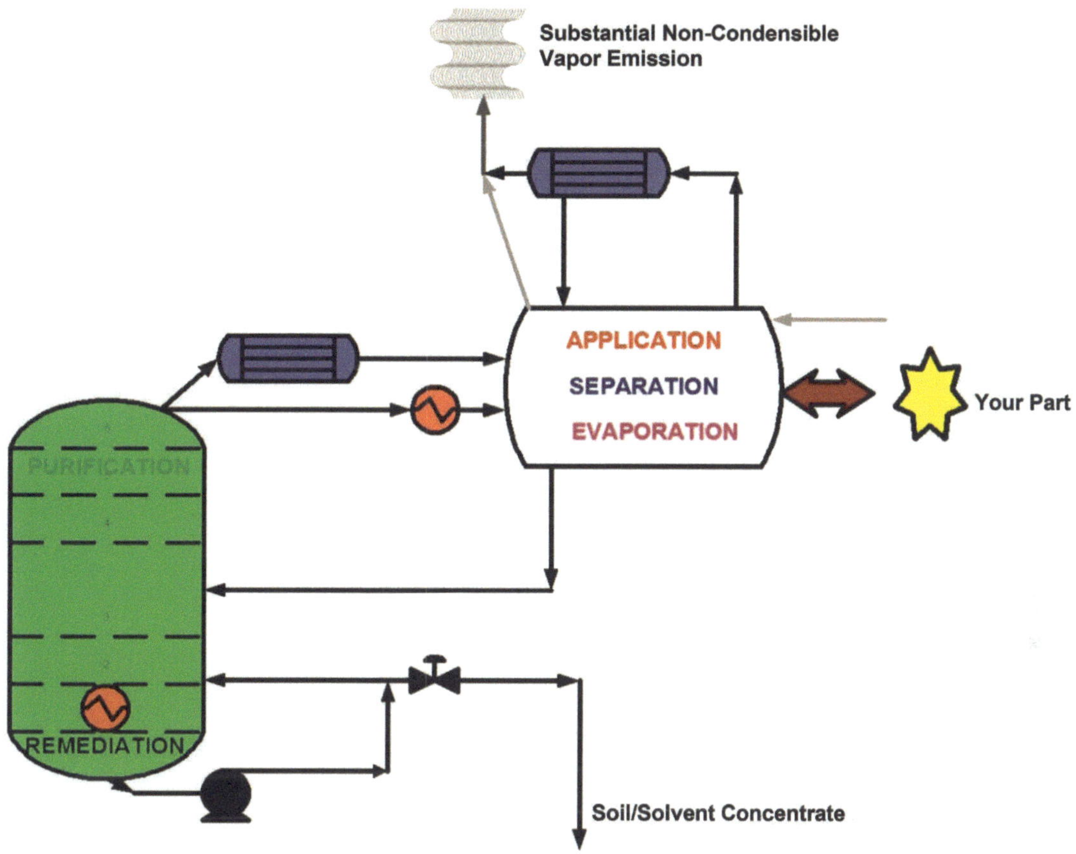

Figure 2.30 Airless Non-Vacuum Enclosed Vapor Degreaser

a multi-course meal in a single pot, or applying two differently colored coats of paint from the same can of paint. To complete many activities, one needs to have at one's disposal the ability to store intermediates – a lay-down area, a shelf, or a storage tank.

Two-sump open-top vapor degreasers don't have internal intermediate storage – the cleaning sump accumulates soil until the charge of solvent must be discarded (or periodically or continuously purified[II]); the rinse sump also accumulates soil as dragout until its charge must be discarded (or periodically or continuously purified). Open-top vapor degreasers are batch machines, with external storage. By analogy, a two-sump open-top vapor degreaser is similar to children each taking a bath in the same bathtub and the last child not being as clean as the first.

Enclosed vapor degreasers offer continuous cleaning – because they have a continuously operating distillation column integral to the apparatus. Solvent storage is internal. By analogy, each of the above children is taking a shower, and the contents of the shower drain are continuously routed to a water treatment facility.

2.11 COMMERCIAL VACUUM DEGREASERS

Global suppliers of vacuum enclosed vapor degreasers number as of 2013 number about one-half dozen. The machines (apparatus) of each supplier manage the contact between parts and solvent, and the storage of solvent(s), in ways which are similar and only slightly different.

Diagrams of the apparatus of two such suppliers are shown in Figures 2.32 and 2.33. The suppliers are described here as "S" and "D" respectively. Both firms equip their apparatus with two storage vessels, a distillation column[25], various heaters and coolers, a work

[25] As noted in Footnote 25, a multi-plate distillation column is not always provided, or needed. Firm "S" generally uses the secondary storage tank (in light grey in Figure 2.32) to provide adequate separation by "flash" (one-plate) distillation between soil components and solvent. Were the cleaning solvents to be less volatile such as the "3A" solvents (Ref. 2, Chapter 4.11), a multi-plate distillation column would almost certainly be a necessity.

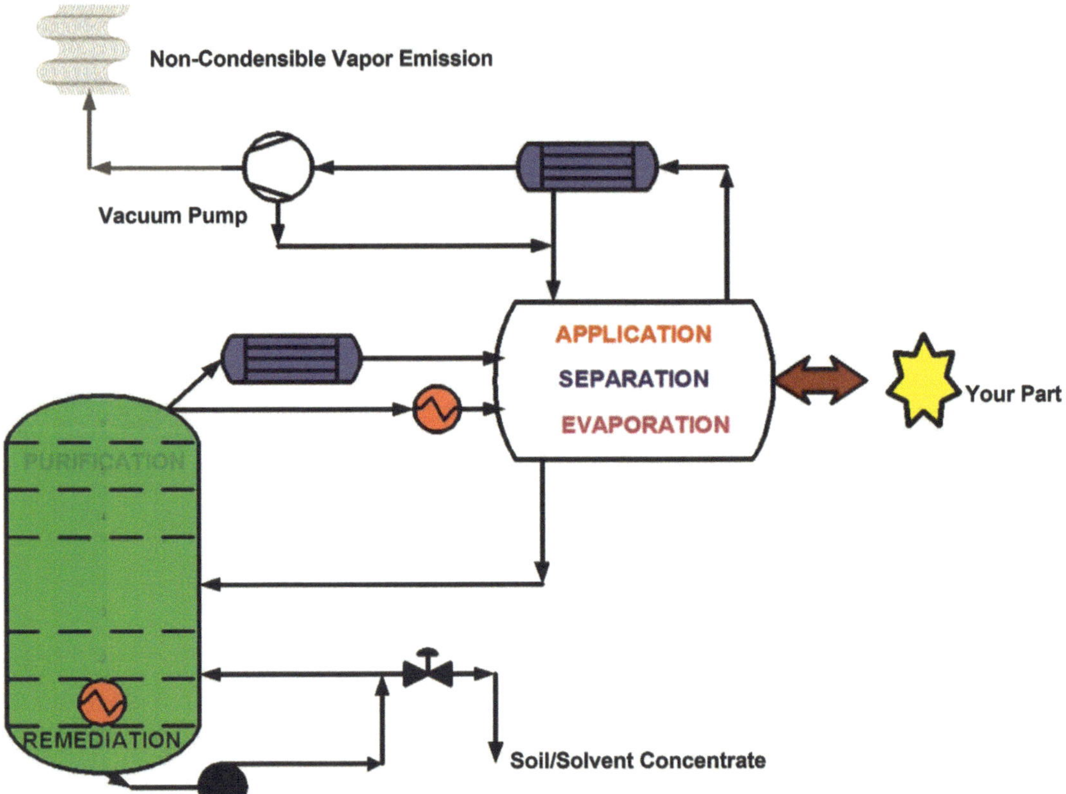

Figure 2.31 Airtight (Vacuum) Enclosed Vapor Degreaser

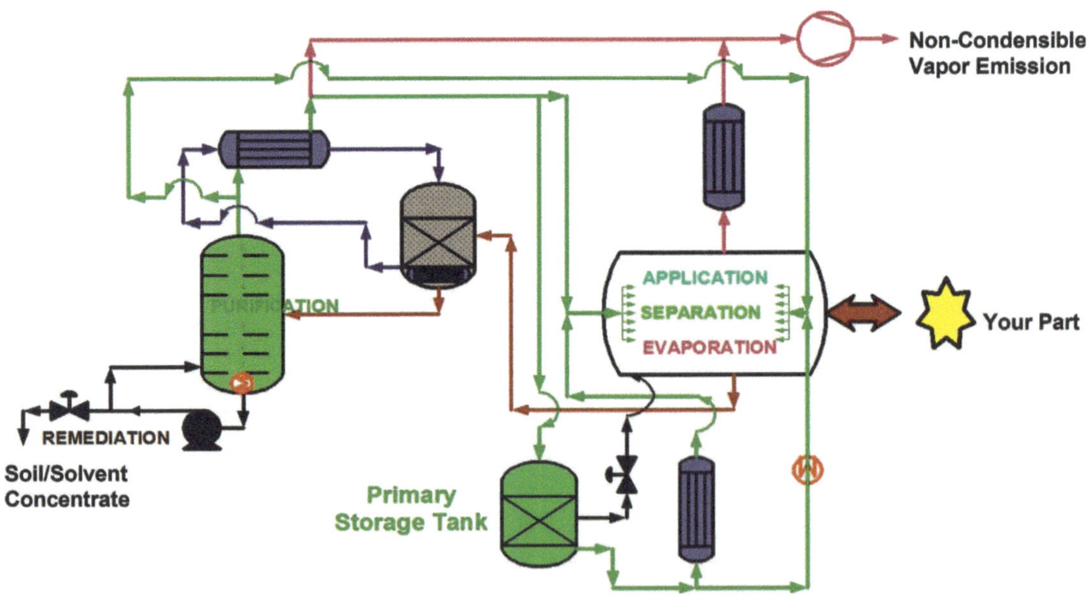

Figure 2.32 The Apparatus of Firm "S"

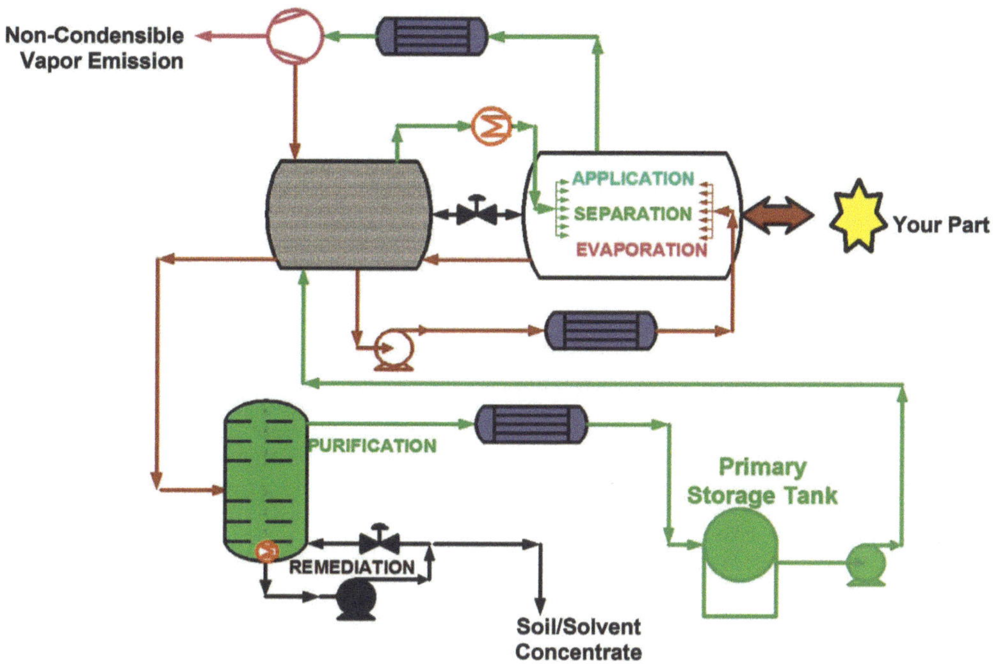

Figure 2.33 The Apparatus of Firm "D"

chamber, a pressure-equalization line between primary solvent storage and the work chamber, and a vacuum pump connected to the work chamber.

One storage vessel in each firm's apparatus is the primary solvent storage tank which receives cooled liquid solvent as product from the distillation column. The second storage vessel in each firm's apparatus contains liquid and vapor solvent which have been used in the work chamber.

The two firms have different visions about their primary approach to cleaning parts. One can see this in the way that they use their second storage vessel.

- The apparatus of firm "S"[JJ,KK] implements the primary approach of spraying hot solvent *vapor* to part surfaces. If the parts are cooler than the hot solvent vapor the vapor condenses, and the parts are rinsed with pristine solvent.
- With a vacuum-capable enclosed vapor degreaser, this relationship is easily arranged. Vapor flows easily within whatever torturous paths there are within the arrangement of parts (Chapter 2.15). Reduction of chamber pressure by engaging the vacuum pump, reduces the temperature of the parts[26] (Chapter 2.3). And as Figure 2.33[LL] shows, heated solvent vapor may be fed to the work chamber[27] – which will raise the pressure.
- In this way, multiple consecutive cycles of rinsing with pristine distilled solvent can be readily achieved. This sequence of rinsing steps is either impossible or very difficult to implement in a conventional open-top degreaser (Figure 1.9), or any enclosed vapor degreaser other than one capable of producing a vacuum. And, hot solvent vapor added after formerly cold parts are heated will dry liquid rinse material from the parts as it contacts them.
- The key apparatus component (other than the vacuum pump) in this sequence is the vessel with a light grey color, known variously as the secondary storage tank, vapor storage tank, vapor supply reservoir, or intermediate solvent storage tank.
- The apparatus of firm "D" is based on the primary approach of immersing parts in hot (or cold) *liquid* pristine distilled solvent, or spraying them with vapor of the same fluids.
- As with the apparatus of firm "S", these immersion treatments may be repeated as desired.

---

[26] Some changes happen faster than others. Pressure, and hence temperature, can be quickly changed by either injecting solvent vapor or implementing the vacuum pump. Temperatures of parts change less quickly. This is because heat transfer rates between vapor and solid surfaces are not rapid.

[27] It is necessary to equalize pressures between the work chamber to effectively make fluid transfers.

Box 2.6 The Volatility of Soils

One must remember that liquid components of an overall soil are both soluble in solvent, and volatile to some extent. Those volatile materials fractionate in intermediate storage vessels. Both liquid and vapor phases of solvent become contaminated by components of the overall soil based on their reactive volatility.

As an example, grease is an amalgam of both low and high volatility components. The former can be base oils, glycerol, rosin oil, mineral oil, and the like. The latter can be low volatility soaps such as stearates, but also non-volatile materials such as clays, tars, metal particulate, and Teflon. Certainly this mixture will fractionate when mixed with solvent in a flash drum or a distillation column – and become a challenge to soil management.

Recycle of liquid solvent from any intermediate storage vessel will contain the low-volatility soil components; hot solvent vapor will contain the high-volatility soil components. Presumably, solid components (such as particulate) of an overall soil are either "movable" or soluble in liquid solvent, but not volatile to any extent – even under vacuum conditions.

- The key component in the apparatus of firm "D" is the vacuum pump. It allows both reduction of pressure in the work chamber, and (if desired) in the distillation column. In addition, the secondary storage tank in the apparatus is also the vessel with the light grey color.

In summary, the two firm's apparatus differ in the way that they manage soil materials, especially those which have a mixture of volatilities (Box 2.6), and how they usually contact parts – with liquid or vapor (Chapter 2.14).

- The apparatus of firm "S" incorporates pressurized drainage or pump-out of liquid materials from the work chamber to the light grey vessel. One has four possibilities for management of the contents of the secondary storage tank (the light grey vessel): (1) discharge to the distillation column for PURIFICATION and REMEDIATION, (2) return to the work chamber as hot vapor, (3) return to the work chamber as cooled (or hot) liquid, or (4) retention.
- Possibilities #2 and #3 can return soil materials to the work chamber as vapor or liquid respectively.
- That can't happen with the apparatus of firm "D". Only purified liquid solvent from the primary storage tank, or hot vapor direct from the distillation column, can be fed to the work chamber. The contents of the secondary storage tank can't be fed as liquid or vapor to the work chamber.
- Obviously, the extent to which this difference matters depends upon the amount and volatile character of the soil materials.

Either apparatus of Figures 2.32 or 2.33 can provide identical contact with solvent: hot vapor spray, hot solvent spray or soak while immersed, cold solvent spray or soak while immersed. In addition, other process steps which can be implemented in either apparatus include: application of pressure waves from ultrasonic or megasonic transducers (Table 1.5 and Figure 1.29), impact by pressurized fluid jets, or cycling of temperature or pressure (Figure 2.3).

Enclosed vapor degreasers from each firm operate continuously, repeating cycles of some cleaning processes (Chapter 2.19), but parts must be loaded and unloaded in batches. As shown in Figures 2.25 and 2.26, two or more work chambers may be used with the same solvent processing facilities and control system to have one work chamber implementing a cleaning process at all times.

Both systems typically recover 99.9% of vapor and liquid solvent from the work chamber during a single cycle of operation. The remaining amount is very expensive of capital and energy to recover. So it is purged from the vessel with air or nitrogen and may be captured by activated carbon (Chapter 4).

The actual internal step of solvent recovery involves more than simple condensation in a heat exchanger. The vacuum pump compresses the vapor (Chapter 2.2.3) removed from the work chamber, which results in condensation. This vapor is then condensed further in a heat exchanger. In the absence of air, heat transfer when condensing a pure solvent vapor can be very rapid.

Neither of the vapor degreasers shown in Figures 2.32 and 2.33 are intended for use where soil levels are high[MM]. A pre-cleaning step in a conventional open-top vapor or liquid (cold cleaning) degreaser, or blast cleaning with some media, should precede use of the enclosed vacuum vapor degreasers. But, vacuum enclosed vapor degreasers are well-suited for removal of soils from crevices (called "features" in semiconductors) with high aspect ratios, as well as to produce a high quality of surface cleanliness because of essentially unlimited opportunities to rinse with pristine distilled solvent.

2.12 MANAGEMENT OF VACUUM (VAPOR)

There is only one[NN] (Footnote 41) reason why the ability to evacuate a work chamber to 759 TorrV (1 Torr absolute pressure) is provided (and desired) by users of many or most vacuum-capable airtight or airless enclosed vapor degreasers.

- That reason is to manage the solvent vapor within the work chamber.

While that may seem an inconsequential task, management of the solvent vapor in a work chamber allows control of essentially all the features of an airtight or airless cleaning vapor degreaser. By analogy, it's like controlling the oxygen in a room full of humans.

Box 2.7 **Management of Vacuum Evacuation**

Note that the *vacuum evacuation* of a chamber and refilling the chamber with another gas is **not** like *flushing* the same volume with the replacement gas. It's considerably more efficient. Two reasons for this are the fluid mechanics of mixing, and to a lesser extent – dragout.

Number of Volume Replacements	Percent Original Contents Remaining
1	36.79
2	13.53
3	4.98
4	1.83
5	0.674
6	0.258
7	0.091
8	0.034
9	0.012
10	0.005

Flow of a fluid (liquid or gas) into any filled volume involves mixing of the replacement fluid with the previous contents. So, at any time in the replacement process, some of both fluids are present. Conventional fluid dynamic theory is that, for a well-mixed chamber, three volumes of replacement fluid are needed to reduce the concentration of the previous contents to about 5 volume percent; seven volumes are necessary to reach less than 0.1 percent (table left).

But, a cylindrical work chamber, filled and emptied from the top, full of parts in a basket which is blocking flow, is the opposite of a well-mixed vessel. So *even more, much more,* replacement fluid is necessary (Ref. 1, Table 1.15).

Furthermore, flushing never displaces the previous contents to the extent that one would predict given the number of chamber volumes of displacement fluid used. Some of the previous contents, liquid or vapor, are always trapped within the confines of the material dwelling within the work chamber or on surfaces of that material. That's what happens with rinsing – another word for flushing (one example is found in Table 2.8, another is found in Chapter 4.9, and Ref. 1, Chapter 1.12 has a more complete discussion).

By contrast, vacuum evacuation produces complete removal of chamber contents – *providing all volume of those contents can communicate with the vacuum pump.*

Vacuum evacuation is not a substitute for use of superheated vapor (in Table 2.6 both process steps are sequentially used). The latter converts liquid to vapor; the former removes the vapor. The processes are complementary.

Several crucial capabilities are enabled or frustrated by management of the solvent vapor in a work chamber.

- One capability is to remove all, or as much as economically justifiable, of whatever vapor is in the work chamber – air, solvent, nitrogen, etc. One does this to prepare for the next step in the cleaning process, such as introducing liquid solvent or opening the work chamber. Relative to atmospheric pressure, a reduction of absolute pressure to 1 Torr (759 TorrV) represents a 99.87% (1/760) removal of the previous vapor contents of the work chamber.
- A second capability enabled by management of solvent vapor in a work chamber is to convert that vapor to liquid by condensation. Essentially this means to make liquid solvent appear and disappear within or upon the configuration of part structures – so that it can dissolve soil. Application of liquid solvent to surfaces is covered in detail in Chapter 2.15.
- The third capability enabled by management of solvent vapor in a work chamber is to dry the parts within it, rapidly and thoroughly, by evaporation of the liquid wetting them. This is the reverse of the second capability. Essentially this means to create more solvent vapor within the chamber. Drying is covered in detail in Chapter 2.18.

2.13 REMOVAL OF SOLVENT VAPOR FROM THE WORK CHAMBER BY VACUUM

The list of process steps in Tables 2.6 through 2.10 provides examples of how a vessel should be purged (Box 2.7)[OO] of solvent fumes when vacuum capability is available. Note that these processes include sequential steps where the work chamber, filled with a solvent atmosphere: (1) has 99.87% of that atmosphere removed by vacuum; (2) is then filled with air (or any inert gas such as nitrogen), and that atmosphere is also removed by vacuum; and (3) finally is filled with dry air before opening.

The first step above is probably not adequate. The work chamber would still contain 1300 ppm of solvent in air (or inert) after removal of 99.87%[28] of the solvent vapor. And that's too high a concentration of solvent-contaminated air to be emitted to the work area (no matter what solvent is used).

That's why the second step is taken – to dilute the solvent-rich gas with an inert. When 99.87% of the diluted

[28] Thirteen hundred ppm is the remainder when 99.87% of 1,000,000 ppm (pure solvent vapor) is removed (1,000,000 x [1 – 0.9987] = 1300).

Table 2.3 Effect of Vacuum Capability on Chamber Evacuation of Oxygen

Actions	With Vacuum Evacuation	Without Vacuum
	ppm Oxygen	
At start, chamber is full of air at 1 atmosphere pressure	219,180 (21.918%)	
After vacuum evacuation to 759 TorrV	$219,180 \times \frac{1}{760} = 288$	NA
After nine displacements with nitrogen (Footnote 29)	$288 \times 0.00012 = 0.034$	$219,180 \times 0.00012 = 26.3$

gas is then removed by vacuum and replaced with additional solvent-free inert, the concentration of solvent in the air *emitted* to the work area has been reduced to about 2 ppm ($1300 \times \frac{1}{760} = 1.7$).

Obviously, dilution of the emitted air with room air will reduce exposure of employees to that high concentration (1300 ppm) of cleaning solvent – while paying the price of increasing the volumetric amount of air pollution. The normal and proper fate of that emitted cleaning solvent is covered in Chapter 4 – tailpipe treatment with a bed of activated carbon adsorbent.

To summarize, vacuum evacuation of a work chamber is generally done to manage the vapor and liquid contents of the chamber so that activities related to cleaning can be efficiently and safely conducted.

2.13.1 Examples of Vacuum Evacuation of the Work Chamber

A comparative example is shown in Table 2.3. Two situations are considered. In one, the work chamber containing air is evacuated from atmospheric pressure to 759 TorrV; then flushed with three chamber volumes of nitrogen, three times in succession. In the other situation, there is no vacuum evacuation – only the same three successive flushings with nitrogen.

At completion, in both cases the chamber is full of nitrogen at 1 atmosphere pressure, but the retained oxygen content in the chamber is considerably different.

In both cases described in Table 2.3, the total mass of oxygen removed from the work chamber is essentially the same: (219,180 – 0.034) vs. (219,180 – 26.3). And since the work chamber contains no solvent, no emissions need to be treated. Obviously, this work defines the employment of an airless vapor degreaser.

Consider a second example – removal of pure solvent. This is a practical example where one has a work chamber "full" of trichloroethylene vapor (molecular weight 131.4). But the pressure is not 1 atmosphere, nor is the temperature ambient either. When the temperature is about 50°C, the pressure in the work chamber should be about 200 Torr absolute (560 TorrV).

The mathematics of dilution (vacuum evacuation) and displacement (purging with an inert) in Table 2.4 are straightforward and allow easy recognition of two principles:

1. One can use either vacuum evacuation or nitrogen displacement to produce the same outcome.
2. A single stage of vacuum evacuation is a powerful treatment. For example, a single application of vacuum evacuation from 200 to 1 Torr absolute produces a similar outcome as does displacement with six vessel volumes of inert gas (nitrogen).

All emission from the actions described in Table 2.4 must be directed to a solvent treatment system (Chapter 4). And the enclosed vapor degreaser in which this solvent is contained could be classified as an airless (if inert is used) or an airtight (if air is used) vapor degreaser.

2.14 SELECTION OF OPERATING TEMPERATURE

The ability to produce and manage vacuum conditions is not similar to pizza, sex, and football where the human belief that "more is better" applies. It isn't similar; it doesn't apply. Cleaning work is not done at low levels of vacuum; it is certainly not done at 1 Torr absolute (759 TorrV or 1 mm Hg). More vacuum is not better in terms of cleaning capability or performance!

Cleaning work is done at a selected temperature, and the absolute pressure (level of vacuum (or level of over-pressure)) is adjusted (with the aid of energy transfer) to produce that condition.

The chamber temperature is selected based on the nature of the soils to be removed (usually the higher

Table 2.4 Effect of Vacuum Capability on Evacuation of Solvent Vapor From Chamber

At start, chamber is full of solvent vapor at 200 Torr pressure and 50°C	**With Vacuum Evacuation**		**Without Vacuum**	
	Trichloroethylene Vapor Mass[PP] in 10 CF Chamber (as ideal gas)			
	$\left\{\left[\frac{131.4}{359}\right]\times\left[\frac{25+273.15}{50+273.15}\right]\times\left[\frac{200}{760}\right]\right\}\times 10 = 0.889\text{lbs}$			
	After vacuum evacuation from 200 Torr to 1 Torr absolute	$0.862 \times \frac{1}{200} = 0.00431\text{lbs}$	Three displacements with nitrogen	$0.862 \times 0.049 = 0.0422$ lbs
	Chamber filled with air	0.00431 lbs	Three displacements with nitrogen	0.0429 x 0.049 = 0.0021 lbs
	After vacuum evacuation from 760 Torr to 1 Torr absolute	$0.00431 \times \frac{1}{760} = 0.000005\text{lbs}$	Three displacements with nitrogen	$0.0021 \times 0.049 =$ 0.00010 lbs
	Chamber filled with air	5.7×10^{-6} lbs	Three displacements with nitrogen	5.3×10^{-6} lbs

the better, in order to reduce the viscosity of the soils) or the fragility of the parts (usually the lower the better, in order to avoid damage). These are two of the normal factors which should always be considered in the design of a cleaning process[QQ].

- Recall the reasons for using vacuum relate less to cleaning with solvents than much more to managing contained solvent plus inerts, and drying of parts.

Management of vacuum level in cleaning operations, for example:

- Frees users from having to accept as a maximum temperature the normal boiling point of the chosen solvent (such as with perchloroethylene, or possibly trichloroethylene), or
- Enables users to use the "3A" solvents which have low volatility and high flash points (such as Isopar L and isopropyl lactate) in boiling service, or
- Enables users to use highly volatile solvents which have low flash points (such as acetone and HFC-43-10mee) in boiling service.

To summarize this subchapter, the normal boiling temperature isn't a limitation when vacuum capability is available.

2.14.1 The Antoine Equation

One reason cleaning work is not done at a pressure of 1 Torr absolute (759 TorrV), is that the solvent temperature at that pressure is too cold for use. For trichloroethylene and n-propyl bromide, their equilibrium temperatures[RR] at pressures of 1 Torr absolute are both below −70°C!

$$Log(Pv) = A + \frac{B}{(T + C)} \quad \text{2.6}$$

To select an operating pressure (managable by a vacuum pump) for the work chamber, given a chosen operating temperature, one must calculate the vapor pressure (P_v) in equilibrium with that temperature. That value of temperature for single solvents can be determined from the Antoine Equation[SS], shown in Equation 2.6. The values of "A," "B," and "C" are coefficients determined via a regression analysis of vapor pressure vs. temperature data[TT].

In this book, only two Antoine constants are used for simplicity[29] – making the value of "C" zero[UU].

Often one wants to know the temperature of solvent in equilibrium with a certain level of pressure in the work

[29] Little accuracy in calculation of vapor pressure is lost by this simplification because the temperature range over which cleaning solvents are used is quite narrow. The gain provided by this simplification is that only two data points (values of temperature and corresponding vapor pressure) must be known for a preliminary estimate of the Antoine coefficients "A" and "B." And often with little-used or newly developed solvents, only two data points are posted in scientific or commercial literature – the normal boiling point and the vapor pressure at 20 or 25°C.

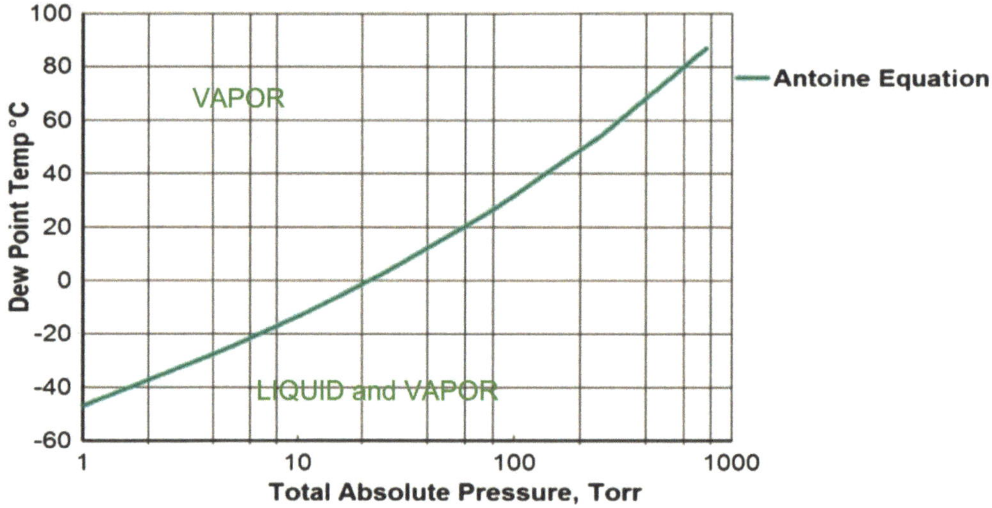

Figure 2.34 Equilibrium Relationship for Trichloroethylene, Using the Antoine Equation

chamber. Equation 2.7 is Equation 2.6 restated for determination of solvent equilibrium temperature.

$$T = \frac{B}{\{Log(Pv - A)\}} - C \quad (2.7)$$

2.14.2 It's Hotter Than You Think

One can estimate with the Antoine Equation (Equation 2.6) the pressure at which solvent vapor condenses into liquid, for an inert-free system, at a temperature chosen for cleaning.

- Higher temperatures, at the same pressure, cause the solvent to be present in the vapor phase. This may be exactly what's wanted if the process step is vapor spray.
- Colder temperatures, at the same pressure, cause the solvent to be present in the liquid phase. Similarly, this may be exactly what's wanted if the process step is liquid immersion and the parts are susceptible to thermal damage.

As noted above, *the use of vacuum capability is chiefly to produce the required pressure in the work chamber to obtain desired equilibrium solvent temperature.*

Calculated values of equilibrium pressure vs. temperature from the Antoine Equation for trichloroethylene are plotted in Figure 2.34[30].

To clean parts with trichloroethylene at about 50°C using Figure 2.34, one can see that the pressure in the inert-free work chamber to produce that temperature should be about 200 Torr absolute (560 TorrV) – as long as there is liquid present.

Note in Figures 2.32 and 2.33 that there are heaters (red symbols) in the lines from storage vessels to the work chamber. Their purpose is to heat vapor or liquid feeds to these higher temperatures if the work chamber had previously been at a lower absolute pressure (hence lower temperature).

To emphasize, vacuum capability to produce a chamber pressure of 1 Torr absolute is provided with some enclosed vapor degreasers for the purpose of controlling the vapor environment or accelerating the rate of drying. It's not provided to clean parts at −70°C!

2.15 IT'S ABOUT WHAT YOU DEW

Unfortunately, the apparent simplicity of Figure 2.34 is misleading. Its monochromatic presentation camouflages a level of complexity that's almost always present.

The relationship (green line Figure 2.34) between absolute static pressure (level of vacuum) and equilibrium temperature for single solvents is more than slightly complicated by the fact that a second component is nearly always present in a work chamber. This is the retained inert purge gas – often air (in an airtight vapor degreaser), but usually nitrogen (in an airless vapor degreaser). It may be inert, but it can't be ignored.

However, the additional complexity is a familiar one. It's found in any weather forecast. The familiarity is that in a weather forecast, and in prediction of operating conditions in a vacuum chamber, one wants to know about "rain" – condensation of liquid from vapor which is diluted by an inert vapor.

The desired parameter is called the dew point – the temperature at which the first drop of water (dew in weather forecasts) condenses from humid air at atmospheric pressure, or the temperature at which the first drop

[30] One can see in Figure 2.34 that if the inert-free line were extended to 760 Torr absolute (0 TorrV), the value of the plotted dew point would be in the mid 80s°C (the normal boiling point is 86°C).

Box 2.8 **Method for Calculation of Dew Points**

The key assumption is that an ideal solution (solvent and nitrogen) is in equilibrium with an ideal gas (nitrogen). Technically, this means to assume Raoult's Law is valid. Algebraically, this means

$$y_i \times P = x_1 \times Pv_i, \text{or } x_i = \frac{y_i \times P}{Pv_i}$$

where y_i is the vapor phase mole (volume) fraction, x_i is the liquid phase mole (volume) fraction, Pv_i is the vapor pressure (from the Antoine Equation) at some temperature, and P is the total pressure – all for the ith component. In this case, there are only two components: solvent and nitrogen. Hence, the liquid phase compositions must sum to 1, or

$$x \text{ solvent} + x \text{ nitrogen} = 1$$

At the dew point, essentially all the mass of the mixture is in the vapor phase, so the vapor phase composition, y_i, is the total composition, z_i, for the ith component. The equation summing liquid phase compositions is the equation from which dew point can be calculated:

$$P = \left\langle \left[\frac{z \text{ solvent}}{Pv \text{ solvent}}\right] + \left[\frac{z \text{ nitrogen}}{Pv \text{ nitrogen}}\right] \right\rangle = 1, \text{ or}$$

$$P = \frac{1}{\left\langle \left[\frac{Z \text{ solvent}}{Pv \text{ solvent}}\right] + \left[\frac{z \text{ nitrogen}}{Pv \text{ nitrogen}}\right] \right\rangle}.$$

Note all compositions are molar or volumetric; weight compositions must be calculated from molar compositions and molecular weights.

The above equation can't be directly solved for dew point temperature. The variable of temperature does not appear in it.

But, each vapor pressure, P_{vi}, is a direct function of temperature – using the Antoine Equation. See Chapter 2.14.1 about Antoine coefficients for cleaning solvents. Values of Antoine coefficients for nitrogen gas can be obtained from NIST, and are: A = 3.7362, B = 264.651, and C = −6.788. See http://webbook.nist.gov/cgi/cbook.cgi?ID=C7727379&Mask=4.

The procedure for calculation of dew point temperature for any two-component mixture is to assume a trial value of temperature; calculate both values of vapor pressure using the Antoine Equation for solvent and nitrogen; and using the above equation, solve for the total pressure. When it equals the specified total pressure, the trial temperature is the dew point temperature. Spreadsheets have a "goal-seeking" function to automate this work.

of liquid solvent condenses in a mixture of cleaning solvent vapor and nitrogen at any pressure (above, at, or below atmospheric). They are the same phenomena[31].

One can calculate[32] dew point from commonly used simplifying assumptions (Box 2.8). For trichloroethylene, the outcomes are plotted in Figure 2.35 – the relationship of saturation temperature vs. total absolute pressure. The varying parameter in Figure 2.35 is the level of nitrogen, in weight percent. The line marked "Inert-Free" (teal color) is for pure trichloroethylene (Figure 2.34) and is calculated using the Antoine Equation (Equation 2.6).

At temperatures above the dew point, by definition all condensable material exists in the vapor phase. At temperatures below the dew point, condensable and non-condensable coexist[VV].

At any absolute pressure, increased amounts of residual nitrogen reduce the temperature at which the liquid cleaning solvent condenses in the two-phase mixture. Note that the relationship between pressure and temperature for trichloroethylene is only slightly affected by retaining 2 or 5 weight percent nitrogen; but larger amounts of nitrogen have considerably larger effects.

Dew point curves for other cleaning solvents which are vapor degreasing solvents are plotted in Figures 2.36 (methylene chloride), 2.37 (perchloroethylene), and 2.38 (n-propyl bromide).

Using Figure 2.36, one can see that to conduct cleaning operations at 50°C with methylene chloride, an airtight (pressure-tight) enclosed vapor degreaser can be used. But it is not a vacuum vapor degreaser because the required pressure is above 760 Torr.

Using Figures 2.37 and 2.38, one can see that to conduct cleaning operations at 50°C with perchloroethylene and n-propyl bromide, the absolute operating pressures in a vacuum enclosed vapor degreaser (nearly inert free) should be about 100 and 500 TorrV respectively.

Some users do or will prefer the "3A"solvents – those which display flash points above 60°C (140°F) so that there is no concern about wastes being hazardous, or electrical connections being rated explosion-proof.

Dew points are calculated for two such solvents: Isopar L (Figure 2.39) and isopropyl lactate (Figure 2.40).

It may be that use of the "3A" solvents is synergistic with enclosed vacuum vapor degreasers – assuming that the

[31] That's why the vertical axis in Figure 2.34 is named dew point temperature.

[32] Seldom is dew point calculated for water in humid air. It's too important and very easy to measure. It is a very important statistic for general aviation pilots, as it is used to calculate the likelihood of carburetor icing and fog. It's measured by using a polished metal mirror which is cooled as air is passed over it. The temperature at which dew forms is, by definition, the dew point. The method for calculation described in detail with cleaning solvents and Nitrogen is given in Endnote XX.

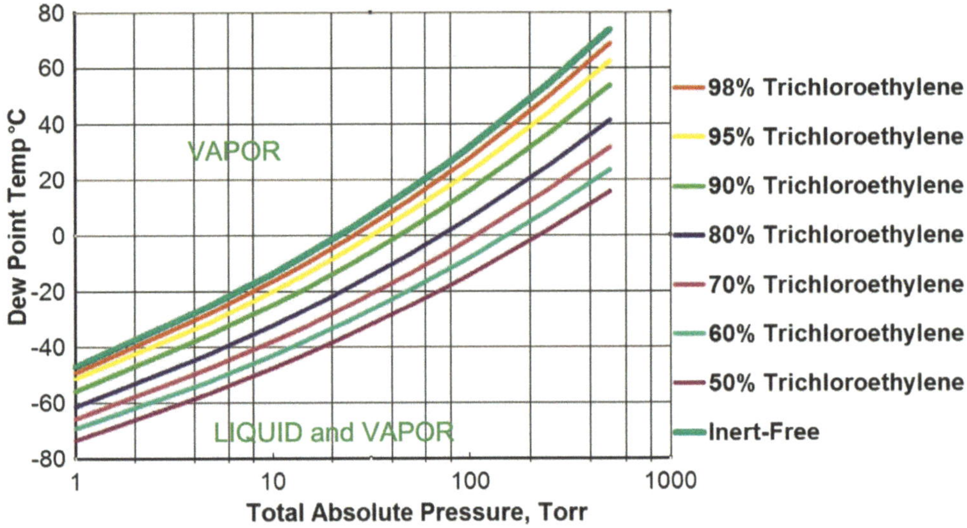

Figure 2.35 Dew Point Estimates by Calculation for Trichloroethylene

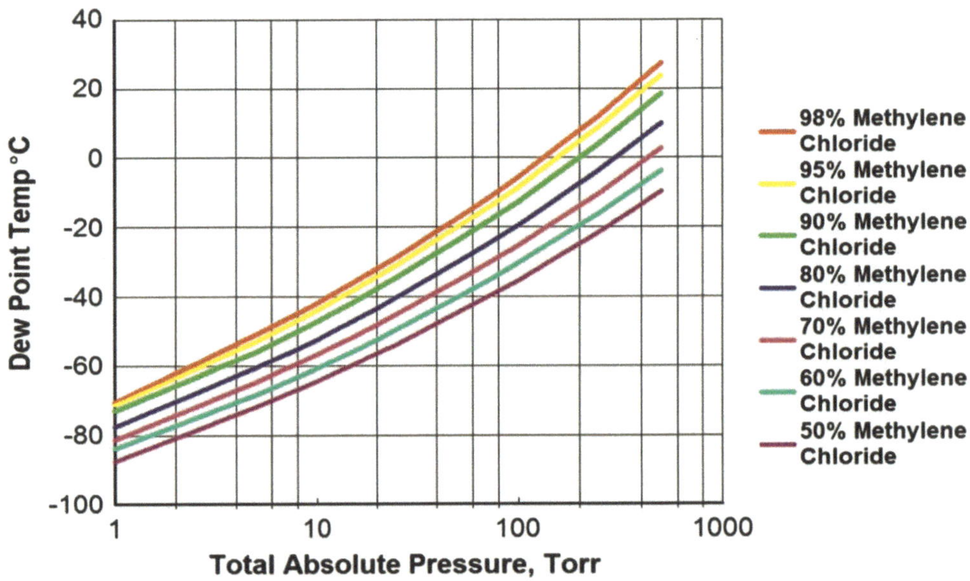

Figure 2.36 Dew Point Estimates by Calculation for Methlene Chloride

solubility parameters of the soils match adequately match those of these different solvents. Using Figures 2.39 and 2.40, one can see that to conduct cleaning operations at 75°C with Isopar L and isopropyl lactate, the absolute operating pressures in a vacuum enclosed vapor degreaser should be about 50 and 130 TorrV respectively – values easily achieved even with an eductor vacuum pump (Chapter 2.2.2 and Figures 2.7 and 2.8).

2.15.1 Colder than Expected

Ignoring the specificity of Figures 2.35 to 2.40, there is an important general point about the presence of inert gases in an active work chamber.

- Inerts[WW] reduce the range of operation for liquid cleaning solvent, and expand it for vapor.

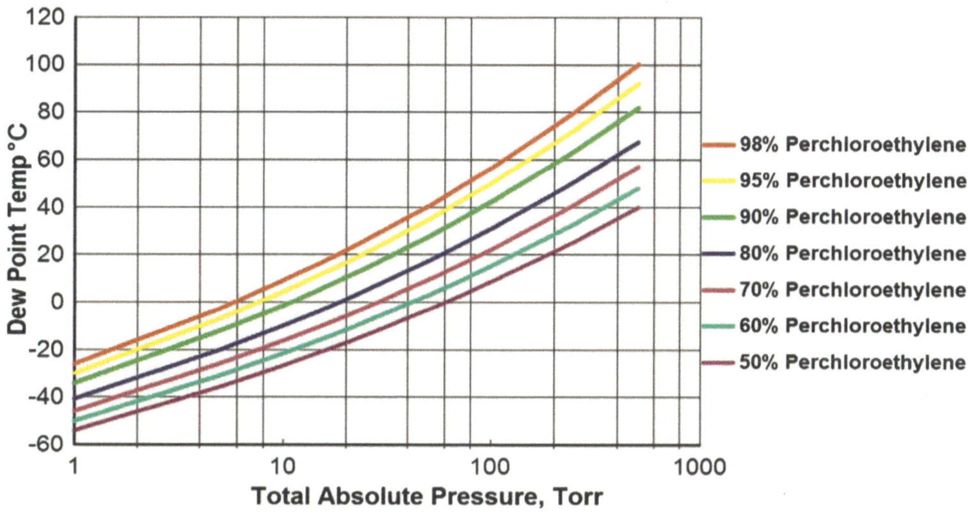

Figure 2.37 Dew Point Estimates by Calculation for Perchloroethylene

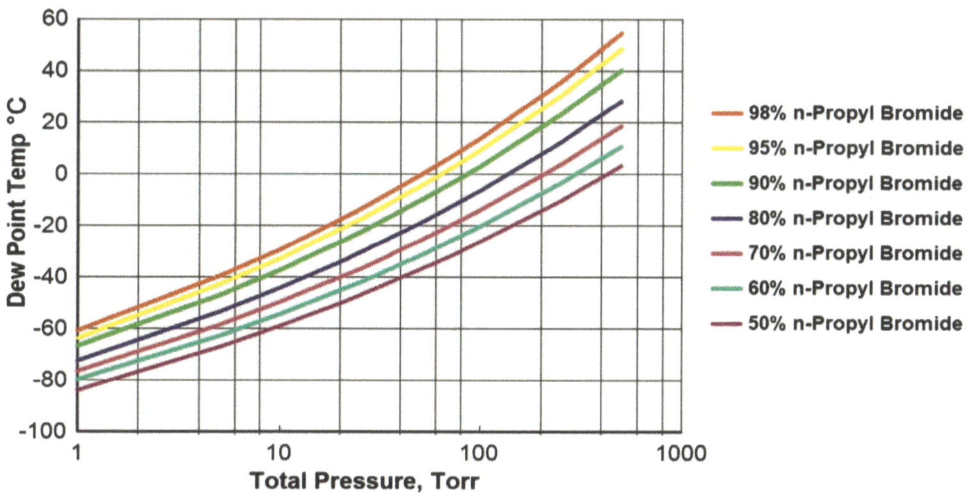

Figure 2.38 Dew Point Estimates by Calculation for n-Propyl Bromide

In the presence of inerts, one must cool the contents of the work chamber by a significant amount (perhaps 50°C) below the level expected from vapor pressure data of pure solvents (the Antoine Equation) in order to *initiate* condensation of liquid solvent.

In other words, the presence of inert gases enables solvent vapor to exist *without condensation* at a temperature considerably lower than would be expected from pure component vapor pressure data. If inerts are not purged from the work chamber, vapor spray can exist at a lower temperature; and condensed liquid will be colder as well – than expected.

2.16 CHAMBER FULFILLMENT

The work chamber is seldom completely filled with liquid solvent as some vapor is trapped within internal sections of parts, and only slowly displaced by moving liquid.

Since the volatility of nitrogen is considerably higher than any cleaning solvent (it's always a gas), the contents of the work chamber are a condensable vapor (the cleaning solvent) and a non-condensable gas[XX] (nitrogen).

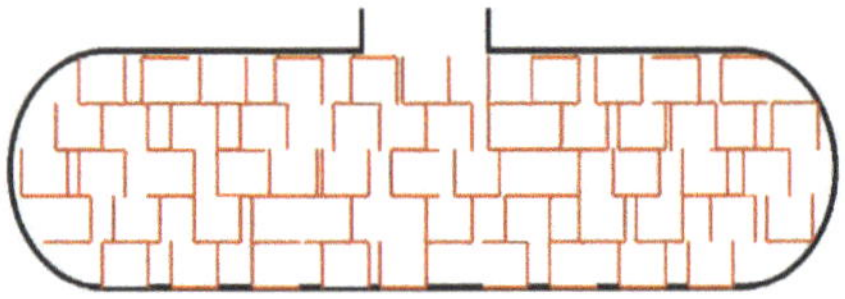

Chamber Empty of Liquid

Imagine a fictitious example of a cylindrical chamber full of thimbles and nitrogen, without any liquid, such as the image above. This represents the work chamber after the parts (the thimbles) are loaded and atmospheric air has been displaced with an inert gas.

Thimble

Imagine pouring water (blue color) through a hole into the top of that chamber. The water flows around and into some of the thimbles, as well as into the free volume of the chamber. But, not all the vapor in the chamber is immediately displaced. This is because the temporary contents of the chamber (the thimbles) trap vapor, and there is a gravitational separation of the low-density vapor from the high-density liquid. An image of the partially filled chamber is shown below. This represents conditions in the work chamber after some time of filling and displacement of vapor.

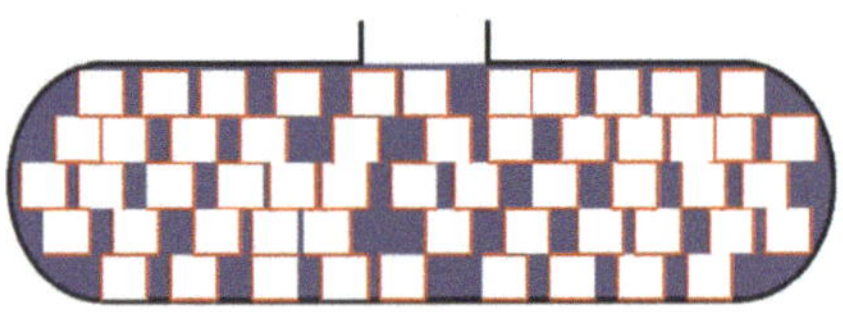

Partially Filled Chamber

Now, imagine trying to fill that container with liquid within a certain time. It would be easier to fill all the volume within the thimbles with liquid if one had twice that amount of time to displace the vapor, still easier if one had even longer, and very easy if one could agitate the container

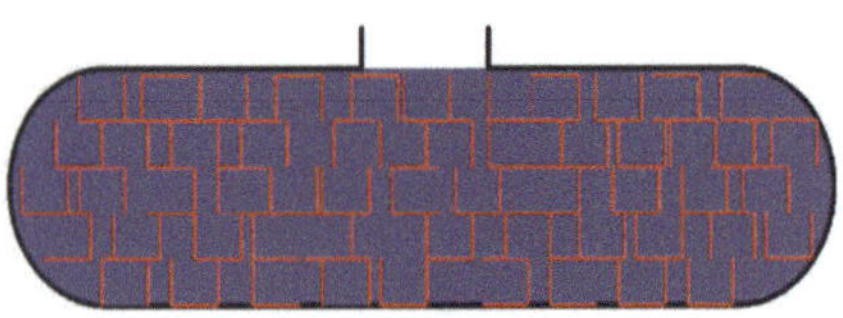

All Vapor Displaced From Chamber

and take as long as one wished. An image of the chamber, completely filled with liquid water (blue color), is shown at the bottom of the left-hand column on this page.

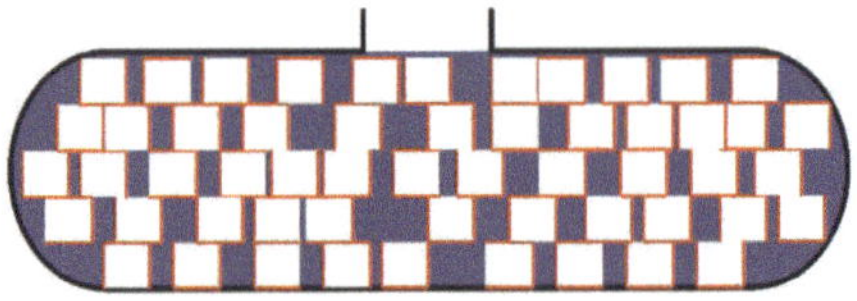

Partially Filled Chamber

It usually takes time to displace vapor with liquid, especially so if there are objects in the vapor volume. And, all cleaning work is done within time constraints. Tables 2.6 to 2.10 are replete with examples of these constraints.

So the outcome of this imagined example is that not all the vapor in the chamber is displaced by liquid. Some vapor, however much, remains. This represents conditions in the work chamber at the close of the liquid fill step of the overall cleaning cycle. The image of the partially filled chamber is repeated above.

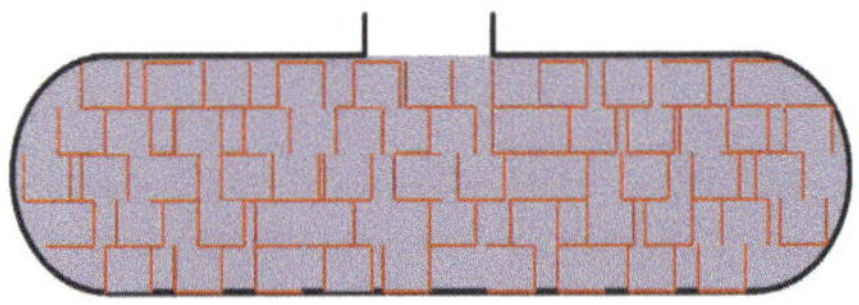

Homogeneous Contents

When filling with liquid is complete, the work chamber isolated, the pressure adjusted as desired, the imagined work chamber would be full of a relatively homogeneous[33] (Box 2.9) mixture of liquid water and some nitrogen gas (usually), as shown above (light blue color).

To summarize this subchapter, one shouldn't overfill the work chamber with complex parts. While a jumble of parts may not interfere with migration of solvent vapor, it is almost certain to do so with liquid solvent. When this happens, the expected cleaning work won't be completed because all the part surfaces won't have been wetted to the extent expected and for the time anticipated.

Box 2.9 A Value of Vacuum

Pressure is easily, and almost immediately, modified and controlled in a crowded work chamber through movement of vapor – by removal of it into a vacuum pump or addition of it via a pump or other device. As long as all volumes of the chamber are in fluidic communication with one another, all of the chamber contents are at the same pressure. That's what is meant by homogeneous.

[33] Pressure control by movement of liquid may not easily, and certainly won't immediately, be effected. Transport properties (viscosity, etc.), physical properties (density, etc.), and surface properties (surface tension) are so different for liquids than vapors.

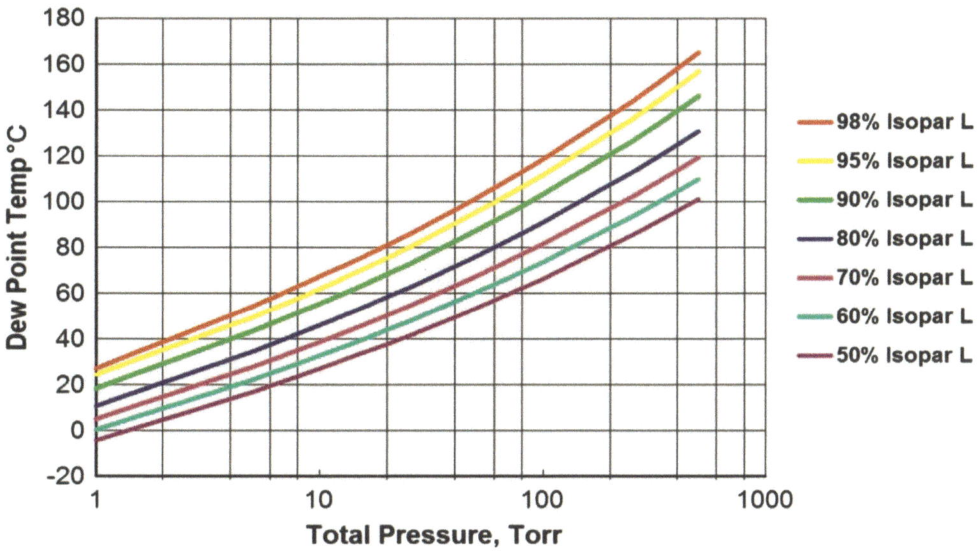

Figure 2.39 Dew Point Estimates by Calculation for Isopar L

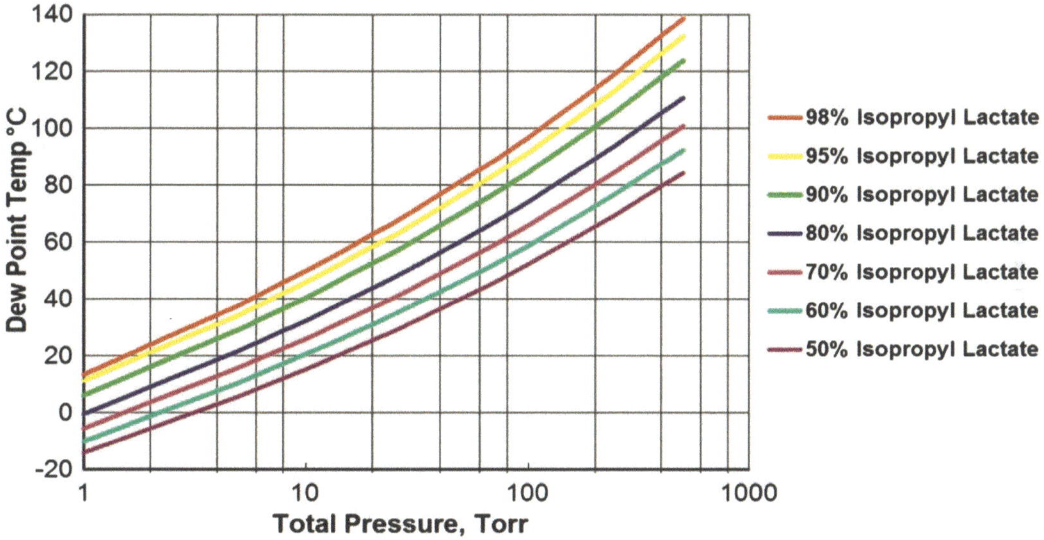

Figure 2.40 Dew Point Estimates by Calculation for Isopropyl Lactate

2.17 MANAGEMENT OF WATER CONTAMINATION

Intrusion of water into open-top cleaning vapor degreasers is unavoidable[YY], and harmful to continuity of operation, quality of cleaning, and integrity of that equipment. Water exists in two places in a vapor degreaser – insoluble or supernatant water, and water soluble in the solvent.

An open-top cleaning apparatus from any supplier will usually include a water separator (Figure 2.41) which performs, based on miscibility, removal of insoluble water from the cleaning solvent.

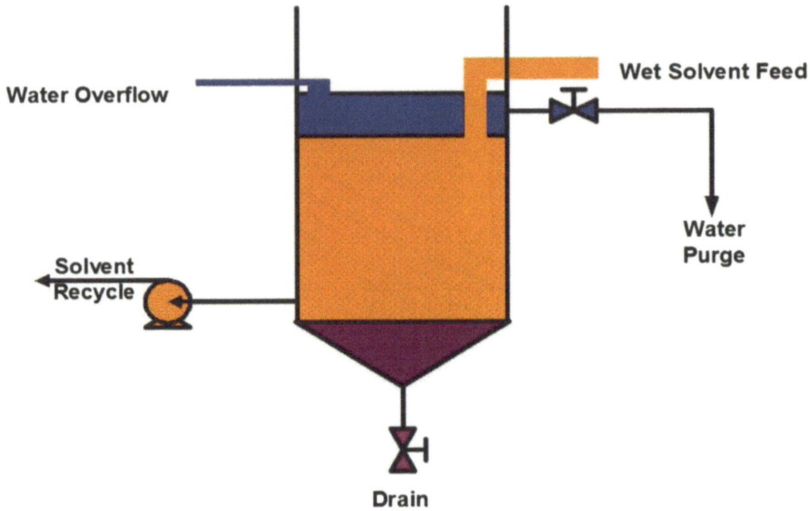

Figure 2.41 Idealized Image of a Water Separator (a Decanter)

Table 2.5 Solubilities of Water and Chlorinated Solvents

Solvent	Solubility of Water, wt% in Solvent	Ratio of Molecular Weights to Water	Reaction Equivalent Concentration of Solvent, wt%	Solubility of Solvent, wt% in Water
trans-1,2-dichloroethylene	No Data	5.4		0.45%
1,1,1-Trichloroethane	0.03%	7.4	0.23%	0.13%
Carbon Tetrachloride	0.01%	8.5	0.11%	0.08%
n-Propyl Bromide	0.50%	6.8	3.40%	0.24%
Chloroform	0.06%	6.6	0.40%	0.82%
Trichloroethylene	0.33%	7.3	2.4%	0.13%
Perchloroethylene	0.01%	9.3	0.09%	0.02%
Methylene Chloride	0.24%	4.7	1.2%	1.85%

Removal of water soluble in solvent is another matter – in any cleaning apparatus. Especially, because there can be plenty of it as shown in Table 2.5, and it can an out-of-proportion harmful effect!

Soluble water is usually consumed by reaction with solvent in open-top vapor degreasers, and then the deleterious effects of its reaction products are negated by further reaction with stabilizers.

But stabilizers are not wanted in enclosed vapor degreasers because: (1) it's difficult for them to leave as other than solid debris on supposedly clean parts, and (2) they have to be purchased and their use controlled. One doesn't want to discard a key value of enclosed airless or airtight machines of any type – that the solvent in them can last a long time with only a tiny amount of stabilizer; or better yet – none.

One does not generally remove water by vacuum-enhanced evaporation – because both water and solvent will be removed, and it may not be easy to separate them by single-plate distillation[34].

One best removes water vapor from a cleaning work chamber by flushing the work chamber with water-free vapor – *after the parts are inserted, but before any solvent is*

[34] For example, the boiling points of trichloroethylene (86°C) and perchloroethylene (121°C) are close enough to that of water (100°C) that they can be dried by distillation, but only with columns with multiple plates.

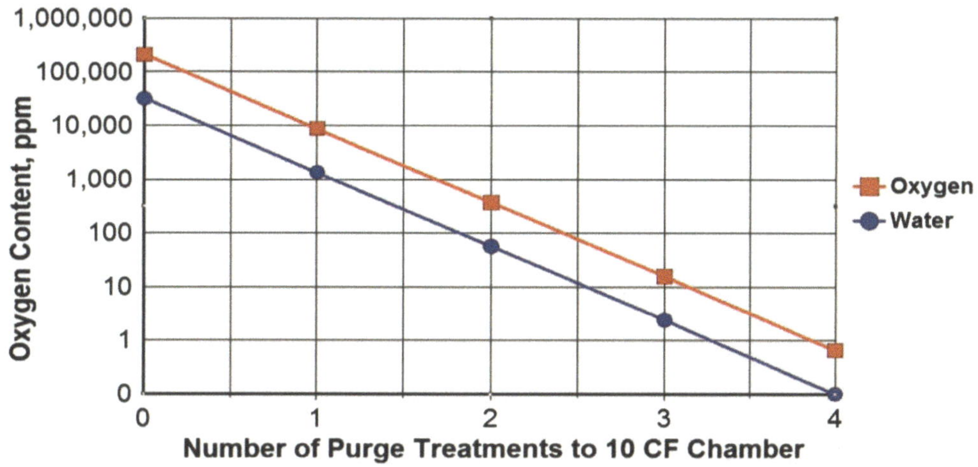

Figure 2.42 Treatment of 125 CFM N_2 flow for 2 minutes

added[35]! This is usually done with nitrogen gas – often heated to enhance evaporation, maintained at a pressure between 500 and 760 Torr absolute.

Multiple flushing (displacement) treatments may be done to further reduce the partial pressure of water, because the work chamber is seldom a perfectly mixed vessel (Chapters 2.15 and 4.9).

This is shown in Figure 2.42 where two displacement treatments reduce the water content from 32,000 ppm (at 32.2°C (90°F) and 100% relative humidity) to 57 ppm.

In other words, one doesn't remove water from the work chamber by evaporating it under vacuum because:

- The latent heat of vaporization of water (about 1000 BTU/lb) is about five times that of solvent. So vacuum-driven evaporation of water is slow and energy intensive, relative to removal of solvent (which should be removed by vacuum).
- The minerals soluble in water don't evaporate, and do become water spots on the cleaned parts.

In still other words, the problem of separation of water dissolved in solvent from solvent never has to be faced because they are never combined! This approach is the correct one for all three types of enclosed machines: externally sealed, airless, and airtight.

[35] One might ask why this approach isn't taken with open-top vapor degreasers. The reason is that purging of volume with inert gas isn't nearly as efficient because of the shape of the volume of the work chamber. The tubular chamber shapes shown in the photographic images of Figures 2.21 to 2.27 are more quickly and efficiently purged than are the "boxy" shapes shown in the photographic image of Figure 2.20 or the diagrams such as those in Chapter 2.15.
Recall that the purging operation is required each time a cleaning cycle is started.Some operators who require low residues have flushed parts entering an open-top degreaser with nitrogen dispensed from a hand-held wand or flooded with a nozzle moved by a robotic arm. This step adds cost, won't purge oxygen and water from deeply nested or complex parts, but it will reduce stabilizer demand.

Finally, this is solvent cleaning technology, not aqueous cleaning technology. The removal of water by purging with inert is not drying of parts, it is protection of solvent.

2.18 DRYING OF PARTS IN ENCLOSED VAPOR DEGREASERS

Water is removed from solvents by purging it from the work chamber (usually before the cleaning work has started); solvents are removed from parts by evaporation. The rate of removal may be enhanced by temperature and/or pressure.

An *externally sealed* enclosed vapor degreaser is an open-top vapor degreaser – in a cocoon, with an airlock. Drying of parts is accomplished as described in Chapter 1.17 – as done in a normal open-top vapor degreaser.

The liquid solvent is vaporized from the parts where the temperature is at the normal boiling point. Then the solvent vapor surrounding the parts is condensed (mostly) in the zone of a vapor degreaser at where the cooling coils have their effect. Drying quality can be excellent.

- Solvent vapor not condensed in the drying step, and not retained within the tall freeboard of the vapor degreaser is contained within the cocoon.

An *airless* enclosed vapor degreaser, where the environment is at ambient pressure but does not contain atmospheric air (with oxygen and water), dries solvent from parts through vaporization into a solvent-free environment.

That environment is usually nitrogen gas, often heated to enhance the rate of vaporization. The gas stream is generally not recycled – thus maximizing the driving force for vaporization – but is treated to remove solvent fumes by a bed of activated carbon (Chapter 4).

Essentially, the driving force for solvent evaporation in the work chamber is that which would occur in the open air. For volatile solvents, such as chlorinated and brominated solvents, HFCs, HFEs, and OSs, drying will probably be rapid and complete.

But for hydrocarbon solvents with a high flash point, drying will be slow in the work chamber and possibly incomplete, unless the drying gas is heated and the drying cycle extended. After drying, the operator will probably have to flush the work chamber with cool dry gas to cool the parts. Drying quality can be excellent with volatile solvents trapped within highly porous and complex structures, and at least acceptable when hot nitrogen gas is used to dry low-volatility hydrocarbon solvents.

- One doesn't select an airless enclosed cleaning vapor degreaser to achieve high drying rates[36]. *One selects an airless enclosed cleaning machine to negate the certainty of solvent decomposition due to contact with oxygen and water.*

An *airtight* enclosed cleaning vapor degreaser may be used to dry the parts under vacuum, even if the parts aren't being cleaned under vacuum conditions. It may be that the only processing step which is completed under reduced pressure is the final one (Table 2.10, page 41).

Often the first step in drying of parts under vacuum is to flush the contents of the work chamber with hot nitrogen at atmospheric pressure – as is done in an airless enclosed vapor degreaser. This can remove most water, avoiding the need to supply its high heat of evaporation.

After evacuation by vacuum to between 1 and 10 Torr absolute, the evaporation rate is raised by the huge partial pressure difference between solvent on the part surface and the solvent in vapor space. Naturally, the partial pressure of solvent in the vapor space is low because the vacuum pump is continually removing all vapor from the work chamber. A second stage of evaporation into hot nitrogen at atmospheric pressure followed by vacuum evacuation may be necessary with complex shapes (Table 2.9 on page 39).

The approach would presumably not be necessary when drying volatile solvents such as those mentioned above. But with the increasing interest in using the relatively non-volatile "3A" solvents, this two-step vacuum-enhanced drying process[ZZ] is a likely solution to an otherwise intractable problem. Some[AAA] have noted that the "oily" hydrocarbons "fly off" the parts.

This technique can also be used with water[BBB].

- Evaporative drying rates are maximized when the temperature is increased and the pressure is reduced. Both can be managed with an airtight enclosed cleaning vapor degreaser, and not in any other.

[36] One may operate an airless enclosed cleaning vapor degreaser under vacuum conditions to achieve the highest possible rates of drying. In that case, the work chamber must be rated to contain full vacuum conditions, and the apparatus noted as being of the airtight type.

2.19 CLEANING PROCESSES IN ALL ENCLOSED MACHINES

The spectrum of available cleaning processes in an *externally sealed* enclosed machine is that of any open-top vapor degreaser. For details of a representative process, see Table 1.5 and Figure 1.29.

In the remainder of this chapter, five examples will be related of how *airless* and *airtight* enclosed cleaning machines may be used to implement various cleaning processes with various part structures.

Recall the three process steps related in Chapter 1.2. They are: (1) immersion in liquid, (2) vapor contact by spray, and (3) liquid spray with force. Any of these steps may, and probably will, be performed with the cleaning solvent at a controlled temperature: cold, heated to sub-boiling, boiling, or superheated above boiling. Nearly all solvent cleaning processes are populated with these steps, and no others.

Each of these examples was developed to demonstrate a solution for a cleaning problem. Only the problem in Chapter 2.20 could also be solved with a conventional vapor degreaser.

2.20 SUPERHEATED SOLVENT VAPOR

Superheated vapor is solvent vapor heated to and beyond its normal boiling point at atmospheric pressure. Its unique capability is to heat materials to above the normal boiling point of the solvent used.

- The prime use of superheated vapor is to vaporize the last traces of liquid solvent retained within part crevices. It's about drying!

That capability is not easy to implement quickly because vapor-to-solid heat transfer is very inefficient. Consequently, the process step of contact by superheated vapor substantially lengthens the overall cycle time.

Honeycomb

Table 2.6 Solvent Soaking and Its Aftermath for Complex Structures

Elapsed Time	Actions
0:01	Parts loaded in racks in chamber
0:15	Chamber sealed. Cycle started
4:45	Hot liquid solvent sprayed on cold parts, then allowed to soak
15:30	Superheated vapor sprayed on hot parts
1:15	Vapors displaced with dry hot air
0:45	Hot air displaced with dry cool air, to cool parts
0:15	Cycle complete. Chamber unsealed automatically

The prime reason for accepting the penalties of a lengthened cycle time and the energy cost of heating solvent vapor beyond its normal boiling point is to be able to remove **all** liquid from parts with complicated structures. Two examples are honeycomb panels (with membranes or the ability to cup liquid solvents) used as structural materials for aerospace applications (at lower left), and porous materials which will be used to fabricate implants for humans.

A schedule of process steps which represents implementation of an extended soak cycle and use of superheated vapor for drying is listed in Table 2.6. A graphical accounting of the time spent in each of the steps for this fictional cleaning process is noted as Figure 2.43.

Notice in Table 2.6 and Figure 2.43[CCC] that there are no scheduled events involving vacuum treatment because drying is done with superheated vapor.

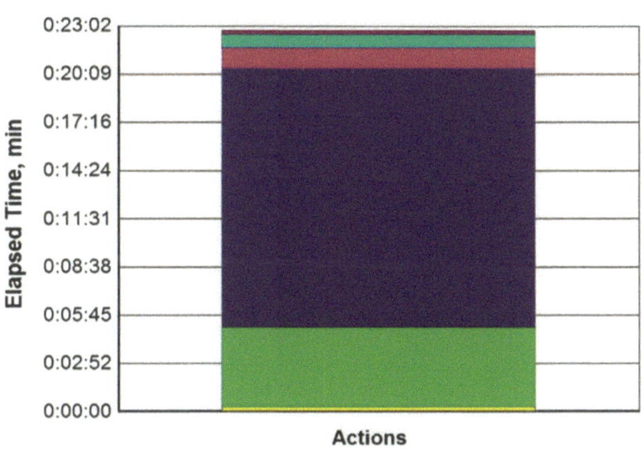

Figure 2.43 Cumulative Graphical Accounting of the Operating Steps in the Soak/Superheat Process of Table 2.6

2.21 THE PARTS DICTATE THE PROCESS DETAILS

The honeycomb part structure requires extensive processing time because of its huge content of surface area. Not all part structures do so, even though the same cleaning process elements are required. A comparative example of such a situation is a part whose structure is simple and whose thermal mass is low. Two of these parts are metal screws and nuts[DDD].

A schedule of process steps which represent implementation of a shortened soak cycle and use of superheated vapor for drying of parts with low thermal mass is shown in Table 2.7. A graphical accounting of the time spent in each of the steps for this fictional cleaning process is noted as Figure 2.44.

Nuts and Screws

Table 2.7 Solvent Soaking and Its Aftermath for Simple Structures

Elapsed Time	Actions
0:01	Parts loaded in racks in chamber
0:15	Chamber sealed. Cycle started
1:30	Hot liquid solvent sprayed on cold parts, then allowed to soak
1:30	Superheated vapor sprayed on hot parts
1:15	Vapors displaced with dry hot air
0:45	Hot air displaced with dry cool air, to cool parts
0:15	Cycle complete. Chamber unsealed automatically

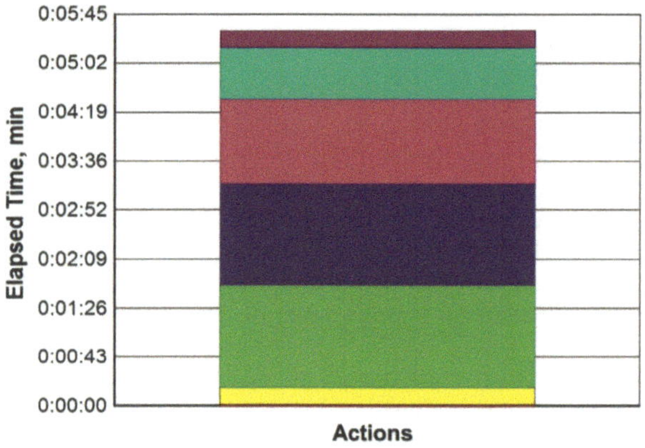

Figure 2.44 Cumulative Graphical Accounting of the Operating Steps in the Soaking-Based Process of Table 2.7

The processes whose schedules are represented in Figures 2.43 and 2.44 have the exact same steps. But the steps are differently "sized" – applied over considerably different lengths of time – because the parts being cleaned have distinctly different characteristics.

The more intricate parts, which have considerably larger surface areas, more void volume for the same mass, and more crevices are the honeycomb structures. They take longer to heat. They hold liquid droplets better. They don't dry well to a high level.

Notice in Table 2.7 and Figure 2.44 that there are no scheduled events involving vacuum treatment because drying has been done with superheated vapor.

The screw and nut parts have the opposite characteristics. It is not difficult to reach their crevices (internal and external threads) with cleaning fluid. They are not difficult to dry, and the need for superheated vapor was only for illustration.

The two lessons of the above subchapters are simple, and not newly expressed:

- Choose the cleaning solvent to match the soil characteristics (Ref. 2, Chapters 2 and 3, and
- Choose the process steps of which the overall process is composed based on the characteristics of the parts.

2.22 VACUUM AS AN AID TO REMOVING NON-VOLATILE SOILS

Cleaning of parts with high levels of non-volatile soils is an example of the application of the second lesson above. The amount of the soil present does matter – not in the choice of solvent (usually), but it matters highly in the choice of how the three process steps are chosen and ordered, as well as the length of time for which they are applied.

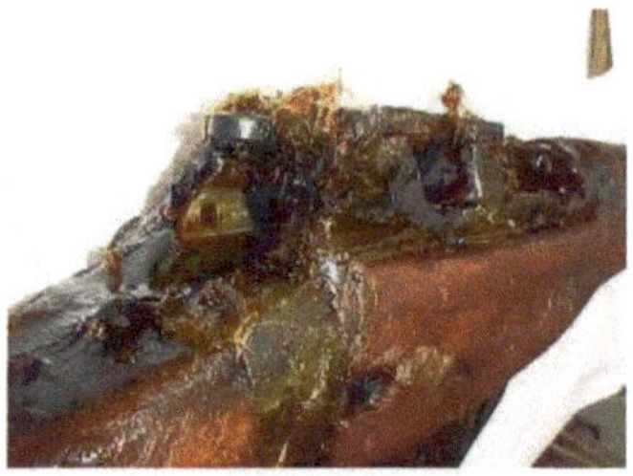

Firearm with Cosmoline

Consider part structures impregnated with grease for lubrication, multiport injectors which may have retained metal chips, and firearms with retained Cosmoline (image above[EEE]).

The process elements selected to clean parts of this nature (well-soiled) will be repeated in a sequence with others because of the quantity of soil to be removed.

But the main characteristic of this process and its cleaning steps is that they are to be conducted under vacuum – to increase volatilization of relatively non-volatile hydrocarbons[37]. So the first major process step is to reduce chamber pressure to 1 Torr absolute (759 TorrV or 1 mm Hg). And the portions of this cleaning process in which there are soil components present will all be conducted at that level of vacuum[38]. Process steps are listed in Table 2.8 and a graphical accounting of them is plotted in Figure 2.45.

Removal of low-volatility and soluble soil components from parts is aided by anything which increases volatility – such as reduction of work chamber pressure, even though the best solvent may have been chosen.

---

[37] One would not use an enclosed cleaning vapor degreaser to remove all Cosmoline from firearms, only the last vestiges. The great majority of cosmoline would likely be removed by soaking the parts for hours in a vat of high-flash (> 93.3°C (200°F)) paraffinic hydrocarbons (~C20). And depending upon the next treatment step in an enclosed vapor degreaser with the parts to provide adequate finish cleaning.

[38] This program of actions for cleaning highly intractable soils under vacuum certainly conflicts with advice previously given in Chapter 2. 12, and other locations in this volume. The reason for this departure from guidance is justified by the unusual character of this cleaning operation, as the soil material (Cosmoline) is: (1) assumed to be present in very large amount, (2) very immobile having a high viscosity, (3) and extremely non-volatile.

So two seldom-used capabilities will be employed to make this unusual cleaning problem less difficult to successfully resolve: (1) a very low pressure environment to enhance volatility of the soil material, and (2) superheated vapor to increase temperature of contact and so reduce soil viscosity. And it may well be that these two capabilities are not sufficient to substantially relieve the known difficulties!

Table 2.8 Removing Non-Volatiles with Vacuum

Elapsed Time	Actions	Comments
0:01	Parts loaded in racks in chamber	
0:15	Chamber sealed. Cycle started	
2:15	Air removed by vacuum pump-down to 1 Torr absolute[FFF]	Elapsed time could be shortened by increased investment for a vacuum pump with a higher capacity
1:30	Cold parts immersed in condensed hot solvent	Used to flush particles, chips, or dissolve Cosmoline. Chamber pressure will be increased afterward
2:00	Solvent vapors are removed by vacuum	
2:00	Parts sprayed with hot solvent vapor	Vapor will penetrate structure, condense, and rinse crevices
2:00	Solvent vapors are removed by vacuum	
10:00	Superheated vapor sprayed on hot parts	Used to enhance removal of residual soil and solvent, over long application time
2:00	**Solvent vapors are removed by vacuum**	**Second application is intended to completely remove all solvent (Chapter 2.17). Alternately, one may choose to omit the second vacuum step**
1:00	**Chamber filled and flushed with air**	
2:00	**Solvent vapors are removed by vacuum**	
1:00	**Chamber filled and flushed with air**	
0:15	**Cycle complete. Chamber unsealed automatically**	

Removal of non-volatile insoluble components (metal chips) is aided by repeated flushing with liquid solvent.

Note that the application of superheated solvent vapor is the most time-consuming process step, and with that the total cycle time approaches one-half hour. Compare this to the example of Table 2.7 and Figure 2.44 where, for parts of simple structure, the total cycle time is barely over 5 minutes. As with solvent selection – structure matter!

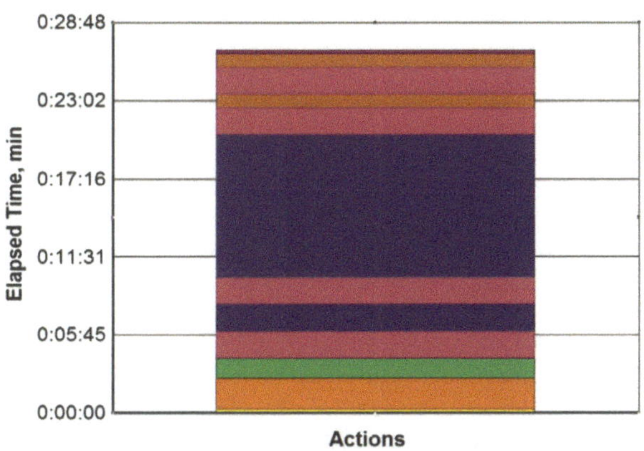

Figure 2.45 Cumulative Graphical Accounting of the Process to Remove Non-Volatiles of Table 2.8

2.23 REPETITION OF CLEANING STEPS

Piston Rings

A television commercial for fast food[GGG] once claimed "parts is parts." What may be true in food service is not so in industrial cleaning: "parts is not parts." It depends upon what they are and what's on them which is not wanted – the soil.

Watch Movement

Consider small parts with a low thermal mass, such as piston rings[HHH], gears for a watch movement, or tips for an atomic force microscope. Since these parts move, any contamination, such as organic films or particulate, will affect their movement.

The surface cleaning operations necessary to remove submicron layers of films or particles from parts are certain to involve repetition of rinsing steps. This is because: (1) there is no better cleaning fluid than pristine distilled (properly chosen) solvent, and (2) cleaning solvent is made dirty by being used and spreads some of that soil material to the surfaces wetted by used solvent.

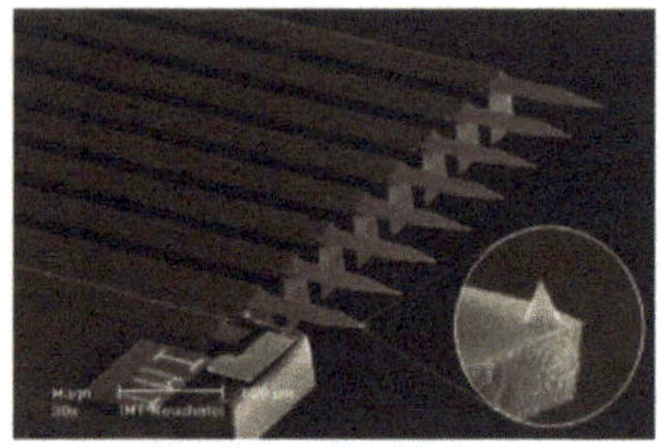

AFM Tips

Process steps are listed in Table 2.9 and a graphical accounting of them is plotted in Figure 2.46.

Table 2.9 Repetition of Cleaning Steps to Achieve a High Level of Soil Removal

Elapsed Time	Actions	Comments
0:01	Parts loaded in racks in chamber	
0:15	Chamber sealed. Cycle started	
2:15	Air removed by vacuum pump-down to 1 Torr absolute	
1;30	Cold parts immersed in condensed hot solvent	Parts are made cold by evaporation of solvent during vacuum operation (Chapter 2.3)
3:00	After draining, parts sprayed with hot solvent vapor	This step removes volatile soil components, leaving only non-volatile ones to be removed by repeated rinse steps
2:00	Solvent vapors are removed by vacuum	
1:30	Cold parts immersed in condensed hot solvent	Rinse step #1
2:00	Solvent vapors are removed by vacuum	
1:30	Cold parts immersed in condensed hot solvent	Rinse step #2
2:00	Solvent vapors are removed by vacuum	
1:30	Cold parts immersed in condensed hot solvent	Rinse step #3
2:00	**Solvent vapors are removed by vacuum**	
1:00	**Chamber filled and flushed with air**	
2:00	**Solvent vapors are removed by vacuum**	
1:00	**Chamber filled and flushed with air**	
0:15	**Cycle complete. Chamber unsealed automatically**	

Multiple rinsing steps dilute the remaining soil. Unfortunately, all soil is never removed from parts and the work chamber[III]. The path towards increased cleanliness with additional rinsing steps is one of diminishing returns[JJJ].

As Figure 2.46 shows, the penalty in cycle time for each rinsing step must include the time to remove the solvent as vapor with vacuum.

Repetition of the same cleaning steps until a desired cleanliness result has been achieved can be completed in all cleaning equipment described in this chapter.

2.24 CRITICAL IMMERSION CLEANING WITH ULTRASONICS

Commonly used in atmospheric open-top vapor degreasers, and enclosed vapor degreaser which operate at atmospheric pressure, ultrasonic transducers[KKK] are seldom found in machines which create a vacuum environment[39,40,41,42]. The reason is perceived cost – of the needed specialty transducers.

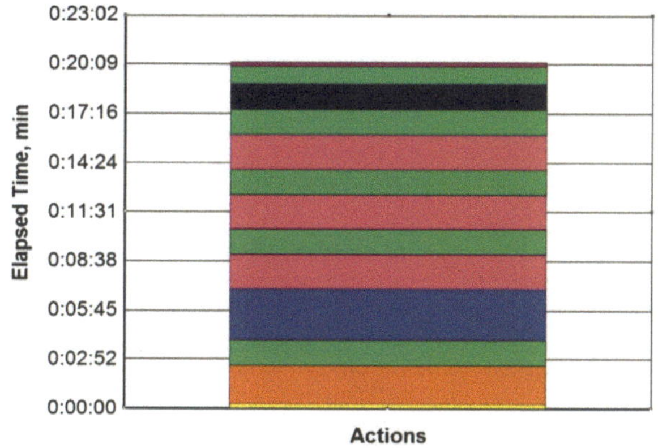

Figure 2.46 Cumulative Graphical Accounting of the Repetitive Operating Steps in the Process of Table 2.9

The problems noted in Footnotes 38 to 41 can add an additional 200 to 300% to the required investment for each package of 500 to 600 watts of ultrasonic power. That increase should not be determinant because a conventional transducer package only costs around $2000 to $3000 – not a significant portion of the investment of any enclosed airtight vapor degreaser. Unfortunately it has become often determinant.

Process steps are listed in Table 2.10 and a graphical accounting of them is plotted in Figure 2.47. Note that this process is one which should be promoted for applications which would be classified as critical cleaning (where the level of residual soil determines the value of the cleaned part).

There are at least four characteristics of this integration of process steps which contribute to a high level of control of cleaning quality: (1) one and one-half minutes of cycle time dedicated to removing oxygen from the work chamber

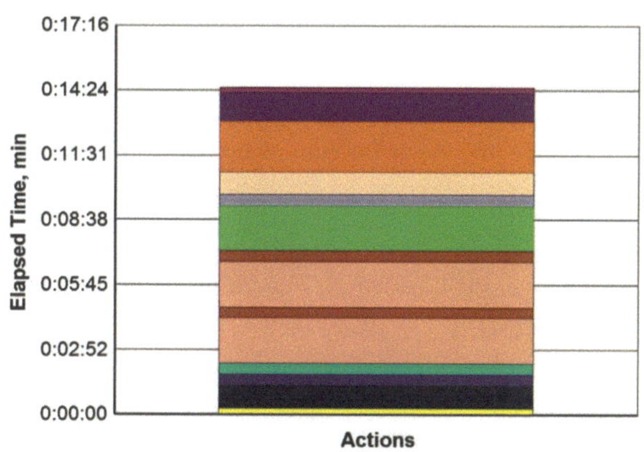

Figure 2.47 Cumulative Graphical Accounting of the Operating Steps in the Process Using Ultrasonic Transducers of Table 2.10

[39] Piezoelectric or magnetostrictive transducers of any frequency are only occasionally used in airtight enclosed vapor degreasers where the pressure is increased above atmospheric. This is because these transducers produce pressure pulses above and below the prevailing static pressure in the work chamber. Hence, each of the dozen or so individual diaphragms within a transducer body at atmospheric pressure must produce additional force during one half of the oscillation cycle – enough to move the body of the transducer the desired distance *and* overcome the increase of the prevailing static pressure above atmospheric pressure. These "overload" transducers can be purchased for a premium price.

Sonic transducers are covered in considerable detail in Ref. 1, Chapter 7, Section 7.10, pages 357 to 374, especially Figure 7.35 and Table 7.6.

[40] Designers of transducers to be used in vacuum environments don't have the problem noted in Footnote 42. They have the opposite problem! It is to only move the transducer body, during one half of the oscillation, the desired distance without having to overcome the normal atmospheric pressure. In other words, each individual diaphragm must be rated to produce a force considerably less than normal, and do so synchronously with all others. These "underload" transducers can also be purchased for a premium price.

[41] Any transducer can be used with solvents rated flammable. The manufacturer must provide the capability to: (1) continually purge the transducer with inert gas, usually nitrogen, (2) completely seal the circumference of the transducer enclosure, with the exception of at least two well-sealed connections for input and output of purge gas, (3) design a power supply and connections to the transducer which are rated for the appropriate flammability classification, and (4) have an independent organization (such as Factory Mutual) certify the entire apparatus as acceptable for that flammability classification. While the latter approval is not mandatory, no user should accept a transducer unit without it.

[42] The situations noted in Footnotes 38 and 39, and the situation noted in Footnote 40, are each interlocked. The reason is that the transducer housing must be totally sealed, to eliminate: (1) contact with flammable solvents, and (2) damage to the equipment. Were the pressure differential across the transducer body be just that applied by the transducer diaphragms, an adjustment in applied force wouldn't be necessary. Unfortunately, that's impossible. Transducer bodies are, and must be, sealed vessels.

Table 2.10 Use of Ultrasonic Transducers in Critical Cleaning

Elapsed Time	Actions	Comments
0:01	Parts loaded in racks in chamber	
0:15	Chamber sealed. Cycle started	
1:00	Chamber flushed with hot nitrogen	To displace all oxygen
0:30	Chamber filled with oxygen-free hot nitrogen	
0:30	Fill with hot solvent liquid	To enable immersion cleaning
2:00	Immersion cleaning with ultrasonics	First cycle
0:30	Drain liquid, flush, and fill with hot solvent	
2:00	Immersion cleaning with ultrasonics	Second cycle
0:30	Drain liquid and flush with hot solvent	First rinse cycle, with concern about dragout
2:00	Continuous flushing with hot clean solvent	Second cycle
0:30	Drain liquid solvent from tank	Concern about dragout
1:00	Chamber flushed with hot nitrogen	To dislodge liquid from crevices
2:15	Vacuum pump-down to 1 Torr absolute	To produce high quality drying
1:15	Flush chamber with clean dry air	Prepare for opening of chamber
0:15	Cycle complete. Chamber unsealed automatically	

to minimize solvent degradation, (2) two cycles of rinsing with one delay to minimize solvent dragout, (3) two cycles of immersion cleaning with ultrasonic transducers to liberate soil, and (4) vacuum drying – after treatment with hot nitrogen. Note that this process does not include vapor degreasing.

This multi-cyclic cleaning process does not have to be conducted in an airtight enclosed vacuum cleaning machine. After all, the first three items noted in the previous paragraph could be conducted in a suitably designed vapor degreaser. The reason why the incremental investment for an enclosed vacuum vapor degreaser is justified is because of the value of complete drying!

2.25 SUMMARY OF AIRTIGHT MACHINES

The three essential process steps included in Tables 2.6 through 2.10 are the same three: liquid immersion, vapor spray, and liquid spray with force. Cleaning parts requires no other. What makes them so powerful is that they can be combined, and implemented at any pressure – from super-atmospheric to vacuum.

Endnotes

A. USP 5,106,404, Grant, D.C.H., April 21, 1992. Emission Control for Fluid Compositions Having Volatile Constituents, and Method Thereof. Assignee is Baxter International.

B. USP 5,343,885, Grant, D.C.H., September 6, 1995. Vacuum Airlock for a Closed-Perimeter Solvent Conversation System. Assignee is Baxter International.

C. The normal boiling point of methylene chloride is 40°C (104°F), and that of perchloroethylene is 121°C (250°F).

D. Perchloroethylene has no hydrogen atoms, doesn't engender hydrogen-bonding intermolecular forces, and so its H-bonding HSP is zero. That for methylene chloride is 7.1 $MPa^{1/2}$.

E. USP 5,232,476, Grant, D.C.H., August 3, 1993. Solvent Recovery and Reclamation System. Assignee is Baxter International.

F. Also USP 5,304,253, Grant, D.C.H., April 19, 1994. Method for Cleaning with a Volatile Solvent. Assignee is Baxter International.

G. Systems that initially remove air with vacuum and later remove solvent with vacuum are referred to as vacuum-to-vacuum systems within the EPA NESHAP literature.

H. Torr is a non-SI unit of pressure. It was chosen to honor Evangelista Torricelli (1608–1647), an Italian physicist and mathematician who discovered the principle of the barometer in 1644. The Torr (capitalized when used as a symbol such as "760 Torr" or small case when used as a name such as "the units of pressure are torr") is identical to the unit of mm of Hg.

Comparison of Pressures which Equal 1 Torr
133.322 pascal
133.322×10^{-3} bar
1.33322 millibar
$1.33322\times10^{+3}$ microbar
1.3158×10^{-3} atm
19.337×10^{-3} psi
1333.22 barye
5.3577×10^{-1} inches of water
3.937×10^{-2} inches of mercury
1000 microns of mercury

There are exactly 760 Torr in one atmosphere. The Torr is a more convenient unit to use in high-vacuum physics and engineering work. Capabilities and limits of enclosed machines are usually specified in the pressure unit of Torr. A benchmark level of pressure (vacuum actually) is 1 Torr.

The SI (International System of Units) unit of pressure is the pascal, noted as Pa, and named to honor Blaise Pascal (1623–1662), a French mathematician and philosopher. One Pa is one newton of force applied over one square meter of area – a tiny exertion. One Torr is equal to 133.322 Pa.

The more useful SI pressure unit is the bar (not capitalized nor named for anyone), which is exactly 100,000 Pa, 100 kPa (kilopascals), or 0.1MPa (megapascals). One Torr is equal to 133.322×10^{-3} bar; hence the pressure unit of bar is more commonly found in high-pressure physics and engineering work. However, various prefixes modify this unit to make it more useful in design of enclosed machines. One Torr is equal to 1.33322 millibar and $1.33322\times10^{+3}$ microbar.

TheEnglishunitofpressureistheatmosphere,orthepoundofforce persquareinch.Theformerisabbreviatedasatmandthelatteris abbreviatedaspsi.Neitherunitiscapitalizedornamedtohonor anyone.Oneatmosphereisequalto14.696psi.OneTorrisequalto1. 3158×10^{-3}atmand19.337×10^{-3}psi.

An unofficial English unit of pressure more useful with enclosed machines is the inch of water column. This unit is an artifact of experimental pressure measurements being made with a manometer where the height of the fluid column was the measurement sought. This unit offers a convenient scale for identifying small values of pressure. One Torr is equal to about one-half inch of water column pressure (actually 5. 3577×10^{-1} inches of water).

Another unofficial but common English unit of pressure commonly used with vacuum pumps is the inch of mercury. This refers to the height of a column of mercury, as measured in inches. One Torr is 3.937×10^{-2} inches of mercury. Full vacuum, 760 TorrV, is taken to be 29.9215 inches of mercury.

The CGS (centimeter gram second) or metric unit of pressure, is the barye (NOT bar), named to honor Antoine-Louis Barye (1796–1875), a French sculptor of animals. The bayre is not capitalized or in common use. It is abbreviated as ba, and is also known as the barad, barrie, bary, baryd, baryed, or barie. One ba is equal to one dyne of force exerted over one square centimeter of area, or 0.1 pascal (an extremely tiny level of exertion). One Torr is equal to 1333. 22 ba; hence the CGS unit ba is seldom found in rating the capacity of enclosed machines.

An unofficial but common CGS unit of pressure is the micron. This also refers to the height of a column of mercury, as measured in microns (one inch is 25,400 microns). One Torr is exactly 1000 microns of mercury pressure.

I. This characteristic of an enclosed vapor degreaser was noted in the 1994 NESHAP for Halogenated Solvents (Chapter 1, Endnote T) as a basis for specifying permissible compliance options.

The US EPA has rated the vacuum-to-vacuum systems as the most efficient way to meet requirements with a near zero emissions resulting in substantial savings of solvent cost.

J. The word vacuum is derived from the Latin term for "empty." A perfect vacuum would be absent of all matter; a volume perfectly empty.

K. The image of Figure 2.6 is courtesy of Vacuum-Guide, http://www.vacuum-guide.com/english/equipment/hw-slidingvanes.htm.

L. The liquid ring vacuum pump was patented in 1914 as USP 1,091,529, to Louis Havelock Nash.

M. Figures 2.8 and 2.11 are courtesy of Gardner Denver Nash, http://www.gdnash.com/.

N. Vacuum levels of 1 Torr were selected to meet the EPA emission and OSHA exposure regulations. They could be achieved with a vacuum package available at a reasonable investment. When using non-NESHAP and low-price solvents such as isopropanol, normal design does not require vacuum levels as low as 1 Torr.

O. For the performance data of steam-driven eductors, see Foisy, E.C. and Munkittrick, M.T., "Energy Comparison Vacuum Producing Equipment – Mechanical Vacuum Pumps Vs. Steam Ejectors," *Proceedings from the Fourth Industrial Energy Technology Conference* Houston, TX, April 4–7, 1982.

P. Systems that initially remove air with vacuum and later remove solvent with vacuum are referred to as vacuum-to-vacuum systems within the EPA NESHAP literature. It is perfectly possible, and some firms have done or are doing so, to use more than one type of vacuum pump in an enclosed cleaning machine. For example, some firms use rotary piston pumps to remove air from the work chamber. These pumps require oil changes and mist eliminators. But they don't contaminate the cleaning solvent with seal oil because solvent never contacts the pump.

For evacuation of solvent vapor from the work chamber, a liquid ring pump is used. The seal liquid is cooled clean solvent. The warmed seal liquid is circulated through a clean solvent holding tank. No oil ever contacts solvent!

Q. There are other types of vacuum pumps, seldom found in cleaning machines. A fifth type of pump is a dry pump, which can be a screw, scroll (high speed), or claw design. Booster pumps are also dry and a generally rotary lobe type.

Booster pumps can be used on larger systems to enhance removal at lower pressures and to attain a final lower pressure.

Dry pumps can produce very high levels of vacuum in a booster role. Some screw pumps can reach less than 0.1 Torr absolute.

R. The image of Figures 2.18 is courtesy of Hyperflo, http://www.hyperflo.com.

S. The images of Figures 2.19 to 2.28 is courtesy of Serec Corporation, http://serec-corp.com/index.html.

T. Specifics about the design of open-top vapor degreasers, and the components from which they are fabricated, are related in Ref. 1, Chapter 7.

U. Other factors beside vessel size determine its cost. They include:

- Vessel openings – the size of the opening is one factor which determines the amount of machining required to create the groove for the gasket. But more than size affects cost. Openings placed on curved surfaces such as those of end caps (Endnote BB) cost substantially more to weld and machine than those on flat surfaces. But or course they are necessary to complete a design.
- Heat exchange capability – electric is cheapest and easiest to install but if flammable solvents are used, steam or heat transfer oils may be required, adding to required investment.

- Ergonomics – method of lifting and removing the lid (cover). There are at least four possibilities: (1) the operator manually lifts it if possible (cheapest), (2) hydraulic means lift it (more expensive), (3) the operator slides it (probably still more expensive but required for heavy lids), and (4) automatic lift and slide (most expensive but required for large lids).
- Orientation of the work chamber (of any size) – vertical orientation is the cheapest if the operator can load or if lifts are available. Horizontal orientation requires additional locking of lid to prevent spills and rails or rollers internally to receive parts. Horizontal orientation is generally required for long parts or heavy parts (Figure 2.19) transported on carts with rollers.

V. In the US the primary code is that developed, published, and supported by the American Society of Mechanical Engineers known as the ASME Boiler and Pressure Vessel Code, Section VIII, Division 1. Within states and regions of the US various jurisdictions have modified the means by which local operators are allowed to implement the national ASME code. The national code, which is more complicated than the naming system for fluorinated chemicals (Ref. 3, Appendix 2), can be purchased together with a training course for its use at http://www.asme.org/products/courses/boiler-and-pressure-vessel-code-section-8-division.

W. Most nations have their own codes which are similar to the ASME code. Though they are based on the same principles, they usually are differently implemented. Two areas in which the US and some international codes differ are whether fabricated vessels must be pressure tested (and how) after fabrication and if an inspection is required (and how) to demonstrate their structural integrity before they are placed into operation.

X. This equation is found in a summary of various design equations in Figure 4.6 of the reference of Endnote V. The equations for calculating the wall thicknesses of non-cylindrical vessel shapes are considerably more complicated and outside the scope of this book.

For those desiring more information, one can download the chapter Section VII – Division 1: Rules for Construction of Pressure Vessels, from the URL http://www.krrao.com/images/ASME_Ch21_p027–138.pdf.

Y. Weld joint efficiency (E) accounts for the proven quality of a welded joint. The longitudinal weld seam often found in a cylindrical vessel is such a major welded joint. Other joints are those between nozzles and the vessel, and often the attachment of the end caps to the vessel.

It is understood that all welded joints may not be treated the same in terms of quality. If each and every weld joint is inspected by radiographic means, the value of E is taken as unity. If only some weld joints are inspected by radiographic means (spot inspection), the value of E is taken as 0.85. If no weld joints are inspected by radiographic means, the value of E is taken as 0.75.

Z. The corrosion allowance is based on the expected annual corrosion rate for the vessel material in the anticipated process environment. That corrosion rate is multiplied by the nominal design life of the vessel (usually 20 years) to determine the total lifetime corrosion allowance. Corrosion allowance is an application-specific judgment call.

Typical values for carbon steel are often ¼ inch (~6 mm), or 0.0625 inches (~1.5 mm), or an intermediate value; and for stainless steel are 0 because it's considered that it is not subject to general corrosion.

One must be careful about the corrosion allowance for pitting corrosion of stainless steel – there should not be one. This is because pitting corrosion is not a continuing phenomenon. When pitting corrosion happens, it happens over a short time which is not the situation envisioned by Equation 2.3.

AA. Some non-cylindrical vessels cannot be designed per ASME code because their shape is not considered in the code.

BB. This most common of austenitic grades contains approximately 18% chromium and 8% nickel (304 SS or "18/8"). This grade is used for chemical processing equipment, food, dairy, and beverage industries; heat exchangers and structural tubing; and less aggressive solvents. The "L" grade (low carbon) is preferred for improved corrosion resistance in welded structures, but is probably not necessary for solvents described in this book. This grade of stainless steel, 304, is almost certain to be the material of construction for any enclosed machine purchased during the useful lifetime of this book. The same is probably also true for purchase of open-top vapor degreasers or cosolvent machines.

CC. This also common grade contains 16 to 18% chromium and 11 to 14% nickel (316 SS). It also has molybdenum added to the nickel and chromium of the 304 material. The molybdenum is added to control pitting corrosion. Type 316 is used in chemical processing; the pulp and paper industry for food and beverage processing and dispensing; and in more corrosive environments. The molybdenum content must be a minimum of 2%.

DD. Either pipe or tubing can be used to create a small vacuum vessel. But tubing is different from pipe, of the same material and fabrication technique. The difference between a pipe and a tube fabricated similarly is chiefly is in how they are measured, and for what they are used.

Pipe is sized by the outside diameter. The inside diameter is different, as the wall thickness is chosen to achieve different levels of mechanical strength to resist or contain stress. Vessels up to 30 inches in outside diameter can be made from pipe. Vessels of larger outside diameter are made from rolled plate whose seam is then welded. The named pipe size is the measured outside diameter for pipe sizes 14 inches and above; smaller named pipe sizes have measured outside diameters slightly larger than the named size. Pipe is usually made from strip which is rolled and the seam welded, although bar stock can be drawn over a mandrel to produce smaller diameter pipe without a welded seam. Seamless stainless steel pipe is seldom available in diameters greater than about 10 inches. So it is too small for general use in fabricating vessels for solvent cleaning.

Tubing is also sized by the outside diameter, with the inside diameter being different so support strength can be modified. The named tubing size is always the outside diameter. Tubing is more often used as a structural material because joints between materials can be fixed and certain. Tubing is often produced by drawing a shape over a mandrel (called DOM) to avoid a weld seam. Seamless stainless steel tubing is seldom available in diameters above 3 inches. So it too is too small for general use in fabricating vessels for solvent cleaning.

The wall thickness of pipe is denominated by its schedule and that of tubing by its gauge. Pipe and tubing with a thicker wall have a larger schedule and a smaller gauge respectively.

Vessels for vacuum service are normally preferred to be fabricated by seamless wall construction to avoid a possible incidence of crevices where particles, soiled solvent, and rinse fluid can be trapped. Yet as noted above, seamless pipe and tubing are not available to fabricate vessels in sizes desired for enclosed cleaning machines. Consequently, these vessels, such as those shown in Figures 2.19 through 2.26, are fabricated not from pipe or tubing but from stainless steel plate which is rolled to shape, seam welded to size, and then precision ground to achieve a desired internal surface finish.

EE. Many of the carbon steel systems sold in the 1990s are still in operation.

The cost of carbon vs. stainless makes carbon steel fabrication worth looking into. It is not a choice for universal application, but for some applications it works well. Under no circumstances would one use carbon steel to fabricate the clean solvent storage tank where there is some air, unless the process design required purging with nitrogen prior to solvent entry. The cost difference between stainless and carbon steel materials is often outweighed by the cost difference in fabrication between the two materials.

FF. Ref. 1, Chapter 1.

GG. Chapter 1.2.

HH. The "normal" boiling point is the boiling point at a pressure of 760 Torr – one atmosphere.

II. In Ref. 1, Chapter 4.12.3 one can read an explanation of why continuous purging of soiled solvent (or aqueous cleaning agent) to a purification (or treatment) unit is a procedure which produces the highest possible quality of cleaning.

JJ. USP 5,469,876, Gray D.J., and Gebhard, P.T.E., November 28, 1995, Cleaning Method and System. Assigned to Serec.

KK. USP 5,538,025, Gray D.J., and Gebhard, P.T.E., July 23, 1996, Cleaning Method and System. Assigned to Serec.

LL. The images of Figures 2.32 and 2.33 were produced by this author from diagrams published in articles and advertisements by two well-known firms. Some license has been taken for simplification and to enable direct comparison. No inference should be made about relative size of components.

MM. However applications involving high levels of soil have been successfully developed by firm "S". The key is to design for and manage the distillation rate at an adequately high level to remove soil from the system at or higher than the rate it enters. One example is removal of "post-quench" oil from drill bits.

NN. Associated USP references are: USP 6,418,942, Gray D.J., and Frederick, C., July 16, 2002, Solvent and Aqueous Decompression Processing System.

This patent was followed by others where the technology (known as Vacuum Cavitational Streaming) was modified and applied to water instead of solvent. They are:

USP 6,743,300, Gray D.J., June 1, 2004, Multistep Single Chamber Parts Proceeding Method.

USP 6,783,601, Gray D.J., and Frederick, C., Aqueous cleaning of liquid residue by etching August 31, 2004, Method for Removing Particles and Non-Volatile Residue from an Object.

USP 6,783,602, Gray D.J., August 31, 2004, Multistep Single Chamber Parts Processing Method.

USP 6,824,620, Gray D.J., and Frederick, C., November 30, 2004, Method for Removing Particles and Non-Volatile Residue from an Object.

USP 7,444,761, Gray D.J., November 4, 2008, Intrinsically Safe Flammable Solvent Processing Method and System

USP 7,846,262, Gray D.J., and Frederick, C., December 7, 2010, Aqueous Cleaning of Liquid Residue by Etching,

USP US 20070107748 A1, Gray D.J., and Frederick, C., May 17, 2007, Vacuum Cavitational Streaming

OO. A mass balance on a continuously fed stirred reactor (CSTR) displacing one fluid with another for nitrogen displacing solvent vapor (or air) produces the classical equation $\theta = \frac{V}{q} ln(\frac{1-X_1}{1-X_2})$. The CSTR is simply a vessel, a volume, a work chamber. It is called a reactor because this equation finds numerous applications – including describing the events within chemical reactions. The core assumption supporting this equation is that all the volume of this vessel is instantaneously and perfectly mixed – something obviously not universally valid but suitably adequate so as to be universally adopted as an estimate of many actual processes. The parameters in this equation are: V (volume of vessel), q (volumetric throughput), θ (the time every element of fluid spends (is held up) in the vessel before being discharged), and X (volumetric fraction of nitrogen) with (1 – X) being the volume fraction of oxygen (= [O]). The subscripts 1 and 2 refer to first and last times respectively between which the volume (= q x θ) of nitrogen has been fed to the vessel.

The CSTR equation can be restated to determine the number of vessel volumes of nitrogen fed for any time interval where a turnover is the number of times the work chamber has been continually filled with nitrogen: $\frac{q\theta}{V} = turnovers = -ln(\frac{1-X_2}{1-X_1})$. The word turnover means that the contents of the work chamber have been displaced or turned over. With [O = (1 – X)] being the oxygen concentration at any time, the above equation can be further restated as $[O]_2 = [O]_1 \times e^{-\{turnovers\}}$. It is this equation which is used to calculate the values in the table of Box 2.5.

PP. The density of an ideal gas is $\frac{Molecular\ Weight}{359}$ in units of lb/CF.

QQ. Ref. 2, Chapter 1.9.2.1 has a more complete discussion about the selection of operating temperature.

RR. This is the temperature at which the vapor pressure of the solvent is 1 Torr absolute.

SS. Use of the Antoine Equation to estimate the flash point of solvent blends is described in Ref. 2, Appendix A2, Section K.

TT. Note that temperature (T) has the units of kelvin, which is the temperature in centigrade +273.15.

UU. The "A" and "B" constants for more than 210 cleaning solvents are tabulated in the General Data of Ref. 2, Appendix A1.

An excellent source of adjustable parameters for the Antoine Equation is the reference Yaws C. L., and Yang H.-C., "To Estimate Vapor Pressure Easily. Antoine Coefficients Relate Vapor Pressure to Temperature for Almost 700 Major Organic Compounds", *Hydrocarbon Processing,* Vol. 68, No. 10, 1989, Pages 65 to 68.

VV. At some temperature well below the point where liquid solvent begins to condense from solvent vapor (the dew point), a second condition may be defined. Not useful in management of solvent cleaning, but only noted here for completeness.

This condition is where the first bubble of nitrogen vapor is formed from liquid nitrogen in a mixture of liquids. One can calculate the mixture bubble point from the assumptions above and a second equation. In this second equation, one sums the estimated vapor phase compositions and seeks to identify the temperature at which the sum is equal to unity. That temperature is the bubble point of the bi-liquid mixture.

The only problem with the specific calculation for the nitrogen and trichloroethylene mixture is that the outcome is below the melting point of trichloroethylene (–73°C)! As stated, the point of this endnote is to recognize the mating calculation of bubble point of solvent mixtures.

WW. Gasses such as nitrogen or carbon dioxide are usually referred to as "inerts."

XX. At the pressures where solvent cleaning is normally done, as its boiling point is –196°C.

YY. Ref. 2, Chapter 11.

ZZ. Some useful patent references associated with this technology are:

USP 5,051,135, Tanaka M., and Ichikawa, T., September 24, 1991, Cleaning Method Using a Solvent While Preventing Discharge of Solvent Vapors to the Environment. Assigned to Tiyoda.

USP 5,106,404, Grant D.C.H., April 21, 1992, Emission Control for Fluid Compositions Having Volatile Constituents, and Method Thereof. Assigned to Baxter International.

USP 5,193,560, Tanaka M., and Ichikawa, T., March 16, 1993, Cleaning System Using a Solvent. Assigned to Tiyoda.

USP 5,449,010, Turicco, T., September 12, 1995, Pressure Controlled Cleaning System.

AAA. Durkee, J.B., and Gray, D., "Enclosed Cleaning Systems," Handbook of Critical Cleaning, Second Edition, April 11, 2011, ISBN 1439828261.

BBB. USP 5,301,701, Nnafzifer, C.P., April 12, 1994, Chamber Cleaning, Rinsing, and Drying Apparatus, and Method Therefor. Assigned to Hyperflo.

CCC. The process steps enumerated in Figures 2.43 through 2.47, and Tables 2.6 through 2.10, are coded by color to make identification of steps simpler. The color code is:

- Red – parts loaded on racks,
- Yellow – chamber sealed and the cleaning cycle is started,
- Dark green – parts flushed with hot clean solvent,

- Dark blue – superheated solvent vapor sprayed on parts,
- Maroon – flush chamber with clean dry air,
- Turquoise – hot air displaced with dry cool air, to cool parts,
- Purple – cycle complete and chamber open,
- Orange – air removed by vacuum pump-down to as low as 1 Torr absolute,
- Dark brown – chamber filled and flushed with air,
- Light brown – immersion cleaning with ultrasonics,
- Light blue – draining or pum-pout of liquid solvent,
- Blue – chamber filled with hot oxygen-free nitrogen,
- Black – chamber flushed with hot nitrogen.

DDD. The images used in Chapters 2.20 and 2.21 are courtesy of Indiamart (http://www.honeycomb.com) and Cutcaster (http://www.cutcaster.com), respectively.

EEE. The image showing Cosmoline on a rifle is courtesy of FirearmsForum (http://www.firearmsforum.com).

FFF. "...This step prepares the chamber for efficient vapor flushing. The presence of a small amount of air can result in the condensing vapor process being controlled by the diffusion of vapor through the collection of air around the part. As the vapor condenses, the air diffusing to the surface of the part begins to accumulate and the vapor now needs to diffuse through this air resistance. The heat transfer coefficient can be reduced 10 times with just 5% air present. The difference is a rapid flushing in an air free environment vs. a gradual condensing process as observed in an open top degreaser. Rapid is better...".

Personal communication, Dr. Don Gray, President, Vacuum Processing Systems, January 12, 2013.

GGG. The claim was made in association with a past television commercial for fast food, asking the question "Where's the beef?"

HHH. The image of piston rings is courtesy of EngineBasics (http://www.enginebasics.com). That of watch components is courtesy of watchmakingblog.com (http://www.watchmakingblog.com). That of an atomic force microscope (AFM) is courtesy of US NASA (http://www.nasa.gov).

III. Ref. 1, Chapter 5.1.2.3.

JJJ. Ibid, Chapter 1.12 and Figures 1.2 and 1.3.

KKK. Also known by some as "cleaning accelerants."

Chapter | 3 |

Cosolvent Machines

A cosolvent (or co-solvent) machine or process is one which uses more than a single solvent.

It's that simple. The cleaning machine could be a vacuum enclosed machine or an open tank without a cover. If the process uses more than one solvent, it's a cosolvent process.

It's just that there are a variety of types of cosolvent machines which use a variety of similar solvents to clean parts. Fortunately, there is a simple classification system which makes it easy to choose which type is most useful for a given application.

Generally, cosolvent systems have only two solvents. There are not more solvent components than two because it's just too difficult to manage three or more solvent components in a process which already is required to manage a soil material which probably has more than one component.

3.1 THE REASON FOR COSOLVENT MACHINES

It's still simple. Cosolvent machines and processes exist because a single solvent isn't the best choice for both cleaning/rinsing, and rinsing/drying of parts.

There is usually just one requirement for a cleaning solvent:

- That it easily dissolves the soil, usually by having its Hansen Solubility Parameters (HSPs) match those of the soil materials[A].

But there is another and different requirement for a solvent with which to dry parts:

- That it rapidly evaporates, leaving clean surfaces dry[B].

Over the last two decades a third and still different requirement has come to dominate:

- That it raises no concerns about safety, health, and environmental (SHE) issues[C].

Almost never do individual solvents satisfactorily fulfill all three requirements. More than occasionally one of these requirements is not well met in a commercial solvent cleaning process; and occasionally more than one is not well met[D]. Said another way, there are no perfect cleaning solvents[E].

The value of 1,1,1-trichloroethane (TCA), trichloroethylene (TCE), or n-propyl bromide (n-PB) as cleaning solvents was/is that they uniquely fulfilled the first two requirements – cleaning, rinsing, and drying – as a single solvent.

Broadly true, it's the legitimate and urgent concerns about SHE issues of these (and like) solvents which are the reasons why cosolvent machines have been created, developed, and used. If TCA were not an ozone depleting compound, if TCE presented no concerns about toxicity or carcinogenicity, or if n-propyl bromide didn't raise the same concerns, the reason for the existence of cosolvent machines would be difficult to identify.

3.2 THE NOMENCLATURE

One of the two solvents is a cleaning agent called the solvating agent, or SA.

The other solvent is a rinsing agent, called the RA. The RA is likely also to fulfill the drying requirement noted above.

In a true vapor degreaser the SA and the RA coexist well because they are the same fluid.

- It is the relationship of the SA to the RA which defines the type of cosolvent process in which both are used.

http://dx.doi.org/10.1016/B978-0-323-22520-5.00003-4

- But more than that, for any type of cosolvent process, a part's cleanliness depends on how the SA and RA coexist with one another in the same cleaning machine, relative to how that was anticipated in the process design.

Any SA may well remove soil from part surfaces, but if the SA remains on part surfaces, the parts aren't clean. That's the sole role of the RA – to remove the SA in some way, and then be removed (or remove itself) from the parts.

3.3 RINSING RULES

Rinsing is necessary to keep clean surfaces clean by removal of the soil-laden SA remaining on parts with clean cleaning solvent (here, called the RA).

There are two mechanisms of rinsing the SA from parts with the RA. Both mechanisms of rinsing parts are involved with specific and different cosolvent processes.

- Rinsing can be *dilution* of the soil-laden SA with fresh clean SA so that the solvent remaining on parts[F] contains as little soil as necessary to meet surface quality standards. Know that the evaporative drying step won't remove either this non-volatile residual soil or the SA because both are relatively non-volatile.
- Rinsing can also be by *displacement* of the soil-bearing SA with an RA which is immiscible with the soil-bearing solvent, *and* evaporates rapidly.

Commercial cosolvent cleaning processes are of both types. But the *displacement* type has become more common than the dilution type. Both will be discussed below.

The chosen RA will have various characteristics: be flammable or not flammable, display a low or high flash point, be miscible or not miscible with the SA, and be less or much more expensive than the SA. But, whatever those characteristics are, it is likely they enable drying of rinsed parts – or that RA won't be chosen.

3.4 THE COSOLVENT CLASSIFICATION SYSTEM[G]

There are four types of cosolvent process. The relationship between the solvating agent and the rinsing agent is the defining factor:

- Class I – *Semi-aqueous;* water is involved as the RA, the SA can be any solvent.
- Class II – *Immiscible* SA and RA. There are two subtypes, with different types of SA (volatile and non-volatile). Water is not involved.
- Class III – *Miscible* SA and RA with neither being water.
- *Azeotropic*[H], where the SA and the RA are irretrievably combined, and neither is water.

Each type is described below, with examples where use is advised and not.

3.5 PROCESS DESCRIPTION: CLASS I – SEMI-AQUEOUS

Developed in the late 1980s, the semi-aqueous process was designed to use low-cost solvents, (terpenes[I] and refined hydrocarbons[J]), and water, in order to replace the cleaning performance of CFC-113 ("Freon") and 1,1,1-trichloroethane (TCA). Unfortunately for the developers, the initial applications have become nearly the only applications through 2013.

Those applications were for removal of acidic flux and flux residues from various printed circuit (wire) boards (PWBs), removal of carnauba wax from optic materials, removal of buffing and lapping compounds from metal substrates, and some others. These applications share at least one characteristic – that high quality, thoroughly repeatable cleaning is essential. And these applications remain highly valued.

Because of concerns about complexity, repeatability, and operating cost, semi-aqueous technology has not been well-adopted by the industries doing other cleaning work such as metal cleaning.

Generally, this narrow spectrum of successful applications involves continuous processing, rather than batch processing. So these application are performed in inline machines (Chapter 1.27 and Figure 1.50) with multiple stages of vapor-liquid contact, and not vapor degreasers with one or two stages of vapor-liquid contact (see Chapter 1.7 and Figures 1.10 to 1.15). In the semi-aqueous process, no vapor is intentionally generated; contact is only with liquids in batch mode.

The semi-aqueous cleaning process is truly well-named. It's constructed around the use of water as the RA (though there is no water in the SA). SHE concerns about solvents in general are a principal reason why the semi-aqueous process was developed using water as one of the two components. In addition to water, there are required contributions from: detergents and solvents, pressurized sprays, and especially management of temperature variations. These are the three basic actions which are found in any cleaning process[K].

Like most cosolvent processes, the semi-aqueous process is not as simple as single-solvent processes – though the cleaning steps alone can make it appear so in Figure 3.1.

But, it's not the cleaning steps which add complexity. It's three other factors which foster complexity: (1) there are multiple or multi-component solvents and multiple needs to separate and reconstitute them for reuse, (2) the key separation is done gravimetrically, based on immiscibility and a difference in fluid density, so that the effectiveness

Figure 3.1 Simple Process Flow Diagram of the Semi-aqueous Process

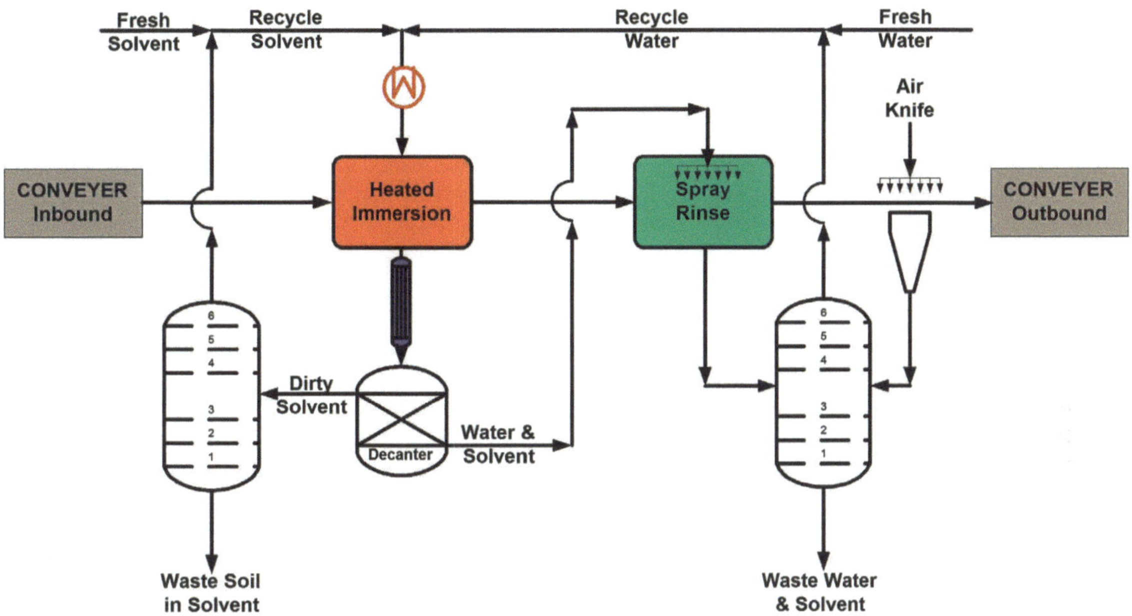

Figure 3.2 Equipment Diagram for the Semi-aqueous Process

of separation is strongly dependent on temperature and composition (as well as soil type and loading), and (3) the number of process control set points has grown from one or none in a vapor degreaser to perhaps at least a half-dozen.

In other words, a semi-aqueous cleaning system is not forgiving of change. And almost none of the complexity is associated with removal of soil from parts; it is associated with the separation steps.

3.5.1 General Semi-Aqueous Complexity

There are many semi-aqueous processes, each slightly different and sponsored by a firm with commercial interest in solving a specific problem.

A simplified process diagram of a general continuous semi-aqueous process is shown in Figure 3.1. Compare that figure to the process equipment diagram of Figure 3.2 to see that the complexity is not within the cleaning baths.

The general semi-aqueous process:

- Uses a mixture of solvents (Chapter 3.5.5) which are miscible with water at an elevated temperature such as 140°F (60°C) to 160°F (71°C)., and immiscible at reduced temperature such as 100 to 110°F (38 to 49°C)[1].
- Uses cleaning agents (SAs) in the cleaning bath that are mostly water. Concentrations of the total solvent mixture in water are usually less than 10 weight percent. The reason for this is that the solvents are expensive and *do perform adequately* at these concentrations. Remember, the solvent/soil mixture dissolved in water is the dragout material left on parts as they leave the

[1] As there is no generic semi-aqueous cleaning process, the above relationship between temperatures may be reversed in some implementations. Many surfactants precipitate from water at temperatures above160°F (71°C) (the situation noted above), and others don't. Some surfactants precipitate from water below 100°F (38°C), and others don't. Generally, one favors cleaning of oil and grease at temperatures closer to 160°F (the situation above) vs. 100°F because the reduction of soil viscosity at higher temperatures makes them more easily removed by contact with sprayed fluid (Figures 1.62 and 1.63).

For an excellent reference about equipment design, cleaner specification, operations, cleanliness testing, and troubleshooting of semi-aqueous cleaning technology, consult the *Post-Solder Semi-Aqueous Cleaning Handbook*, IPC-SA-61A, 2005. Contact IPC, 2215 Sanders Road, Northbrook, Illinois, 60062–6135, (847)-509–9700.

Temperatures are primarily specified in this chapter in °F because that's how they are generally specified by the suppliers of goods in the US (though not necessarily by those who supply them in Europe).

cleaning bath. That expensive solvent product is both lost when diluted in the rinse water (the RA), and left on the parts as residue.

- Does cleaning by immersion[2] where the solvent mixture and water are miscible. Contact times between parts and the water/solvent mixture can be as little as 30 seconds but are more commonly 5 minutes or more. Agitation by under-immersion or in-air fluid spray is often incorporated.
- Cools the soil-laden water/solvent-mixture brew, then feeds it to a decanter such as that shown in Figure 2.41, which allows it to phase-separate. The heavier water phase, by at least 0.15 to 0.2 g/cc, usually contains soluble solvent components, but little soil. The lighter solvent mixture phase contains solvent, little water, and hopefully all the soil (Box 3.1).

> **Box 3.1 Spray Cleaning by the Semi-aqueous Process**
>
> Depending on the specific components of the SA mixture, a distillation column may be used instead of a decanter. So, the purposes and methods of separation of water from solvents from soils are not the same.
>
> 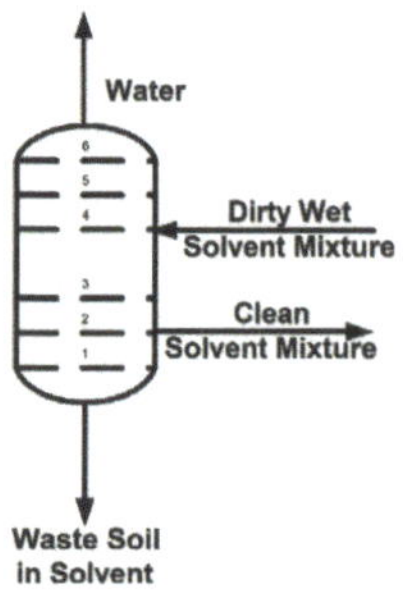
>
>
> This distillation column shown above is not the one shown in Figure 1.8, which separates solvent mixtures from soil – as the water has already been removed in the decanter. This distillation column is designed based on the understanding that all components of the SA mixture are less volatile than the RA (water). So, the distillation column shown above separates water from the solvent mixture, and then as much of the SA mixture as possible from all of the soil components.
>
> Certainly it's true that a simple flash distillation will be less forgiving of changes in soil or solvent composition than will the multi-plate distillation column above. But it will have a different energy cost, and impose a different kind of complexity.

- Fluid holdup time in the decanter is not short. It may be 5 to 30 minutes – probably closer to the latter. The decanter is a simple tank, though likely with cove corners (to avoid fluid retention), and certainly without agitation (to avoid mixing the phases together). Since the decanter contains two phases, there will be an interface. An interface monitor, often optically activated, controls pumps which keep the interface within a range of decanter heights by periodically (usually) pumping away a batch of aqueous (RA) material or soil-laden SA material as needed[3].
- Without using any capability for mixing, the process designer will strive to make control of the temperature in the decanter as continuously close to a laboratory-determined goal as possible. Temperature variation from goal means that the integrity of phase separation will vary from goal as well[4].
- Easily segregates soil materials (oil, grease, wax, and particulates) for concentration by distillation. Present in the solvent mixture are either higher-boiling materials or those insoluble in solvent. Distillation allows the solvent mixture to be recovered and reconstituted for reuse.
- Imposes a cost because some of the costly solvent mixture is lost as the soils are disposed as waste, and also because the dilute rinse water is treated to recover the water. Because the cleaning agent is a multi-component solvent mixture, recovery and reconstitution of it from the waste soil concentrate is seldom attempted. Incineration of the soil concentrate is a common disposal technique.
- May produce an emulsion[5] of the solvent mixture in the water effluent from the decanter. This is done by design in many cases[6]. A low concentration of the solvent

---

[2] Some semi-aqueous cleaning is done by spray cleaning. In this case, the various solutions which are part of the semi-aqueous technology are successively applied to surfaces (probably large ones) and later collected. Then the solvent and water processing inherent to semi-aqueous technology is practiced in separate facilities.

[3] Should operation of the interface within the decanter become compromised, it is likely that operation of the entire process will be compromised. Consider the decanter to be the "brain" of the semi-aqueous process, and compromise of the interface as a "stroke." Unintended release of material rich in solvent mixture and soil into the cleaning spray of emulsion (Figure 3.3) will convert partially cleaned parts into re-soiled parts. Unintended release of material rich in water through the distillation column into the immersion cleaning vessel will reduce the concentration of solvent mixture and so the quality of cleaning. And without continuous monitoring of the soil levels in the emulsion and immersion vessels, one doesn't know that a semi-aqueous cleaning machine has suffered a "stroke" until too late.

[4] This was a hard-earned lesson for developers of the semi-aqueous process. Unless soil materials are familiar in nature and quantity, laboratory tests are essential to commercial success with the semi-aqueous process.

[5] An emulsion is a mixture of two (or more) immiscible fluids. Emulsions are unstable, do not form spontaneously, and their formation is aided by addition of mechanical energy to the fluid mixture. Over a time span of minutes to hours, emulsions generally tend to revert to the stable state of the phases comprising them.

[6] In one low-cost variant of semi-aqueous technology, the parts are cleaned with an emulsion (½ to 2%) of solvating agent (SA) in water followed by water rinses (RA).

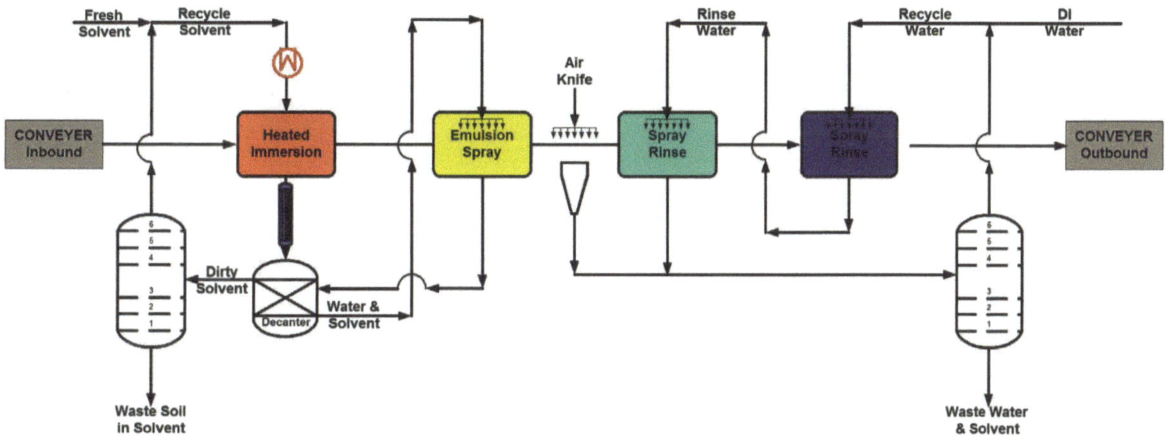

Figure 3.3 Equipment Diagram for a Semi-aqueous Process Used to Clean Flux from PWBs

mixture emulsified in water can become a superior cleaning agent[7] because the emulsion is a weak one (making the solvent available) and the bulk water dilutes the soil.

- Offers the opportunity to recover the water (RA) from the dilute rinse fluid. A distillation column is shown for this purpose, but because of its high energy costs, reverse osmosis membrane systems have often been used for the same purpose.

3.5.2 Cleaning PWBs with the Semi-Aqueous Process

The semi-aqueous process, when used to clean flux (organic acids) and flux residues from PWBs necessarily becomes even more complex (Figure 3.3).

The reason for this complexity, and additional equipment cost, is that the PWBs must be as completely free of residue as possible, including the SA, acidic flux or residue (the soils), water (the RA), or anionic or cationic ionic material (impurities in the RA).

The process of Figure 3.3 adds the following features to the basic semi-aqueous cleaning technology:

- The SA and the RA are engaged in a circular relationship: (1) a weak oil-in-water (Box 3.2) emulsion of solvent mixture (SA) in water (RA) is formed, (2) the emulsion is used as a second or "polishing" spray cleaning step, and (3) the emulsion is recycled back[8] to the decanter to recover the component values of the solvent mixture and remove the soil in the distillation column.
- A high-velocity spray of air, from air knives, is blown across the PWBs after the emulsion cleaning stage in

> **Box 3.2 The Basics of Emulsions**
>
> There is a general form in which an emulsion exists – one phase is dispersed within another. The former is called the dispersed phase; the latter is called the continuous phase. Emulsions produced within the semi-aqueous process are of oil (soil) dispersed into water. The mechanical energy applied to the fluids in the semi-aqueous process is that of mixing the fluids together with spray nozzles; in food preparation the mechanical energy applied to olive oil and vinegar from a moving whisk produces a familiar salad dressing.
>
> Often an emulsion is formed, or made more stable, by addition of a third fluid. Called the emulsifier, in semi-aqueous cleaning it is the surfactant added to make the solvent (SA) and water (RA) form a single fluid temporarily in the immersion chamber. In general, the Bancroft rule applies – the continuous phase (the water) is the one in which the emulsifier (the surfactant) is most soluble.

[7] The emulsion can become something else – a foam. If the spray nozzles are not properly aimed (and sized) so that they act directly on collections of liquid and not on part surfaces, foam can be generated. Foam is a fearsome enemy in all cleaning systems. Foam is air and chemicals. When foam breaks in the rinse vessel by the air being removed, the cleaning chemicals remain on part surfaces as chemical-rich residues. Usually, when foam is noticed within the semi-aqueous process, the operating machine must be shut down and the system flushed with fresh water – else soiled parts will continue to be produced.

[8] The rate of emulsion draw off to the decanter, ie. the spray rate of emulsion over the PWBs, controls the equilibrium concentration of cleaning agent (SA) in the emulsion.

Some soil components may be water soluble. In that case, if water-soluble soils build up in the recycled water, a small stream may be purged to waste treatment.

The liquid level in the emulsion stage is maintained by feeding either fresh (not DI) water or water from the first cascade rinse.

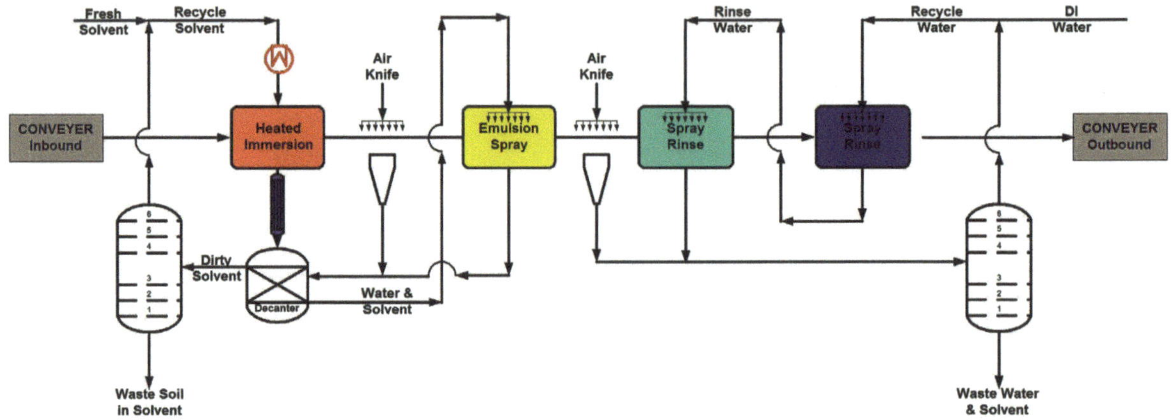

Figure 3.4 Equipment Diagram for a Semi-aqueous Process Used to Clean Flux from PWBs, Using Two Blowoff Stations

order to remove a controllable (Chapter 3.5.3) amount of liquid dragout.

- Multiple stages of rinsing with water (RA) to dilute dragout (dilution rinsing). Note how the final contact of fluid with the PWBs is with deionized (DI, or ion-free) water. As with vapor degreasing, this step is practiced so that the final fluid contact with the parts is with the cleanest possible fluid[9].

Note also how the version of the semi-aqueous process used to clean PWBs is parted in half where the air knife separates the cleaning operations from the rinsing operations. Upstream of this blow-off station, the solvent-water mixture (SA) is used for both immersion and spray cleaning; downstream of the blow-off station, the RA is used to dilute dragout containing SA and dissolve non-volatile residues.

Not shown in the above figures is any method by which parts are dried. This is because no specific method of drying is inherent to the semi-aqueous process. Yes, the liquid removed by drying is water. But it may be partially removed by impact with high-velocity air, nearly completely removed by evaporation, or almost totally removed by displacement with hydrophobic surfactants (Chapter 3.5.5).

Semi-aqueous cleaning technology is an excellent choice where: (1) the solvent mixture is matched to soil components, (2) high levels of removal of small amounts of soils are required, (3) soil materials are not changed, and (4) the site is populated by staff educated, trained, and experienced in reproducibly operating complex chemical processes[10].

These conditions are found in the application of cleaning PWBs.

3.5.3 Don't Blow-Off Process Control

In some embodiments of the semi-aqueous process, the high-velocity air spray chamber is located after parts leave the immersion tank but before they enter the emulsion spray vessel.

In other embodiments, it is located as shown in Figure 3.3 – after the emulsion spray rinse. And in still other embodiments, in both places (Figure 3.4); and in others, neither (Figure 3.2).

Surprisingly, the air knife apparatus (as shown in Figure 3.5[L]) can act as a "chemical control" device, by adjustment of the operating pressure to remove more or less water. One might want do that because the blow-off effect of the air knife can be controlled – both in advance of operation, and during operation.

Moving the air knife closer to and further away from the target part surface increases and decreases, respectively, the mechanical impact upon the target.

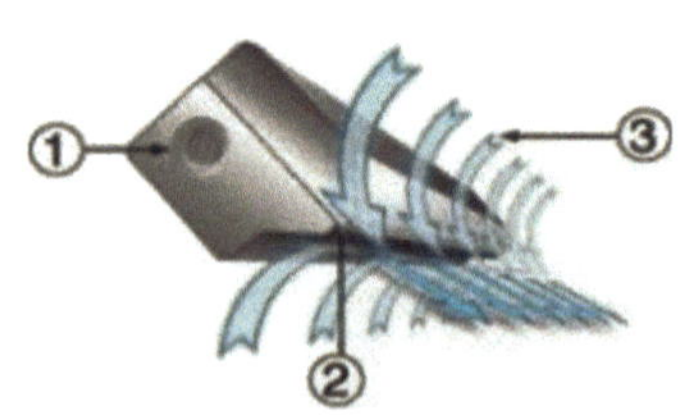

Figure 3.5 Air Knife Powered by Compressed Air

[9] The DI water used for final rinsing is also the makeup water for the overall process, replacing that lost by evaporation and with solvent.

[10] Semi-aqueous cleaning systems are not forgiving of mediocre skills of process control.

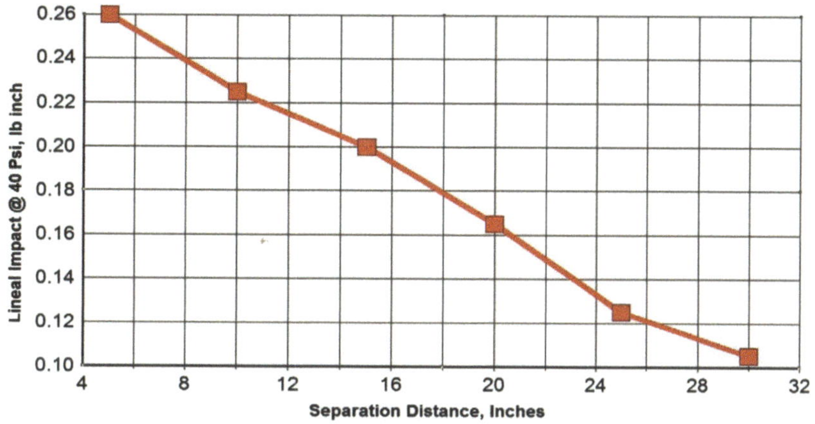

Figure 3.6 Adjustable Mechanical Impact on PWBs by Movement of the Air Nozzle

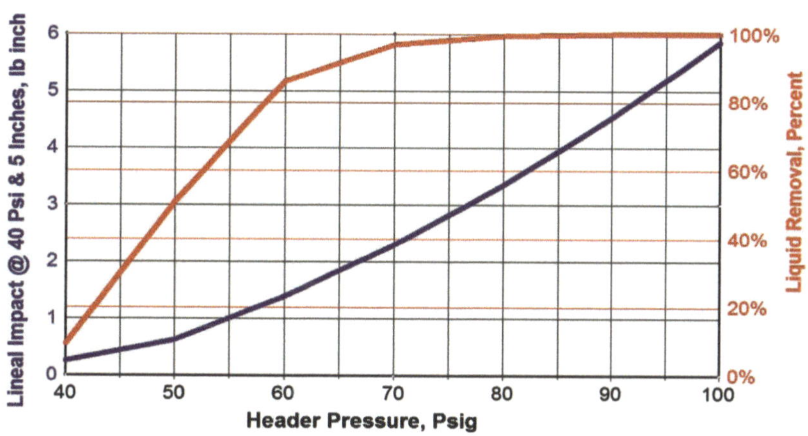

Figure 3.7 Substantial Adjustable Mechanical Impact on PWBs by Change of Header Air Pressure

Data[M] plotted in Figure 3.6[11] show that the position of the air knife relative to the target substrate has a general effect which can be controlled prior to and usually not during operation. But, this effect is not one that could be used in process control because it has only a modest gain[12] (response/stimulus).

Inlet air pressure (and implicitly discharge air flow rate) to the air knife, however, can be adjusted on-line. That adjustment can be part of a process control scheme because it provides a much more substantial gain.

Consider Figure 3.7 which shows: (1) the huge extent to which mechanical impact upon a target surface can be increased by increasing the supply pressure to the air header (blue curve), and (2) how mechanical impact can "sweep" essentially all liquid dragout from surfaces (red curve). The latter is a high-gain opportunity – by increasing header pressure from 40 to 60 psig (27.5 to 41.4 N/cm^2).

So, increased mechanical impact, from about 0.2 in-lb to more than 6 inch-lb, removes substantially more liquid (dragout) from the PWBs after they exit the immersion vessel, but before they enter the emulsion spray vessel. This liquid is returned, with the soil it contains, to the decanter.

Soil belongs in the decanter. There it fractionates so that the vast majority of it is fed to the distillation column within the solvent mixture (Figure 3.4), where it is concentrated for disposal and the solvent mixture is recovered.

[11] The numeric labels in Figure 3.6 represent the following capabilities: (1) compressed air at pressures between ~40 and ~100 psig enters a manifold or distribution chamber which directs air flow to a slotted orifice whose aperture has been precisely formed, (2) a uniform sheet of high velocity air flow is directed outward along the full length of the air knife, and (3) the air moving at a high velocity entrains ambient air at a ratio of 30/1 to 50/1 producing a well-defined sheet of air moving in laminar flow and impacting a surface with significant force. Essentially, an air knife transforms the pressure energy of compressed air to kinetic energy of motion.

[12] Roughly, the gain reported in Figure 3.6 is to only double the low level of impact when the air knife is relocated by a huge distance of 2 feet closer to the target surface.

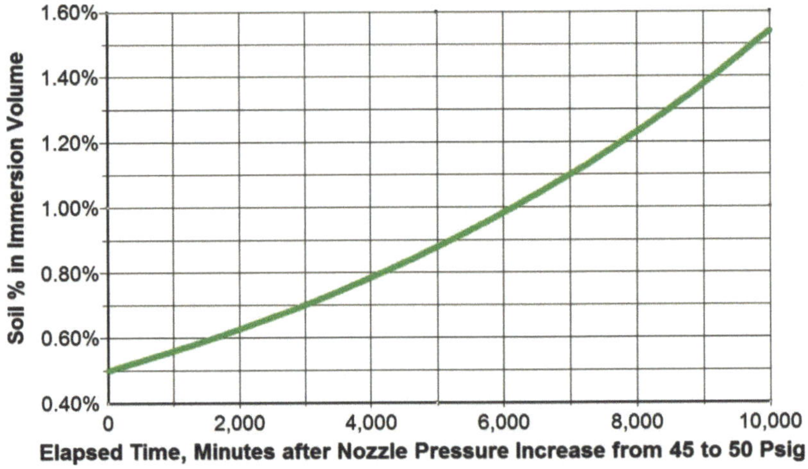

Figure 3.8 Example of Deterioration Over Time - Soil Buildup in Solvent with Increase of Blowoff Effect

But some soil in the decanter always fractionates into the aqueous phase with the water-soluble or emulsifiable portions of the solvent mixture. That soil is fed to the emulsion rinse vessel. There it can infect the supposedly cleaned parts. An increased amount of soil found in any vessel downstream of the primary cleaning vessel is never a good idea, in any cleaning process.

The calculated[13] effect of a modest increase of nozzle pressure is shown in Figure 3.8[N]. The plotted effect is on the concentration of soil in the water/solvent mixture used for the emulsion rinse. Note that after about one week (10,080 minutes), a probably undetected increase of nozzle pressure from 45 to 50 psig has tripled the soil concentration in water fed to the emulsion rinse vessel.

The point of Figure 3.8 isn't that there is soil in the rinse fluid (there will always be some). The simple point is that the concentration of it substantially increases during a modest time period due specifically to a tiny (and probably unmonitored) change in the feed pressure of the air blow-off system. Operators of semi-aqueous processes find it essential to manage the blow-off system with care and concern.

- But, the broader point of this subchapter is that the semi-aqueous process is metastable. Cleaning effectiveness can be excellent; reproducibility and repeatability of that outcome can be acceptable, or unacceptable.

Fortunately, process control technology is simple, mature, available[O], and organized to improve the reproducibility and repeatability of operations. With it, the semi-aqueous process can be rendered stable. There are two keys for success: (1) identify *all* the appropriate aim points to validate correct process operation, and (2) always use appropriate process control technology to audit operation is on-aim.

3.5.4 Equipment Used in Semi-Aqueous Cleaning

There is nothing uncommon about the equipment components used in the semi-aqueous process. They are the pumps, filters, tanks with cove corners, spray nozzles, etc., noted as "Best" in Chapter 7[P] of Ref. 1. The image of Figure 3.9[Q] shows one such apparatus in which that process is practiced.

Figure 3.9 Semi-aqueous Multistage Console

[13] The calculation is based on the information plotted in Figures 3.6 and 3.7, supplemented by the following assumptions: a small system with an immersion vessel volume of 12 gallons, a large decanter volume of 5 gallons; the decanter is pumped to the distillation column or the emulsion rinse vessel sufficient to maintain an equal volume of each phase; initial soil levels in the decanter and rinse are 8 and 0.5 wt%, respectively; the parts are two-sided PWBs, each 4" x 6", containing 75% of additional surface area within electrical components, and an average soil level of 10 mils of unspecified soil material. A batch of 100 boards is cleaned every 10 minutes.

Figure 3.10 Cutaway View of Multi-stage Semi-aqueous Cleaning Machine

After all:

- The immersion tank is a tank in which aqueous or solvent immersion cleaning can be done,
- The vessel in which emulsion is generated and sprayed is an open tank (or volume) with pressurized spray nozzles, such as found in an aqueous or solvent cleaning line,
- The vessel(s) in which spray rinsing is done are also open tanks with pressurized spray nozzles, such as found in aqueous or solvent cleaning lines,
- The semi-aqueous decanter is a more precisely controlled tank in which immiscible fluids are reproducibly separated than is the decanter found in an aqueous or solvent cleaning line (Figure 2.41), and
- A hoist/transport is a hoist/transport for lifting and moving baskets of parts, such as found in aqueous and solvent cleaning lines.

Said another way, semi-aqueous cleaning apparatus are not the most expensive facilities for cleaning parts, but they are certainly not inexpensive. Automation, necessary to practice the semi-aqueous process, is a crucial component which probably at least doubles the purchase price of new equipment.

Facilities to practice the semi-aqueous process would not be purchased on the internet as a commodity machine, without specific laboratory testing. They would be purchased in a highly competitive environment because applications are highly customized.

Another example of equipment is shown in cutaway view in Figure 3.10[14] – an apparatus marketed as an "optimum console," capable of being customized for either aqueous or semi-aqueous service.

Key capabilities needed to practice the semi-aqueous process not normally present in the equipment found in aqueous cleaning lines are to:

- Separate wastes and recover solvent mixtures at a very high level of efficiency (and reproducibility) to conserve expensive chemicals,
- Treat rinse water in large volumes because the semi-aqueous process is practiced in applications where large

[14] The apparatus of Figure 3.10 is a cutaway view courtesy of Crest Ultrasonic (http://www.crest-ultrasonics.com/). Facilities found in both aqueous and semi-aqueous cleaning apparatus include:

1. Custom fixtured baskets.
2. Automated load stations, which sequence baskets through the process.
3. Power supplies for ultrasonic transducers.
4. Immersion wash, enhanced by agitation generated by ultrasonic transducers.
5. Multiple stages of spray or immersion rinse where immersion rinse is enhanced by agitation generated by ultrasonic transducers.
6. Final spray or immersion rinse with DI water.
7. Hot air dryer with up-flow design (to avoid reinfecting drying surfaces with water droplets).
8. HEPA (high-efficiency particulate air) filter to remove particles and oil from pressurized air.
9. Electrical connections, accessible at end of machine rather than under or within machine.
10. Sliding dryer cover to both contain heated air, and make parts available to transport.
11. Operator controls.

volumes of soiled rinse water are generated because of a need to produce nearly pristine surfaces, and

- Manage measurable operating conditions (temperatures[15], interface positions, liquid flow rates, soil concentrations[16], solvent concentrations, water conductivity and hardness, air pressures, time of parts basket(s) in various positions (holdup times), overall cycle time, and of course, cleanliness) as well as non-measurable (normally) operating conditions (positions and numbers of parts in racks, soil composition, amount and distribution, and unanticipated upstream or downstream events) as compared to the point of aim for each condition.

The semi-aqueous process also forces the consideration of two management issues not associated with aqueous cleaning technology:

- Although the base solvents found in semi-aqueous solvent mixtures are normally of low volatility, flammability of aerosols[R] is an issue. Solvent-air aerosols are created by expansion of emulsion through pressurized nozzles into air. Consequently, the emulsion vessel must be managed to safely contain flammable materials. The presence of water does not innoculate against this hazard[S].
- Water quality of the final rinse fluid (DI water) must be managed. This fluid is usually produced on-site from facilities which might include a distillation column, or a reverse-osmosis membrane system, or an ion-exchange column.

3.5.5 Solvent Mixtures Used in Semi-Aqueous Cleaning Processes

Semi-aqueous cleaners usually are formulated for very specific applications. No user considering cleaning with the semi-aqueous process should adopt it without a legitimate demonstration of operation of the solvent mixture/water decanter without the actual soil components present. Of secondary but significant importance is a demonstration of the immersion and spray cleaning operations with actual soil components.

Examples of these formulations are listed in Table 3.1a.

One can't help but notice a characteristic absent in these formulations. They are water-based, but there are no alkaline components – as there are with aqueous cleaning technology. So, corrosion of parts or machine components is not an issue, as it is with aqueous cleaning technology.

3.5.6 Summary of Semi-Aqueous Applications

The semi-aqueous cleaning process *could* be named the nearly aqueous process. Yes, it *is* a solvent-based process, being based on the solution of soils by solvents. But, the purchased mixture of solvents (and detergents) is always diluted 1/10 or more with water. The fluids being managed within the process are the water/solvent mixture, or a dilute emulsion of the solvent mixture in water, or DI water, or rinse waters, or waste water; *or* a concentrate of soils in the solvent mixture.

The real advantage of semi-aqueous processes is that they provide acceptable-to-excellent cleaning for heavy oils, greases, flux and flux residues, waxes, and tars[17] – especially compared to aqueous[18] cleaning technology. Yes, the cleaning quality with volatile halogenated solvents is at least equivalent. But those solvents raise concerns about SHE issues (Ref. 3, Chapter 20), and solvents used in the semi-aqueous process do not.

The real disadvantage of the semi-aqueous processes is that it requires a superior level of willingness and capability to manage chemical processes on site.

There are other issues of less moment that raise both positive and negative concerns.

- The consumption of solvent is less than that of a vapor degreasing process, and the consumption of water is less than that of an aqueous cleaning process.
- The process and equipment must be engineered to constrain flammable aerosols, dry relatively non-volatile water from parts, provide water of DI quality, not require excessive floor space, dispose of both water and solvent wastes, and *especially* integrate significant instrumentation and control elements.

[15] When removing high-melting soils as waxes, one must implement a pre-wash step. This can be done in several ways: by dipping the parts again in the wax tank (containing pure wax) to dissolve build-up, by dipping parts in hot water (180°F (82°C)), or by melting in an oven. The purpose of this pre-wash is to remove as much wax as possible before the solvent-based cleaning is started, by reducing wax viscosity via raising its temperature.

[16] Soil concentration in the immersion cleaning bath and emulsion rinse vessel can be a significant issue. This is because the soil removal capability of a solvent mixture and a surfactant with semi-aqueous technology is considerably greater than when only a surfactant is employed with conventional aqueous technology. So soil concentration is 5 to 10 wt% vs. only ½ to 1 wt%. Hence, management of rinse quality is essential to success. That's why on-line direct or indirect measurements of soil concentration are (or should be) commonly done.

[17] Note that these soils, and the main group of cleaning agents (SAs) listed in Table 3.1a or Table 3.1b, are similar in that neither group displays high levels of polar or hydrogen-bonding intermolecular force. In other words, these soils are cleaned with solvents (SAs) which are compatible with them.

[18] The basic reason for this relative level of performance is the considerably higher ratio of solvent (SA) to soil used in semi-aqueous technology vs. the same for conventional aqueous cleaning technology. In the former, the solvent mixture in water can contain as much as 10 wt% soil; in the latter, the loading of soil in water is perhaps ½ to wt% soil.

Table 3.1a Types of Semi-Aqueous Cleaning Agents (SAs)

Base Solvent	Components other than Water (Chiefly Acting as Cosolvents or Surfactants)	Application
Aliphatic (non-aromatic) linear and cyclic hydrocarbons	Tergitol 15-S-15 (surfactant) $[C_{12-14}H_{25-29}O[CH_2CH_2O]_xH]$	Metal cleaning
	DBE[T] (cosolvent)	Oils, greases, and high-melting waxes (including carnauba) from metal, glass, and plastic substrates
Aliphatic (non-aromatic) linear and cyclic hydrocarbons (though this cleaner is touted as being "non-petroleum")	Diethylene glycol butyl ether[U] (cosolvent)	Engine degreasing
D-Limonene (in order to conceal identity, also referred to as (R)-4-isopropenyl-1-methylcyclohexene; i.e., D-limonene)	Methyl dodecanoate (cosolvent)	Cerita wax, Cosmoline, greases, high melt point soils, lubricants, oils, Rigidax wax, Rosswax, Tolber wax, fixturing wax in optics and wafer industries
Orange terpenes (in order to conceal identity, also referred to as (R)-4-isopropenyl-1-methylcyclohexene; i.e., D-limonene)	Nonylphenoxypolyethoxyethanol (surfactant)	Organic soils such as tar, asphalt and bitumen
	Ethylene glycol butyl ether, sodium dodecyl benzene sulfonate (surfactants) Buffers and builders	All types of oil, wax and grease contamination
	Methyl laurate (surfactant)	Rosin, no clean and lead-free solder flux residues
D-Limonene (only)	Ethoxylated alcohols, coconut diethanolamide (surfactants)	Flux residues on PWBs
N-Methyl-2-Pyrrolidone	Alcohols, glycol ethers, ketones, and aromatic/chlorinated hydrocarbons (surfactants)	Photoresist stripper, defluxing, graffiti remover

- The consumption of operating cost, investment, and energy is more than that of a solvent or aqueous cleaning process used for the same purpose.
- Semi-aqueous solvents are VOCs, and probably have odors which may be objectionable.

3.6 COSOLVENT PROCESS DESCRIPTION: CLASS II – IMMISCIBLE

Class II is probably the most commonly applied variant of all of the four cosolvent technologies.

In Class II cosolvent machines, the SA and the RA are immiscible with one another.

3.6.1 Immiscibility Is Not New

This relationship among solvents has been long used to dry (and rinse) water from parts in what was formerly called a "kerosene dryer," whereby water-wet surfaces would be washed with kerosene. Since they are mutually immiscible (Figure 3.27), the water was displaced from the parts by the kerosene. When the kerosene evaporated the parts would be left dry of water. Obviously, this technology predated recognition of smog-forming chemicals such as VOCs.

- Essentially, this is how the rinsing (Chapter 2.21) and drying sections of any Class II cosolvent machine work, except that kerosene and water[V] are not involved.

So the rinsing mechanism in all Class II cosolvent processes is displacement – because the two fluids are immiscible. A significant difference in density ($\geq$0.15 to 0.25 g/cc) between the two fluids affords the best rinsing of the liquid solvating agent (SA) by the liquid rinsing agent (RA). In this case, the combination of surface forces due to immiscibility and gravity forces due to density difference makes the separation between phases on the parts positive and lasting.

3.6.2 The Basis of a Subclassification

Class II cosolvent machines can have SA or RA materials which are either easily ignited by a spark in oxygen, or not. That distinction is the basis for an important subclassification of Class II cosolvent processes:

- If the immiscible SA or RA has a flammability classification by OSHA/DOT[W] of Flammable or Combustible Class II, the cosolvent process using those solvents is known as Class II A.
- If the immiscible SA *and* RA have the favorable flammability classification of Combustible Class IIIA (or Ignitable/NOS), the cosolvent process using those solvents is known as Class II B.

More specifically, this distinction:

- For the *Class II A* cosolvent processes, is earned in the US[X] by *either* of the chosen SA or RA having a flash point less than 140°F. Technically, that distinction forces the above flammability rating by US OSHA/DOT[Y], and a waste classification by the US EPA of hazardous.
- For the *Class II B* cosolvent processes, is earned by *both* of the chosen SA and RA having a flash point greater than 140°F. Technically, that distinction forces the above flammability rating by US OSHA/DOT, as well as the absence of a waste classification by the US EPA of hazardous.

Practically, for Class II A cosolvent processes that distinction means: (1) equipment design must account for the normal presence of a flammable liquid[Z], and (2) all solvent wastes must be classified as hazardous[AA]. For Class II B, neither is true.

In all Class II cosolvent machines, the SA is often a VOC (in the US), and the RA can be VOC exempt. When the latter is so, the volatile VOC-exempt RA can act (because of immiscibility) as a barrier to keep the VOC SA from being emitted[BB]. However, where either (or both) an SA or an RA cosolvent which are also VOC exempt (anywhere) can be chosen for optimum performance, that choice is normally[19] made (Chapter 3.6.3).

[19] Fluorinated VOC-exempt (in the US) solvents, HFCs, HFEs, HFOs, etc., are priced one-to-two decades higher than are non-fluorinated VOC-exempt halogenated solvents, and oxygen-bearing solvents such as acetates or acetone, which are also VOC-exempt. VOC-exempt OS solvents are priced closer to VOC exempt fluorinated solvents. In other words, there isn't always a premium commanded for the VOC exemption.

But, where a premium is commanded, this author's opinion is that the price is justified considering that the former also don't raise concerns about flammability and the latter do.

3.6.2.1 Class II A (Flash Point <140°F, in the US)

A user chooses Class II A cosolvent machines for one of two reasons: (1) the cosolvent (SA) whose HSPs best match the HSPs of the soil mixture has a flash point <140°F and there is no SA cosolvent with similar HSPs which has a flash point >140°F, or (2) the user is required by specification[CC] to use an SA which happens to have a flash point <140°F.

- No user rationally chooses to use an SA or an RA which has a flash point <140°F. That choice is made out of necessity because the solvent "works" or is "specified." Said another way, the fact that users choose to use the Class II A cosolvent process is a credible endorsement[20] of the design, functionality, and inherent safety of these solvent cleaning machines.

3.6.2.2 Class II B (Flash Point >140°F, in the US)

A prudent user would choose Class II B cosolvent machines vs. Class II A in every case where performance and specification allows.

3.6.3 Method for Selection of Class II A or Class II B Cosolvent Processes

Consistent with guidance found in other chapters of this book, the following approach to selection of Class II cosolvents (by far the most common) is recommended, and implemented below:

A. Obtain a database of properties of cleaning/rinsing/drying solvents of interest. Properties for each solvent should include: the three HSPs, flash point, density, boiling point, surface tension, and any outstanding SHE concerns[DD].
B. Identify[EE], to the extent possible the three HSP values which most closely represent the soil composite.
C. Calculate the R_A value (separation distance[FF]) between each SA cosolvent of interest and the soil mixture. Use Equation 2.8 from Ref. 2 as noted in Endnote FF. Sort these cosolvents into a list based on the calculated values of R_A, with the lowest ranked first. For each SA cosolvent, note also the flash point, density, and surface tension.
D. Using the three items noted above, select the preferred SA, based on (in order): (1) lowest value of R_A[21],

[20] There is a substantial business in producing Class II A cosolvent cleaning machines capable of safely cleaning with isopropanol, and to a lesser extent, ethanol (Ref. 3, Chapter 12). The investment required for such machines is justified by the high value of the cleaned objects.

[21] It is noted in Ref. 2, Chapter 3.28 that solvents with R_A values >8 $MPa^{1/2}$ should be avoided. Such a condition suggests that the soil and solvent do not adequately share values of intermolecular force for there to be good cleaning by dissolving soil into solvent.

(2) highest flash point or absence of same, (3) lowest density[22], (4) highest boiling point[23], and (5) lowest surface tension. The first two items are of the highest priority; the third and fourth, of significance; the fifth, not so much.

An example of such a sorted list is Table 3.1b with the soil being assumed to be cottonseed oil.

Selection of the SA cosolvent for cottonseed oil soil from Table 3.1b is straightforward (given the above priorities), but some compromise and multiple opinions are normal. In this case, tripropylene glycol n-butyl ether (TGBE) is a useful choice. There are others[24]. From Table 3.1b:

- The R_A for this SA cosolvent is not the lowest at 7.69 $MPa^{1/2}$, but is within the goal limits of 8.0. Actually, no SA cosolvent is an excellent match for cottonseed oil soil.
- The SA cosolvent does have a flash point, but its value is well above any value of concern.
- The density value of the SA cosolvent is below unity, meaning that it is likely that an RA cosolvent can be found with a density value of at least 0.15 g/cc higher.
- The volatility of the SA cosolvent is very low, with a boiling point of 530°F (277°C).
- The surface tension of the SA cosolvent is relatively high, at about 30 dynes/cm. While not determinant of dissatisfaction with cleaning quality, no other suitable choice of SA cosolvent has a significantly lesser value.

Selection of the preferred RA cosolvent is not done on an absolute basis. It is done *relative to the SA cosolvent selected* to clean cottonseed oil, tripropylene glycol n-butyl ether (TGBE).

- Calculate the R_A value (separation distance) between each RA cosolvent of interest *and the previously chosen SA* – also using Equation 2.8. Sort these RA cosolvents into a list based on the calculated values of R_A, with the *highest*[25] ranked first. For each RA cosolvent, note also the flash point, density, and surface tension. An example of such a sorted list is Table 3.2, which has validity only relative to the choice of SA cosolvent in Table 3.1b.

Selection of the RA cosolvent, for cottonseed oil soil, from Table 3.2 is not straightforward (given the above priorities)[26]. It comes with risk. HFC-43-10mee (Ref. 3, Chapters 22 and 26) is this author's choice – chiefly because it is VOC exempt (in the US).

- Of greatest and significant risk is that its R_A, relative to TGBE, is quite[27] low at ~10.1 $MPa^{1/2}$. This value is at (or below) the lowest considered for efficient and effective separation between immiscible fluids in the rinse sump[GG]. So this risk produces the concern (or likelihood) that rinsing of the SA will be extremely poor.
- HFC-43-10mee has no flash point which mandates that its OSHA/DOT flammability classification be NOS.
- The density value of HFC-43-10mee is above 1.5 g/cc meaning that it is very likely to separate effectively from the chosen SA cosolvent in the rinse sump[28] – even though the level of immiscibility with TGBE is not high[29].
- The volatility of HFC-43-10mee is high, with a boiling point value of 130°F (54°C) – considerably less than that of water. That high volatility means: (1) TGBE is separated very adequately from HFC-43-10mee in the vapor zone above the sumps so that the calculated concentration of the SA cosolvent in the RA cosolvent vapor is less than 0.1 volume percent, and (2) that condensation and recovery within the cooling coils should be excellent[30].

---

[22] The preferred SA cosolvent will have a density value adequately lower than the preferred RA cosolvent so that it will float in the rinse sump, and the RA cosolvent will sink – quickly and completely.

[23] The preferred SA cosolvent will not be very volatile, having a high boiling point. One wants little SA material vaporized into the RA cosolvent above the rinse sump. The vapor above the rinse sump will ultimately be condensed as the parts are dried when lifted above the cooling coils. Retained easily condensed SA will be non-volatile residue on the part surfaces.

[24] The choice of TGBE is made because this SA cosolvent is commonly available. It is used as a component of hydraulic brake fluid, as a solvent in paint stripping formulations, and as a dye carrier for textile dye processes. TGBE is a low-volume commodity solvent.

Useful choices other than TGBE for the SA cosolvent are also available. Three are similar molecules: propylene glycol ethyl propyl ether (PGEPE), propylene glycol methyl propyl ether (PGMPE), and tripropylene glycol monomethyl ether (TGEE, Ref. 3, Chapter 14.3.2). Another is Texanol (Ref. 3, Chapter 12, Table 12.2). The RA cosolvent best matched to each of these four SA cosolvents is HFC-43-10mee. All four choices would be components of a Class II B cosolvent process (because neither the SA nor the RA are flammable).

[25] The highest value of R_A, relative to the chosen SA cosolvent, ensures the greatest immiscibility between the chosen SA and the to-be-chosen (from Table 3.2) RA cosolvent. For a Class II cosolvent process, one wants the chosen SA and RA cosolvents to most efficiently and effectively separate from one another in the rinse sump – representing the highest value of R_A between the SA and the RA.

[26] Selection of SA and RA cosolvents is normally straightforward. Cottonseed oil soil was chosen for illustration of situations where some compromise must be made.

[27] There is little direct evidence for judging what level of R_A is necessary for the RA cosolvent to be separated from the SA cosolvent. One estimate is about 10 to 11 $MPa^{1/2}$. This is based on generally successful commercial experience of using Soygold 1000 as the SA and HFE-7200 as the RA cosolvent. Obviously, higher values will produce greater immiscibility which is desirable.

[28] The density separation from the information in Tables 3.1 and 3.2 is 0.65 g/cc (1.580 – 0.934 = 0.646), which far exceeds a typical minimum value for use in decanters in the semi-aqueous cosolvent process.

[29] This author would make another choice of RA cosolvent were the choice "possible." Today, no one in their right mind would choose carbon tetrachloride as a rinsing aid (solvent). Formerly used as a dry cleaning solvent, a pesticide, and a refrigerant, carbon tetrachloride was banned in 1970 within the US for use in consumer products. Both an ozone-depleting and a global warming compound, exposure to "carbon tet" at various concentrations for various times affects the central nervous system, degenerates the liver and kidneys, causes cancer, and will react under temperature to produce phosgene.

[30] That's the good news. The bad news is that recovery must be excellent because this solvent, among other fluorinated solvents, is highlypriced.

Table 3.1b Selection of SA Cosolvent for Cottonseed Oil Soil

Possible SA Cosolvents	δd (MPa$^{1/2}$)	δp (MPa$^{1/2}$)	δh (MPa$^{1/2}$)	R_A MPa$^{1/2}$	Flash Point, °F	ρ g/cc	γ dynes/cm	Boiling Point, °F
Stoddard Solvent	16.1	1.6	2.1	5.55	102.2	0.780		318.2
Isobutyl Acetate	15.1	3.7	6.3	6.19	69.8	0.872	23.1	242.6
Methyl Isobutyl Ketone	15.3	6.1	4.1	6.44	60.8	0.801	23.6	243.3
Ethylene Glycol Diethyl Ether (Ethyl Glyme)	15.4	5.4	5.2	6.44	80.6	0.842		249.8
Butyl Glycol Acetate	15.3	7.5	6.8	6.53	183.2	0.972	26.0	377.6
n-Propyl Propionate	15.5	5.6	5.7	6.61	76.0	0.881	24.7	253.4
n-Propyl Acetate	15.3	4.3	7.6	6.64	57.2	0.888	23.8	214.9
Dipropylene Glycol Dimethyl Ether	15.6	7.0	6.5	6.94	149.0	0.902	26.3	347.0
Dibutyl Ether	15.2	3.4	3.2	6.95	77.0	0.768	22.4	284.4
n-Butyl Propionate	15.7	5.5	5.9	7.01	89.6	0.874	25.9	292.5
Tripropylene Glycol n-Propyl Ether	15.0	2.6	8.6	7.04	266.0	0.938	29.3	501.8
Diethylene Glycol Dimethyl Ether (Diglyme)	15.7	6.1	6.5	7.05	134.6	0.945	27.0	323.6
Texanol	15.1	6.1	9.8	7.08	248.0	0.948	28.9	471.2
VM&P Naphtha	15.0	2.7	2.7	7.09	40.3	0.730		273.5
Propylene Glycol Ethyl Ether Acetate	15.6	6.3	7.7	7.10	129.2	0.941		318.2
Diethylene Glycol Diethyl Ether (Ethyl Diglyme)	15.8	5.9	5.6	7.21	194.0	0.908	27.2	372.2
Diethylene Glycol Dibutyl Ether (Butyl Diglyme)	15.8	4.7	4.4	7.41	246.2	0.881	27.0	492.8
OS-30 Decamethyltetrasiloxane	12.2	1.0	0.0	7.48	350.0	0.854	17.3	381.0
Dipropyl Ketone	15.8	7.6	4.9	7.49	120.2	0.800	0.0	293.0
Diethyl Ketone	15.8	7.6	4.7	7.51	55.4	0.814	24.7	215.1
n-Butyl Acetate	15.8	3.7	6.3	7.52	80.6	0.882	24.9	259.7
OS-20 Octamethyltrisiloxane	11.7	1.0	0.0	7.55	84.0	0.820	16.5	307.0
Amyl Acetate	15.8	3.3	6.1	7.63	71.6	0.871	25.2	287.6
Tripropylene Glycol n-Butyl Ether (TGBE)	15.8	5.7	8.5	7.69	491.0	0.934	29.7	528.8
Tripropylene Glycol Monomethyl Ether	15.3	5.5	10.4	7.75	249.8	0.965	29.0	491.8
Triethylene Glycol Dimethyl Ether (Triglyme)	16.0	7.3	6.3	7.77	235.4	0.986	29.4	420.8

Table 3.1b Selection of SA Cosolvent for Cottonseed Oil Soil *(continued)*

Possible SA Cosolvents	δ d	δ p	δ h	R_A	Flash	ρ	γ	Boiling
	$MPa^{1/2}$			$MPa^{1/2}$	Point, °F	g/cc	dynes/cm	Point, °F
2-Ethyl Hexyl Acetate	15.8	2.9	5.1	7.79	159.8	0.872		390.2
Methyl Isoamyl Ketone	16.0	5.7	4.1	7.79	96.8	0.810		293.0
Methyl Propyl Ketone	16.0	7.6	4.7	7.90	44.6	0.809	23.3	216.1
Propylene Glycol Methyl Ether Acetate	15.6	5.6	9.8	7.91	122.0	0.967	28.2	295.7
Propylene Glycol Mono n-Butyl Ether	15.3	6.1	10.8	8.00	113.0	0.870	24.4	303.8
Diisobutyl Ketone	16.0	3.7	4.1	8.06	120.2	0.805	25.5	334.4
Propylene Glycol Monopropyl Ether	15.8	7.0	9.2	8.08	118.4	0.855	25.4	300.2
Diethylene Glycol Monobutyl Ether Acetate	16.0	4.1	8.2	8.16	240.0	0.980	30.0	447.8
Methyl Amyl Ketone	16.2	5.7	4.1	8.18	102.2	0.811	26.1	304.7

Table 3.2 Selection of RA Cosolvent – Relative to Tripropylene Glycol n-Butyl Ether (TGBE) SA Cosolvent

Solvent	δ_d	δ_p	δ_h	R_A	Flash Point, °F	Cosolvent Class II Type	Boiling Point, °F	ρ g/cc	γ, dynes/cm
	$MPa^{1/2}$								
Methanol	14.7	12.3	22.3	15.52	53.6	A	148.5	0.791	22.1
Acetonitrile	15.3	18.0	6.1	12.61	35.6	A	178.9	0.782	28.7
Ethanol	15.8	8.8	19.4	11.39	55.4	A	172.8	0.790	22.0
Carbon Tetrachloride	17.8	0.0	0.6	10.47	None	B	168.8	1.593	26.4
Hexane	14.9	0.0	0.0	10.32	−9.4	A	155.7	0.659	17.9
SBP 5	15.3	0.0	0.0	10.21	−0.4	A	197.6	0.707	
Cyclohexane	16.8	0.0	0.2	10.20	−0.4	A	177.3	0.779	24.7
SBP 80/110	15.5	0.0	0.0	10.18	10.4	A	188.6	0.710	
HFC-43-10mee	14.2	1.1	0.0	10.12	None	B	130.0	1.580	14.1

- The surface tension of HFC-43-10mee is 14.1 dynes/cm, which is certain to aid in wetting soiled surfaces with rinsing cosolvent.

Note that this analysis proposes the choice of a Class II B cosolvent process – involving two solvents classified by US OSHA/DOT as not flammable or combustible.

Suitable ingredients of a Class II A cosolvent process with cottonseed oil soil would be OS-30 (decamethyltetrasiloxane) for the SA cosolvent and dimethyl carbonate for the RA cosolvent[31]. The former SA solvent has a flash point of 350°F; the latter RA solvent has a flash point of 64.4°F – making the cosolvent classification clearly be II A. The chief reason one

[31] The R_A between cottonseed oil and OS-30 is an acceptable value of 7.5 $MPa^{1/2}$ (<8 $MPa^{1/2}$). The R_A between OS-30 and dimethyl carbonate is a superior value of 14 $MPa^{1/2}$ (>~10–11 $MPa^{1/2}$).

would choose this Class II A cosolvent process is that both components are VOC exempt (in the US).

- And that reason may be determinant. The drawback of a flammability classification can be overcome by: (1) well-known and proven equipment capabilities which avoid exposure of flammable solvents to both oxygen and ignition sources, and (2) implementing a somewhat higher level of care and effort, and accepting a somewhat higher level of expense for waste disposal[HH]. But in return, the precious VOC exemption (in the US) for both SA and RA cosolvents means that no emissions have to be permitted legally, controlled, measured, or reported. Managers are paid to make the choice between flammability and exemption.

It is imperative that the selection of RA and SA be done in two sequential steps: first, selection of an SA cosolvent to dissolve the soil; and second, selection of an RA to be immiscible in the rinse sump with the chosen SA cosolvent. The first choice depends upon the nature of the soil to be cleaned; the second choice depends upon the first choice.

One doesn't simply use a Class II A or Class II B cosolvent process on a whim. One makes that choice based on the nature of the expected soil materials. As with the semi-aqueous cosolvent process, if the soil materials are known to have changed, the choice of the two cosolvents should be reviewed. With two-solvent processes, one can gain flexibility but trade simplicity for complexity.

This example with cottonseed oil soil clearly shows that selection of the two cosolvents for any Class II cosolvent process is requiring of experience, prejudice, judgment, and the ability to compromise. The analysis done for Tables 3.1 and 3.2 is based on the data of Appendix A1 and use of a spreadsheet.

3.6.4 Selection of Cosolvents for Various Soil Materials

The method described above[32] was applied for various single chemicals which are used commercially and could be soils or components of soils. The results are noted in two tables: Table 3.3 for Class II A (flammable) and Table 3.4 for Class II B (non-flammable) cosolvent processes.

In both evaluations a strong preference is made for an RA that is VOC exempt (in the US).

Note that even though the SA cosolvents chosen for all soils in Table 3.3 are non-flammable, the chosen RA cosolvents are flammable. Thus the classification of Class II A.

All cosolvents chosen to clean the soils in Table 3.4 either manifest no flash point or one well above values raising concern about flammability (140°F). That's why all cosolvent pairs are classified as II B.

The selection method above was followed for each soil material. The author's stated preference for RA cosolvents was implemented for use of solvents which are VOC exempt (in the US), don't have a measured flash point, and do have a high density. For every soil material in Tables 3.3 and 3.4:

- The calculated difference in R_A between the chosen SA cosolvent and the soil is less than 8 $MPa^{1/2}$ (except for natural rubber where there was no lower value), so solution cleaning by immersion should be effective.
- The calculated difference in R_A between the chosen SA cosolvent and chosen RA cosolvent is greater than 10 to 11 $MPa^{1/2}$, so that the SA cosolvent and the RA cosolvent should be immiscible with one another.
- The RA cosolvent is considerably denser than the SA cosolvent, and the difference in solvent density between the SA cosolvent and RA cosolvent greatly exceeds 0.15 g/cc, so that: (1) the more dense RA cosolvent will sink and the less dense SA cosolvent will float in the rinse sump[33], and (2) separation between them will be very efficient and effective.
- The calculated volumetric concentration of SA cosolvent in the RA vapor is quite low so that NVR after rinsing should be quite low.

One can't fail to observe a soil common to both Tables 3.3 and 3.4. It's sperm oil (a waxy liquid obtained from sperm whales), which is printed near the bottom row of each table with a grey background. It's found in both tables because it can be cleaned with either type of cosolvent process – with flammable materials (Class II A), or with those not flammable (Class II B)[34]. This is frequently the case, and is one reason why the selection process is made with the aid of a spreadsheet – so that various possibilities can be explored.

The general character of the SA and RA cosolvents in Tables 3.3 and 3.4, respectively, is that the former are low-volatility, low-flammability VOC solvents (in the US) with modest levels of polar and hydrogen-bonding intermolecular force (HSPs), and the latter are high volatility, high density solvents with very low levels of polar and hydrogen-bonding intermolecular force[35].

The reason for the preferred values of HSPs for the SA is for it to engage well with soil materials – hence favoring good solution cleaning. The reason for the preferred values of HSPs for the RA is so that it will not engage well with the SA – hence favoring immiscibility between the two cosolvents.

[32] For *each* result in Table 3.3 or 3.4, an equivalent of Table 3.1b and Table 3.2 was prepared, and evaluated.

[33] And be easily decanted from the rinse sump over a weir to the wash sump.

[34] No satisfactory pair of non-flammable cosolvents could be found to clean the soil materials listed in Table 3.3 (Class II A).

[35] It is the latter character which often enables the classification of VOC exempt (in the US).

Table 3.3 Selection of Cosolvent Pairs with Various Soils for Class II A Cosolvent Processes (Immiscible, Flammable)

Soil	Soil δ d	Soil δ p	Soil δ h	SA Cosolvent	Soil vs. SA, R_A	Flash Point °F	ρ g/cc	Mole % SA in Vapor	T boil °F	RA Cosolvent	SA vs. RA, R_A	Flash Point °F	ρ g/cc	T Boil °F	Δ ρ g/cc
	MPa$^{1/2}$														
ASTM Fuel 'B'	15.30	0.40	0.60	OS-30 Decamethyl-tetrasiloxane	6.26	350.0	0.854	0.07%	381.0	Dimethyl Carbonate	13.97	64.4	1.064	194.0	0.210
Auto Transmission Fluid	14.30	0.40	0.60	Isopar L	1.40	145.4	0.770	0.13%	374.0	Dimethyl Carbonate	13.02	64.4	1.064	194.0	0.294
Mil-H-5006 (Petroleum)	14.70	0.80	0.60	Isopar L	1.08	145.4	0.770	0.13%	374.0	Dimethyl Carbonate	13.02	64.4	1.064	194.0	0.294
Mineral Oil (White, Refined)	14.50	0.00	0.00	Isopar L	0.80	145.4	0.770	0.13%	374.0	Dimethyl Carbonate	13.02	64.4	1.064	194.0	0.294
Motor Oil SAE 20W	14.70	0.40	0.40	Isopar L	0.69	145.4	0.770	0.13%	374.0	Dimethyl Carbonate	13.02	64.4	1.064	194.0	0.294
Sperm Oil	14.30	2.12	2.70	Isopar L	3.64	145.4	0.770	0.13%	374.0	Dimethyl Carbonate	13.02	64.4	1.064	194.0	0.294
Silicone Oil	13.80	5.00	1.20	Isopar L	5.59	145.4	0.770	0.13%	374.0	Dimethyl Carbonate	13.02	64.4	1.064	194.0	0.294

Table 3.4 Selection of Cosolvent Pairs Cosolvent Soils for Class II B Cosolvent Processes (Immiscible, NOT Flammable)

Soil	Soil δ d (MPa$^{1/2}$)	Soil δ p (MPa$^{1/2}$)	Soil δ h (MPa$^{1/2}$)	SA Cosolvent	Soil vs. SA, R_A	Flash Point °F	ρ g/cc	Mole % SA in Vapor	T boil °F	RA Cosolvent	SA vs. RA, R_A	Flash Point °F	ρ g/cc	T Boil °F	Δ ρ g/cc
Auto Brake Fluid	15.80	6.10	10.20	Ethylene Glycol Diacetate	1.66	210.2	1.104	0.31%	375.6	HFC-43-10mee	11.18	None	1.580	130.0	0.476
Bitumen	17.82	4.60	4.05	Propylene Glycol Diacetate	6.13	187.0	0.966	0.25%	376.0	HFC-43-10mee	10.79	None	1.580	130.0	0.614
Castor Oil	13.62	6.04	10.46	Diethylene Glycol Monobutyl ether	4.86	172.0	0.954	0.17%	424.4	HFE-7200	12.10	None	1.430	168.8	0.476
Cholesterol	20.40	2.80	9.40	DBE	7.11	212.0	1.092	0.05%	384.8	HFE-8200	11.88	None	1.430	168.8	0.338
Coal Tar Pitch	18.70	7.50	8.90	Propylene Glycol Diacetate	5.54	187.0	0.966	0.25%	376.0	HFC-43-10mee	10.79	None	1.580	130.0	0.614
Cottonseed Oil	12.20	5.77	5.77	Tripropylene Glycol n-Butyl Ether	7.69	491.0	0.934	0.03%	528.8	HFC-43-10mee	10.12	None	1.580	130.0	0.646
Linseed Oil	13.90	3.50	3.70	Propylene Glycol Diacetate	7.26	187.0	0.966	0.25%	376.0	HFC-43-10mee	10.79	None	1.580	130.0	0.614
Mil-H-8446 (Silicate)	14.50	6.10	7.60	Texanol	2.51	248.0	0.948	0.02%	471.2	HFC-43-10mee	11.15	None	1.580	130.0	0.632
Mil-L-7808 (Ester)	14.30	2.90	3.10	Tripropylene Glycol n-Butyl Ether	6.72	491.0	0.934	0.03%	528.8	HFC-43-10mee	10.12	None	1.580	130.0	0.646

Mineral Spirits	15.46	3.26	3.99	Tripropylene Glycol n-Butyl Ether	5.10	491.0	0.934	0.03%	528.8	HFC-43-10mee	10.12	None	1.580	130.0	0.646
Natural Rubber	20.80	1.80	3.60	Isophorone	9.91(!)	194.0	0.921	0.35%	410.9	HFC-43-10mee	10.20	None	1.580	130.0	0.659
Neets Foot Oil	14.30	2.90	3.70	Tripropylene Glycol n-Butyl Ether	6.25	491.0	0.934	0.03%	528.8	HFC-43-10mee	10.12	None	1.580	130.0	0.646
Olive Oil	15.90	1.20	5.40	Ethylene Glycol Diacetate	5.65	210.2	1.104	0.31%	375.6	HFC-43-10mee	11.18	None	1.580	130.0	0.476
Palm Oil	17.70	3.50	3.70	Propylene Glycol Diacetate	6.54	187.0	0.966	0.25%	376.0	HFC-43-10mee	10.79	None	1.580	130.0	0.614
Pine Oil	12.13	5.43	5.80	Tripropylene Glycol n-Butyl Ether	7.80	491.0	0.934	0.03%	528.8	HFC-43-10mee	10.12	None	1.580	130.0	0.646
Sperm Oil	14.30	2.12	2.70	Tripropylene Glycol n-Butyl Ether	7.38	491.0	0.934	0.03%	528.8	HFC-43-10mee	10.12	None	1.580	130.0	0.646
Trioctyl Adipate	16.70	2.00	5.10	Tripropylene Glycol n-Butyl Ether	5.27	491.0	0.934	0.03%	528.8	HFC-43-10mee	10.12	None	1.580	130.0	0.646
Turpentine	16.40	1.40	0.40	Solvesso 150	3.08	147.2	0.903	0.14%	363.2	HFE-8200	10.19	None	1.430	168.8	0.527
VM&P Naphtha	15.03	2.68	2.68	Tripropylene Glycol n-Butyl Ether	6.66	491.0	0.934	0.03%	528.8	HFC-43-10mee	10.12	None	1.580	130.0	0.646

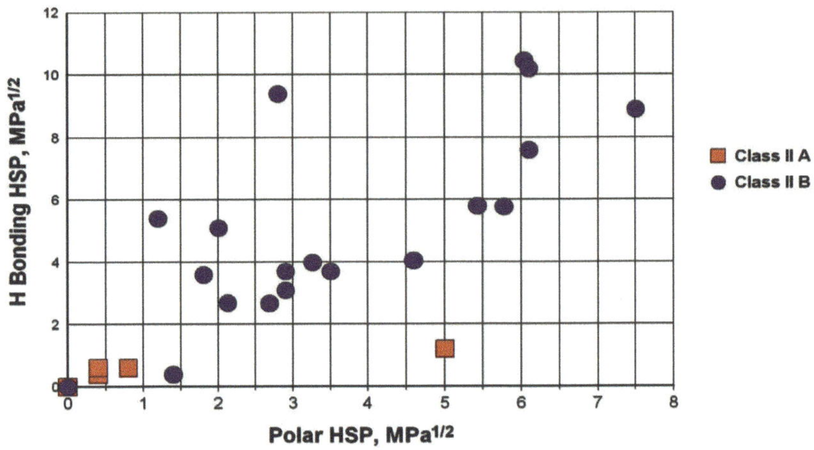

Figure 3.11 A Broad Range of Soil Materials can be Cleaned with Class II Cosolvent Processes

3.6.5 Selection of Cosolvents in General and for Substitution of Solvents

The polar and H-bonding HSP values of the 25 soils included in Tables 3.3 and 3.4 are plotted in Figure 3.11[36].

Two points are easily seen. First, a broad spectrum of soils can be cleaned with the SA and RA cosolvents used Class II cosolvent machines.

Second, soils best cleaned by Class II A (flammable) cosolvent machines generally display low levels of polar and H-bonding intermolecular force. And it is the soils best cleaned by Class II B (non-flammable) cosolvent machines which display high to moderate levels of polar and H-bonding intermolecular force. That segregation is an outcome, not a specification.

- SA cosolvents useful with soils having high levels of polar and hydrogen-bonding intermolecular force must share those characteristics. They are necessary to produce good solvency.
- The associated RA cosolvents will *not* have high levels of polar and hydrogen-bonding intermolecular forces. That's necessary for their SA cosolvent to be immiscible with the RA cosolvent, and enable separation of the SA and RA cosolvents in the rinse sump.

Both points taken together suggest that Class II B cosolvent pairs are useful to replace single solvents about which there are SHE concerns – which display moderate levels of polar and H-bonding intermolecular force – as shown in Table 3.5.

[36] Those soils best cleaned in a Class II A process are plotted with red square symbols. Those soils best cleaned in a Class II B process are plotted with blue circular symbols in Figure 3.11.

3.6.6 The Meaning of Cleaning

Solvent cleaning is valued for at least three characteristics, including the capability to: (1) completely remove a soil from a surface by taking it into solution, (2) rinse a wetted surface with pristine fluid, and (3) dry a wetted surface by rapid evaporation of the fluid wetting it. Item 2 is what's meant by the phrase "vapor degreasing." Item 3 is normally provided in any vapor degreasing process.

None of the three capabilities is normally provided by aqueous cleaning technology.

All of these three capabilities were provided by single solvents whose manufacture is now banned (TCA[JJ], Freon (CFC-113), HCFC-141b), or are provided by single solvents whose use raises significant concern about SHE issues (TCE, PCE, n-PB).

So to where should a user needing *all three* capabilities turn?

There are three options involving solvent technology; all are covered by this author:

- Substitute *another single solvent* for use in an open-top vapor degreaser. This approach is covered in detail in Ref. 2, where two independent approaches are implemented.

  The conventional approach is solvent substitution by matching HSP characteristics, which is covered in Ref. 2, Appendix D1. A forward-looking approach is solvent substitution by identifying solvents which generate less ozone when emitted[37] and also have similar HSP characteristics. This approach is explained

[37] This means to engender lower values of MIR (maximum incremental reactivity).

Table 3.5 Selection of Class II B Cosolvent Pairs to Replace[ll] Single Solvents Having SHE Concerns

Solvent to Be Replaced	Solvent δ d (MPa$^{1/2}$)	δ p	δ h	SA Cosolvent	Replaced Solvent vs. SA Cosolvent, R_A	Flash Point °F	ρ g/cc	Mole % SA in RA Vapor	T boil °F	R_A Cosolvent	SA vs. RA, R_A	Flash Point °F	ρ g/cc	T Boil °F	Δ ρ g/cc
n-Propyl Bromide	16.40	7.90	4.80	Propylene Glycol Diacetate	4.96	187.0	0.966	0.25%	376.0	HFC-43-10 mee	10.80	None	1.580	130.0	0.614
Trichloroethylene	18.00	3.10	5.30	DBE	5.20	212.0	1.092	0.05%	384.8	HFE-8200	11.90	None	1.430	168.8	0.338
Perchloroethylene	18.30	5.70	0.00	Isophorone	6.09	194.0	0.921	0.97%	410.9	HFE-7200	10.30	None	1.430	168.8	0.509
Methylene Chloride	18.20	6.30	6.10	DBE	4.71	212.0	1.092	0.28%	212.0	HFC-43-10 mee	10.80	None	1.580	130.0	0.488
Methyl Ethyl Ketone	16.00	9.00	5.10	DBE-6	2.92	235.4	1.064	0.08%	235.4	HFC-43-10 mee	10.80	None	1.580	130.0	0.516

and applied in Ref. 2, Appendix D2. In addition the relationship between measurable solvent characteristics and solvent structure is related in Ref. 2, Chapter 1.

Three drawbacks to the substitution option are: (1) that the substituted solvent isn't identical to the replaced solvent, and won't perform the same functions in the same ways – compromise in selection is essential, (2) that a suitable substitute may not even exist[KK], and (3) vapor degreasing with the replacement solvent may not be possible because of its high volatility (low boiling point).

- Use a multi-step cleaning process within an *enclosed vacuum machine* (see Chapter 2), and a suitable apparatus for control of "tailpipe" solvent emissions of the (see Chapter 4). This option allows any solvent (including the solvent being considered above for substitution) to be used at any temperature, with exceedingly limited emissions or exposure to workers. All three of the above capabilities are provided. The drawbacks to this option are: (1) increased investment required to purchase such a machine, and (2) increased investment and attention to purchase and operate the necessary[38] emission management system.
- Use a *Class II B cosolvent process* (Chapter 3.6.2.2) with two suitably chosen cosolvents chosen to match the performance of single solvents, (Table 3.5, and Chapter 3.6.7), or matched to the soils to be cleaned (Tables 3.3 and 3.4, Chapter 3.6.5). The drawbacks to this option are: (1) that the base of supporting operating experience is lean – it isn't commonly practiced outside a few industries[39], (2) there is a relatively low temperature limit to which parts can be immersed – it's the boiling point of the RA cosolvent[40], and (3) management of two solvents is more complicated than management of one.

All three options have little or no threat from current or likely future environmental regulations – especially the two options covered in this chapter.

3.6.7 Substitution Options

Information in Table 3.5 offers options to replace single solvents which raise SHE concerns. These are good options where: none of the Class II B SA cosolvents has a flash point which engenders a flammability classification of concern, none of the Class II B RA cosolvents has a measured flash point, all of the Class II B RA cosolvents are VOC exempt (in the US), none of the Class II B SA and RA cosolvents raises concerns about SHE issues, and all of the Class II B SA cosolvents have small R_A values relative to the single solvent under consideration for replacement.

These are not the same options produced by substituting one solvent for another single solvent[LL], or produced by changing one piece of equipment for another – such as described in Chapter 2 where a vacuum chamber allows options not available at ambient pressure.

The options developed in Chapter 3.6.5 and listed in Table 3.5 are based on a change of the cleaning process – within in the capabilities of an open-top vapor degreaser.

Table 3.6 is where the options produced by the three different approaches are compared for five often-used solvents about which there are SHE concerns.

One must note that the specifications for a Class II B SA cosolvent and its companion RA cosolvent replacements are not the same specifications as those for selection of a single solvent as a replacement solvent.

- Specifications in common are similarity of intermolecular forces (low R_A) between soil and SA cosolvent, and the OSHA/DOT flammability classification.

  But –

- Low vapor pressure matters when selecting an SA cosolvent to reduce its concentration in the rinse vapor phase, and high vapor pressure matters when selecting the RA cosolvent to enhance drying rate.

---

[38] The emission management system is necessary because the SHE issues which compromise use of the solvent being replaced must be resolved. Solvent escaping the work chamber will raise those SHE issues again.

[39] Removal of wax from optic surfaces and removal of flux/residues from PWBs are two broad-based applications in which both Class I and Class II cosolvent machines have been used.

In the early 1990s, the Class II B cosolvent process was relatively popular for other applications. Marketed as the AVD (Advanced Vapor Degreasing) process, the SA was a mixture of aliphatic hydrocarbons supplemented by an ester of dibasic acids (Ref. 3, Chapter 19). The RA was either PFC-5060 (perfluorinated hexane), or later HFC-43-10mee. Figure 3.15 is an image of an operating AVD vapor degreaser.

Regulators and customers gave a failing grade to this then-new process, for two reasons: (1) PFCs were known to be highly inert and have a high level of persistence in the stratosphere (becoming global warming or greenhouse gases), and (2) the SA and either RA had too similar levels of intermolecular force (HSP values) to allow efficient and effective rinsing of the SA with the RA. The R_A between the SA and the RA was $< 2\ MPa^{1/2}$! So, the vocal concern expressed by customers about rinsing of the SA by the RA could have been foreseen (looking backwards).

[40] Since the two cosolvents are contained within the same equipment envelope, the temperature of one cosolvent must be the temperature of the other cosolvent. And as long as the RA cosolvent hasn't been completely boiled away, the temperature of the RA can't exceed its normal boiling point. So also cannot the temperature of the SA cosolvent exceed that value.

There is no inherent reason, other than cost, to locate the SA and RA cosolvent sumps into separate equipment envelopes. This arrangement is perfectly feasible, though not produced commercially. It is shown for Class III (miscible) cosolvent machines in Figure 3.33. The patent claims (described in Endnote GG) refer to a Class II cosolvent machine with separate sumps but don't specify whether they must be located in separate equipment envelopes.

Table 3.6 Comparison of Options for Solvent Replacement

	Solvent Substitution	**Machine Substitution**	**Process Substitution**
Item	**Single Solvent Substitution (Ref. 2 as Noted in Endnote KK)**	**Solvent Containment in an Enclosed Vacuum Machine (Chapter 2.1)**	**Substitution of Two Solvents in a Class II B Cosolvent Machine (Chapter 3.6.3)**
Replacement Solvent (SA)	Dipropylene Glycol Methyl Ether	n-Propyl Bromide	Propylene Glycol Diacetate
R_A vs Solvent, $MPa^{½}$	2.49	0	4.96
OSHA/DOT Flammability Classification	Combustible Class IIIA	Flammable Class IB	Combustible Class IIIA
Can Vapor Degrease?	NO	YES (with n-propyl bromide)	YES (with HFC-43-10mee RA)
Drying Effectiveness	Poor	Excellent	Excellent
Process Equipment	Open Tank	Enclosed Vacuum Machine	Modified Vapor Degreaser
Required Investment	1X	4 to 6X	2X
Replacement Solvent (SA)	DF-2000 (replacement dry cleaning fluid)	Trichloroethylene	DBE
R_A vs Solvent, $MPa^{½}$	5.76	0	5.02
OSHA/DOT Flammability Classification	Combustible Class IIIA	Flammable Class IC	NOS
Can Vapor Degrease?	NO	YES (with trichloroethylene)	YES (with HFC-43-10mee RA)
Drying Effectiveness	Poor	Excellent	Excellent
Process Equipment	Open Tank	Enclosed Vacuum Machine	Modified Vapor Degreaser
Required Investment	1X	4 to 6X	2X
Replacement Solvent (SA)	DBE	Methylene Chloride	DBE
R_A vs Solvent, $MPa^{½}$	4.71	0	4.71
OSHA/DOT Flammability Classification	NOS	Flammable Class IA	NOS
Can Vapor Degrease?	NO	YES (with methylene chloride)	YES (with HFC-43-10mee RA)
Drying Effectiveness	Poor	Excellent	Excellent
Process Equipment	Open Tank	Enclosed Vacuum Machine	Modified Vapor Degreaser
Required Investment	1X	4 to 6X	2X

(*Continued*)

Table 3.6 Comparison of Options for Solvent Replacement *(continued)*

	Solvent Substitution	Machine Substitution	Process Substitution
Item	**Single Solvent Substitution (Ref. 2 as Noted in Endnote KK)**	**Solvent Containment in an Enclosed Vacuum Machine (Chapter 2.1)**	**Substitution of Two Solvents in a Class II B Cosolvent Machine (Chapter 3.6.3)**
Replacement Solvent (SA)	DF-2000 (Replacement Dry Cleaning Fluid)	Perchloroethylene	Isophorone
R_A vs Solvent, $MPa^{1/2}$	7.67	0	6.09
OSHA/DOT Flammability Classification	Combustible Class IIIA	Combustible Class II	Combustible Class IIIA
Can Vapor Degrease?	NO	YES (with perchloroethylene)	YES (with HFC-43-10mee RA)
Drying Effectiveness	Poor	Excellent	Excellent
Process Equipment	Open Tank	Enclosed Vacuum Machine	Modified Vapor Degreaser
Required Investment	1X	4 to 6X	2X
Replacement Solvent (SA)	Dipropylene Glycol Monomethyl Ether	Methyl Ethyl Ketone (MEK)	DBE-6
R_A vs Solvent, $MPa^{1/2}$	2.62	0	2.92
OSHA/DOT Flammability Classification	Combustible Class IIIA	Flammable Class IA	NOS
Can Vapor Degrease?	NO	YES (with MEK)	YES (with HFC-43-10mee RA)
Drying Effectiveness	Poor	Excellent	Excellent
Process Equipment	Open Tank	Enclosed Vacuum Machine	Modified Vapor Degreaser
Required Investment	1X	4 to 6X	2X

- VOC status (in the US) matters when selecting a single solvent as a replacement, and not so when selecting an SA cosolvent (because the more volatile RA cosolvent is the one emitted as it overlays (covers) the volume of the SA sump).
- Low liquid density is preferred when selecting an SA cosolvent to increase the rate of separation (rise) in the rinse sump.
- Immiscibility relative to an SA cosolvent, a VOC exemption (in the US), a high liquid density, and no (or a high) measured flash point matters when selecting an RA cosolvent.

The approach of changing cleaning processes while retaining the same (or similar) process equipment adds complexity with management of two chemicals being required. But it is still the same three process steps[MM] of liquid immersion, vapor spray, and liquid spray with force that are involved in the revised two-solvent process.

The simple process diagram of either a Class II A or Class II B cosolvent machine is shown in Figure 3.12. Part surfaces are dried as they contact hot RA vapor, which is then condensed as shown in Figure 1.14 and described in Chapter 1.8.

The second "line" of process steps reveals the possibility of multiple immersion treatments in the SA followed by rinsing by immersion in the RA[41].

[41] Note in Figure 3.12 that if a second step of cleaning with an SA cosolvent and its associated step of rinsing with an RA cosolvent is to be completed, both are done before part surfaces are dried by contact with hot vapor. Once parts are lifted into the freeboard zone above the condensing zone where surface are dried, they are hardly ever returned to the immersion sumps.

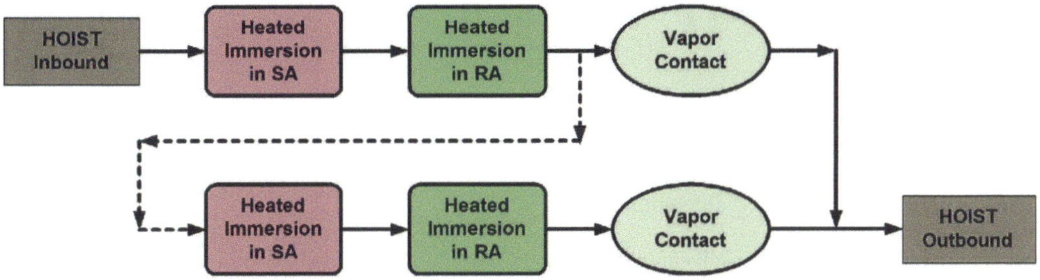

Figure 3.12 Process Diagram for Multistage Class II Cosolvent Technology

3.6.8 Temperature Management in a Class II A or II B Cosolvent Machine

With a different solvent in each of two sumps in a Class II cosolvent machine, at what temperature is cleaning/rinsing/drying done?

- All these operations are done at the boiling temperature of the most volatile cosolvent[42], or at some lower value.

If there are no thermal barriers between the two liquid sumps, and if there is a suitable[43] condensing coil above the sumps, and if there is a straightforward path for the condensed liquids to be returned to the sumps, and if the pressure is ambient; when heat is applied to either sump the temperature in either sump can't exceed the normal boiling point of the most volatile solvent[44]. It simply can't[45]!

- Should there be no pool of condensed RA liquid, continually applied heat will increase the vapor temperature to approach the boiling point of the SA liquid. This might occur if the condensing capability is lost[46] or if the cleaning machine has been constructed to superheat the part surfaces above the boiling point of the RA cosolvent in order to achieve improved quality of drying[NN].

The calculations of vapor composition in Table 3.1 and 3.2 are based on operation at the normal boiling point of the RA cosolvent.

Note in Tables 3.3 and 3.4 that one generally cannot heat parts in the cleaning (or rinsing) sump to the boiling point of water.

- For Class II A cosolvent machines (flammable cosolvents), the chosen RA cosolvent from Table 3.4 is dimethyl carbonate whose boiling point is 92°C (194.0°F).
- For Class II B cosolvent machines (non-flammable cosolvents), the chosen RA from Table 3.4 is HFC-43-10mee whose boiling point is only 54°C (130°F)[OO].

So removal of greases, synthetic oils, and waxes from surfaces may be limited by the degree to which their viscosity can be reduced by being heated in a Class II B cosolvent machine[47].

Since the RA cosolvent is boiled *and* condensed, vapor degreasing (rinsing of surfaces with pristine RA cosolvent) can be done in *either* class of cosolvent machines. Furthermore, surfaces can be rapidly dried because a high difference in temperature can be maintained between the primary cooling coils and the vapor zone above the coils.

---

42 For the temperature to be further increased, all of the more volatile RA cosolvent would have to be vaporized and not returned as condensate to the liquid sumps. Simply, both cosolvents can't boil at the same time. Evaporate yes, boil no.

43 A suitable condenser (Table 1.4) is one which can provide a cooling load to remove the heat of vaporization and some of the sensible heat from the more volatile RA cosolvent at whatever rate it is heated and evaporated. Operation of such a heat exchanger insures there will always be liquid of both cosolvents in the two sumps.

44 Heat added to a liquid raises its temperature, up to its boiling point. Additional energy supplies the heat of vaporization to enable a phase transformation. But, the liquid temperature can't rise above the boiling point until all the liquid is vaporized. Then, when no more liquid is present, further additions of heat raise the temperature of the vapor.

But, if vapor is condensed and the condensate is returned to the sumps, the complete phase transformation of all the liquid RA cosolvent to vapor is never completed, and the temperature in the both sumps is never increased beyond the boiling point of the RA cosolvent.

45 So, if the temperature in either type of Class II cosolvent machine can't exceed the temperature of the most volatile solvent, it also can't ever reach the boiling temperature of the least volatile solvent! In other words, while Class II cosolvent machines are capable of vapor degreasing and rapid drying of surfaces, they aren't capable of cleaning in boiling solvent (the SA).

46 If the capacity of the boilup heater (Chapter 1.15) substantially exceeds the heat removal capacity of the primary condenser, then two outcomes are likely: (1) the liquid RA cosolvent will be fully evaporated and so the temperature in the SA and RA sumps can rise, and (2) the secondary condenser will be overloaded with RA vapor and so the vapor degreasing unit will likely expel RA solvent (Chapter 1.19.1).

47 One with such a need might well elect to use the Class II A cosolvent machine because its commonly selected VOC-exempt (in the US) flammable RA cosolvent, dimethyl carbonate, has a considerably higher boiling point than does a commonly selected non-flammable RA cosolvent, HFC-43-10mee.

3.6.9 The Market for Class II Cosolvent Technology

Vapor degreasers are like automobiles, spaghetti sauce, laptop computers, history textbooks, and television "situation comedies." Within a category, they are all the same; and yet they can be constructed, formulated, designed, written, and programmed completely differently – thus producing different outcomes.

A cornerstone upon which Class II cosolvent technology rests is the interests of the firms who have been its primary developers. Almost without exception, in the US[PP], the primary developers of the Class II cosolvent process of Figure 3.12 have been those whose interests were to supply the cleaning agents (Footnote 42) – rather than those whose interest was to supply equipment[48]. After all, success for suppliers of cleaning agents produces a long-term stream of earnings for them.

This fact has several significant consequences for users:

- Without outside input, selection of proven cosolvent pairs will be limited to those SA or RA cosolvents offered by firms who are either developers or supporters[49] of the Class II cosolvent process, and the specific application(s) they target.
  - No other outcome should be expected. There is nothing wrong with self-interest[50].
- Without input from developers supporting use of just a single pair of cosolvents in some family of applications, selection and testing of alternate pairs of cosolvents for use in other applications is the burden and responsibility of the user.
  - One purpose of this book is to provide guidance in both regards.
- Commercialization of Class II cosolvent technology has progressed at a discontinuously slow pace in the US because of the limited options for customization (Footnote 52).
  - One purpose of this book is to provide a sound approach and reliable basic data which would support an expansion of commercialization.

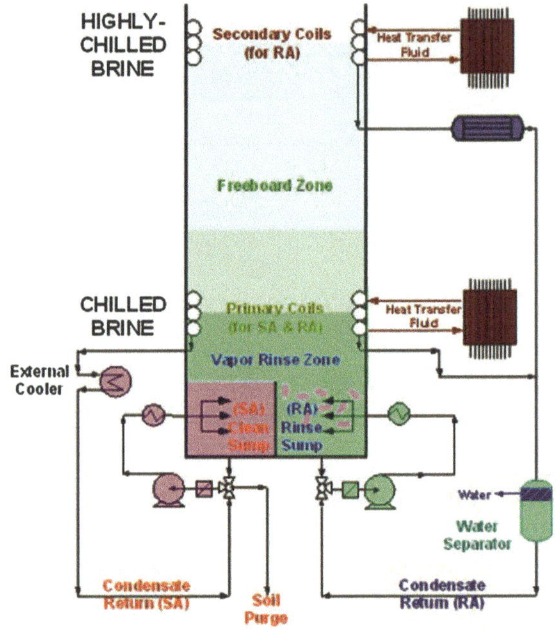

Figure 3.13 Either Class II A or II B Cosolvent Machine

3.6.10 Class II Cosolvent Process Equipment

The same process steps (Figure 3.12) are involved in operating either a Class II A or a Class II B cosolvent machine – yet not all machines of the same class are the same.

The general cosolvent process as shown in Figure 3.12 can be implemented[51] in a conventional vapor degreaser as shown in Figure 3.13 – with the necessary modifications noted below. For comparison, a previously shown diagram of a vapor degreaser is repeated as Figure 3.14.

An aspect of design, common to both II A and II B classes of cosolvent machines, is the selection of refrigeration type and capability:

---

[48] One reason for this situation is that less change of process equipment (Figures 3.13) is required to replace vapor degreasing using single solvents with two Class II cosolvents, than is change required to replace a single solvent with two solvents with different characteristics.

[49] Such is the nature of the chemical industry that it is highly unlikely that both cosolvents are manufactured by the same firm. The cosolvents preferred in Chapter 3.6.4 are simply too unlike one another.

Hence partnership arrangements are normal – one firm having more experience in development of the Class II cosolvent processes provides one of the two cosolvents, and a second firm provides the other. This partnership is a one-size-fits-all arrangement with little opportunity for customization as done with the approach of Chapter 3.6.3 and the information in Tables 3.1 and 3.2.

[50] There is nothing wrong with self-interest, but it can lead to the situation described by the humorous aphorism "when you are a hammer everything looks like a nail." Seeking applications for chemicals which are existing products is not the same as seeking the chemical which provides the optimum benefit in use.

[51] Implementation by every supplier of the process steps of Figure 3.12 within the apparatus of Figure 3.13 leads to an apparatus which is furnished differently by each supplier. Examples are a different instantaneous and overall capacity of the boilup heater; excess instantaneous condensing capacity vs. the boilup heater, metallurgy, and design of primary condenser; capacity of the secondary condenser; location of thermocouples in the rinse sump and freeboard zone; layout and especially the thermal mass of the parts basket; location of soil drain in the washing sump; configuration of tank bottoms so as to not trap debris; freeboard height; capacity and orientation of pump/nozzle systems; tank dimensions for the same volumetric liquid capacity; design of weirs and gratings; as well as control systems and programmed logic.

By far, for short term performance, maintenance of the overall heat balance on a continuing basis is of most importance – without regard to which specific facilities are implemented to achieve it. Characteristics of commercially supplied components of which a cleaning machine is assembled are described and ranked as "good; better; best" in Ref. 1, Chapter 7.

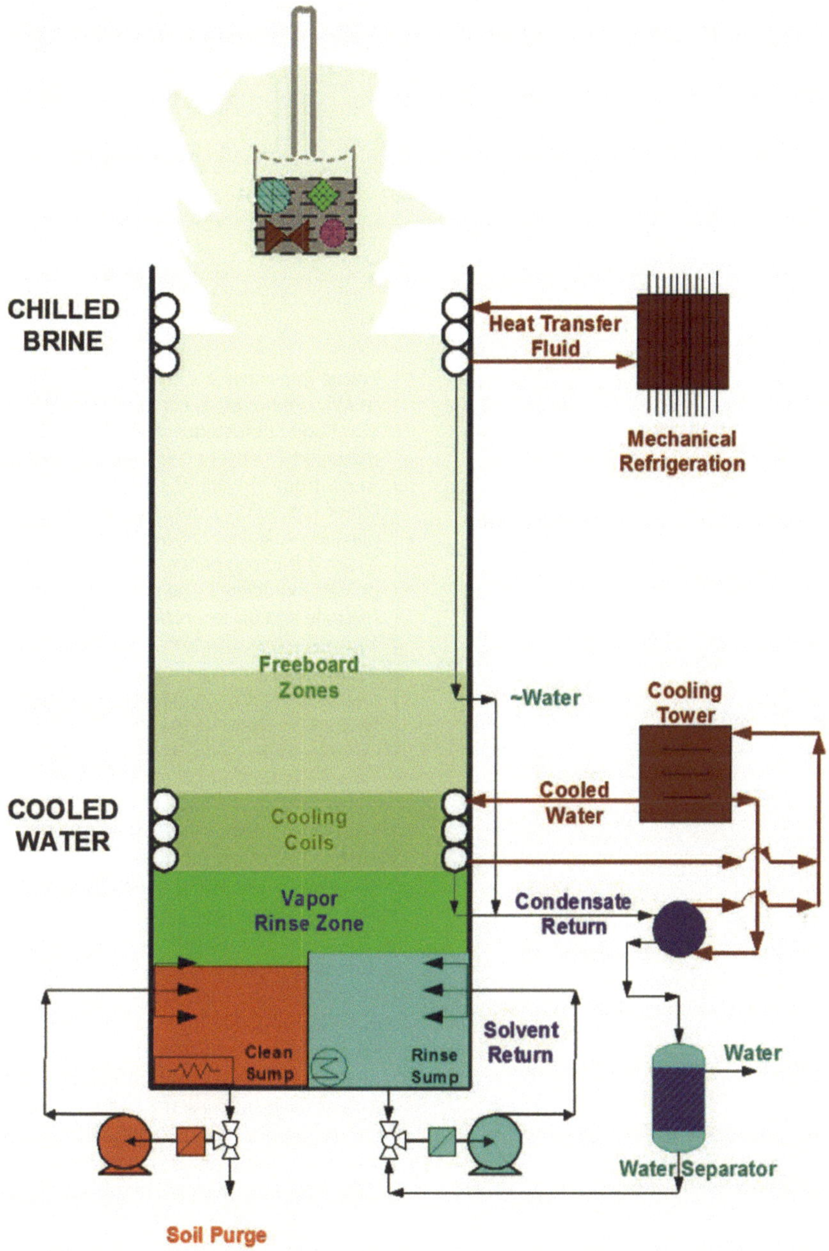

Figure 3.14 Single Solvent Open-Top Vapor Degreaser

- A condenser with chilled brine as a coolant is used in the primary cooling coils. Water from a cooling tower is not generally adequate[52], as it is with some single vapor degreasing solvents such as perchloroethylene, to either condense the expensive and highly volatile HFC-43-10mee, or rapidly condense the somewhat less volatile and less expensive residual dimethyl carbonate solvent from rinsed part surfaces.

A second aspect of difference between the vapor degreaser used with single solvents and that used with cosolvents is the management of fluid transfer between the two sumps.

- In a Class II cosolvent machine there is no fluid transfer.
  - Any RA cosolvent that reaches the cleaning sump, sinks into the SA cosolvent (because it is immiscible and more dense) but evaporates from the cleaning sump (because it is more volatile).
  - Any SA cosolvent that reaches the rinse sump floats on top of the RA as a discrete phase for the same reasons (Figure 3.13 – note the pink "blobs" of SA solvent in the rinse sump).
- In a single-solvent vapor degreaser (Figure 3.14), there is fluid transfer from the rinse sump to the cleaning sump as condensate overfills the rinse sump and flows over a weir (Figures 1.3 and 1.5) into the boil sump.

A third aspect of difference is that multiple heat sources (or sinks) are required[53] with a Class II cosolvent machine, instead of just one with a single-solvent degreaser.

- The reason is that different temperatures are involved in the same apparatus – the RA cosolvent must be heated to its normal boiling point while the SA cosolvent must be heated (or cooled) under control to whatever temperature is desired for parts washing – no warmer than the boiling point of the RA cosolvent.
- In a single-solvent vapor degreaser only one temperature is involved – the normal boiling point of the solvent – and one need not control it because it cannot be exceeded unless the boil sump is empty of fluid.

3.6.11 Management of Water

Another aspect of difference is that there are no stabilizers needed (Ref. 2, Chapter 9). This is because neither cosolvent is a halogenated hydrocarbon which can react with water to produce a corrosive mineral acid. So no monitoring for stabilizer content is necessary.

As with most solvent cleaning machines of any type, water is a tramp impurity (Box 3.3) entering with the parts as humidity in entrained air or absorbed on part surfaces.

Retained water is a significant problem with halogenated solvents in any cleaning machine. These solvents react with the water and produce corrosive acids. That's why stabilizers are provided with halogenated solvents[QQ].

But a hydrolysis reaction between tramp water and the solvents used in Class II cosolvent machines doesn't occur, because there are no chlorinated or brominated solvents[54] present. In these machines, water is not a reactant, but a diluent. Water simply accumulates unless it is removed.

Fortunately, it is easy to remove water from Class II cosolvent machines, because water is highly immiscible with the RA cosolvent[55]. The displacement-based water separator shown in Figure 3.15 is that which must

Box 3.3 **Out of Control – Tramp Impurities**

The Little Tramp

Tramp impurities are those out of control. Their source is usually unknown, as is their frequency of appearance, their duration of persistence, and their concentration is surely not managed. In solvent cleaning operations, water and oxygen (both from air) are the most common tramp impurities. Other are color, odor, stains, particulate, NVR, or unexplained peaks in a UV or other scan.

Some tramp impurities are products of degradation of process ingredients. The presence of tramp impurities can frustrate and tax technical support organizations because it is almost never clear whether or not their effects are a curiosity or a determinant of significant outcomes. While risky and often painful to experience, a fruitful strategy for management of technical resources is to ignore tramp impurities.

[52] Despite this concern, some users of Class II cosolvent machines do not use chilled brine. Obviously, these users will experience reduced energy costs and required investment, and also experience increased costs of solvent loss. Certainly, chilled brine is not required for effective cleaning. In Table 1.6 is tabulated a comparison of cooling tower performance vs. that of mechanical refrigeration of "brine." The choice between chilled brine and cooling tower water is simply one of optimization.

[53] A minor but potentially significant aspect of this difference is that the cutoff switch for the heaters on the SA cosolvent should be set to not exceed the normal boiling point of the RA cosolvent. The reason is to avoid boiling away all the RA cosolvent.

[54] Fluorinated chemicals used as RA solvents are too inert to react with water and create corrosive acids under the conditions found in Class II cosolvent machines (Ref. 3, Table 22.2). Actually, fluorinated solvents are quite immiscible with water and are used to displace it from surfaces.

[55] In general, water is retained as a soluble impurity in the SA cosolvent. The solubility of water in the SA cosolvents noted in Tables 3.3 through 3.5 is typically around 5 weight percent. However, values can range from essentially zero with Isopar L to 16 percent with ethylene glycol diacetate.

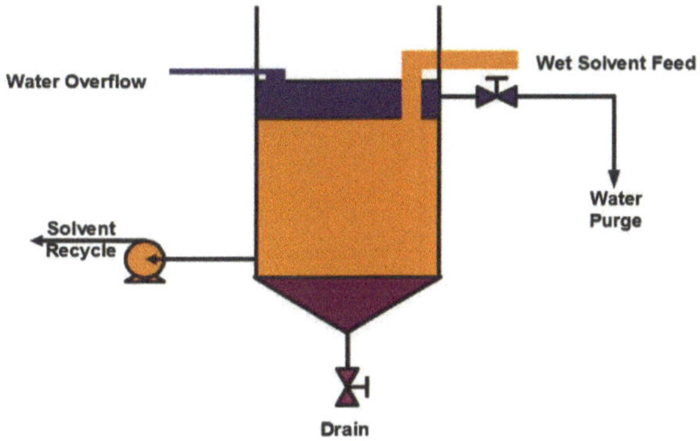

Figure 3.15 Open-Top Vapor Degreaser

be installed in a Class II A or Class II B cosolvent machines where the RA cosolvent is immiscible with water[56].

3.6.12 The Reason for the Subclassification of Class II A and B Cosolvent Machines

Figure 3.12 is a process diagram of either Class II cosolvent machine. It's not a photograph of a Class II B working machine (as is Figure 3.15[57]). And even if it were a photograph, the necessary differences between Class II A and Class II B cosolvent machines wouldn't be evident!

For the solvent selections within Tables 3.3 to 3.5, the differences among cleaning machines are hidden, as are the physical attributes of men and women sometimes hidden by their choice of clothing, Trojan malware hidden in downloads of software, or the taste of a "lite beer" hidden after being poured into a glass before serving.

There are two general differences between Class II A and Class II B cosolvent machines. One is significant; the other is dominant[58]. The significant difference is about cosolvent physical properties. The dominant difference is about electro-mechanical/safety design issues.

The most significant heat transfer property of Class II A and Class II B cosolvents is very different for HFC-43-10mee vs. that for dimethyl carbonate:

- The heat of vaporization per *volume*[59] for HFC-43-10mee (Class II B) is less than half of that for dimethyl carbonate (Class II A)[60]. So a cosolvent vapor degreaser must be designed with a different boilup and primary condensing capacity (perhaps double capacity) for the Class II A cosolvent machine than for the Class II B cosolvent machine.

Use of flammable materials imposes a dominant, and usually legal[61], burden. It's the burden of preventing ignition of the flammable materials, but more especially of proving that one has done so. This burden falls on the designer (Box 3.4) and then the users of machinery in

Box 3.4 Insurance Requirements for Ignition Control

In the US, at least two agencies have jurisdiction over the acceptance or approval for installation of specific equipment as being in compliance with NFPA, and OSHA/DOT codes. They are:

- Underwriters Laboratories, Inc. (UL) is probably the best-known agency. UL is a for-profit agency that serves the insurance industry by determining that products meet certain minimum safety standards. UL does not approve any item for use in any specific application. They simply list items that meet their minimum safety standards, don't refer to those which don't meet them, and don't approve anything. This is an important distinction vs. FM (below). UL has not approved this book.
- Factory Mutual Research Corporation (FM) is also an insurance rating company (for profit). FM rates and approves heater systems, other components, and assembled machinery for use in industrial applications insured by the Factory Mutual Insurance Company. Unlike UL, FM approves products for use in facilities insured by their parent company (and other insurance companies).

This author believes that any user contemplating the purchase of a cosolvent or mono-solvent machine intended to use flammable solvents should insist that the purchased machine be approved by FM.

56 Both dimethyl carbonate (Class II A) and HFC-43-10mee (Class II B) machines are not miscible with water because their intermolecular forces (HSPs) are completely unlike those present in water.

57 Figure 3.15 is an image of an operating AVD Class II B cosolvent machine (Footnote 42).

58 Simply put, one can't purchase a vapor degreaser designed and outfitted for use with a specific non-flammable Class II B cosolvent and successfully use it with any flammable Class II A cosolvent. The reverse may or may not be true; but is a waste of money if purchased as new.

59 The contents of the cleaning sumps are denominated by volume because they are always full of liquid.

60 It's not the same with heat-up time to reach the boiling point, as volumetric heat capacities for HFC-43-10mee and dimethyl carbonate are within 10% of one another.

61 The connection in the US between a specific user and a permit to use for or a legal citation against a specific site goes through what the NFPA identifies as a local "Authority Having Jurisdiction" (AHJ).

Consulting engineers sometimes refer to this body as the "Questioning Authority." The AHJ may be identified as a commission, Fire Marshall (Chief), council, or board. In any case, the AHJ is empowered by local voters to define local plumbing, building, safety, and fire codes. Usually, but not always, these codes (laws) are based on standards independently developed by the NFPA or another like agency, supplemented by local bias or needs. But the local AHJ can accept or reject any NFPA standard because NFPA has no empowering legal authority.

It is the expectation of this author that a user in any US locality must inform their: (1) insurance carrier, and (2) AHJ of an intent to use flammable solvents on site in a cleaning machine. Surely, this is one of the advantages of a Class II B cosolvent machine (no flammable solvents) vs. a Class II A cosolvent machine, and a reason for the subclassification.

Box 3.5 The Meaning of 'Explosion-Proof'

Many think that the phrase "explosion-proof" refers to the environment external to potential ignition sources, and the provision of a barrier between that environment and those sources. Actually, that's not correct. Explosion-proof refers to containment of the outcome of an ignition – assuming it occurs.

The technical definition of explosion-proof is taught by the National Fire Protection Association (NFPA) in its Standard No. 70, the National Electrical Code (NEC). In Article 100, the Code offers this general definition of an explosion-proof apparatus:

> ***"...Apparatus enclosed in a case that is capable of withstanding an explosion of a specific gas or vapor that may occur within it and of preventing the ignition of a specified gas or vapor*** surrounding ***the enclosure by sparks, flashes, or explosion of the gas or vapor within, and that operates at such an external temperature that a*** surrounding ***flammable atmosphere will not be ignited thereby...".***

The NEC is covered more fully in Ref. 1, Chapter 3.22.

This definition does not require the apparatus to be designed so that an internal explosion cannot occur. The only requirement is that any such explosion be confined to the apparatus interior.

However, explosion-proof design of electrical connections and wiring does also what one intuitively expects – it acts as a barrier to keep flammable materials and oxygen from an obvious type of spark source.

which the flammable materials are contained. One often, but not always, characterizes this burden as being explosion-proof (Box 3.5).

3.6.13 Explosion-Proof Facilities for Class II A Cosolvent Machines

Should a flammable material meet an ignition source in the presence of oxygen, only the low strength (energy level) of the ignition source or the low (or high) relative concentration[62] of either the flammable material or oxygen can prevent an ignition. Since, without sophisticated measurement equipment, one can't identify the strength of an ignition source, all real and potential ignition sources are to be contained – by so-called explosion-proof design.

Examples of explosion-proof design are shown in Figures 3.16 (terminal box), 3.17 [RR] (temperature sensor), 3.18 (cable connector or gland), and 3.19 (flowmeter housing with external control buttons and touchscreen).

Safety engineers identify the approach of using explosion-proof hardware as being "containing." This approach assumes that an ignition can occur and seeks to eliminate damage and hazard when it does.

Figure 3.16 Class II B Cosolvent Machine – Boiling and Cleaning

Figure 3.17

Figure 3.18 Explosion-Proof Temperature Sensor

[62] These concentrations are the lower explosive limits (LELs) and upper explosive limits (UELs).

Figure 3.19 Cable Connector or Gland

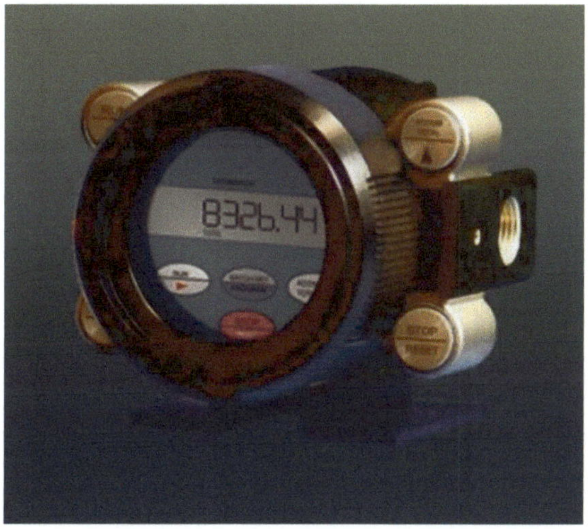

Figure 3.20 Flowmeter Housing with External Control Buttons and Touchscreen

3.6.14 Intrinsically Safe Facilities for Class II A Cosolvent Machines

There is a second approach – a proven (and low-cost) method by which one can safely use flammable materials in cleaning machines, and not require expensive explosion-proof electrical connections.

This method is based on simultaneous elimination of *two* (Box 3.6) of the three factors necessary to produce an ignition. One (1) constructs an apparatus with no electrical connections, other than those deemed "intrinsically safe," *and* (2) purges all volumes where flammable materials are located with an inert gas – usually nitrogen, but occasionally carbon dioxide – to a level that is deemed safe.

For example, a cleaning apparatus with no electrical hazards and no excessive concentrations of flammable materials would:

- Use pumps with motors powered by compressed air (Figure 3.20) rather than motors powered by electricity even though they were rated explosion-proof (Figure 3.21).
- Use steam, steam condensate, or hot water for supply of heat for boilup.
- Eliminate the possibility of sparks generated by friction between metal parts, components, or tools by using "non-sparking"[63] materials where necessary[64].
- Use instruments providing pneumatic or low-voltage outputs. The former are known as "suck and blow" and are well-proven over decades of use. The latter are known as "intrinsically safe" (Box 3.7).
- Maintain conditions everywhere so that the concentration of flammable material in air never exceeds 25[65] volumetric percent of the minimum amount needed to support combustion (the lower explosive limit (LEL)) in air. One does this by purging internal volumes such as the area above the sumps with nitrogen, and providing excessive ventilation with air in external areas. Then, one proves that one has done so effectively by measuring with an

> **Box 3.6 To Prevent Ignition**
>
> Actually, removal of only one of the three factors is all that's necessary to "guarantee" that an ignition can't be created by circumstance. But, normal, proper, safe, and conservative design should require removal of two of the three.

[63] "Non-sparking," "spark reduced," "spark-resistant" or "spark-proof" materials are names given to objects made of soft metals such as brass, bronze, Monel metal (copper-nickel alloy), copper-aluminum alloys (aluminum bronze), or copper-beryllium alloys (beryllium bronze).

Actually, non-sparking materials can generate sparks sometimes referred to as "cold sparks". These cold sparks have a low energy level and do not ignite carbon disulfide, which has the lowest ignition point of any common solvent.

Replacement materials can be plastic, elastomeric, or the softer materials noted above.

[64] In a vapor degreaser objects of concern include: overhead cranes often remote to but occasionally servicing the apparatus, moving components of the hoist and its guides/rails, connections of the movable cover to the degreaser body, tools, and cellular telephones. These metal objects are normally made of 304 or 316 grades of stainless steel in order to provide a long life in potentially corrosive environments. Vanes, casings, bushings, and shafts of pumps powered by compressed air typically are made from non-metallic materials as to not defeat their purpose of use.

[65] The offset of 25% is a 4/1 safety factor.

Figure 3.21 Pneumatic Motor (aka Compressed Air Motor)

> Box 3.7 **Intrinsically Safe Systems**
>
> In an intrinsically safe system, all possible equipment is designed and installed in such a way that it does not have enough energy to cause ignition of the potentially explosive gas mixture, *even in a fault (spark) condition*. One uses low-power devices, such as those powered by 24 VDC. Nearly all sensors are available as intrinsically safe models.
>
> However, heavy-duty motors, motor starters, circuit breakers, etc. can't be powered by 24 VDC. Where devices requiring higher energy levels are necessary, designers use a safety barrier between the "safe area" and the "hazardous area", so that any fault that generates a high energy level (and sparks) cannot get carried over to the hazardous area. The spark (fault) is stopped at the safety barrier. Safety barriers can be classified as Zener barriers (passive devices) and Isolators (active devices). Functionally, they achieve the same objective, i.e. prevention of high energy levels from reaching the hazardous area (potentially containing flammable materials).

explosimeter local concentrations of the flammable materials relative to 25% of the LEL (Chapter 1.35, and Figure 3.22).

Safety engineers identify the approach of using "intrinsically safe" hardware as being "preventive." This approach assumes that an ignition can be prevented and seeks to do just that.

The debate about "containing" vs. "intrinsically safe" design is for safety engineers, and outside the scope of the book. This author has successfully used both approaches. Fires in vapor degreasers using flammable solvents are non-existent, or occur with an extremely low incidence, or are well-concealed.

Figure 3.22 Motor Rated 'Explosion-Proof'

3.7 COSOLVENT PROCESS DESCRIPTION: CLASS III – MISCIBLE

Class III cosolvent processes are opposite to and different from the Class II cosolvent processes. With Class III cosolvent processes, both cosolvents are miscible[66], and the flammability rating of the mixture is not specified[SS].

3.7.1 The "Failed" Cosolvent Process

One might explain the Class III cosolvent process by noting that it is a "failed" Class II cosolvent process.

- In any proper Class II cosolvent application, the SA and RA cosolvents are separated from one another in "HSP space" by *at least* 10 to 12 $MPa^{1/2}$ (or more) so that *im*miscibility is maintained. In all Class III cosolvent processes, the separation (R_A[TT]) is less than 10 $MPa^{1/2}$; hopefully, the R_A between Class III SA and RA cosolvents is as close to zero $MPa^{1/2}$ as possible – so that complete miscibility is maintained.
- In a proper Class II B (not flammable) cosolvent application, the RA is immiscible with and very dense relative to the SA cosolvent, and is relatively inert to ignition. Hopefully, it is also relatively inert to reactions with UV light in the atmosphere, and a fairly volatile solvent. Because of the latter inertness, it may often be classified as VOC exempt (in the US).

66 Technically, this means that the R_A between the SA and RA cosolvents is greater than 10, and hopefully greater than 12 $MPa^{1/2}$.

Table 3.7 Solvents with Similar HSPs but Different Volatility

SA Cosolvent	Boiling Point,°C	RA Cosolvent	Boiling Point,°C	R_A Relative to Specific SoilMPa$^{1/2}$
Stoddard Solvent	159	HFE-7100 (VOC exempt)	60	2.14
Diethylene Glycol Dibutyl Ether (Butyl Diglyme)	256	HFC-365mfc (VOC exempt)	40	4.46
Diethylene Glycol Monobutyl Ether Acetate	231	Tetrahydrofuran	66	2.27
D-Limonene	176	Diethyl Ether	35	4.20
2-Ethyl Hexyl Acetate	199	Diethyl Ether	35	1.18
Tripropylene Glycol n-Propyl Ether	261	Ethyl Acetate	92	3.44
Triethylene Glycol Dimethyl Ether (Triglyme)	216	Tetrahydrofuran	66	1.40
Butyl Glycol Acetate	192	Acetone (VOC exempt)	56	2.66
Isododecane	178	HFC-365mfc	40	0.40
Isopar L	190	2-Methyl Pentane	17	2.10
Diethylene Glycol Monobutyl Ether Acetate	231	Methylal	42	3.07
Dipropylene Glycol Dimethyl Ether	175	*trans*-1,2 Dichloroethylene	49	3.94
Mineral Spirits	200	HFC-365mfc	40	2.34
D-Limonene	176	Cyclopentane	49	3.47

- In all Class III applications, none of that matters. In a proper Class III cosolvent application, the SA cosolvent must have similar intermolecular forces to the RA cosolvent – because they are miscible. Further, the blend (both SA and RA cosolvents) must have intermolecular forces similar to those of the soil – because the SA and the soil also are adequately miscible for good solution cleaning.
- In a proper Class II A (flammable) cosolvent application, the RA is immiscible with and probably not dense[67] relative to the SA cosolvent, not relatively inert to ignition, probably not inert to reactions with UV light in the atmosphere, and a fairly non-volatile solvent. Because of its lack of inertness to ignition, it is likely to be classified as flammable. Because of its absence of inertness in the atmosphere, it is not likely to be classified as VOC exempt (in the US).

[67] The reason for this outcome is that flammable solvents, the RA solvents in a Class II A cosolvent process, are not heavily fluorinated. So their density is normally less than unity. If they were heavily fluorinated, they would still be volatile but display a density well above unity.

As above, none of that matters in all Class III cosolvent applications.

3.7.2 Rationale for Class III Cosolvent Processes

It's simple – to gain a difference in volatility (boiling point), within the HSP "confines" of a single solvent.

That is the drawback to Class II cosolvent systems – the maximum operating temperature is limited to the normal boiling point of the most volatile solvent (Chapter 3.6.8, Footnote 47). The Class III cosolvent systems repair that "flaw" (if indeed it is a flaw).

A key point with Class III cosolvent systems is that solvent volatility and HSP values are not directly related. To practice the technology, one wants two Class III cosolvents which are fully miscible (have similar HSP values), but one is relatively non-volatile (the SA cosolvent) and the other is highly volatile (the RA cosolvent).

That this can be done is demonstrated by the pairing of cosolvents in Table 3.7. Between each pair of SA and RA cosolvents there is little difference in R_A relative to a specific soil, but great differences in volatility (normal boiling point).

Figure 3.23 Explosimiter for Personal Use

Any of the selected pairs of SA and RA cosolvents listed in Table 3.7 could be chosen to clean one of seven different soil materials[68].

The point of Table 3.7 is not to select solvent pairs to clean specific soils[69] but to show there are multiple pairs of solvents which can contribute in a Class III cosolvent process – miscible cosolvents which are broadly separated in volatility (normal boiling point).

That point is augmented with Figure 3.24, where the polar and H-bonding HSP values for the seven quite different soils (from Footnote 76) are plotted. Clearly there is a considerable range of soils which can be cleaned with these seven pairs of miscible Class III cosolvents.

Furthermore, the HSP values for three commonly used halogenated solvents are also plotted in Figure 3.24, showing that these Class III cosolvent pairs might well be expected to dissolve soils cleaned by these halogenated solvents.

3.7.3 The Uniqueness of Class III Cosolvent Processes

The process diagram of Class III cosolvent machines is similar to that of Class II cosolvent machines (Figure 3.12). But it differs somewhat as the parts can't be returned from the RA sump to the SA sump without passing through a zone of vapor contact (Figure 3.25).

Class III cosolvent processes address a limiting problem inherent with Class II cosolvent systems – that the SA cosolvent can only be heated to the boiling point of the RA cosolvent[UU].

Imagine! With Class III cosolvent systems, it's like having a single solvent with two boiling points: a high temperature value (probably) for when it's performing the role of an SA cosolvent – removing a heavy (viscous) grease; a low temperature value (probably) when it's performing the role of an RA cosolvent – rapidly evaporating. Since the SA and the RA cosolvents are quite miscible, each cosolvent must have similar solubility parameters.

The effect of dual boiling points is implemented by physical separation of the SA cleaning sump from the RA rinsing sump.

This is shown in Figure 3.26 (compare to Figure 3.13, which is not to the same scale).

- The SA cosolvent usually will (or can) have a high boiling point (150 to 200°C, or more). At that temperature, substantial reduction of the viscosity of waxes, buffing compounds, greases, Cosmoline, and heavy oils can be achieved[70].
- SA cosolvents whose volatility is this low will normally have a flash point above, and possibly considerably above, the value where flammability is not of concern (140°F).
- The RA cosolvent, chosen in order to enable part surfaces to dry rapidly, is heated to its considerably lower normal[71] boiling point.
- Each open sump operates as a single-sump vapor degreaser, with a heater, primary and secondary condensing coils, and a freeboard zone to contain emission of solvents and to dry parts.
- Parts in baskets are shifted by hoist from the SA sump to the RA sump after immersion cleaning. The only direct fluid contact[72] between sumps is the SA dragout material[73] adhering to the parts and basket when immersed in the RA sump.

[68] Cosolvent SA and RA pairs with the same background color could be selected to clean the same soil.

The R_A value in Table 3.7 is between the SA and RA cosolvents. For example, the intermolecular forces manifested by 2-ethyl hexyl acetate and diethyl ether are surprisingly similar – separated by an R_A value of only 1.18 $MPa^{1/2}$. These cosolvents would be a good choice to remove linseed oil (light yellow color in Table 3.7) – if one was not concerned about flammability or stability of the ether.

Notice that because these pairs of solvents have very similar levels of intermolecular forces it is not necessary to specify the mass ratio of each in the pair.

[69] One should not believe that there are only two pairs of Class III cosolvents which can clean each of the seven soil materials noted in Table 3.7. In fact there are usually several pairs; and many more soil materials than seven which are suitable for cleaning by Class III cosolvents. The selections in Table 3.7 were arbitrarily made to demonstrate the concept.

[70] It is not necessary to operate with the SA being boiled in a Class III cosolvent process. Obviously, cold cleaning may be done with the SA in either Class II or Class III cosolvent processes.

[71] Note in Figure 3.26 that both sumps are open to the ambient atmosphere as cosolvent processes are implemented in relatively inexpensive open-top machines.

[72] Certainly, the sumps will be physically located within the confines of the same apparatus, and adjacent to one another. Heat exchange and other facilities will be shared. This process has been developed by one US firm and units sold under the trade name "dual chamber."

[73] This is illustrated by the visual (but wholly imaginary) form of SA droplets in the RA cosolvent rinsing sump as shown in Figure 3.26. There can be no such droplets because the cosolvents are fully miscible with one another.

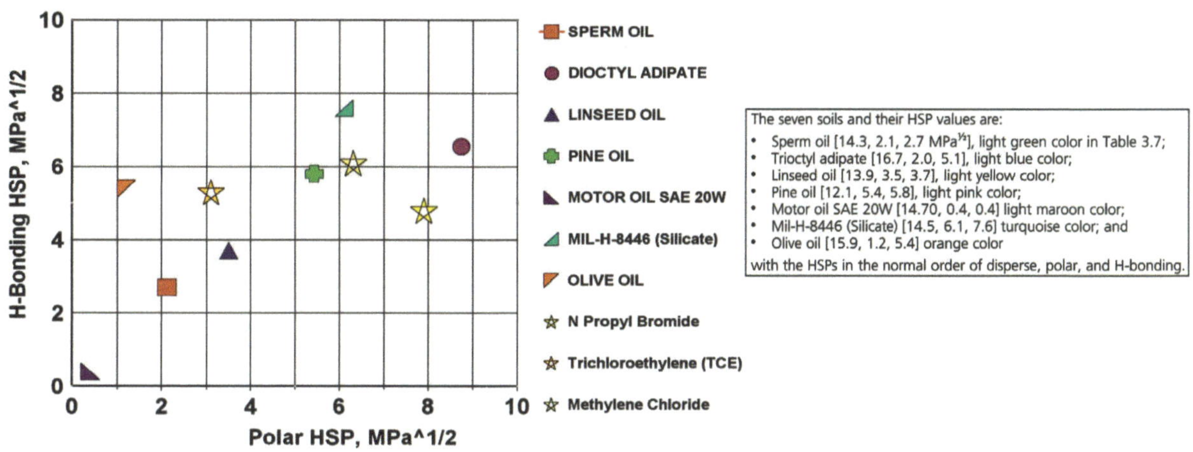

Figure 3.24 Range of Soil HSP Values which can be Cleaned by Class III Cosolvent Pairs

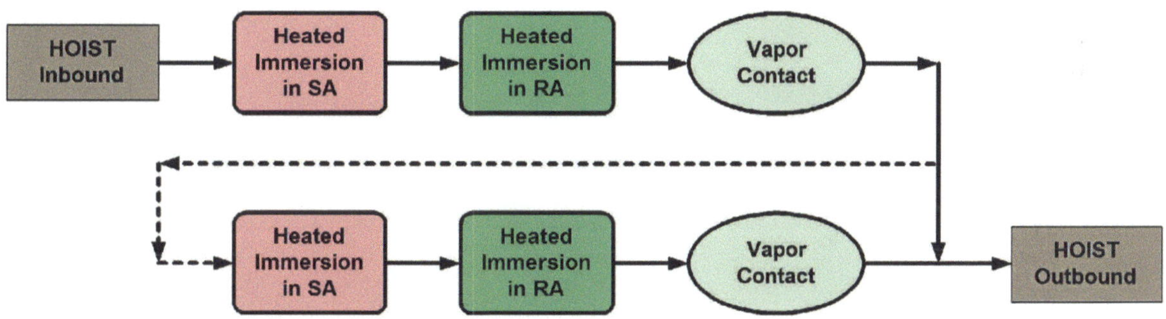

Figure 3.25 Process Diagram for Two-Sump Class III Cosolvent Technology

Thus the chief drawback of Class II cosolvent processes can be overcome – by use of separate sumps[74] and two compatible and different solvents.

[74] There is little reason for any cosolvent process to exist if incremental investment has been provided for facilities to control working pressure. Should, with an enclosed machine, it be desired to:

- Treat with a second solvent, simply attach a pressure-capable tank to the work chamber to store, dispense, and recover the second solvent. With an enclosed machine, there is no reason for the semi-aqueous process as low-cost hydrocarbon and terpene-based SA solvents can be used, contained, and evaporated at any temperature. If in the unlikely event that the second solvent must be water, one should use construction materials of 316 stainless steel for protection against corrosion, and implement a desiccant dryer (Chapter 3.7.7).
- *Operate at a high temperature*, simply do so as the work chamber contains the excess pressure above ambient. The situation is the same for a reduced contact temperature.
- *Produce very dry parts*: flush with nitrogen, evaporate under vacuum; and repeat.
- *Treat with a single solvent*. Use an airtight enclosed machine to produce vacuum and volatilize most any low-volatility cleaning solvent so that part surfaces are free of liquid (Chapter 2.8).

3.7.4 Rinsing Can Be That Important

Actually, there is a second reason to use Class III cosolvent machines.

One may choose Class III cosolvent processes not just because they allow high-temperature solvation cleaning, but because of the type of rinsing their use provides!

- Class II A and II B cosolvent processes are founded on the conventional idea that rinsing in a solvent cleaning process should be done by *displacing* a low-density solvent with high-density solvent. Both of these solvents are to be *im*miscible with one another.
  - The ultimate limitation on rinsing by displacement is the level of immiscibility between the RA cosolvent and the SA cosolvent. This is shown for various Class II cosolvent processes in Tables 3.3 and 3.4 (the column labeled SA vs. RA, R_A). The average R_A

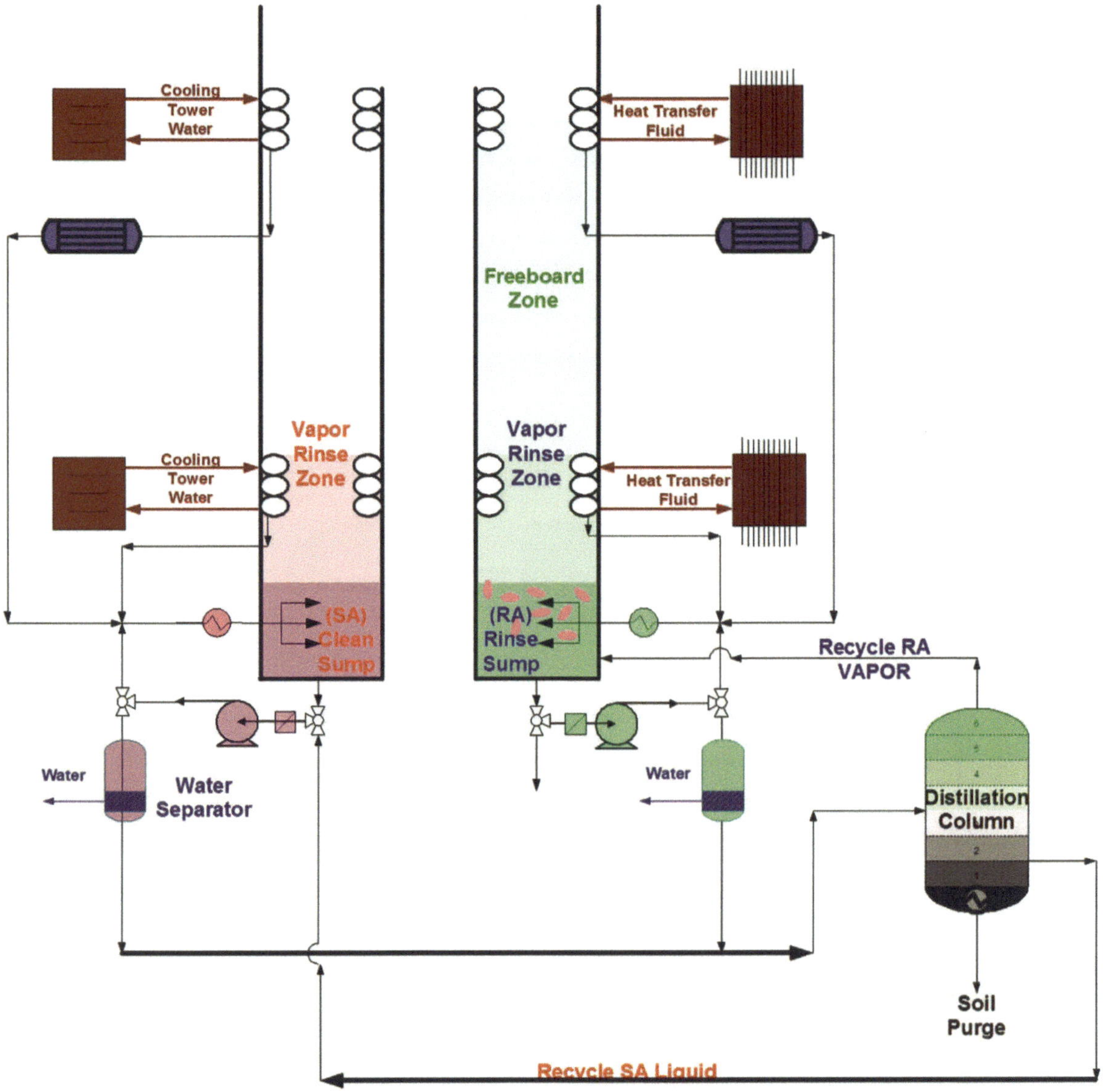

Figure 3.26 Dual Boiling Point Apparatus Exploiting Capability of Class III Cosolvent Systems

between SA and RA[75] for these cosolvent pairs is 11.24 MPa$^{1/2}$.

For comparison, the R_A between SA and RA for water and kerosene (Chapter 3.6.1 about kerosene dryers) is 24.6 MPa$^{1/2}$. A rotated three-dimensional plot of the relationship between water and kerosene in "HSP space" is shown in Figure 3.27. Note that there is no overlap between the two colored spheres[76].

[75] This R_A value is a measure of immiscibility between the SA and RA cosolvents. A higher value represents less similarity between the intermolecular forces of the cosolvents; a lower expectation that the RA can be dissolved in the SA.

[76] In Figure 3.27, the dark green sphere represents water, the light green sphere represents kerosene. In Figure 3.28 the dark green sphere represents Soygold 1000; the light green sphere represents HFE-7200. Both images are rotated to show the separation of the two cosolvents to best advantage. The images of Figures 3.27 and 3.28, and those of Table 3.9, were produced with the software identified in Hansen, C.H., Abbott, S., and Yamamoto, H., *Hansen Solubility Parameters in Practice, 2012*; available for purchase including an e-handbook at http://www.hansen-solubility.com.

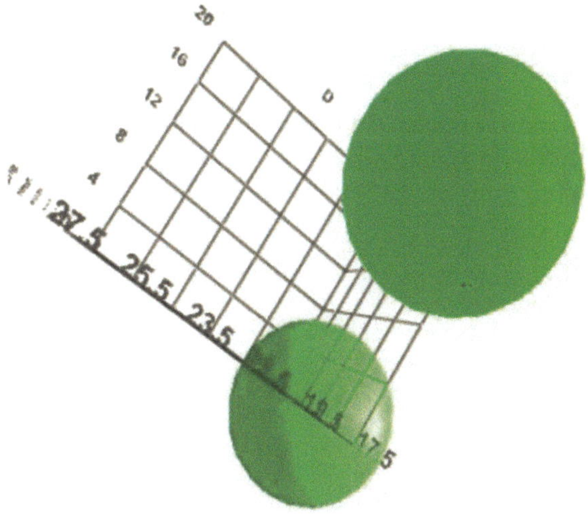

Figure 3.27 Ingredients in a Kerosene Dryer, $R_A = 24.6\ \text{MPa}^{1/2}$

A commercial cosolvent cleaning process using Soygold 1000 (Ref. 3, Chapter 6, Table 6.1) as the SA and HFE-7200 (Ref. 3, Chapter 22, Table 22.4) as the RA cosolvent has a degree of immiscibility represented by an R_A of 11.21 $\text{MPa}^{1/2}$. A rotated three-dimensional plot of the relationship between HFE-7200 and Soygold 1000 in "HSP space" is shown in Figure 3.28. Compare the level of overlap to that in Figure 3.27. In summary, one might expect (at best) fair-to-good but not excellent

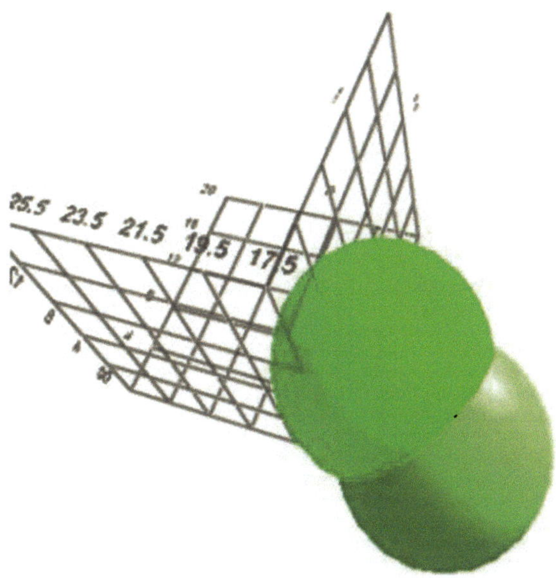

Figure 3.28 Soygold/HFE-7200 Class II B Cosolvent Process, $R_A = 11.2\ \text{MPa}^{1/2}$

rinsing quality in a Class II cosolvent process with these materials.

- Class III cosolvent processes are based on an analogy with the conventional idea that rinsing in an *aqueous* cleaning process should be done by *diluting* the soil/surfactant-laden water with pure water. Assuming perfect miscibility and mixing, one can calculate how many repetitions of dilution contact (stages) are necessary to dilute the concentration of soil/surfactant-laden water (which is in contact with part surfaces) to a specified level of water quality[VV].
  - Dilution rinsing with aqueous cleaning technology is expensive of floor space and investment. Each stage of dilution rinsing requires a separate treatment volume (a tank), and its associated facilities.
  - In the Class III cosolvent processes, all dilution rinsing of low-volatility SA cosolvent is done in the same vessel (the rinse sump) with a continuing flow of pristine distilled high-volatility RA cosolvent produced by an adjacent distillation column. The operating cost paid to implement dilution rinsing with Class III cosolvents is basically the cost of the energy needed to operate the distillation column and condense the solvent vapor in the primary condensing coil above the RA sump (Figure 3.26).
  - Inherently, dilution rinsing can and normally does provide a higher quality of surface treatment than does displacement rinsing[77] as part of a cosolvent cleaning process[78]. Though, treatment time for dilution rinsing is probably longer than that for displacement rinsing (Chapter 3.6.11).

In other words, one may choose a Class III cosolvent machine in order to either operate the wash sump at a high temperature to remove viscous soils, *or* to produce superior rinse quality – or both!

[77] Another factor affecting the capability to rinse complex part structures are the forces of surface tension. Films of a solvent with a *relatively* high value of surface tension may block flushing of complex surface structures. In other words, an SA cosolvent which is perfectly miscible (Class III) or immiscible (Class II) with an RA cosolvent, can't either dilute or displace, respectively, the RA cosolvent if surface forces block /limit its passage to meet the SA cosolvent.

This is one reason why supercritical CO_2 is considered for solvent cleaning – not because supercritical CO_2 has HSP values coincident with those of soil materials (it doesn't), but because supercritical CO_2 has essentially zero surface tension (and zero viscosity).

The RA cosolvents (with an exception) chosen in Table 3.3 all have low values of surface tension (< 20 dynes/cm) – one reason why they were selected by this author. Unfortunately, the surface tension of dimethyl carbonate is not low (31 dynes/cm).

[78] One might enquire about the practice of Class II cosolvent technology in a degreaser system with two separated sumps (as shown in Figure 3.26). The reason for not doing so is that the practitioner of Class II cosolvent technology has chosen to value displacement rinsing over dilution rinsing.

3.7.5 Fluid Management with Class III Cosolvent Processes

A study of Figure 3.26 shows that there are several instances where the two cosolvents must be properly managed, and treated in a Class III cosolvent process. The five primary issues about treatment are: (1) separation of soil materials from the SA cosolvent and elimination of the soil materials from the cleaning process, (2) separation of the SA and RA cosolvents from one another so each can be reused, (3) achievement of the desired rinse quality, (4) use of energy in an efficient manner, and (5) removal of water from both cosolvents:

A conventional distillation column is used to complete the first four management tasks.

Separation of the two cosolvents from one another is entirely straightforward because they are miscible and, by design, their volatilities are very different.

Separation and elimination of soil components may not be straightforward by distillation. If their volatility is low, and similar to that of the SA cosolvent, it will be difficult to separate the one from the other by distillation. Similarly so, if the soil components have the same high volatility (less likely) as the RA cosolvent[79].

Achievement of the desired rinse quality is a matter of time – in which the parts are held above the RA sump, exposed to pristine distilled solvent (vapor from the distillation column condensed by the primary condenser above the RA sump). This material dilutes, not displaces, the soil-laden cosolvent from the part surfaces.

3.7.6 Energy Management with Class III Cosolvent Processes

The Class III cosolvent process requires considerably more energy than a Class II cosolvent process. Energy is also used in a different way, because it's a different process.

Use of energy in an efficient manner is not essential because it is good practice, it's essential because there are two energy-intensive operations to be performed: (1) heating of the SA cosolvent to approach its high normal boiling point[80], and (2) containing the expensive and volatile RA cosolvent. There are several parts to a successful energy management strategy:

- Use of cooling tower water (CTW) as the primary and secondary coolant above the SA cosolvent sump. This is the least expensive method (Table 1.6) of heat removal from the cosolvent system. A CTW should easily contain and condense the high-boiling SA cosolvent.
- Implementation of a "hold" time as the parts basket is withdrawn and located within the secondary coils above the SA cosolvent sump. The purpose is to cool the parts basket with the low-cost cooling capability of CTW so that parts can be directly rinsed by condensing hot vapor above the RA cosolvent sump (vapor degreasing).
- Continuous feed of hot RA vapor from the distillation column directly to the parts held adjacent to the primary condensing coil above the RA Cosolvent sump. This vapor is condensed on the part surfaces using the primary condenser to remove heat. In this way, the parts are flooded with pristine RA cosolvent as a dilution rinse fluid[81].
- Primary, secondary, and tertiary condensers are used with an expensive brine refrigerant to contain the expensive RA cosolvent above the RA cosolvent sump (Figure 3.26).

3.7.6.1 Latent Heat Isn't Sensible (Heat)

In vapor degreasing with a single solvent, heat energy must be supplied to heat the liquid solvent from ambient temperature[82] to its normal boiling point, and then to change its phase to vapor. The former is known as sensible heat; the latter is known as latent heat. It's thus with Class II and III cosolvent systems. The individual properties of each solvent affect the outcome heat balance.

It's expected that Class III cosolvent systems will require more heat energy than do Class II cosolvent systems. After all, two major characteristics of the Class III cosolvent processes are that their SA sump can be raised to an elevated temperature, and that the RA cosolvent is a volatile and expensive material.

---

[79] In either case, cleaning with a single solvent may be the appropriate process. That single solvent must be chosen to not only share HSP values with the soil components, but to also have a volatility intermediate between the SA and RA cosolvents being considered for a Class III cosolvent process.

[80] It is unlikely that vapor degreasing will be done above the SA cleaning sump because of the high boiling point of the Class III SA cosolvent. If cold cleaning is to be done, certainly the bill for energy will be reduced.

[81] When using the Class III cosolvent process described in Figure 3.26, it is neither necessary nor desirable to immerse the parts basket into the RA sump. This is because the liquid in the RA sump below the primary condenser is less clean of soil and SA cosolvent materials than is the pristine distilled RA cosolvent vapor. The RA sump exists to collect the liquid dragout material (SA cosolvent and dissolved soil materials) rinsed from the parts and basket surfaces, as well as RA cosolvent. This collection is not retained in the RA sump but is immediately fed to the distillation column. Immersion in the RA cosolvent sump would dilute one of the advantages of the Class III cosolvent process.

[82] !?A3B2 twb=.4w?>Actually, all the liquid solvent isn't heated and vaporized for every cycle of cleaning. But, significant heat energy must be supplied every cleaning cycle to heat the incoming cold parts. They enter the cleaning sump at ambient temperature and must be heated to a temperature approaching the normal boiling temperature of the SA cosolvent.

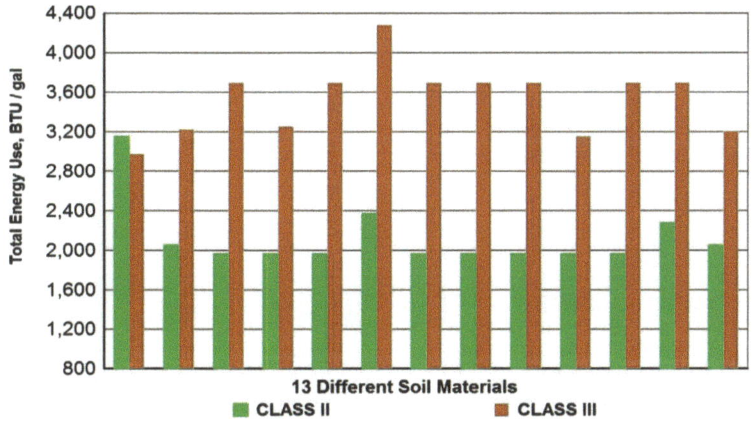

Figure 3.29 Comparison of the Energy Requirements for Operating a Class II vs a Class III Cosolvent Machine

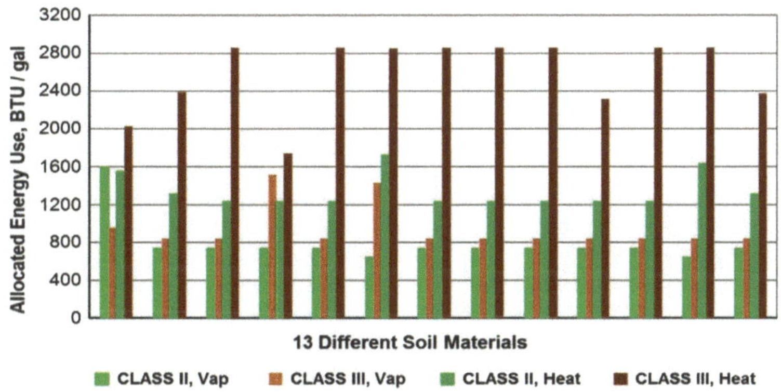

Figure 3.30 Comparison of Allocated Energy Use for Both Cosolvent Vaporization and Heatup

While process designs will be system specific (soil, solvent, and cleaning quality), general calculations[83] show that roughly 50% more total energy is required for a Class III (brown bars in Figure 3.29) vs. a Class II (green bars) cosolvent process. This can be recognized for 13 different soil materials in Figure 3.29.

The need for an increase in energy consumed in a Class III cosolvent process, above that for a Class II cosolvent process, can also be seen to be that expected – to provide the additional *sensible heat* to raise the SA cosolvent to its high boiling point. This can easily be recognized by examination of Figure 3.30[84].

[83] The basis for the calculation is simple: the sensible heat is calculated by multiplying the appropriate specific heat for each SA and RA cosolvent, times the temperature difference between the normal boiling point for the RA cosolvent and an assumed ambient temperature of 70°F, times the appropriate solvent density (to express the energy demand in energy units per volume). The latent heat is calculated by multiplying the latent heat of vaporization for the RA cosolvent times the solvent density (for the same reason). The latent contribution for the RA cosolvent is doubled because of the belief that excellent rinsing is required. The total energy use is the sum of the two. The SA cosolvent is assumed to not be vaporized in Class III cosolvent processes, and can't be vaporized in both Class II cosolvent processes.

The calculation is made for 13 pairs of solvents listed in Table 3.8. Each pair has been chosen to clean a specific soil material with both the Class II and Class III cosolvent processes.

For reference, total energy values for single solvents in vapor degreasing are 1530 BTU/gal (n-propyl bromide), 1690 (trichloroethylene), 1830 (perchloroethylene), 1790 (methylene chloride), and 1780 (MEK) – somewhat less than for Class II cosolvents.

[84] The label "heat" in Figure 3.30 refers to sensible heat needed for the SA cosolvent, and the label "vap" in Figure 3.30 refers to the heat of vaporization needed for the RA cosolvent.

Compare the dark colored bars, green (Class II) and brown (Class III) vs. the light colored bars (green (Class II) and brown (Class III), to assess the effect on energy requirements of increasing the volatility difference between SA and RA cosolvents.

Note that, for each soil material, the selected Class II and Class III cosolvents are not the same solvents, because there are different selection criteria for each class of cosolvent process. That's why the values are different for each soil in Figures 3.29 and 3.30.

3.7.7 Simultaneous Removal of Soluble and Insoluble Water

Water enters Class III (and Class II) cosolvent machines as it does all single solvent cleaning machines[WW] – with the parts.

Water can be soluble or insoluble in either cosolvent. Consequently the decanter of Figure 3.14 can't be the default device for water removal as it only removes water insoluble in solvents.

Another method for water removal must be sought. That usually is a method which doesn't depend upon the phase in which the water resides – the preferred method is a bed of desiccant (Box 3.8).

Box 3.8 Desiccants – Their Types and Uses

A desiccant is a substance that induces or sustains a state of dryness by removal of general or specific liquids. There are two general types: sorbents, and reactants. In this case, desiccants which are hygroscopic (attractive of water) are wanted.

Sorbents are solid materials which trap water through absorption or adsorption. Drying of water from solvents by *ab*sorption means that the water is incorporated throughout the sorbent. Drying of water from solvents by *ad*sorption means that the water doesn't penetrate the sorbent material but remains on the surface of the sorbent. Absorption is a volume-controlled process; adsorption is an area-controlled process.

Notwithstanding the similarity of names, confusion in understanding use of the two types of sorbents can arise because the commercially useful sorbents have huge amounts of internal surface area. When highly magnified, the internals of an adsorptive desiccant look like the insides of an English muffin. Popped popcorn kernels adsorb butter and salt; watermelons absorb vodka.

Three types of sorbents are commonly used to dry cleaning solvents. All are sold as granular 1/8 inch diameter spheres, are highly porous, and can hold at least 20% of their gross weight as water.

- Activated alumina – a form of aluminum oxide with high surface area which adsorbs liquids and gases without any change in form. Surface area is about 83 x 10^{+6} SF/CF. Water can be released by heating to at least 175°C (350°F) for two hours or more.
- Silica gel – a non-crystalline but not gelatinous form of silica made synthetically from sodium silicate which adsorbs water. There is a volume change which can cause cracking on absorption of sufficient water. Water can be released by heating to 120°C (250°F) for about two hours.

3 Å Mol Sieves

- Molecular sieves ("mol sieves") – a form of aluminosilicate derived from crystalline and non-crystalline naturally occurring clays. These minerals have a formula such as Al_2SiO_5. They are also commercially synthesized for specific uses. A major use is to *filter* and *ad*sorb water from solvents, or compressed air. It's a two-step treatment: (1) some molecules are blocked because of their large size from entering pores which function as gateways to internal surfaces, and (2) *ad*sorption of the molecules which do diffuse through the pores onto the huge amounts of internal surface. Surface area is about twice that of activated alumina. Water can be released by heating to at least 250°C (480°F) for two hours or more. The image above is courtesy of http://www.ceramic-honeycombs.com (Chapter 4, Footnote 70).

Reactants are both liquids and solids which trap water through chemical bonding. Solid sodium and potassium hydroxide, and liquid sulfuric acid, all aggressively trap water through chemical bonding. For reasons of safety, none are used neat to dry water from cleaning solvents. However, caustic materials can be compacted with walnut shells to form briquettes. These products are used to remove water from oil byproducts in refining operations, and generally are not reused. Useful solid desiccants which don't present serious hazards are lithium, sodium, magnesium, and calcium sulfates; and lithium, calcium and zinc chlorides.

All sorbent desiccants release heat (exothermic) when water is captured. Similarly, desorption requires addition of heat energy (endothermic). Reactant desiccants can be either exothermic or endothermic, depending upon the nature of the chemical reaction. This is also known as chemisorption. Other pertinent details are noted in Endnote C and Footnote 40 of Chapter 4.

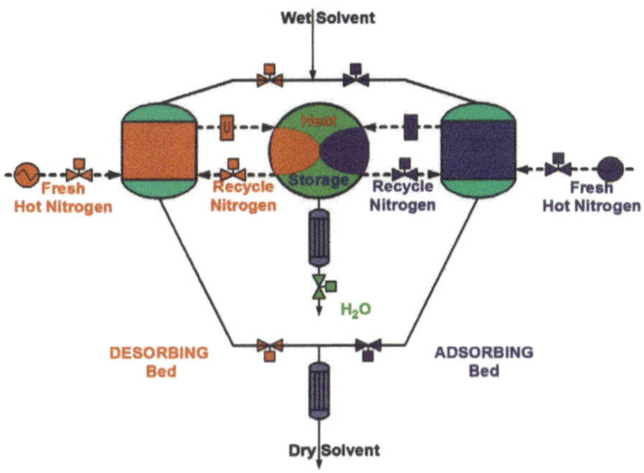

Figure 3.31

The preferred desiccant, a molecular sieve[85], is contained (packed) within cylinders. Desiccant dryers are generally made up of a pair of identical cylinders (light blue objects in Figure 3.31).

The desiccant contents of each cylinder is called a bed. One bed continuously *ad*sorbs water[XX] from liquid solvent being used in the cleaning process.

When the solvent emitted from the desiccant bed in one cylinder (the dark blue zone in Figure 3.31) becomes moist, the flow of wet solvent is transferred to the other cylinder (the red zone in Figure 3.31) while the first bed is regenerated (desorbs water).

Regeneration is done by treatment of the bed of desiccant with heated gas – usually hot nitrogen. *Ad*sorbed water is released from the desiccant internal surface because the attractive polar forces which bond water to the mineral are overpowered by the increased kinetic energy of the water molecules excited at the higher temperature. Said another way, the increase of temperature increases the solvent vapor pressure, making it more difficult for extra-molecular forces (polar bonding) to bind moisture to mineral. The gas is simply the agent which raises the temperature of the mineral, and has no additional effect on *ad*sorption.

Because gases hold so little heat (low specific heat), considerable volumes of hot nitrogen are required to desorb any quantity of water from a desiccant bed. A heat wheel[YY] or similar device (the green symbol in Figure 3.31[ZZ]) is commonly used to store and recover energy values. One does this to reduce both operating costs and the time to regenerate a bed of adsorbent. An adsorption/desorption system is shown in Figure 3.32.

Water evaporated from the desiccant by the hot gas saturates the gas and can be removed as a liquid by cooling.

3.7.7.1 Water, Water Everywhere

The outstanding question is how much water can be retained with the Class III cosolvent – 0.1%, 1%, 10% – without initiating harm during repeated cycles of cleaning?

Water not purged from the cosolvent cleaning system via *ad*sorption in a desiccator bed[86] can react with the cosolvents, corrode metal parts/baskets/cleaning equipment, consume heat transfer capability in the primary cooling condenser and distillation column[87], or ultimately dilute the cleaning power of the cosolvents.

With halogenated[AAA] solvents, only supernatant (immiscible) water can be efficiently removed, and *all* water reacts with the solvents[BBB]. Management of the retained water level devolves to management of added stabilizers which sequester the reaction products of the water and the halogenated solvent – corrosive mineral acids. So the answer to the question above with halogenated solvents is to use acid acceptance test kits and always maintain the positive presence of at least 0.2% NaOH equivalent[CCC].

For non-halogenated solvents, the obvious answer to the above question is that it depends – on the cosolvents, and how they are being used[88]. But there is one key difference between management of residual water level in non-halogenated solvents and that in halogenated solvents.

- It is that it is possible to lower the retained water level in non-halogenated solvents to nearly nil, by use of the desiccator bed noted above[89], and in Figure 3.31 [DDD].

[85] While all three types of *ad*sorbent can be and are being used to dry solvents in cleaning processes, the use of the somewhat more expensive mol sieves is strongly recommended because of their excellent selectivity for water. The 3 Å (3 Angstrom or 0.3 nanometer) size is preferred to accept water and reject all organic solvents a small kinetic diameter of 2.65 Å (0. 265 nm). The 4 Å size can allow some polar organic solvents to pass through pores and consume retention capacity for water. The 13 Å size can allow polar solvents, non-polar solvents, and water to pass through pores. Kinetic diameters for alcohols are 3.8 Å (methanol), 4.3 Å (ethanol), 4.6 Å (isopropanol), and 5.4 Å (isobutanol). Kinetic diameters for paraffins are 5 Å for butane and 5.8 Å for hexane. The 5 Å size (actual kinetic diameter of 4.3 Å) is used in a commercial process to separate smaller linear paraffins from larger un-branched-chain paraffins and cyclic hydrocarbons. The kinetic diameter of trichloroethylene is about 6.2 Å (0.62 nm) and that of methylene chloride is 4.0 Å (0.4 nm).

It is well known that temperature has a sizable effect on molecular sieving. The aperture size decreases with temperature decrease. Thus, for some molecules, molecular sieving occurs only below certain temperatures.

[86] Note in Figure 3.26 that water can be purged with the soil materials removed from the distillation column. This isn't an effective technique because the volatility of water much more resembles that of the RA cosolvent (highest volatility) than it does that of the soil materials. Said another way, water enters with the parts, is returned to the cleaning sumps by condensation in the primary condenser and by recycle of the RA (especially) and SA cosolvents from the distillation column, and so accumulates with both cosolvents.

[87] Recall that the latent heat of vaporization for water is roughly five times that of organic cosolvents.

[88] Some cleaning solvents are fully miscible with water (alcohols and glycols). Others are miscible only to a limited extent of a few percent (ketones and esters). Still other cleaning solvents are not miscible to any significant extent (hydrocarbons, HFCs, HCFCs, HFEs, and OSs).

[89] Desiccator beds are almost never used to dry halogenated solvents. The reason is simple – it's cheaper to purchase (and manage) stabilizer materials to control the reaction products of water than it is to purchase, manage, and pay the energy cost of operating a desiccator bed system to totally remove water.

Figure 3.32 Two Stage (Adsorption/Regeneration) System for Drying of Solvents

In general, most oxygenated solvents don't react with water. Those most likely to react are esters – to produce the organic acids and alcohols[90] from which they were produced[EEE].

All organic reactions are enhanced by an increase of temperature. So use of oxygenated solvents in a Class II cosolvent process raises less concern about reaction with water than does use of the same solvents in a Class III cosolvent process.

At operating temperatures above ~212°F (100°C), which are unlikely to occur in a Class II cosolvent process, and more than likely to occur in a Class III cosolvent process, reactivity of any solvent with water should not be surprising. But two other factors must be considered:

- Most importantly, the propensity for chemical reaction between water and cleaning solvent (and soils) increases when there are metal oxides present which can be catalysts. Metal oxides can play another reactionary role – they can be the reactant with water instead of the catalyst for its reaction with water.

Metal oxides exist throughout a metal cleaning operation[FFF], on the surfaces of the parts basket, the parts within it, or the cleaning apparatus – especially the heater. And heated metal oxides are the main material present in the desiccator bed[91].

Of somewhat lesser importance is that the propensity for chemical reactions between water and cleaning solvents increases, often in a non-linear fashion, when the amount of water present increases.

- Water can form azeotropes – many of them[92]. It forms blends (mixtures) with oxygenated and non-oxygenated[GGG]solvents which boil as if they were single solvents[HHH]. Of these, the vast majority are more volatile than expected[93]. The increased volatility makes it likely that these cosolvent/water blends will be recovered by condensation in the distillation column or the primary condenser.
- But, the Class III two-solvent soil-removal system will become more complex when water azeotropes must be managed. Instead of a simple mix of chemicals with two cosolvent components and a single soil component, each of these three components may be azeotropic with water – potentially doubling the number of components which must be managed.

This author's answer to the above question is that a desiccator bed of 3.0 Å mol sieves *must* be used because there is no better way to remove water, and that when the *ad*sorbing bed has broken through (reached capacity to hold water as detected by a simple in-line conductivity monitor), it should be taken off line and replaced with a freshly regenerated bed (as shown in Figure 3.31). In other words, the level of water which should be maintained in working cosolvents is zero. Nothing good can happen because of the presence of water – there is no reason to tolerate the presence of a potential for harm by using the desiccator bed[III] on a part-time basis.

3.7.8 About Oxygen

Oxygen and water molecules enter the cleaning bath with the parts, as a component of the air entrained with the parts in their basket. Whereas water is condensible, oxygen is not. Both are unlikely reactants.

Oxygenated solvents can be chemically oxidized. Alcohols are oxidized to aldehydes, aldehydes to ketones, and ketones to acids – the starting and finishing ingredients being the ingredients which produce an ester solvent.

But, these reactions don't just happen. They are catalyzed by other ingredients (such as acids or bases), and still other ingredients participate in the oxidation reaction. It is these other ingredients which provide the oxygen atoms for the oxidation, and to no surprise they are called oxidizing agents.

In other words, molecular oxygen, present as a component of air, isn't of concern as an oxidizing agent. Molecular

[90] Esters are not hydrolyzed to any significant degree without a catalyst, but the list of catalysts is long. Weak acids reversibly hydrolyze low molecular weight esters (Ref. 3, Chapter 14); weak bases hydrolyze the same esters irreversibly. Higher molecular weight esters (Ref. 3, Chapters 16 and 19) are hydrolyzed with strong bases. Soap is made from hydrolysis with concentrated sodium hydroxide of esters found in animal and vegetable fats and oils. Soap is the metal salt of the acid formed from hydrolysis. Metal oxides generally don't catalyze hydrolysis of esters.

[91] The tendency towards chemical reactions involving water and cosolvents on the hot desiccator bed surface is mitigated by the atmosphere within the bed being nearly free of oxygen as pure nitrogen is used to desorb the bed.

[92] Water is the most common azeotropic component. In the "bible" of azeotropic information, the reference given in Ref. 3, Appendix 2, there are 64 pages devoted to lists of binary azeotropes with water. Each page has approximately 50 entries of different azeotropes with water as one component. For comparison, there are 30, 35, 10, and 13 pages for methanol, ethanol, n-propanol, and isopropanol, respectively.

[93] The blend boils at a temperature less than that of either component, i.e., minimum-boiling.

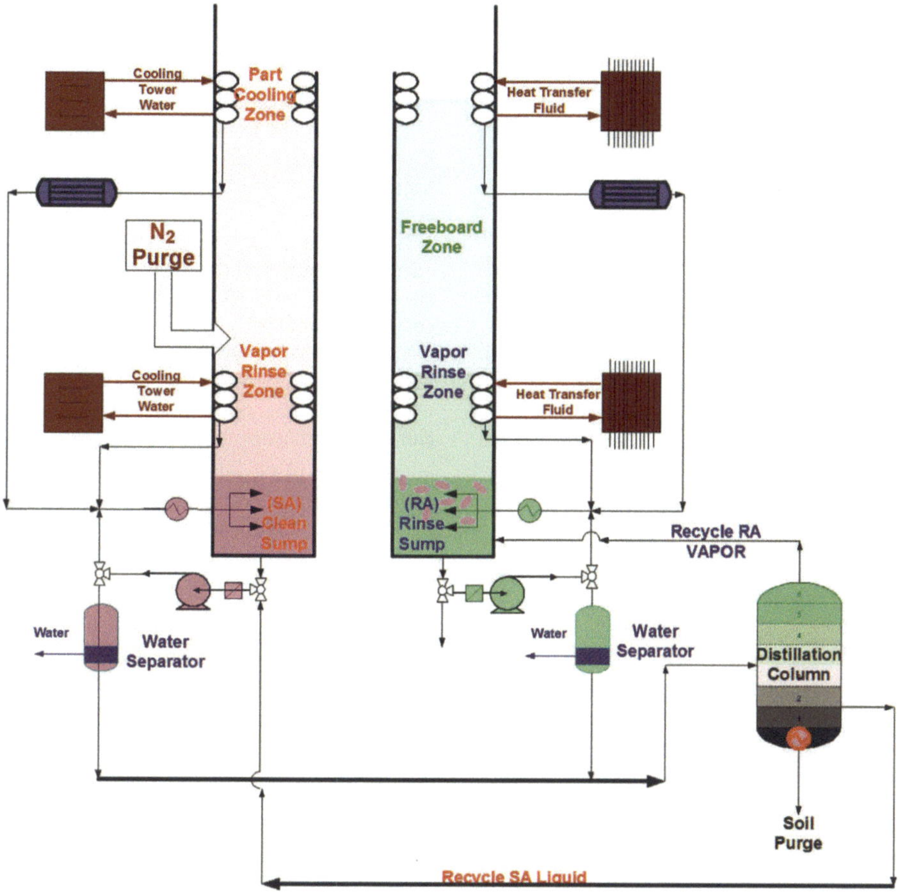

Figure 3.33 Dual Boiling Point Apparatus Exploiting Capability of Class III Cosolvent Systems – with Nitrogen Purge to Eliminate Oxygen Intrusion

oxygen doesn't oxidize oxygenated cleaning solvents[JJJ], except for ethers[KKK].

Nonetheless, if one is to practice the Class III cosolvent process with some SA solvents at their normal boiling points (which may range up to 400 to 500°F (~200 to 250°C)), prevention is the order of the day.

Above the SA cleaning sump, one is advised in this circumstance to establish a "barrier zone" of nitrogen. The incoming parts basket would have to pass through this zone in the freeboard volume before becoming immersed in the SA cleaning sump.

The oxygen-bearing air entrained with the parts will be diluted with impure[LLL] nitrogen and so become somewhat less vulnerable to unwanted thermal decomposition reactions in which oxygen molecules participate.

Practically, this "barrier zone" is produced by directing a small flow of nitrogen upward against the downward motion of the parts basket. This precautionary arrangement is shown in Figure 3.33.

3.7.9 Selection of Class II and Class III Cosolvents

This author has four clear preferences[94] for selection of cosolvents with both classes of cosolvent machines. Repeated and summarized they are:

- A minimum value of R_A between soil materials and the SA cosolvent to achieve the highest level of solvency of soil materials from surfaces,
- Only "Sanitary" emissions – those which don't raise concerns about flash point, being VOCs (in the US), meeting exposure low limits, and controlling odors,

[94] These four criteria play the crucial role in selection of the RA cosolvent, and hardly any role in selection of the SA cosolvent. This is because in the Class II cosolvent processes the RA cosolvent is the only fluid in direct contact with the ambient environment; in the Class III cosolvent process the same is so, plus the SA cosolvent is chosen for minimum volatility so fugitive emissions will be low.

- A maximum level of immiscibility (largest R_A) between the SA cosolvent and the RA cosolvent for *both* Class II cosolvent processes, and
- A minimum level of immiscibility (smallest R_A) between the SA cosolvent and the RA cosolvent for all Class III cosolvent processes.

These preferences are reflected in the previous selections of Class II cosolvents and they are extended to Class III cosolvents for a broad range of 25 soil materials in Table 3.7 and Figure 3.24. Obviously, one can make other selections using different and necessary preferences.

3.7.10 Comparison of Class III with Class II Cosolvent Systems

In Table 3.8 one can compare Class III with Class II cosolvent processes, making the comparison based on the same soil materials being cleaned by either process.

The selection of cosolvent pairs is at its core a compromise. One would prefer to find values of R_A between Class II cosolvents above 16 $MPa^{1/2}$. Similarly one would prefer to find values of R_A between Class III cosolvents approaching 0 $MPa^{1/2}$. One would also prefer to be rich.

The cosolvent pairs chosen for each soil and each process are the best which could be found using the extensive solvent data of Appendix 1, and the ability to engender compromise of the author. Some selections encourage selection of a process with its chosen cosolvent pairs; other selections encourage selection of another technology to clean the soil under consideration. These outcomes mirror the lessons about compromise taught by life (Chapter 3.7.10.2).

There are three areas of comparison: (1) limits on cleaning temperature, (2) quality of rinsing; and (3) overall practicality.

It is the first area in which the capability of the conventional Class II cosolvent cleaning process is severely limited, and where the capability of the Class III cosolvent cleaning process can provide superior capability. Solution cleaning temperature is a critical variable, not because it affects HSP but because it thins highly viscous soils and makes them less intractable (Ref., Chapter 1.9.2.1).

3.7.10.1 The Range of Solution Cleaning with Class II and III Cosolvent Processes

The normal operating temperature of the SA cosolvent in Class II cosolvent processes cannot be high, as it's limited[95] to the value of normal boiling point for the RA cosolvent in the Class II cosolvent processes. However, that for the SA cosolvent in Class III cosolvent processes is nearly unlimited. These two values are in the two columns printed in italic font in Table 3.7.

The increment between them is very large – averaging more than 154°C across the 25 soils listed in Table 3.7. Matched pairs (for those common soil materials) of temperature values are plotted in Figure 3.34.

It is this temperature increment which can be employed in solution cleaning of the same soil if one chooses a Class III cosolvent process vs. a Class II cosolvent process[96].

One may thus craft a Class III cosolvent process which removes a low molecular weight water-soluble wax used as a protectant for optic components – a highly valued cleaning operation.

- First, in a physically isolated cleaning chamber, heat the optical materials in water at a sub-boiling temperature. Then, in a physically unconnected cleaning chamber, dry all the water (the SA) by exposure to an HFC or HFE cosolvent (the RA) which boils at 120 to 140°F (50 to 60°C)[97]. The wax will be removed, the thoroughly dry optics won't be damaged by heat.

Or one may craft another Class III cosolvent process which removes coking residue (coal tar pitch) from semi-porous electrodes used in an electrolysis process – an extremely difficult but financially necessary cleaning task.

- Selected cosolvents are listed in Table 3.8 – the SA is triethylene glycol dimethyl ether (triglyme) whose boiling point is within the boiling range of the soil material[98]; the RA is HFC-245fa[99] whose boiling point is within the range of ambient temperatures, and whose surface tension value is less than 15 dynes/cm.
- First, in a physically isolated cleaning chamber, immerse the electrodes in the very hot SA cosolvent – possibly even conducting vapor degreasing operations at temperatures more than twice the boiling point of water! Mechanical agitation and long treatment cycles will likely be involved because of the high viscosity and softening temperatures of these soil materials.

---

[95] Recall that the SA cosolvent in the cleaning sump can't be heated above the normal boiling point of the RA cosolvent in a Class II cosolvent process because there is always (by intent) liquid RA cosolvent present – and the two sumps are physically connected. Recall also, that the SA and the RA cosolvent sumps are physically not connected in a Class III cosolvent process – so any desired temperature can be achieved.

[96] Within the capabilities of a Class III cosolvent process, there is no mandate to heat parts and soils to the maximum temperature for which the process is capable. One might only do cold cleaning in the SA cleaning sump. Alternately, one may do vapor degreasing at temperatures over 250°C (approaching 500°F) should that be necessary to dislodge thick viscous tar materials from rugged industrial parts.

[97] Note that this is the opposite of the Class I semi-aqueous cosolvent process, where water is used as the RA cosolvent.

[98] Coal tar (pitch) is produced by the carbonization, or coking, of coal. It's a black viscous liquid (or semi-solid), with a naphthalene-like odor (think moth balls), and reportedly a sharp burning taste. It's obtained by the destructive distillation of bituminous coal in coke ovens. One ton of coal yields 8.8 gallons of coal tar. Since it's a mixture, coal tar has a boiling range – typically from 190 to 400°C.

[99] A solvent better known as an agent for blowing foams. The rinsing mechanism will be dilution, not displacement because of the bare compatability of HFC-365mfc with triglyme.

Table 3.8 Comparison of Class II and Class III Cosolvent Pairs

	Class II Cosolvents (from Tables 3.3 and 3.4)						Class III Cosolvents					
Soil Material[MMM]	SA Cosolvent	Solubility R_A	T Boil °C	RA Cosolvent	Immiscibility R_A	*T Boil* °C	SA Cosolvent	Solubility R_A	*T Boil* °C	RA Cosolvent	Rinsing R_A $MPa^{1/2}$	T Boil °C
ASTM Fuel 'B'	OS-30 Decamethyl-tetrasiloxane	6.26	193.9	Dimethyl Carbonate	13.97	*90.0*	2-Ethyl Hexyl Acetate	5.24	*199.0*	HFC-365mfc	5.57	3.2
Auto Brake Fluid	Ethylene Glycol Diacetate	1.66	190.9	HFC-43-10mee	11.18	*54.4*	OS-30 Decamethyl-tetrasiloxane	4.28	*189.0*	HFE-7100	3.38	60.0
Auto Transmission Fluid	Isopar L	1.40	190.0	Dimethyl Carbonate	13.02	*90.0*	OS-30 Decamethyl-tetrasiloxane	4.28	*193.9*	HFE-7100	3.38	60.0
Bitumen	Propylene Glycol Diacetate	6.13	191.1	HFC-43-10mee	10.79	*54.4*	2-Ethyl Hexyl Acetate	4.50	*199.0*	HFC-365mfc	5.57	3.2
Castor Oil	Diethylene Glycol Monobutyl ether	4.86	218.0	HFE-7200	12.10	*76.0*	Diethylene Glycol Dibutyl Ether (Butyl Diglyme)	7.58	*256.0*	HFE-7100	6.00	60.0
Cholesterol	DBE	7.11	196.0	HFE 8200	11.88	*76.0*	Benzyl Alcohol	6.84	*205.3*	Methyl Formate	7.42	31.7
Coal Tar Pitch	Propylene Glycol Diacetate	5.54	191.1	HFC-43-10mee	10.79	*54.4*	Triethylene Glycol Dimethyl Ether (Triglyme)	6.02	*216.0*	HFC-245fa	6.42	14.9
Cottonseed Oil	Tripropylene Glycol n-Butyl Ether	7.69	276.0	HFC-43-10mee	10.12	*54.4*	Diethylene Glycol Dibutyl Ether (Butyl Diglyme)	7.41	*256.0*	HFE-7100	5.95	60.0
Linseed Oil	Propylene Glycol Diacetate	7.26	191.1	HFC-43-10mee	10.79	*54.4*	Diethylene Glycol Dibutyl Ether (Butyl Diglyme)	7.41	*256.0*	HFE-7100	6.00	60.0

(*Continued*)

Table 3.8 Comparison of Class II and Class III Cosolvent Pairs *(Continued)*

	Class II Cosolvents (from Tables 3.3 and 3.4)						Class III Cosolvents					
Soil Material[MMM]	SA Cosolvent	Solubility R_A	T Boil °C	RA Cosolvent	Immiscibility R_A	*T Boil* °C	SA Cosolvent	Solubility R_A	*T Boil* °C	RA Cosolvent	Rinsing R_A $MPa^{1/2}$	T Boil °C
Mil-H-5006 (Petroleum)	Isopar L	1.08	190.0	Dimethyl Carbonate	13.02	*90.0*	OS-30 Decamethyl-tetrasiloxane	5.04	*193.9*	HFE-7100	3.38	60.0
Mil-H-8446 (Silicate)	Texanol	2.51	244.0	HFC-43-10mee	11.15	*54.4*	Diethylene Glycol Dibutyl Ether (Butyl Diglyme)	4.35	*256.0*	HFE-7100	6.00	60.0
Mil-L-7808 (Ester)	Tripropylene Glycol n-Butyl Ether	6.72	276.0	HFC-43-10mee	10.12	*54.4*	Diethylene Glycol Dibutyl Ether (Butyl Diglyme)	3.73	*256.0*	HFE-7100	6.00	60.0
Mineral Oil (White Refined)	Isopar L	0.80	190.0	Dimethyl Carbonate	13.02	*90.0*	OS-30 Decamethyl-tetrasiloxane	4.71	*193.9*	HFE-7100	3.38	60.0
Mineral Spirits	Tripropylene Glycol n-Butyl Ether	5.10	276.0	HFC-43-10mee	10.12	*54.4*	Diethylene Glycol Dibutyl Ether (Butyl Diglyme)	1.64	*256.0*	HFE-7100	6.00	61.0
Motor Oil SAE 20W	Isopar L	0.69	190.0	Dimethyl Carbonate	13.02	*90.0*	OS-30 Decamethyl-tetrasiloxane	5.05	*193.9*	HFE-7100	3.38	61.0
Neets Foot Oil	Tripropylene Glycol n-Butyl Ether	6.25	276.0	HFC-43-10mee	10.12	*54.4*	Diethylene Glycol Dibutyl Ether (Butyl Diglyme)	3.57	*256.0*	HFE-7100	6.00	61.0

Olive Oil	Ethylene Glycol Diacetate	5.65	190.9	HFC-43-10mee	11.18	*54.4*	2-Ethyl Hexyl Acetate	1.74	*199.0*	HFE-7100	5.91	61.0
Palm Oil	Propylene Glycol Diacetate	6.54	191.1	HFC-43-10mee	10.79	*54.4*	Diethylene Glycol Dibutyl Ether (Butyl Diglyme)	4.05	*256.0*	HFE-7100	6.00	61.0
Pine Oil	Tripropylene Glycol n-Butyl Ether	7.80	276.0	HFC-43-10mee	10.12	*54.4*	OS-30 Decamethyl-tetrasiloxane	7.29	*193.9*	HFE-7100	3.38	61.0
Natural Rubber	Isophorone	9.91	210.5	HFC-43-10mee	10.20	*54.4*	Solvesso 150	6.50	*184.0*	HFC-43-10mee	7.35	54.4
Silicone Oil	Isopar L	5.59	190.0	Dimethyl Carbonate	13.02	*90.0*	OS-30 Decamethyl-tetrasiloxane	5.26	*193.9*	HFE-7100	3.38	61.0
Sperm Oil	Tripropylene Glycol n-Butyl Ether	7.38	276.0	HFC-43-10mee	10.12	*54.4*	2-Ethyl Hexyl Acetate	3.92	*199.0*	HFC-245fa	5.68	14.9
Trioctyl Adipate	Tripropylene Glycol n-Butyl Ether	5.27	276.0	HFC-43-10mee	10.12	*54.4*	Diethylene Glycol Dibutyl Ether (Butyl Diglyme)	3.32	*256.0*	HFE-7100	6.00	61.0
Turpentine	Solvesso 150	3.08	184.0	HFE 8200	10.19	*76.0*	Solvesso 150	3.08	*184.0*	HFC-43-10mee	7.35	54.4
VM&P Naphtha	Tripropylene Glycol n-Butyl Ether	6.66	276.0	HFC-43-10mee	10.12	*54.4*	Diethylene Glycol Dibutyl Ether (Butyl Diglyme)	3.06	*156.0*	HFE-7100	6.00	61.0

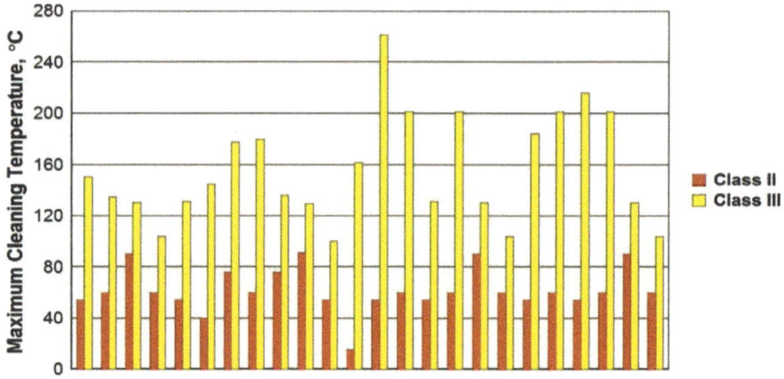

Figure 3.34 Maximum Cleaning Temperature for Common Soils in Various Cosolvent Processes

Transfer the somewhat-cooled wet porous electrodes to another physically isolated cleaning chamber[100], at a location adjacent to the primary cooling coils in the chamber (called the vapor rinse zone in Figure 3.33). Flush the electrodes with RA cosolvent vapor[101], which easily condenses to liquid in the primary cooling coils. The SA cosolvent and RA cosolvents are adequately miscible, the exceptionally low surface tension of the RA cosolvent will aid in flushing the porous structure of these electrodes.

These two examples illustrate the broad range of cleaning applications which can be considered by cosolvent processes – especially Class III cosolvent processes.

3.7.10.2 The Practicality of Solution Cleaning with Class II and III Cosolvent Processes

To gain the most value from Table 3.8 one needs a sense of the ridiculous, and/or of the practical. After all, a few of the choices made to clean certain soils with either Class II or Class III cosolvent processes do force recognition of the ridiculous. Not all of the choices listed in Table 3.8 are good, some are poor; a few are impractical. The reason all are included is to illustrate use of the selection method. After all, when confronted with an impractical choice one must conclude either another choice is necessary or that the selection method doesn't apply to the situation being considered – just as in life.

Examples from Table 3.8 are:

- Cholesterol is a soil not well suited to be cleaned by *either* type of cosolvent process. A Class II cosolvent process will not have adequate immiscibility between the cosolvents (producing poor rinsing, which is impractical) and a Class III cosolvent process will involve an exceptionally flammable cosolvent (whose use is ridiculous).
- In fact, all of the soil materials in Table 3.8 will be poorly rinsed in any Class II cosolvent process (all R_A values for the RA vs. the SA are well less than 14 to 16 $MPa^{½}$)[102]. So the Class II cosolvent process is an impractical choice for these many soils.
- When the Class III cosolvent process is considered for use with some soils, the RA best matched to the SA is HFC-365mfc. To use a cosolvent whose boiling point is 89°F (32°C) in an open-top machine is ridiculous. But a dual-chamber machine with one chamber enclosed might be an entirely practical choice for certain parts.

Other examples from Table 3.8 stimulate no sense of the ridiculous and appear quite practical. They include:

- Soil materials such as auto brake fluid, auto transmission fluid, MIL-H-5006 (petroleum), mineral oil (white refined), motor oil SAE 20W, pine oil, and silicone oil offer excellent potential for cleaning by the Class III cosolvent process. The R_A values between soil and SA cosolvent are acceptable, the R_A values between RA and SA cosolvent are also

[100] It is likely that the rinsing chamber for this application is an enclosed machine because of the exceptionally low boiling point of the RA cosolvent. It would be impractical to conduct this operation in an open-top machine.

[101] The RA cosolvent vapor should contain little carbonaceous soil or SA cosolvent because of the more than 200°C temperature difference between the overhead product of the distillation column and the fluid material at the bottom of the distillation column.

[102] With but a single exception, there are NO circumstances where an RA cosolvent can be chosen to be acceptably immiscible with a suitable SA cosolvent for any Class II cosolvent process. This statement is based on the requirement for acceptability noted in Figure 3.35, that the R_A separation between SA and RA cosolvents be 14 to 16 $MPa^{½}$ at a minimum.

The author used the HSP data of 210 cleaning solvents (from Ref. 2, Appendix A1), the 25 soil materials listed in Table 3.27, a spreadsheet, and an automated analysis scheme.

No suitable (R_A separation value less than 8 $MPa^{½}$ from any soil material) SA cosolvent could be found, which was separated in "HSP space" by 14 to 16 or more $MPa^{½}$ from any RA cosolvent. Nearly all values of separation distance were between 11 and 12 $MPa^{½}$.

The exception is water, being used as an SA cosolvent with a hydrocarbon RA cosolvent (such as kerosene). This is the opposite of the Class I cosolvent process, where water is used as the RA cosolvent (Chapter 3.7.11). Unfortunately, water (without surfactants) is not a suitable SA cosolvent for many soils.

acceptable, and both RA and SA cosolvent are VOC exempt (US).

- Soil materials thought to be difficult to clean such as natural rubber, coal tar pitch, Mil-H-8446 (silicate), sperm oil, and Mil-L-7808 (ester) appear to be excellent applications for the Class III cosolvent process.

In summary, no cosolvent system is "better" than another, as no solvent is "stronger" than another. Class II cosolvent systems can be excellent choices where there is adequate immiscibility between cosolvents; Class III cosolvents can be excellent choices where there is adequate miscibility between cosolvents. And both can be ridiculous or impractical choices where neither is true.

3.7.11 What Immiscibility Really Means

It means that Class II cosolvent processes won't be recognized for providing part surfaces free of the SA cosolvent. This is because the RA cosolvent (the rinse fluid) is not adequately immiscible with (too soluble in) the SA cosolvent. Experientially, these processes aren't recognized for the quality of their rinsing of SA cosolvent by RA cosolvent.

To rinse one fluid from a surface by *displacing* it with another fluid, the two fluids must not be soluble in one another. Otherwise, the surface won't be free of the first fluid until it has been fully diluted with the second fluid. And that, theoretically at least, will require an infinite time and volume of the second fluid! Where possible, rinsing by displacement is far quicker, more efficient with volume of rinse fluid, and highly capable of producing surfaces more free of the undesired fluid than is rinsing by dilution.

- The ultimate limit on the effectiveness of rinsing is the relative compatibility of the fluid being rinsed with the fluid used *for any method of rinsing*. For displacement rinsing, the least compatibility is preferred; for dilution rinsing, complete compatibility is preferred.

One can study the incompatibility between solvents produced by correlation of their intermolecular forces as characterized by and enumerated with Hansen Solubility Parameters[NNN]. The considerable data of water solubility in many cleaning solvents is a useful database[OOO]. It can be studied in terms of the separation between water and solvent in "HSP space." The illustrations[PPP] contained in Table 3.9 allow the compatibility of water with specific cleaning solvents to be visually examined with three-dimensional images[103], and also the solubility of water in many cleaning solvents to be recognized by measured data (Figure 3.35).

The information in Table 3.9 allows one to place Figures 3.27 (kerosene and water, kerosene dryer, $R_A = 24.6\ MPa^{1/2}$) and 3.28 (Soygold 1000 and HFE-7200, Class II B cosolvent process, $R_A = 11.91\ MPa^{1/2}$) in context.

- Kerosene should be (and has been) an excellent fluid to displace water from surfaces, and
- HFE-7200 should be (and is) too miscible with Soygold 1000 to acceptably and reproducibly displace it from cleaned surfaces as a rinsing step.

3.7.11.1 The Effect of Immiscibility

The second comparison between Class II and Class III cosolvent processes can be made for some[104] of the soil materials listed with candidate cosolvent processes in Tables 3.3 (Class II A – flammable) and 3.4 (Class II B nonflammable). Calculated solubility parameters for both classes of cosolvent processes when used with the common soil materials are found in Table 3.9[105]. At least three observations can be made:

- Cleaning quality should be acceptable and equivalent among both cosolvent classes[106].
- *Displacement* rinsing in Class II cosolvent processes certainly won't be excellent, probably won't be acceptable[107], and may not even be attempted.
- *Dilution* rinsing in Class III cosolvent processes may be as acceptable as desired[108].

The information in Ref. 2, Chapter 3.29.1 about the maximum limitation of R_A for acceptable solvent cleaning (which produced a limiting value of 8 $MPa^{1/2}$) and the information in Table 3.9 (which suggests that R_A for displacement rinsing should be at least 14 to 16 $MPa^{1/2}$, or more) form the basis for design of solvent cleaning systems.

These relationships and capabilities are shown in Figure 3.36. From this illustration, and the above text, one should not conclude all Class II cosolvent processes will produce unsatisfactory outcomes. The proper conclusion is that *displacement* rinsing is likely not to be

[103] Each image represents two chemicals, water and a selected solvent.

[104] The reason that not all of the soils noted in Tables 3.3 and 3.4 are listed in Table 3.9 is that cosolvent pairs could not be found for them which meet the requirements of a Class III cosolvent process.

[105] Remember the differences in the Class II and Class III cosolvent processes. The former is designed to operate at lower cleaning temperatures with immiscible SA and RA cosolvents; it's the opposite for the SA with the latter.

[106] The R_A value between soil materials and the SA cosolvent for both classes averages about 5 $MPa^{1/2}$, with the Class III cosolvent process being about ½ unit better (lower R_A) than the Class II cosolvent process.

[107] In the Class II B cosolvent process with Soygold 1000 as the SA cosolvent and HFE-7200 as the RA cosolvent, the R_A between these "immiscible" SA and RA cosolvents is only 11.2 $MPa^{1/2}$. This cosolvent process is not known for excellent rinsing. Figure 3.28 shows where the HSP overlap between solvents is insubstantial compared to that of a kerosene dryer in which excellent rinsing is completed (see Figure 3.27).

[108] If the SA cosolvent is miscible with the RA cosolvent, the limitation on rinsing quality is the amount of contact provided by the user. And the separation between SA and RA cosolvents for Class III cosolvent processes listed in Table 3.9 averages about 5.5 $MPa^{1/2}$.

Table 3.9 Two Illustrations of the Solubility of Water in Cleaning Solvents

2-Ethyl Hexyl Acetate
R_A =13

Texanol
R_A = 14

Methylene Chloride
R_A = 14

Triethylene Glycol
Dimethyl Ether, R_A = 15

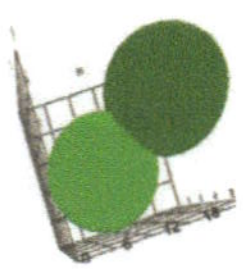
Acetone
R_A = 13

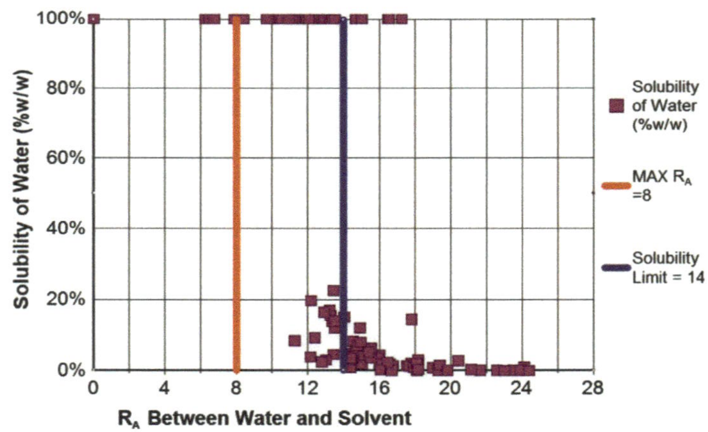

Figure 3.35

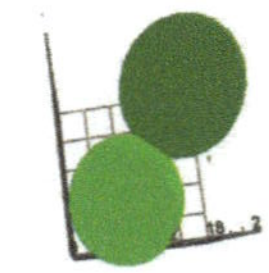
Tripropylene Glycol
n-Butyl Ether, R_A =15

Isopropanol
R_A = 12

The solubility of water in cleaning solvents is determined by their compatibility with water.

Compatibility can be characterized by the separation or intersection in "HSP space" of the solvents and water.

Visually, compatibility is recognized by the intersection (overlap) or separation of solubility spheres, each representing one material.

By independent solubility measurements, compatibility is recognized by the value of R_A being just greater than about 14 $MPa^{1/2}$ (Figure 3.35)

Practically, they both represent the same phenomena — immiscibility.

Methyl Propyl Ketone
R_A = 16

Triethylene Glycol
Methyl Ether, R_A = 11

Hexylene Glycol
R_A =10

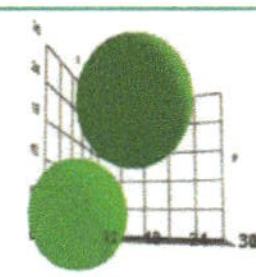
Isobutyl Acetate
R_A = 18

1,4-Dioxane
R_A = 17

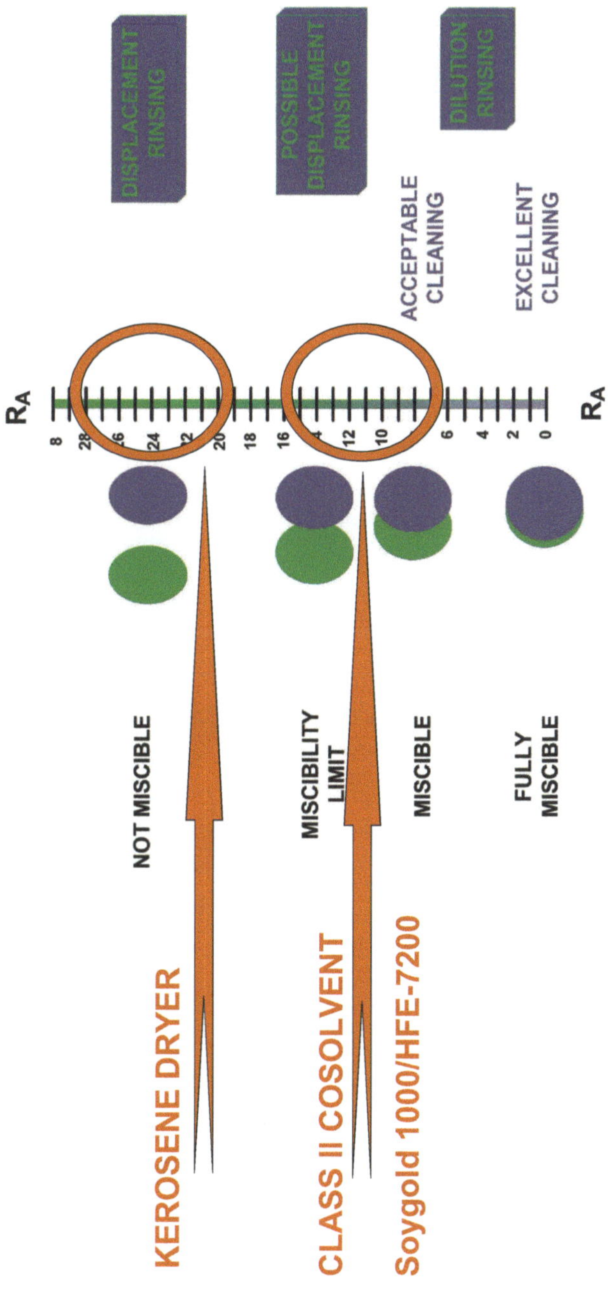

Figure 3.36 Summary of Compatability Relationships Between Miscible and Immiscible

satisfactory for the two cosolvents noted in the figure because their mutual immiscibilities are unlike that of the kerosene dryer.

3.8 PROCESS DESCRIPTION: CLASS IV – AZEOTROPIC

Class IV cosolvent processes are simple. Cleaning work is done with them as if only a single solvent were involved. And, the Class IV cleaning process may[QQQ] be conducted in a standard NESHAP-compliant open-top vapor degreaser[109]. That's simple!

Actually, while the above statements are correct, they do not fully describe Class IV cosolvent processes. The additional needed information is that the integrity of the azeotropic composition can be lost over time of use. And when the integrity of an azeotropic blend is lost it becomes only a mixture of chemicals.

Surprisingly, integrity of the azeotropic composition must be maintained external to, not internal to, the vapor degreaser. That's done with an on-site distillation column – a device commonly found in solvent cleaning processes.

3.8.1 A Misunderstanding about Binary Azeotropes

As one uses the other classes of cosolvent processes to achieve a specific aim, so some use Class IV cosolvent processes to achieve one of two specific but different aims. However, only one of them is achievable; the other is illusory.

The two aims are to: (1) clean soils with a solvent blend[110] whose HSP values can't be matched by any single solvent which doesn't raise SHE or other concerns[RRR], or (2) clean multi-component soils with a blend of solvents where at least one of the solvents preferentially solubilizes one of the soil components in a simple process.

Aim 2 is illusory.

These two aims are quite different. They engender very different visions of what an azeotrope is and how it can be used.

- The first vision is illustrated in Figure 3.37, for the azeotrope of cyclohexane and isopropanol (entry 397 in Ref. 3, Table A 2–2) with 67% alcohol. The vision is that the two solvent components are not capable of being separated into those components (because they are locked together).
- A binary azeotrope is a bulk blend of two solvent components which acts as if it were a third solvent. The blend boils at a temperature different from that of either solvent, displays HSP values intermediate to those of either solvent, and the azeotropic components cannot be separated by means of simple multistage distillation. In their volatility properties and behavior they are locked together.
- The second vision is that some components of an azeotropic blend can be attracted to specific soil components, and so act to solubilize them. That would involve "breaking"[111] the binary azeotrope and enabling one or both of the components to act as an individual solvent.
- The second vision is illustrated in Figure 3.38, which is derived from Figure 3.37.
- The "lock" is shown as having been opened.
- So, each of two solvent components is "allowed" to interact within a solution with other materials.

Figure 3.37 Visual Representation of the Solvent Interrelationships of a Binary Azeotrope

[109] As noted in Chapter 3.7.3 and Figure 3.26, Class III cosolvent processes use solvents which could be present as a single phase, but are conducted in separated (split or dual) chambers so that cleaning and rinsing can be done at different temperatures.

[110] While azeotropes with as many as five components are known, only binary azeotropes are commonly used in solvent cleaning. There are a few three-component cleaning azeotropic solvents patented (and fewer still commercialized).

But simplicity rules! Management of an unknown number of soil components, plus management of more than two components within the cleaning solvent, is generally thought of as being "a bridge too far."

[111] Azeotropic (or zeotropic) components cannot be separated in simple multistage distillation processes, or generally by addition of "special" components. In other words, the azeotrope can't be "broken" or "unlocked." The noun "broken" is conventionally used to mean that the azeotropic components can be separated by simple multistage distillation; that the components of the "broken" azeotropic blend display the normal boiling points and individual HSP values of each component. *None of that is so!*

Binary azeotropes can be and are commercially disrupted (not "broken"). But, that requires designed chemical processes involving additional solvent components and very specific conditions to do so. Individual azeotropic solvent components can't be liberated from an azeotrope by attraction of some soil components.

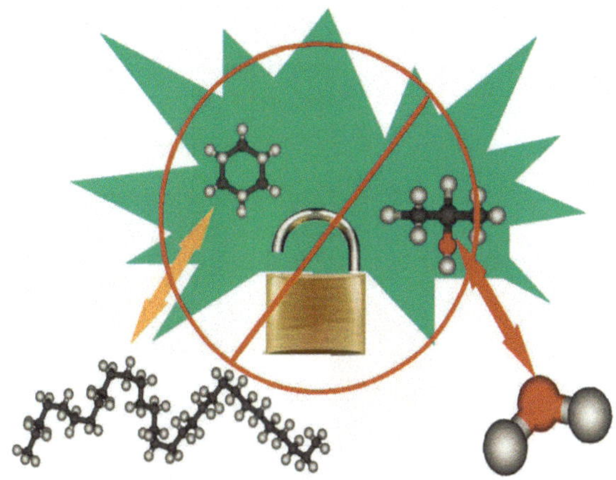

Figure 3.38 Visual Representation of What a Binary Azeotrope is NOT

- The cyclohexane is expected to share disperse electrostatic forces with other saturated hydrocarbons, such as an oil. As a single solvent, it is often used for that purpose.
- The isopropanol[SSS] is expected to share polar and hydrogen-bonding intermolecular forces with water. As a single solvent, it is highly soluble in water[112].

What makes Figure 3.38 a mirage[TTT] is that it's not scientifically possible.

The "lock" doesn't open in that way. Soil components don't attract individual components of an azeotrope to the extent that the integrity of the azeotrope is threatened. That's the red "**NO**" symbol in Figure 3.38. There are commercial examples[UUU] of how specifically tailored chemical processes with additional ingredients and specialized facilities are needed to disrupt (deconstruct) commercially significant binary azeotropes.

In the 1990s, three factors drove a rush to adopt binary azeotropes for cleaning multi-component soils, with each azeotropic component targeted at one or more soil components: (1) *urgency*, the need to comply with the US EPA's phaseout schedule[VVV] for ozone-depleting chemicals (ODCs) when proven options were scarce or unavailable; (2) an *incomplete understanding*, within the parts cleaning industry, about the chemical nature of azeotropes – because there was little experience in their use; and (3) the incontrovertible fact that there were *successful applications* of surface cleaning where it was believed, without proof, that both azeotropic components had contributed to a positive outcome[WWW].

Unfortunately, this second vision, actually a misunderstanding, was the basis for design, construction, purchase, and use of a considerable number of Class IV cosolvent machines.

[112] In fact, isopropanol and water form a binary azeotrope (entry 791 in Ref. 3, Table A 2–2, 87.7 % alcohol).

3.8.2 Management of Water, and other Uncertainties

They are called cosolvent processes because more than one solvent is used to clean parts in the same application (Chapter 3.1). Some of these processes are chameleons in that they appear externally to be conducted in conventional vapor degreasers – having a two-sump immersion cleaning apparatus and an attached distillation column.

But cosolvent facilities aren't designed and operated as are conventional vapor degreasers having only a single cleaning solvent. In a Class IV cosolvent process, with two solvents locked within a binary azeotrope, the integrity of that azeotrope must be managed. That's done by providing some unusual attention to the: distillation column, vapor degreaser, and selection of the binary azeotrope.

Another perspective on the differences between a single-solvent process and the Class IV cosolvent azeotropic process is that facilities and care are needed to manage an uncertainty that is certain to occur.

- The uncertainty is what tramp solvent components will contact the azeotrope, what will be the effect of those components on the binary azeotrope, and how will the integrity of the azeotrope be managed within the distillation column to account for that contact. The only certainty is that there will be tramp solvent components!

Water is one such tramp component. It enters chiefly as humidity in the air around the parts and basket[XXX]. The uncertainty is the fate of both the water and the azeotropic blend. Depending upon the specific azeotropic blend being used, the water:

- May dissolve in the azeotropic blend[113],
- Be immiscible with the azeotropic blend (Chapter 2.16),
- Integrate with the binary azeotropic blend to form a ternary azeotrope[YYY], or
- Do something else[ZZZ,114].

[113] Solution of water, or any tramp solvent, into an azeotrope is governed by the relationship between the values of solubility parameters manifested by the azeotrope and those of the tramp solvent. For single solvents and water, the relationship (separation distance) between their solubility parameters (R_A) and solubility is plotted in Figure 3.35. Shown in that figure is a sharp separation between the solubility and insolubility of water in single solvents which occurs about an R_A value of 14 to 16 $MPa^{½}$.

[114] One may inquire about how "breaking" of binary azeotropes by soil components (such as water) – which does not happen – is distinct from disruption of binary azeotropes by tramp contaminants (such as water) – which does happen. The difference is in the outcome. The former presupposes that the binary azeotrope of cyclohexane and isopropanol liberates cyclohexane to dissolve oils and then the binary azeotrope is reconstituted when the oil is separated. The latter recognizes that the binary azeotrope of ethanol and toluene forms a stable ternary azeotrope of ethanol, toluene, and water; or that ethanol, heptane, and water do the same, or that the binary azeotrope of isobutyl acetate and isobutyl alcohol upon contact with water is disrupted to form two separate and immiscible binary azeotropes.

Whatever chemical contamination is encountered by the binary azeotrope can usually be removed within and around the distillation column associated with a Class IV cosolvent process.

3.8.3 Management of Azeotrope Integrity with a Distillation Column

The distillation column associated with a Class IV cosolvent process is not that found within a conventional single-solvent vapor degreasing process.

- That in the single-solvent vapor degreasing process only operates when needed (batch mode), and produces a different quality of overhead (OH) product throughout the time span of processing a batch of soiled solvent. The quality is initially rich in the volatile solvent components, and finally rich in the least volatile soil components[AAAA].
- That in the Class IV cosolvent process operates continuously, and produces the same quality of overhead (OH) product throughout its use.

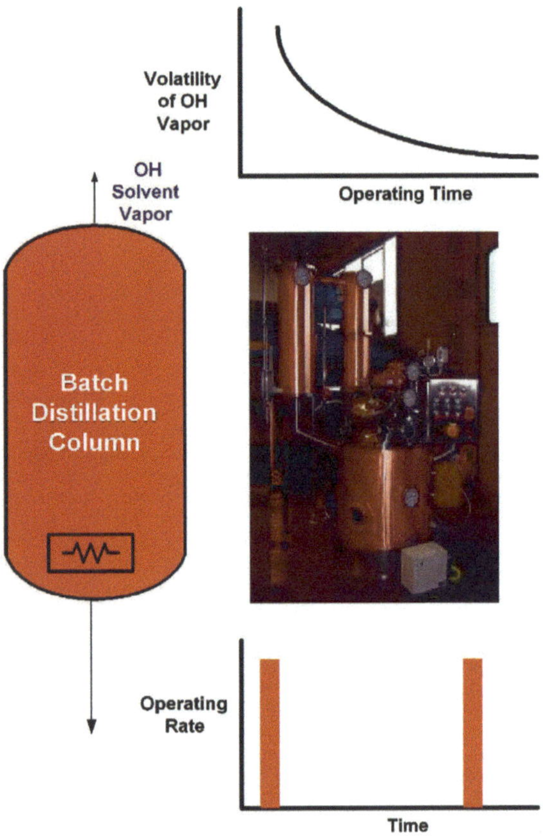

Figure 3.39 Batch Distillation Column

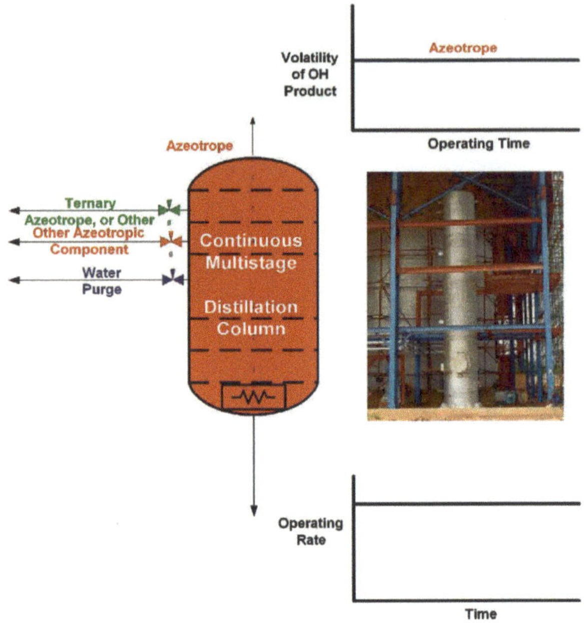

Figure 3.40 Continuous Distillation Column

These capabilities can be compared visually, first by comparing the two types of distillation columns in Figures 3.39 (batch[BBBB]) and 3.40 (continuous); and second by comparing the full installations of distillation column and vapor degreaser in Figures 3.41 (single solvent) and 3.42 (Class IV cosolvent).

The batch-use distillation column exists to separate single cleaning solvents from soil components. The fluid distilled is a single cleaning solvent.

Water is removed from cleaning systems that use single solvents in two ways: (1) insoluble water is removed in a decanter (Figure 3.14), and (2) soluble water is indirectly but effectively removed by reaction with solvent (halogenated) which reaction product then is consumed by reaction with sacrificial stabilizer chemicals[CCCC].

The continuous-use distillation column exists to recover the pure binary azeotrope for further use, and to exclude all other contaminant(s) in the cleaning fluid – including soil components. Identification of non-soil contaminants, as noted in Endnote UUU and Footnote 124, is uncertain. But almost certainly they include water and the products of its interaction with the binary azeotrope or its components (Box 3.9); in addition to any excess of one azeotropic component[115]; and occasionally some unexpected materials.

Integration of the batch-use distillation column into a single-solvent vapor degreaser and the continuous-use

[115] It is nearly impossible not to find an excess of one of the two azeotropic components. For example, mistakes are common in formulation of blends by users.

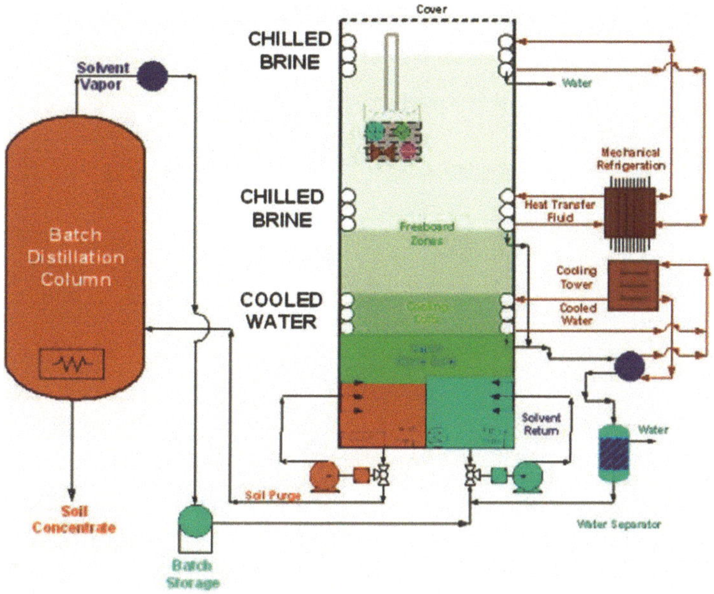

Figure 3.41 Single Solvent Vapor Degreaser

Box 3.9 **Management of Multiple Azeotropic Components**

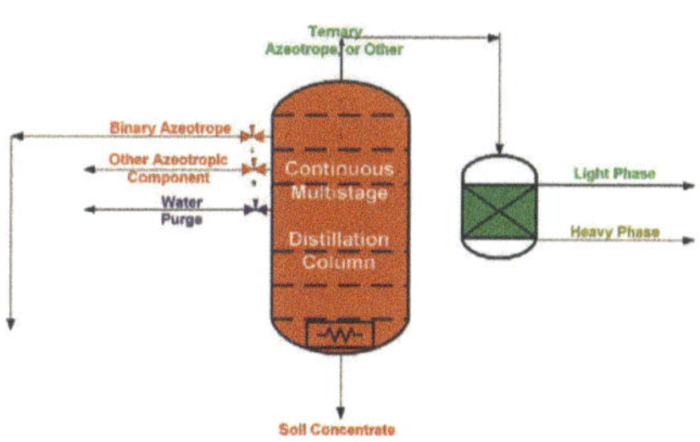

In Figure 3.40 a ternary azeotrope is shown as being slightly less volatile than the binary azeotrope – it is recovered nearer the top section of the distillation column. That assignment was made for visual convenience in this book.

It may be, in actual cleaning operations, that ternary azeotropes or multiple binary azeotropes (both are reported formed in the reference of Endnote TTT) are the most volatile materials, and the binary azeotrope used for cleaning operations is not the overhead (OH) product of the distillation column.

In that case the binary azeotrope being used for cleaning is recovered for reuse as a side draw (product) from the continuous distillation column. Water is the likely tramp impurity producing this conversion of the binary azeotrope to other coordinated solvent assemblies.

In such situations, the material values of the overhead product must be recovered. The image above shows one method of doing this (albeit incompletely) – phase separation, a side draw to recover the intact binary azeotrope for reuse in cleaning operations, and another side draw to remove the water (presumed to be less volatile than the binary azeotrope or its more volatile products of conversion.

distillation column into a Class IV azeotropic cosolvent vapor degreaser is shown in Figures 3.41 (and Ref. 2, Figure 7.27[116]) and 3.42, respectively.

Note in Figure 3.41 that: (1) there are three refrigerated zones within the vapor degreaser– the third one at the top being used to minimize entry of water, (2) the distilled solvent cannot be returned to the vapor degreaser as vapor – to save energy – because it is produced in batch mode and provision for storage of vapor is impractical, (3) cleaning quality can't be constant because the soil level in the cleaning sumps varies as soil accumulates prior to a batch purge to the distillation column, and (4) distilled solvent is added to the sump where the solvent is most clean – the rinse sump[117].

[116] In both Figures 3.31 and Ref. 2 Figure 7.27 a storage vessel is shown so that the production from the batch still may be continuously separated in time from operation of the vapor degreaser.

[117] As is done with aqueous cleaning technology when the cleanest rinse fluid is used as the last fluid to contact the parts.

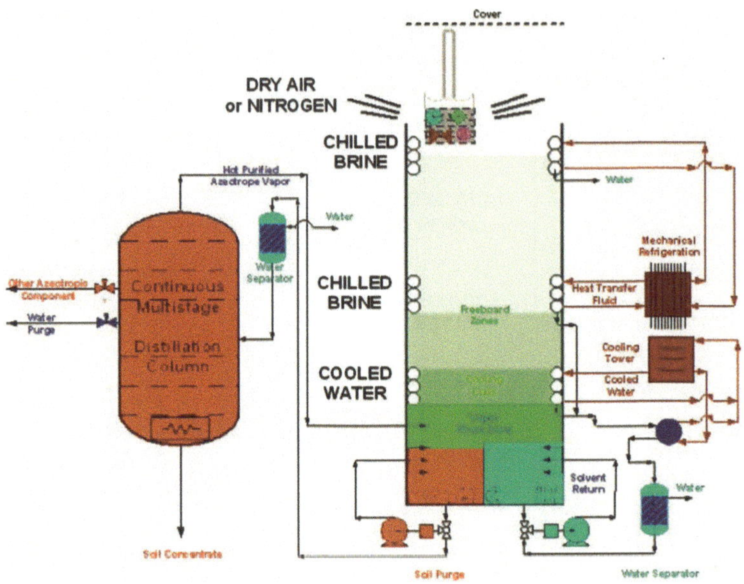

Figure 3.42 Class IV Cosolvent Degreaser

Integration of the continuous distillation column into a Class IV cosolvent vapor degreaser is shown in Figure 3.42.

Note in Figure 3.42 that there are four facilities employed to remove water, because of the above concerns: (1) a top refrigeration coil, (2) a purge station to displace humid air with dry air or nitrogen from the parts basket prior to its entry into the freeboard area of the degreaser, (3) a decanter-type water separator to remove any water condensed in the two lower refrigeration coils, and (4) a second decanter-type water separator to remove any insoluble water prior to the distillation column[118].

Further note in Figure 3.42 that: (1) it is assumed (as most commonly occurs) that there is no interaction of the binary azeotrope with water or any other tramp impurity, (2) azeotrope vapor can be returned to the vapor rinse zone of the vapor degreaser if desired (to save energy), (3) it is necessary through experience to learn the locations of positions in the continuously operating distillation column from which purge streams must be taken and when and how much[119], and (4) cleaning operations in the vapor degreaser should be conducted normally.

While necessary, and possibly appearing to be complex, these facilities are commonly and straightforwardly used in the production of all cleaning solvents. They should be familiar to any reputable supplier of vapor degreasers.

3.8.4 Selection of Binary Azeotropes for Class IV Cosolvent Processes

The characteristic of boiling at a constant[120] temperature and composition[121] is not a random event. One can predict which few compositions will be azeotropic (Box 3.10) from the principles of thermodynamics, and other methods.

Compositions of many binary azeotropes are listed in Ref. 2, Chapter 8, especially Tables 8.1 through 8.8 which are provided to suggest binary azeotropes suitable for replacing single solvents about which there are SHE concerns. Many binary azeotropes have only VOC-exempt (in the US) components (Table 8.9) and many are based on water as one component (Table 8.10).

---

[118] This latter separator may prove to be unnecessary.

[119] Solvent values from these two streams are easily recovered. First, only a single purge stream may be necessary. Second, water can be removed as a contaminant by passing the cooled stream through a bed of desiccant such as mol sieves (molecular sieves, Footnote 93). Third, dried solvent materials may be returned to the tank from which the binary azeotrope is formulated.

[120] Usually, binary azeotropes selected for cosolvent cleaning happen to be minimum-boiling azeotropes. This means that their boiling point is less than those of either solvent component. Probably three quarters of the more than 30,000 identified binary azeotropes are of the minimum-boiling type; most of the balance are of the maximum-boiling type; and a few of the balance are neither as their boiling point is between those of their constituents. And probably a similar fraction are homogeneous blends; the balance being heterogeneous.

One can't independently specify the characteristic of minimum (or maximum) boiling point and the two solvent components of the azeotrope. That would be like choosing a spouse and wanting to change their height, date of birth, fingerprints, or previous life experiences.

[121] Not only does this pair of solvents boil at the same temperature, they boil at a constant composition – as long as some of each component is present (Chapter 3.6.10). Then the boiling temperature becomes that of the remaining azeotropic component.

A further resource is Ref. 3, Appendix 2, Table A 2–2, in which more than 800 binary homogeneous azeotropes are listed. For each azeotrope the following information is supplied: the type of homogeneous azeotrope (boiling point suppression, elevation, or neither); its weight composition, boiling point, calculated density, calculated HSP values; notation of an absence of measured flash point; and some literature or patent references. The table is sorted by total solubility parameter (Hildebrand).

Writ large throughout this book is the message that cleaning solvents are to be matched to soil material components through Hansen Solubility Parameters (HSPs), and chosen to minimize concerns about safety, health, and environmental (SHE) issues. Certainly, those guideposts line the path from application development to operational success.

But there is a third factor necessary to select binary azeotropes for Class IV cosolvent processes – it is their compatibility with water[122]. One needs to know this to properly design and operate the continuous-type distillation column. It is unlikely this factor can be evaluated through the scientific literature; a simple laboratory-scale experiment[123] is necessary.

Though water is the single tramp impurity guaranteed to exist, others are possible, and likely in some industries.

Box 3.10 Deviations from Ideality

The formation of azeotropes, or constant-boiling blends, is known to be a function of the non-ideality of the azeotropic solution, combined with the difference in boiling point between the two pure components. In general, the deviations from ideality can be attributed to contributions from dispersion interactions, dipole/dipole interactions, dipole-induced dipole interactions and hydrogen bonding, this last being one of the most important contributors to azeotrope formation. Ref. 3, Appendix 2 has much more details.

"Deviations from ideality," as enumerated above, are the characteristics of single solvents which determine their solubility parameters (Ref. 2, Chapter 2). See also Ref. 2, Chapter 8.

3.9 COMPARISON BETWEEN COSOLVENT PROCESSES

There is a function provided by each cosolvent of the pair of solvents which comprises a cosolvent application:

- Class I – Primarily used to implement solvent cleaning technology with non-volatile hydrocarbon-based cleaning solvents, whose HSPs match those of soil components, and yet be able to rinse those solvents with water off the surfaces of certain parts.
- Class II A and II B – Primarily used to implement solvent cleaning with both high and low-volatility cleaning technology agents; whose HSPs match those of other soil components, and some of which are VOCs in the US; while restricting emission of some of those cleaning agents as air pollution by displacement of them from part surfaces with immiscible VOC exempt (in the US) solvents. Simply, the VOC-exempt RA cosolvent acts as a "blanket" or barrier to block emission of VOC SA cleaning cosolvents[124]. Nearly all Class II cosolvent

122

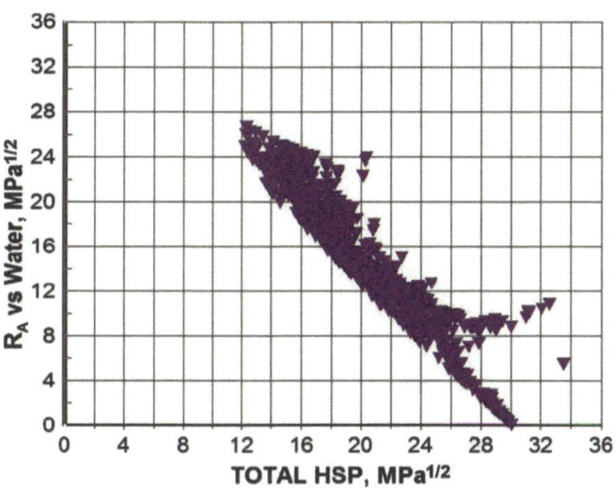

Of the more than 800 binary azeotropes listed in Ref. 3, Appendix 2, slightly less than half (377) are likely to not raise concern about compatibility with water as they either don't contain water or their R_A values relative to water are more than 16 $MPa^{1/2}$ (Figure 3.36). Of the remainder, 103 do contain water (VOC exempt in the US), and an additional 379 have R_A values relative to water less than 16 $MPa^{1/2}$ and are likely to be miscible to some extent with water. It is these latter 379 which have more (rather than less) potential to interact with water.

Solvents compatible with water, i.e. those with low values of R_A relative to water, are those which display a high level of total HSPs. Not surprising this outcome because water has a high level of total HSPs (24.9 $MPa^{1/2}$) and so would solvents compatible with water be expected to also have a high level of HSPs. This is clearly shown in the figure above which is based on all data from Ref. 3, in Appendix A1.

123 The straightforward laboratory experiment would involve boiling the binary azeotrope being considered with ~10 weight percent water added under total recycle, measuring the internal temperatures in the boiling apparatus, and observing/analyzing the resultant mixture. Measurement of a changed boiling point indicates new azeotropes were formed; observation of interfaces indicates immiscible materials were formed. If neither outcome is measured or observed, concern for unusual operation of the continuous-type distillation column is warranted.

124 As this book is published, interest in the US in Class II cosolvent machines is apparently located in California. These machines are a practical way of conducting useful solvent cleaning operations while remaining in compliance with highly restrictive regulations about emissions of VOCs.

Table 3.10 Differences among Cosolvent Processes and How They Work

Process Step	Vapor Degreasing	Cosolvent Processes				
		Class I	Class II		Class III	Class IV
			Immiscible			
		Semi-Aqueous	A (Not Flammable)	B (Flammable)	Miscible	Azeotropic
Example Cleaning Agent	Trichloroethylene	Terpene Detergent Surfactant	"3A" Solvents (Flash Point > 140 F); hydrocarbons, glycols, glycol ethers, esters, dibasic esters	Isopropanol	Glycol Glycol Ether Glyme (Glycol Ether) Ester	Cyclohexane Isopropanol
Example Rinsing Agent		Water	PFC, HCFC, HFC, HFE, OS etc.		Isopropanol Acetone	
VOC Exempt (in US)	NO	NO	Probably not	Probably	Probably not	Possibly
Cleaning	Solvation	Solvation	Solvation	Solvation	Solvation	Solvation
Rinsing	Dilution	Dilution	Displacement	Displacement	Dilution	Dilution
Miscible	Yes	Yes	No	No	Yes	Yes
Purification of Rinse Agent	Distillation	Treatment	Distillation	Distillation	Distillation	Distillation
Flammability of SA	Flash point of 90°F	Aerosol of emulsion can be Flammable Class I	Combustible	Flammable	Combustible	May be flammable
Flammability of RA		Non-flammable	Non-flammable	Non-flammable	Flammable	
Non-Volatile Solvent in Rinse	No	Initially	No	No	Initially	No
Cleaning Agent within Rinse Agent	Yes		No	No	Yes	Yes
Drying Process	Evaporation of volatile rinse agent	Evaporation of water	Evaporation of volatile rinse agent (RA)			
Key Value Provided	Rapid effective cleaning and drying	Low-cost non-toxic solvents, water rinse	Safe heated solvent, Rapid drying	Selected heated solvent, Rapid drying	HSPs matched to soils	Simplicity, HSPs matched to soils

applications are Class II B (non-flammable) for obvious reasons.

- Class III – Primarily used to implement solvent cleaning technology with a pair of cleaning agents with different volatilities (that are VOCs in the US) whose HSPs match those of other soil components, and one of that can be heated to elevated temperatures
- Class IV – Primarily used to implement solvent cleaning technology with a pair of volatile cleaning agents whose blended HSPs match those of other soil components, and whose properties don't raise an inordinate level of concern about SHE issues.

These four classes are compared on an overall basis in Table 3.10.

This author believes that cosolvent machines offer substantial value to those who wish to clean parts with solvents but don't wish to be concerned about the SHE issues associated with familiar halogenated solvents.

Endnotes

A. There is considerably more detail in Ref. 2, Chapters 2 and 3.
B. Ibid Chapter 1.8.3.1.
C. Ibid Chapter 5.
D. Ibid Chapter 1.8.
E. Durkee, J.B., "What Would the 'Perfect' Cleaning Solvent Be?" *Cleaning Times column in Metal Finishing Magazine,* November 2005, p. 61.
F. This solvent residue is called dragout. It's nature, effects, and causes are discussed in detail in Ref. 1, Chapter 1.12, Table 1. 4, Figure 1.5.
G. This classification system was proposed by this author in 1995, by presentation at that year's Precision Cleaning Conference as "Equipping for Cosolvent Cleaning, Rinsing, and Drying." It was published starting on page 26 of the proceedings of that event. The paper can be downloaded from http://www.p2pays.org/ref/09/08458.pdf.
In addition, cosolvent machines and this classification system were also discussed at that year's Precision Cleaning Conference as a symposium titled "Cosolvent Machines Symposium." It was published on page 1 of the proceedings of that event. The paper can be downloaded from http://www.p2pays.org/ref/01/00863.pdf.
H. Ref. 2 Chapter 8.
I. Ref. 3 Chapter 7.
J. Ibid, Chapter 3.
K. There is considerably more information in Chapter 1.2 of the reference of Footnote 1.
L. The image of Figure 3.5 is courtesy of Exair (http://www.exair.com). In Figure 3.5, the features represented by the numbers are: (1) compressed air fed to the air knife and directed to a precisely machined slotted orifice, (2) the combination of both air streams directed as a sheet at a target surface, and (3) room air entrained by expansion of the compressed air. This device uses the energy of pressure within a small volume of air to collect and accelerate a much larger volume of air producing a high-velocity low-pressure stream which can do mechanical work in a focused area. The larger volume of directed air may be as much as 40 times larger than the volume of compressed air.
M. The data upon which Figures 3.6 and 3.7 are based is courtesy of Spraying Systems Corporation. Their URL is http://www.spray.com/applications/pdf/c20c.pdf.
N. The image of Figure 3.8 is based on the same information as Figure 3.7.
O. Chapter 4 of the reference of Ref. 1, Chapter 4 has considerably more information.
P. Tables 7.2 through 7.5.
Q. The image of Figure 3.9 is courtesy of Interlab (http://www.interlab-inc.com).
R. Ref. 2, Chapter 10 and Figure 10.11.
S. An illustration with ethanol and water is plotted in Ref. 2, Chapter 9, Figure 9.5.
T. Ref. 3, Chapter 19.4.
U. Ref. 3, Chapter 14.3
V. Water is not involved in commercial cosolvent machines of Class II, III, or IV. The example of the kerosene dryer was noted to provide a simple explanation and a practical example of Class II cosolvent machines.
Actually a more relevant example of Class II cosolvent machines, and of the kerosene dryer, is the fluorocarbon-based dryer for removal of water from part surfaces. Here the substrate is to be dried by displacement. Drying in this manner avoids an energy consuming drying step (evaporation of water) and can greatly expedite subsequent processing of the substrate component. No thermal stresses are applied to components on surfaces.
The R_A between water and common volatile fluorocarbon solvents such as HFE-43-10mee, HCFC-225ca/cb, HFE-7100, and HFC-365mfc is quite large (24.5, 22.6, 23.5, and 24.3 $MPa^{½}$ respectively). These values are well above the threshold value for immiscibility of about 14 $MPa^{½}$ or greater (Table 3.9). So displacement between water and these (or similar) solvents should be complete.
Yet, some commercial fluorocarbon-based displacement drying fluids are augmented with surfactants whose function is to chelate with the water so that it does not form emulsions with the fluorocarbon. Surfactant-fluorocarbon compositions have been patented to alleviate this condition, especially on non-metallic surfaces. Useful surfactants are:
- Quaternary ammonium salts of a phosphate acid ester (USP 4,618,447, Surfactant-based Solvent System for Dewatering Different Substrates),
- Nonionic surfactant, fatty alcohol ethoxylates (USP 5,256,329, 1,1-dichloro-1-fluoroethane Dewatering Systems),
- Quaternary ammonium salts with fluoroalkyl amines (USP 5,610,128 and USP 5,868,799,, Surfactants and Drying and Drycleaning Compositions Which Utilize Said Surfactants)

In addition both methods and processes are claimed for drying surfaces using similar (but apparently different) technology. They include USPs 5,980,642 and 6,096,240 which involve use of a (per)fluoropolyether coupled with an non ionic additive, and USP 5,733,416 which involves a process wherein water is displaced and entrained in a water-displacement composition comprising an organic liquid and a surfactant with the water evaporated at a rate exceeding the entrainment rate.
W. Flammability classifications by US DOT/OSHA are explained in Ref. 1, Chapter 3.16.2.2.
X. The classification of a solvent as flammable is not based on the same value of flash point in every country (Ref. 2, Chapter 6.13 and Table 6.7).
Y. Unfortunately, there is not a single system for rating ignitability of chemicals. The internationally recognized National Fire Protection Association (NFPA), the US Occupational Safety and Health Administration (OSHA) and Department of Transportation (DOT) (see 29 CFR 1910.106), and the National Electrical Code (NEC) each have different terminologies for rating of ignitability. They are compared in Ref. 1, Table 3.42. Internationally, there are additional terminologies for rating of ignitability.
Z. This situation is explained in more detail in Ref. 1, Chapter 3.22 and illustrated in Ref. 1, Figure 3.28.
AA. The rationale and meaning of this classification are discussed in Ref. 2, Chapter 7.7.1.

BB. For example, the South Coast Air Quality Management District (SCAQMD) has recognized dimethyl carbonate as being VOC exempt, and notes in Rule 102 in 2009 that such solvents can be used in cleaning operations.

CC. An example of the latter is use of isopropanol as the SA to clean or disinfect surfaces of high value.

Isopropanol is specified to clean electronic devices such as contact pins in connectors as well as the connectors, magnetic tape and disk heads, lenses of lasers in optical disc drives, and LCD/LED and glass computer monitor screens. Isopropanol is also used for removal of thermal (solder) paste from integrated circuit packages, and restoration of vinyl records.

Further, isopropanol is specified for the above specific applications and is also frequently specified for others in military, aerospace, pharmaceutical, medical, and biological research (DNA is insoluble in isopropanol) work.

A disinfectant, such as a water solution of isopropanol, destroys some but not all organisms. Spores, for example, aren't destroyed by disinfectants. Sterilants are capable of completely eliminating or destroying all forms of microbial life, including spores. Isopropanol is not a sterilant; isopropanol is a good and versatile disinfectant. For chemical sterilization, ethylene oxide is frequently specified.

DD. Such a database is found in Ref. 2, Appendix A1.

EE. The method for doing this is described with examples in Ref. 2, Chapter 3.

FF. One uses Equation 2.8 from Ref. 2, to make this calculation. This equation is reproduced below. The units of R_A are square root of pressure, and are commonly recognized as $MPa^{1/2}$. Please see equation below.

second solvent [the RA] comprises a VOC exempt solvent...." respectively in separate vessels, and only combined in a recovery tank (a decanter such as in Figure 3.14). Immiscibility is enhanced by cooling the tank contents.

HH. Ref. 2, Chapter 7.7.

II. The replacement proposed in Table 3.5 includes both a change in cleaning chemicals and a change in cleaning processes. The single solvents are normally used in an open-top vapor degreaser (Chapter 1) or an enclosed machine (Chapter 2).

The cosolvent pair which replaces the single solvents is proposed to be used in two different processes. The SA cosolvent is proposed to be used in a cold cleaning process conducted in a sump at the boiling point of the RA cosolvent. The RA cosolvent is proposed to be vaporized from another sump and used to rinse the SA cosolvent from part surfaces and dry them by evaporation as is done in an open-top vapor degreaser. The proposed process is seen in Figure 3.12.

Nonetheless, clean parts should be produced without the use of a solvent, such as n-propyl bromide, whose uses raises significant SHE concerns.

JJ. The abbreviations for solvents used here are TCA for 1,1,1-trichloroethane, Freon for CFC-113, TCE for trichloroethylene, PCE for perchloroethylene, meth for methylene chloride, MEK for methyl ethyl ketone, and n-PB for n-propyl bromide. Other abbreviations used here are OTVD for open-top vapor degreaser, HSPs for Hansen Solubility Parameters, and MIR for Maximum Incremental Reactivity.

KK. Results of this approach are tabulated in Ref. 2, Appendix D1, Tables D1-11 through D1-15 (n-PB), D1-16 (TCE), D1-18 (meth), D1-19 (PCE), and D1-20 (MEK).

LL. See Endnote KK.

$$R_A = \sqrt{(\langle 4 \times [\{\delta_{disperse\ 2} - \delta_{disperse\ 1}\}^2]\rangle + [\{\delta_{polar\ 2} - \delta_{polar\ 1}\}^2] + [\{\delta_{H-bonding\ 2} - \delta_{h-bonding\ 1}\}^2])}$$

GG.

A Bi-Solvent Unit

A method involving what is a modified Class II cosolvent process is claimed in USP 7,604,702, Mouser, W.L., Manchester, R., Barrett, W., and Bergman, F., Method, Apparatus, and System for Bi-solvent Based Cleaning of Precision Components, October 20, 2009; assigned to Crest Ultrasonics Corp. The inventors apply the nomenclature "Bi-Solvent" to their process. A commercial unit in which this process is practiced is shown in the image above, courtesy of Forward Technology.

Two immiscible solvents are employed in this method – both are VOC exempt (in the US). The SA is based on soy-derived chemistry (Ref. 3, Chapter 6). The SA and the RA are used to clean and rinse precision parts.

An apparently similar technology is claimed as a system in USP 8,210,189, Mouser, W.L., Manchester, R., Barrett, W., and Bergman, F., Method, Apparatus, and System for Bi-solvent Based Cleaning of Precision Component, July 3, 2012; assigned to Crest Ultrasonics Corp. Claim 3 of this patent requires "...at least one of the first solvent [the SA] and the

MM. These steps are enumerated and described in Chapter 1.2.

NN. Operation of superheat in open-top vapor degreasers is explained in Chapter 1.20.

OO. For reference, solvent cleaning operations with trichloroethylene and n-propyl bromide are conducted at their boiling points – 86°C (187°F) and 71°C (160°F), respectively.

PP. It is this author's experience that Class II cosolvent processes have found more interest in Europe than in the US. The chief application is removal of flux residues from printed wire boards (PWBs).

The evolution of Class II and Class III cosolvent technology is illustrated by the claims and supporting material found in three U.S. patents, all credited to the same claimants: Hayes, M.E., Hosman, D.P., Hrebenar, K.R., and Sell, R.D. All patents were filed between June 1992 and June 1995. The patents, in order of publication (granting) are USP 05679175B1, USP 05716457B1, and USP 06187729B1. As of publication, none are in force.

USP 05679175B1and USP 05716457B1 are process patents, with a claimed range of binary cosolvent components and compositions. The general sequence of process steps is that of Figure 3.12. The equipment in which these processes could be practiced is that of Figure 3.13. The claimed SA and RA cosolvent components are those of Tables 3.1b, 3.2, and 3.3 – with the addition of the RA PFC 5060 (perfluorohexane). Since the atmospheric lifetime of PFC 5060 is more than 3,000 years (CFC-113 is around 100, and HFC-43-10mee is less than 10), US EPA has in the mid-1990s reserved approval for PFCs to only use sectors where no other option is available. In neither patent are HSP used to characterize similarity of cosolvents, or soils.

In USP 05679175B1 the separation in HSP "space" between representative claimed SA and RA cosolvent components has been calculated by this author to be low.

Example compositions and their RA values are: D-Limonene & HFC-365mfc, 4.93 MPa½; Soygold 1000 & HFC-365mfc, 5.77 MPa½; Soygold 1000 & CFC-113, 6.06 MPa½; and D-Limonene & CFC-113, 6.59 MPa½. In addition, relative volatilities of the claimed SA and RA cosolvents were specified in detail.

USP 05716457B1 represented progress in selection of cosolvent components which were less miscible – in HSP "space." Example SA and RA compositions and their RA values are: D-Limonene & 1,1,1-Trichloroethane, 3.49 MPa½; Soygold 1000 & CFC-113, 6.06 MPa½; Soygold 1000 & PFC 5060, 10.52 MPa½; D-Limonene & PFC 5060, 11.40 MPa½; N-Methyl Pyrrolidone & HFC-365mfc, 14.61, MPa½; N-Methyl Pyrrolidone & PFC 5060 18.63 MPa½, and water & HFC-365mfc, 24.28 MPa½. Use of water as an RA was claimed; though use as an SA was excluded. There was also growth in specification of claimed process conditions and capabilities. The primary application of interest was removing rosin flux from PWBs. And Class III cosolvent applications were claimed (though not with that nomenclature)!

USP 0617729B1 is not a process patent, but claims a composition of matter. The components are indirectly limited by their vapor pressure. For the SA, it is to not exceed 40 Torr; for the RA, it should approach 760 Torr at room temperature. Immiscibility is both preferred (but not quantitatively defined) in some implementations, and not preferred in others. Water may be an SA, but not an RA. The claimed SA and RA cosolvent components are those of Tables 3.1b, 3.2, and 3.3 – with the addition of the RA PFC 5060 (perfluorohexane).

QQ. Ref. 2, Chapter 11.

RR. The source for replacement Figure 3.9 should be CC Hydrosomics (http://www.cchydrosonics.com/ultrasonic-aqueous-cleaning.php)
The source for replacement Figure 3.16 should be ATEX (http://www.aet.fr/en/produit-ATEX_Junction_box_GRP,321.html)
The source for replacement Figure 3.17 should be Reed Sensors (www.reed-sensor.com/)
The source for replacement Figure 3.18 should be Direct Industry (http://www.directindustry.com/cat/cables-connectors-CB.html)
The source for replacement Figure 3.20 should be Mannesman-Demag (http://www.mannesmann-demag.com/en/air-motor.html)
The source for Figure 3.22 should be Mara Industrial Supply (http://www.maraindustrial.com/cart/test-equipment/msa-explosimeter-model-2a-89220-combustible-gas-used-shelf-549-ca.html)

SS. Why that is so is a legitimate question when the Class II cosolvent processes are segregated by level of concern about flammability.

The answer has two parts; it's not needed and wasn't considered. First, almost without exception the SA cosolvent in a Class III cosolvent process will have low volatility so parts can be heated to a high temperature and so there is no concern about flammability, and often the RA cosolvent will be a volatile fluorinated material without a measured flash point so there is also no concern about flammability. Second, when this author created the classification scheme in the early 1990s, the focus was on Class II systems and the author did not adequately give Class III systems due consideration.

TT. One uses the equation noted in Endnote FF, which is Equation 2.8 from Ref. 2.

UU. Chapter 3.6.8, Footnote 47.

VV. This topic is covered in Ref. 1, Chapter 1.12 on page 20 through page 33; especially Chapters 1.12.3.1 and 1.12.3.2.

WW. Ref. 2, Chapter 11.8, and Chapter 3.6.11 this book.

XX. An excellent, and free, reference about adsorption is Yang, R. T., *Adsorbents: Fundamentals and Applications*, John Wiley & Sons, 2003. It can be downloaded by an internet search on its title, or from, http://lib.freescienceengineering.org/view.php?id=30226.

YY. Ref. 1, Chapter 7.12.7.1.1 and Figure 7.62 describes heat wheels which are solid rotary passive energy recovery heat exchangers. The heat exchange matrix is normally made from aluminum or alloys.

ZZ. All of the facilities shown in Figure 3.31 are represented by the single vessel in Figure 3.26 labeled for simplicity as "water separator."

AAA. The adjective halogenated is not used to mean all solvents containing a fluorine, chlorine, bromine, iodine (Ref. 3, Chapter 25.12.1), or astatine atom (Ref. 3, Chapter 22, Footnote 3). Solvents containing fluorine atoms are generally too stable to react with water (Ref. 3, Figure 22.2 and Table 22.1). Solvents containing iodine and astatine atoms are too unstable relative to UV light for use in general vapor degreasing service.

BBB. Ref. 2, Chapter 11.8.

CCC. Ref. 2, Chapter 11.16 and Table 11.4

DDD. The image as Figure 3.32 is courtesy of Mellcon Engineers Pvt. Ltd (India), http://www.mellcon.com/.

EEE. Ref. 3, Equation 16.2

FFF. USP 4,806,658 speaks to cleavage of glycol ethers by hydrolysis in the presence of aluminum oxide, activated alumina, porous aluminosilicates and porous clays at temperatures from 170°C to 350°C (340°F to 660°F).

GGG. Water can form binary azeotropes with a few halogenated solvents, but none of them containing cleaning solvents are of consequence.

HHH. Ref. 2, Chapter 8.3.

III. An excellent brief on design of *ad*sorption desiccator beds can be downloaded for free from http://www.adsorption.com/publications/AdsorberDes2.pdf.

JJJ. More information can be found in the reference: Taube, H., "Mechanisms of Oxidation with Oxygen," *The Journal of General Physiology,* Vol. 49, No. 1, 1965, pages 29 to 50. This article can be downloaded for free at http://www.ncbi.nlm.nih.gov/pmc/articles/PMC2195455/pdf/29.pdf.

KKK. Ref. 3, Chapter 11.5 describes oxidation of ether solvents and the mechanisms by which these reactions occur and can be prevented.

LLL. It is certainly not necessary to purchase "hi-purity grade," "research grade"or" instrument grade" nitrogen gas for this use. A perfectly appropriate product is "industrial grade" or "welding grade." This material is usually 99.5% (or less) pure nitrogen. Alternately, depending upon local availability and pricing, waste carbon dioxide can be used. Price should be the dominant factor affecting purchase.

MMM. Some HSP values are tabulated below. The reference is Hansen, C.M., *Hansen Solubility Parameters: A User's Handbook*, Second Edition, CRC Press, June 15, 2007, ISBN-13: 978–0849372483.

HSP Values of Soil Materials

Soil Name	δ Disperse (MPa½)	δ Polar (MPa½)	δ Hydrogen-Bonding (MPa½)
ASTM Fuel 'B'	15.3	0.4	0.6
Auto Brake Fluid	15.8	6.1	10.2
Auto Transmission Fluid	14.3	0.4	0.6
Bitumen	17.8	4.6	4.1
Castor Oil	13.6	6.0	10.5

(Continued)

(continued)

HSP Values of Soil Materials

Soil Name	δ Disperse	δ Polar	δ Hydrogen-Bonding
	MPa$^{1/2}$		
Cholesterol	20.4	2.8	9.4
Coal Tar Pitch	18.7	7.5	8.9
Cottonseed Oil	12.2	5.8	5.8
Linseed Oil	13.9	3.5	3.7
Linseed Oil (White Refined)	12.1	6.1	6.1
Methyl Oleate	12.8	6.4	6.2
MIL-H-5006 (Petroleum)	14.7	0.8	0.6
MIL-H-8446 (Silicate)	14.5	6.1	7.6
MIL-L-7808 (Ester)	14.3	2.9	3.1
Mineral Oil (White Refined)	14.5	0.0	0.0
Mineral Spirits	15.5	3.3	4.0
Motor Oil SAE 20W	14.7	0.4	0.4
Neets Foot Oil	14.3	2.9	3.7
Olive Oil	15.9	1.2	5.4
Palm Oil	17.7	3.5	3.7
Pine Oil	12.1	5.4	5.8
Natural Rubber	16.0	4.0	6.0
Silicone Oil	16.4	0.0	7.8
Sperm Oil	14.3	2.12	2.7
Trioctyl Adipate	16.7	2.0	5.1
Turpentine	16.4	1.4	0.4
VM&P Naphtha	15.0	2.7	2.7

NNN. Ref. 2, Chapter 2.

OOO. Ref. 2, Appendix A1. Here water is simply a chemical whose solubility in water has been well-studied. Water is not a cosolvent.

PPP. See Footnote 84.

QQQ. Such compliance is not necessary, merely advisable to reduce costly emissions of expensive solvent blends.

RRR. Ref. 2, Chapter 5.

SSS. 2-Propanol in Ref. 3, Table A 2–2.

TTT. This scientific misunderstanding became recognized as truth in the late 1980s and early 1990s – the era of the phaseout of CFC solvents because emissions of them provided a catalyst for reactions which depleted ozone from the stratosphere.

A popular replacement for CFC-113 was one of several binary azeotropes:

- Cyclohexane and isopropanol (entry 397 in Ref. 3, Table A 2–2, 67% alcohol),
- Cyclohexane and ethyl acetate (entry 167 in Ref. 3, Table A 2–2, 54 % ester), and
- Isopropanol and methyl ethyl ketone (MEK) (entry 494 in Ref. 3, Table A 2–2, 30% alcohol).
- Various azeotropes of hexane and primary alcohols were also used.

The scientific misunderstanding was: implemented in equipment capable of handling low-flash point solvents; adopted chiefly in the microelectronics (solder residues on printed wire boards (PWBs)), aerospace (beryllium subassemblies in inertial guidance instruments), and optic (wax coatings on laser components) products; and reported in non-peer reviewed scientific publications. As an example of the latter, consider the following:

"...Cyclohexane/IPA is an effective azeotrope commonly used in both the ... systems because it combines two complementary attributes: the degreasing action of cyclohexane with the more polar IPA, which is effective for inorganic residue, salts, and particles..."

[source, Precision Cleaning Magazine, March 1995, pages 21 to 28].

Literature references from the time of the CFC phaseout which reflect use of binary azeotropes to clean multi-component soils include:

- Davis, P.G., "Cleaning with Alcohol and Cyclohexane," Precision Cleaning Conference, 1994.
- Arnold, G.S. and Uht, J.C., "Nonvolatile Residue Solvent Replacement," Aerospace Report No. TR-95(5448)-1, March 1995.
- Kanegsberg, B., "Cleaning Systems for Low-Flashpoint Solvents," Precision Cleaning Magazine, March 1995, pages 21 to 28.

Yet, the implementation of binary azeotropes of saturated hydrocarbons and primary alcohols was believed (and occasionally is still believed) to provide adequate, if not improved over the alcohol, performance.

Today, the above scientific understanding of binary azeotropes appears to be implemented in the choice of cleaning machines: single low-flash point solvents such as cyclohexane ("...for removing particle contamination or organic films such as oil or grease...") or isopropanol ("...for removal of partial contamination and inorganic films such as salts, fingerprints, and highly activated fluxes...").

UUU. Section K of Ref. 3, Appendix 2.

VVV. Ref. 1, Chapter 2, pages 44 to 56.

WWW. The binary azeotropes generating strong interest as of publication in surface cleaning operations manifest no measured flash point, such as those noted in Ref. 3, Table A 2–2. They include:

- Heptane and HFE-7100 (entry 34, 99.5% hydrocarbon),
- HFC-365mfc and HFE-7100 (entry 47, 60% HFC),
- OS-20 and propylene glycol n-propyl ether (entry 74, 74% siloxane),
- n-propyl bromide and hexane (entry 124, 35% hydrocarbon),
- n-propyl bromide and ethyl acetate (entry 240, 70% ether),
- n-propyl bromide and propylene glycol butyl ether (entry 253, 75% ether),
- n-propyl bromide and 2-butanol (entry 423, 92.5% alcohol),
- HFE-7100 and methanol (entry 536, 96% fluoroether), and
- n-propyl bromide and methanol (entry 691, 80% alcohol).

XXX. Ref. 2, Chapter 11, Footnote 98.

YYY. While there is no direct literature or specific experience bearing on the contamination in cleaning operations by water with binary azeotropes, there is peer-reviewed scientific literature describing the formation of ternary azeotropes and immiscible two-phase mixtures through solution of some binary azeotropes into water. The ternary azeotropes (with water) usually boil at slightly higher temperatures than do the binary azeotropes (without water). Some example literature references are:

- Gomis, V., Font, A., and Saquete, M.D., "Homogeneity of the Water + Ethanol + Toluene Azeotrope at 101.3 kPa," Fluid Phase Equilibria, Vol. 266, 2008, pages 8 to 13. R_A between the binary azeotrope and water is 11 $MPa^{1/2}$.
- Chafer, A., de la Torre, J., Morton, J.B., and Lladosa, E., "Liquid–liquid Equilibria of the Systems Isobutyl Acetate + Isobutyl Alcohol + Water and Isobutyl Acetate + Isobutyl Alcohol + Glycerol at Different Temperatures," Fluid Phase Equilibria, Vol. 265, 2008, pages 122 to 128. This research reports the conversion of the isobutyl acetate and isobutyl alcohol azeotrope (entry 346 in Ref. 3, Table A 2–2, 107.4°C, 45% alcohol) into two immiscible azeotropic liquid phases on contact with water. The phases are water and isobutyl acetate (87.5°C, 80.5% acetate), and water and isobutyl alcohol (79.9°C, 88.3% alcohol). Both immiscible azeotropic phases are more volatile than the original anhydrous binary azeotrope. R_A between the binary azeotrope and water is 15 $MPa^{1/2}$.
- Gomis, V., Font, A., and Saquete, M.D., "Vapour–Liquid–Liquid and Vapour–Liquid Equilibrium of the System Water + Ethanol + Heptane at 101.3 kPa," Fluid Phase Equilibria, Vol. 248, 2006, pages 206 to 210. This research reports that a ternary azeotrope is formed when the binary azeotrope (entry 456 in Ref. 3, Table A 2–2, 48% alcohol) is contacted with water. The boiling point of the ternary azeotrope is 3°C less.

ZZZ. Everson, R.C., and van der Merwe, B.J., "The Effects of Selected Solvents on the Relative Volatility of a Binary System Consisting of 1-Octene and 2-Hexanone," Fluid Phase Equilibria, Vol. 143, 1998, pages 173 to 184. This research is an attempt to generalize the effect on a single binary azeotrope of 31 specific solvents. The selected azeotrope, octene-1 and methyl butyl ketone (2-hexanone) (112.3°C, 31% ketone), is unique. The R_A between its two components is small, only 4.8 $MPa^{1/2}$. But the internal electrostatic properties of the olefin and ketone are quite different; dipole moments are not similar (0.34 D vs. 2.66 D) nor are the dielectric constants (2.08 and 14.56) respectively.

Reported experimental findings were muddled. But the solvent properties with the most significant effects upon the relative volatility of the azeotrope were those with extreme values of dielectric constant (Ref. 3, Table 10.3 and Chapter 12.5.4) and dipole moment (Ref. 2, Appendix C5, Chapter C1). This finding engenders the proposition that the tramp contaminant, water, could have a substantial effect on the vapor-liquid equilibrium of binary azeotropes because of its extraordinarily high value of dielectric constant (80) and relatively high value of dipole moment (1.84 D).

AAAA. Ref. 2, Chapters 7.14.1 and 7.14.3, especially Figures 7.29 through 7.32.

BBBB. Ibid. The batch distillation column is also described in Ref. 2, Chapter 7.14 and other representations are illustrated in Figures 7.29, 7.30 and 7.32.

CCCC. Ref. 2, Chapter 11.

Chapter 4

"Tailpipe" Treatment of Fugitive Solvent Emissions

Aptly named, vapor degreasers (VDGs) produce vapor which has to be "degreased" (cleaned) before it can be emitted.

The vapor is air (or an inert gas such as nitrogen), enriched with vaporized cleaning solvent.

There are, generally, three sources of solvent-laden air within and around any type of vapor degreaser. They are:

- The lip vent attached to an open-top vapor degreaser[1] (Chapter 1.21 and Figures 1.38 and 1.39),
- The off-gas produced when an enclosed cleaning machine[2] (Table 2.1) is purged with air (or inert gas) to relieve pressure, and
- Unintentional leaks of volatile liquid or vents of vapor[3] (Box 4.1).

The volumes of solvent-laden vapor escaping from these three sources (especially the latter) are called fugitive

1

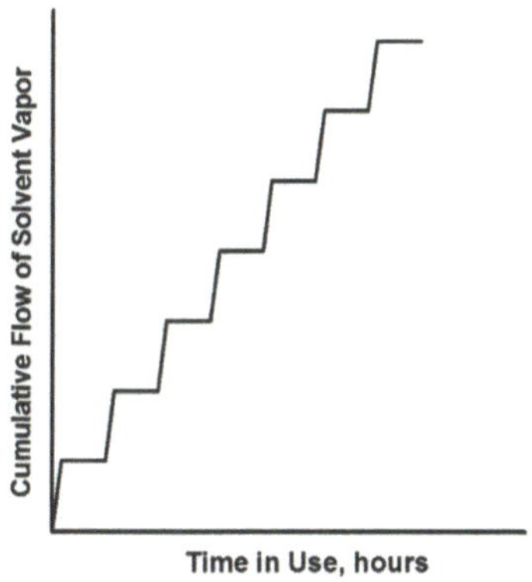

For open-top vapor degreasers with their cover closed, the lip vents can only release solvent vapor which is not condensed by the upper cooling coil and has then risen to the top of the machine. That volumetric emission rate should be very low because there is little inward leakage of air to provide fluid to entrain the uncondensed solvent vapor into the lip vents.

However, when the cover is opened to allow removal of the parts basket, the lip vents draw air from the room into the top of the machine. That action entrains solvent vapor out the lip vents (Figure 1.39). This staged pattern of emission is shown in the figure above.

2

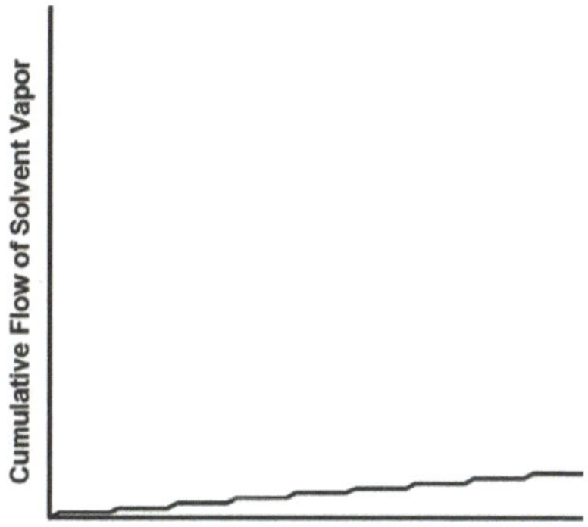

For enclosed machines, there are two times when there are substantial and significant emissions of solvent in air: (1) when the chamber contents of pure vapor at reduced pressure are purged with either clean dry air or nitrogen (Tables 2.6 to 2.10) at ambient pressure, and (2) when the machine is opened to allow removal of the parts basket, and so the vapor contents of the cleaning chamber, diluted of solvent vapor, are released. Otherwise, emissions are only leaks of liquid and vents of vapor. This staged pattern of emission is shown in the figure above, and can be compared with the emission pattern shown in Footnote 1.

3 Leakage of liquid solvent from pump seals, joints of tubing (or pipe), and drains of storage vessels represents abnormal and haphazard, careless, and slipshod operation. As well do vapor vents from the same sources. Depending upon the volatility of the cleaning solvent being used, the former may immediately become the latter. Either emission adds cost, pollution, and may raise concern about a fire.

http://dx.doi.org/10.1016/B978-0-323-22520-5.00004-6

Box 4.1 Pit-it-ful' Location of Vapor Degreasers

One of several reasons for locating vapor degreasers in sub-ground pits or lowered areas (Figure1.22) is to enable a spill of liquid solvent to be contained away from normal shop traffic. Notice in this figure that the open-top vapor degreaser is placed into a catch pan or basin – to collect liquid spills. Liquid spills should be removed immediately as they are recognized (after the cause of their release is stanched) using a liquid vacuum cleaner powered by compressed air.

The low level location should be well ventilated with the exhaust air directed to the same duct (plenum) where the lip vent exhaust, or the suction point adjacent to the cover of the enclosed machine, is directed. Excellent ventilation can be an effective method of keeping workers from being exposed to concentrations of solvent in excess of the posted (by a standard-setting authority or the manufacturer) exposure limit. But it raises the cost of solvent use.

Concentrations of solvent vapor, which is heavier than air, adjacent to the vapor degreaser and in the work area should be continuously monitored using an explosimeter (Figure 3.22). The standard is that these volumetric concentrations are to be no more than 25% of the posted lower explosive limit (LEL).

emissions. Their escape is usually not controlled within the cleaning machine, but these emissions have to be captured and treated to avoid legal penalties for pollution or harmful exposure to employees.

Since these emissions are not controlled or limited within the cleaning machine, treatment of them is called "tailpipe" control.

4.1 USE OF BLACK HOLES

Like fugitives from prison, the existence of these emissions is unscheduled; the volumetric amount of them released is inconsistent and usually unknown; and their concentration in air is uncertain.

That's why capture and treatment/recovery of these fugitive emissions is best done with an apparatus which is in effect a black hole[A] in terms of capturing solvent emissions.

4.1.1 The Comfort of a "Black Hole"

In the analogy above, a "black hole" (Figure 4.1) is a device which can function at any time with any volume of solvent emitted to it at any rate, at any concentration in air, *and* not fail to retain essentially all solvent vapor it receives.

In terms of protecting workers and complying with environmental permits, that capability provides significant comfort to managers as shown diagrammatically in Figure 4.2. Managers need only manage the cleaning work in their cleaning machine – until the "black hole" becomes full – then they "replace" it! See Chapter 4.18 and Figure 4.42 for what that involves.

Figure 4.1 **Computer-Generated Image of a 'Black Hole'**

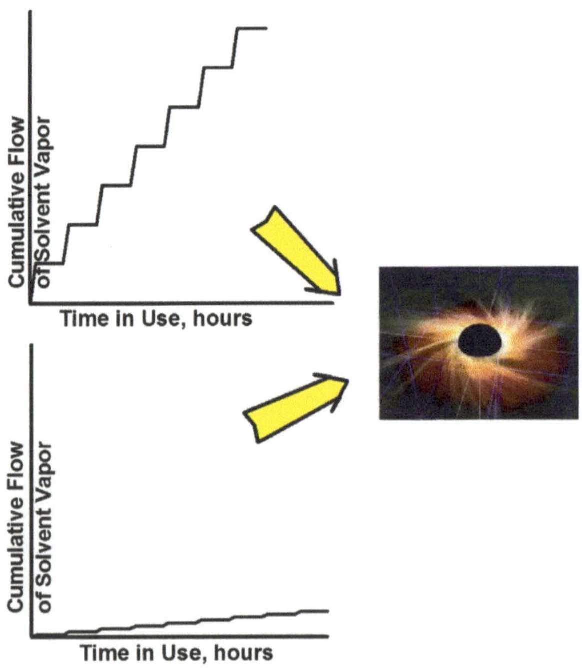

Figure 4.2 **The Meaning of 'Black Hole' Containment/ Treatment**

This approach is almost always taken at plants using solvent vapor degreasers of any type[4]. Where multiple vapor degreasers are employed at a plant, a common "black hole" services all units – even if multiple solvents are involved.

4.2 THE NATURE OF THE ADSORBENT – ACTIVATED CARBON

The actual device providing that comfort is a bed of *ad*sorbent (Box 4.2). Fugitive solvent diffuses (Figure 4.3) into and within the bed, and then is *ad*sorbed onto the pore surfaces[5] (Box 4.3) of the activated carbon[6].

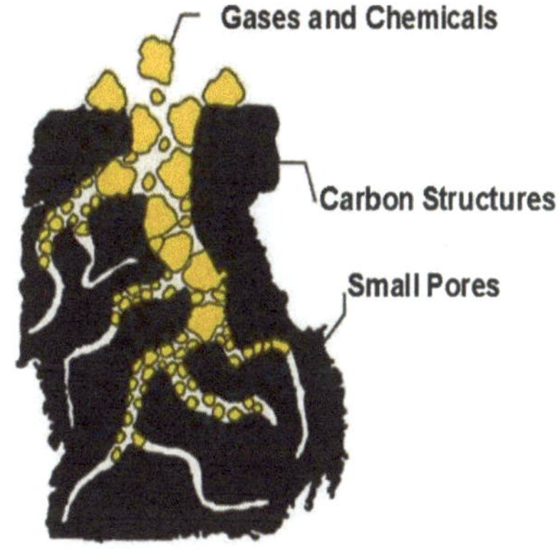

Figure 4.3 **Use of Pores**

Box 4.2 **Desiccants – Their Types and Uses**

Recall from Chapter 3, Footnote 93 that adsorption is the adhesion of atoms and molecules to surfaces of a material; absorption is the penetration of surfaces by molecules and retention of them within the material. It matters not whether the adhesion is to a material's internal or external surface.

What makes activated carbon so useful as an adsorbent is the extremely high amount of surface area within those pores – ~50 to 100 acres[7] of surface area contained within one pound of activated carbon.

Amazing views (Figures 4.4 through 4.6[B]) are found when the physical structures within activated carbon are examined under high magnification.

Activated carbon is an extensive population of semi-polar surface sites which can attract and hold solvents, housed within a network of pathways which collect the solvent vapor and channel it to where it is in close proximity to those sites.

The action by which a bed of activated carbon removes solvents is depicted in Figure 4.3. It is that vaporized solvents diffuse into the vast volume of pore structure and then have a veritable lunar landscape of activated area upon which to adsorb.

Activated carbon can hold up to at least half of its weight in adsorbed solvent. The amount depends upon the specific types of each, operating temperature, and the concentration of solvent present in the inlet air stream[C].

4.2.1 Both Names of the Product

The product name is activated carbon.

No one wants carbon substrates which haven't been activated. It matters not the extent of the area included within the pore structure of the proposed adsorbent – if it

[4] Other approaches and devices (thermal or catalytic oxidation, scrubbers, precipitation, and particulate filters) have been evaluated for general treatment of fugitive emissions of solvent. Some have proven useful in specific cases. But adsorption into activated carbon remains the standard treatment method to be used as a "black hole."

[5] An adsorption site is the area where one solvent molecule may reside. In general a micropore (major diameter < 2 nm) will accommodate many solvent molecules. However, a site may also be a volume as more than one solvent molecule may be attached to a single area within a pore.

It's not *ab*sorption when solvent vapor molecules diffuse to the internal volume of activated carbon and attach to internal surfaces. Its *ad*sorption because the attachment is to the activated surface. Solvent vapor molecules never penetrate into the carbon matrix (printed in black in the two figures above) – that would be *ad*sorption.

Information in Chapter 3, Footnote 94 describes use of mole sieves for drying of water from solvents: the opening of micropores for adsorption of chlorinated solvents is relatively large. Mole sieves rated at 5 Å (0. 5 nm) will allow water to enter, and reject all organic solvents. So, trichloroethylene with a kinetic diameter of 6.2 Å (0.62 nm) will easily enter a pore with a major diameter of 20 Å (2.0 nm). So also will water enter, and compete for precious activated surface.

[6] Activated carbon is an adsorbent as described in Chapter 3, Footnote 93, and in Appendix A2. One of the first large-scale uses of activated carbon as an adsorbent was in military gas masks where complete removal of contaminants was essential. Another essential application is removal of mercury contained in emissions from coal-fired power stations, medical incinerators, and in produced natural gas at wellheads. A value provided in cleaning operations by activated carbon adsorbent is that multiple solvent components can be captured within the same apparatus.

[7] In metric units that is 500 square meters (SM) of surface within a single gram of adsorbent (or nearly 2.5 million SF/pound).

This high level of surface area makes activated carbon an excellent site for catalysis of chemical reactions – whether they are wanted or not.

Box 4.3 **Some Pores Are Poor Adsorbents**

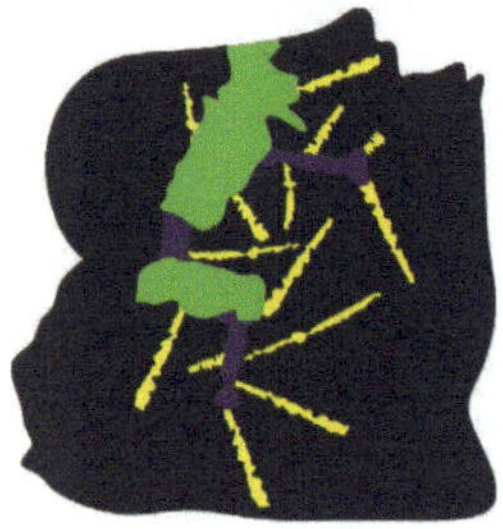

There are three kinds of pores – each differentiated by size. Those with a major diameter smaller than 2 nm (20 Å or 20 Angstrom) are called micropores. Those with a major diameter between 2 and ~50 nm are called mesopores. Still larger than 50 nm in major diameter are macropores. See the conceptualized figure above, where the different-sized pores are delineated by color (green = macropores, blue = mesopores, and yellow = micropores).

The *ad*sorption process in micro pores is very much stronger than that on the relatively flat (relative to the size of a solvent molecule) surfaces of meso- and macro pores (Figures A1-15 to A1-17). The adsorbate solvent molecule in a micropore is closely surrounded by pore walls. Due to the force-distance characteristics of adsorption forces in small pores, the adsorbate solvent vapor (not liquid) interacts with a larger number of solid wall atoms, as compared to larger pores, where the distance of solvent to walls means much weaker interactions with other walls (demonstrated by use of Equation A2-24, and shown in the scaled image below with pore diameters of 2 and 10 nm).

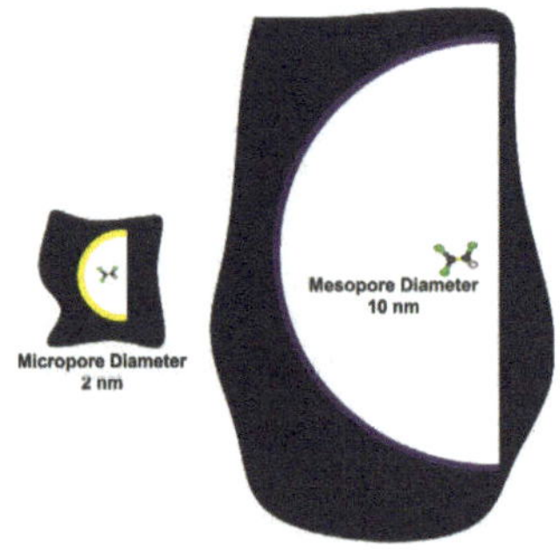

hasn't been activated (Box 4.4). Activation makes the solvents "stick" to the plethora of internal surface. Unactivated carbon is simply debris.

The outcome of activation is a surface which radiates intermolecular forces. Depending upon the method of activation, they can be London dispersion or van der Waals forces or forces with a more polar character associated with oxygen atoms[D]. Surfaces which radiate those forces become targets for pollutant molecules which also radiate the same forces (Figure 4.7 is a hypothetical image of such a surface covered densely with oxygen molecules). The situation is analogous to solvent cleaning where soils are dissolved in similar solvents.

The other half of the product name is carbon – a synonym for commonly available at a low price. It matters

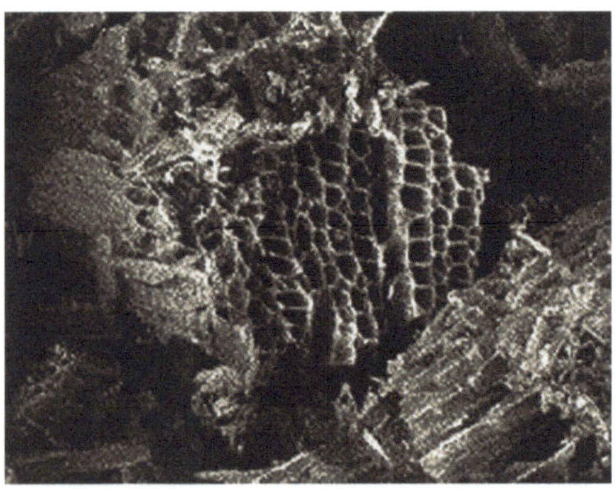

Figure 4.4 Unusual Structure Produced by Activation of Carbon

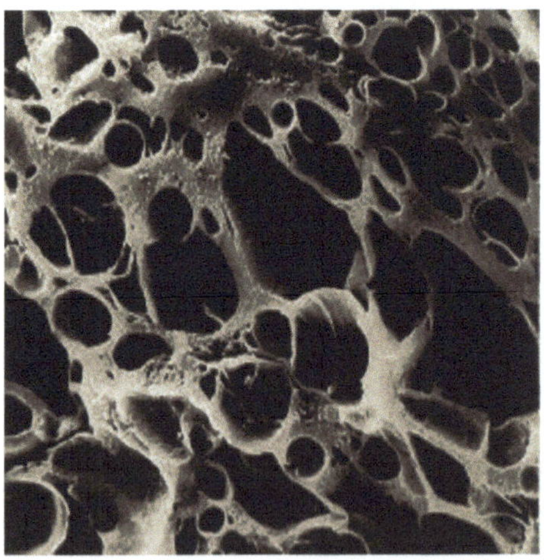

Figure 4.5 Unusual Structure Produced by Activation of Carbon

Box 4.4 Activation Activities

Manufacture of activated carbon requires two steps: carbonization, and then activation of the all-carbon mass.

Carbonization is done by pyrolysis (drying, heating to drive off byproducts, and further heating the material to over 900°C [1650°F] in an oxygen-free atmosphere that cannot support combustion). The purpose of the carbonization step is to convert all materials present, which are often multi-component waste products, to elemental carbon (called char). This step also initiates development of the desired pore structure.

Both tasks can be done in one of two ways. One is called physical (or thermal); the other is called chemical. Both involve the practice of chemistry.

The "physical activation" is done by oxidation with air or carbon dioxide at high temperatures (600 to 900°C) followed by treatment with steam at temperatures above 250°C. Oxidation burns off the pore-blocking structures created during the carbonization phase. A longer period of oxidation results in larger pore sizes.

The "chemical activation" is done at a lower temperature with specific chemicals to achieve specific ends – including modification of the pore structure and minimization of non-porous tar structures. Phosphoric acid, potassium or sodium hydroxide, and zinc, calcium, or aluminum chloride populate the formulary of the chemists who manage this step.

Produced by the above means, probably in a confidential scheme, the activated carbon must have the following characteristics: (1) adequate mesopores (cavities) which constitute pathways for large and small molecules to access the inner porous structure, (2) a surfeit of micropores (less than 2 nm in major diameter) which contain the primary adsorption sites, and (3) nearly complete modification of the internal surface structure to make it capable of attracting and bonding to chemicals via van der Waals forces (weak electrostatic forces between molecules), not the covalent or ionic forces bonding the atoms of molecules together.

not if the activation process produces a "sticky" surface for targeted pollutants if that untreated surface is unique and valuable. Only debris is commercially and economically suitable for activation for adsorption of solvent vapors.

Figure 4.6 Unusual Structure Produced by Activation of Carbon

4.3 CHARACTERIZATION OF AN ACTIVATED CARBON ADSORBENT

Activated carbon is a globally-available product[E,F]. There is an actual universe of substrates[G,H] which are distressed products and can be converted to carbon, and then activated via a great variety of processes.

The reason for this abundance is that there is an actual universe of applications for activated carbon. This includes removal of at least H_2S, SO_2, NO_X, HCN, HCOH (formaldehyde), NH_3, Hg, and CO_2 from large volumes of waste air streams such as emissions from power plants or refineries, as well as controlling

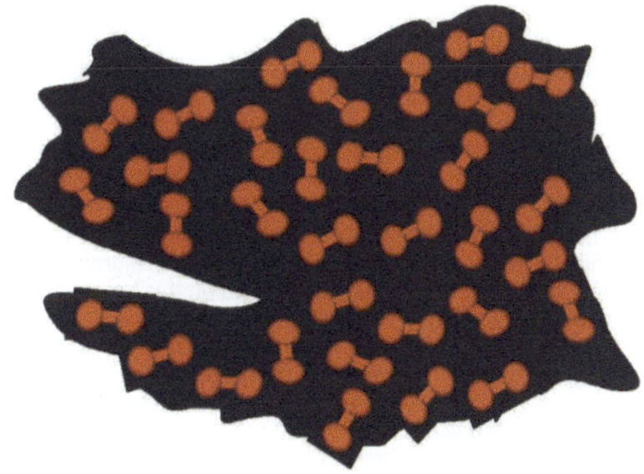

Figure 4.7 Hypothetical Two-Dimensional Image of an Activated Surface

"tailpipe" emissions from cleaning operations. In addition to applications involving cleanup of air with GAC, there are probably more applications involving cleanup of water using PAC.

Each of the adsorption applications mentioned above is best done with a different activated adsorbent. One may choose from a product with a different: smallest pore diameter, distribution of pore sizes and volumes, total amount of surface area per volumetric charge[8]; and especially, the method of activation[I].

For the user, this is good news. There are plenty of capable, experienced, and competitive suppliers of activated carbon[9]. There is no reason to work with a firm not appearing to be so.

The bad news is that preparation, use, and recovery of activated carbon is a business whose global volume is atleast 4 billion dollars – and the amount used in solvent remediation or recovery from or for cleaning operations is negligible in that business volume.

To the extent that treatment of vapor emissions for recovery of solvents is non-negligible in that business volume, it is (or has been) associated with emissions from gasoline stations – especially in California, recovery of solvents used in the tape industry (sealing tape), gravure printing, and cellulose packaging for cigarettes.

The major global business volume for activated carbon is removal of chloramines from water to make it suitable for drinking.

4.4 SPECIFICATION OF ACTIVATED CARBON

Purchase[10] of activated carbon for removal of chlorinated solvents from vapor emissions is similar to purchase of cleaning solvents. One may purchase virgin material (solvent or carbon) from a manufacturer or distributor. Or one may purchase recycled (reprocessed) material (solvent or carbon) from a firm performing that operation. Reprocessed material is almost never that which was earlier used by the purchaser[11].

As this is published, competitive prices for "Bulk Virgin Vapor[12] Phase Carbon" are around $1.75/lb for virgin material and $1.50/lb for processed material. In most cases, the specifications for both virgin and processed (actually reprocessed) types are the same or similar. So, processed material with good specifications is often a low-risk choice to obtain modest savings.

The most important specification about activated carbon is its capacity to capture solvents[J]. This is typically measured in a standard test (ASTM 3467-99), and denominated by the weight percent amount of carbon tetrachloride (Carbon Tetrachloride Number, CTC[13]) held by a charge of carbon under standard conditions. The minimum acceptable value should be at least 60%[14]; 65% is better.

An implicit specification is surface area – probably measured by the BET method (Appendix A2). Typical values are more than 500 and less than 1000 SM/gram (roughly 2.5 to 5 MM SF/lb or 70 to 180 MM SF/CF).

- The amount of internal surface area isn't as significant as the amount of solvent which can be adsorbed.

Other specifications of significantly lesser concern are:

- Hardness Number – This measurement indicates if the external surface of the adsorbent shape can be worn away by mechanical action. One selects a weight of adsorbent shapes and exposes them to the action of moving steel balls for a specified time. Whatever material worn away is captured on a #8

[8] Activated carbon adsorbents are purchased and specified by weight, as are cleaning solvents. But, similarly, they are most often used on a volumetric basis – gallons to fill a cleaning sump, cubic feet to fill an adsorption column (vessel). Since each type and charge of adsorbent has a different amount of surface area per weight, and a different bulk density, the mass amount of adsorbent surface charged to an adsorption column is certain to be different for every use. Note, it is the amount of surface which is valued per charge of an adsorption column. So, adsorbents should be compared and purchased on this basis – minimum or estimated surface per volume (SF/CF, or SM/m^3). This quantity is the product of surface area/mass (SF/g) times by apparent density (g/cc).

[9] Some firms specialize in providing and reprocessing activated carbon used in water purification applications. This product is generally not suitable for cleanup of vapor streams.

[10] Virgin or processed adsorbent is purchased and used in working containers (canisters) holding hundreds of lbs, or supersacks (Figure 4.10) holding 1100 lbs, or drums holding 55 gallons of solid (~225 lbs) activated carbon.

[11] Spent carbon is re-carbonized and re-activated in batches of tens of thousands of lbs. A minimum quantity of several thousand lbs is required in order to have one's spent carbon sequestered for toll-processing.

The word toll means that ownership of a quantity of goods (activated carbon in this case) is retained while those goods are processed for a negotiated fee by some second firm to achieve negotiated specifications. The fee is the toll. Usually it is a quantity of goods which are processed and not specifically identified goods. Cleaning solvents are often toll-processed, in which case the process is filtration and distillation.

In addition, there are set-up fees charged by the processor to learn the physical character and residue profile of the material to be processed. These fees usually total ~$1000. The price of re-carbonized and re-activated carbon is usually $0.10 to $0.25/lb less than virgin material.

The processing firm should be licensed to follow EPA's testing and processing procedures – because processing usually produces waste. Furthermore, if the activated carbon contains a Hazardous Air Pollutant (HAP) (methylene chloride, trichloroethylene, etc.), EPA shipping requirements (and related costs) must be observed for shipping hazardous materials.

[12] This assignment is made to distinguish the product from activated carbon used to treat water.

[13] Due to the harmful effects on humans and the environment by use of carbon tetrachloride, a similar test (ASTM 5228-92) with butane is often used. The minimum acceptable value for activated carbon in the butane test is about 23 weight percent.

[14] Some firms offer a specification of 55% as a minimum level of CTC. That may be perfectly suitable, depending upon the price of the activated carbon but it does not manifest a high standard of performance.

Box 4.5 **Measurements of Particle Size Distributions**

Frequently, a stack of multiple screens are used to obtain a particle size distribution – with screens having a larger spacing between wires located higher in the stack. So results are reported as "minus the smallest sieve size number", and "plus the next largest sieve size number;" or as – the smallest sieve size number (widest separation between wires) × next largest sieve size number (next smallest separation between adjacent wires). For example, −4 +6 (passed through the #4 screen and remained on top of the #6 screen), or 4 × 6.

Unfortunately, there is not a standard for sieve sizes (spacings between wires). That commercialized by the W.S. Tyler Company has the most currency. It's based on counting the number of wires of a standard diameter per inch of sieve grid, where each sieve size uses a different wire diameter. A schedule of sieve sizes with less currency is referred to as US Standard (it's not). Fortunately, for those interested in characterizing the size distribution of activated carbon, sieve numbers for 4 through 8 have the same meaning in both systems.

The spacings for both the Tyler and US Standard sieves are: #4 (4.75 mm or 0.0187 inches), #6 (3.35 mm), #7 (2.81 mm), and #8 (2.38 mm). For the Tyler sieves, the spacings are #9 (2.00 mm), #10 (1.68 mm), and #12 (1.40 mm).

One can use the information produced by a screen size analysis to calculate a weight average particle diameter, by knowing the wire spacing and the weight amount on each screen (sieve).

Tyler sieve[15] (Box 4.5) and weighed. Twice the weight is the Hardness Number. Values should range from 90 to less than 100[K].

- Abrasion Number – This measurement is made around the same exposure which produced the Hardness Number. One measures the particle size distribution using a set of Tyler sieves before and after the wearing exposure. The Abrasion Number is the ratio of the final mean particle diameter to the initial mean particle diameter, expressed as a percentage. Change of mean diameter should not exceed 30%. Both hardness and abrasion resistance are functions of the original source of carbon. Nut shells produce carbon more resistant to abrasion. Lignite and coal produce carbon less resistant.
- Apparent Density – This is the bulk density of the adsorbent shapes. It is simply the total weight of adsorbent shapes held within a volume, measured in g/cc or more likely, lb/CF. This measurement indicates the weight quantity of adsorbent shapes necessary to fill the adsorption chamber. This value should be at least 0.44 g/cc (27.5 lb/CF) and perhaps as much as 0.58 g/cc (36.2 lb/CF) – depending upon the chosen size of the particle shape, and the methods of carbonization and activation[16,17].
- Shape Size – One prefers a uniform distribution of sizes (4 × 6 vs. 4 × 10) so that the pressure drop of air flowing across the bed is minimized for the desired flow of solvent-laden vapor. Larger granule sizes (4 × 6) are preferred vs. smaller ones (8 × 10) for the same reason.
- Shape Density – This measurement can be of significant moment to adsorption performance despite it being relevant to only a single granule. This is because, for a given granule shape and size, shape density depends upon porosity of the activated carbon. For a cylindrical shape as 4 × 6 Tyler mesh size, a typical value is 1.4 g/cc (87 lb/CF) to 1.5 g/cc (94 lb/CF).
- Ash – Ash is non-performing material – wasted space which doesn't contain adsorbing surface. It can be salt residue from an activation operation, or more likely mineral originally present in the original source of carbon. Values should not ever exceed about 8 weight %.
- Moisture – This is another non-performing material. It too is wasted space[18]. It may represent surface not available for adsorption of the pollutant solvent. Values can be as low as 2% and as high as 15%.

A preferred purchase specification, which can be met by several firms in most any global market, is listed in Table 4.1.

4.5 THE SIMPLICITY OF TAILPIPE TREATMENT

There are three simple steps to operating the "black hole" for tailpipe treatment of fugitive solvent emissions. All are covered in this chapter.

[15] Granules and particles are characterized by their size, and possibly size distribution. This character is denominated by sieve or mesh size. Larger particles are recognized by being associated with smaller sieve size numbers.

This is easily measured. One takes a charge of material, places it on a wire screen, and shakes the screen. Material larger than the separation between adjacent wires remains on the screen; smaller material passes through the spacing between wires, and is not retained on the screen. Results are reported as percent of the charge greater than the spacing.

[16] The average of these two values is a common characteristic of useful product.

[17] A possibly significant consequence of this characteristic is described in Chapter 4.20.1.

[18] Note that a purchase of activated carbon is based on CTC or BET values which should be determined using "as received" material – activated carbon containing ash, moisture, or other impurities.

Table 4.1 Specifications for Granular Activated Carbon

Parameter	Specification
Mesh Size Rating (Tyler or US)	4×6 (5% > 4, 5% < 6) mesh
CCT (wt. % Carbon Tetrachloride)	> 60 wt. % (Iodine Number > 1100; ASTM D460)
Hardness Number	> 98%
Moisture (when packaged)	< 2%
Total Ash	< 3%
Bulk (Apparent) Density, g/cc	0.48 to 0.52 (30 to 32.5 lb/CF)

- First, collect the solvent emissions and direct them to the bed of activated carbon (an adsorbent) – Chapter 4.6.
- Second, operate the bed of adsorbent until one notes that the bed capacity has been filled with adsorbed solvent vapor – Chapters 4.7 to 4.10.
- Third, regenerate or replace the solvent-filled adsorbent in the bed so it is charged with fresh adsorbent – Chapters 4.11 to 4.13.

Simplicity in the management of disposal of fugitive cleaning agents can be clearly and favorably contrasted with the complexity of management of disposal of spent aqueous cleaning agents.

4.6 COLLECTION OF FUGITIVE SOLVENTS

As noted in Footnote 3, perfect operation of a solvent cleaning machine, without leaks, spills, or unintentional vents of solvent, would still produce *some* fugitive solvent emissions. Design of these cleaning machines makes it inevitable that there will be emission of solvent vapor into room air – at a large or small rate, depending not at all on the quality of design or dedication of operators. The two major unavoidable emissions are:

- Solvent vapor withdrawn from the top of an open-top cleaning machine either through lip vents (Figure 1.39) or by entrainment with parts baskets when removed from the vapor degreaser. Though undesirable, this exhaust does unavoidably exist in some/many/most open-top vapor degreasers[L] — because the parts basket must be removed, and it is surrounded by air containing some solvent.

Figure 4.8 **Powdered Activated Carbon (PAC)**

Figure 4.9 **Granular Activated Carbon (GAC)**

- Solvent vapor emitted when the cleaning chamber of any enclosed cleaning machine is opened after being purged with clean dry air (Chapter 4.8.1.1 and Figure 4.18) — because the work chamber must be evacuated before parts are removed, and it is surrounded by air containing some solvent.

The proper tool for collection of all fugitive solvent emissions is an air amplifier (air mover powered by compressed air). This device is figuratively shown in use in Figure 4.11[M], where emissions from *all* sources around an open-top vapor degreaser are pulled (sucked) into an air

Figure 4.10 **The Commercial Product — Activated Carbon**

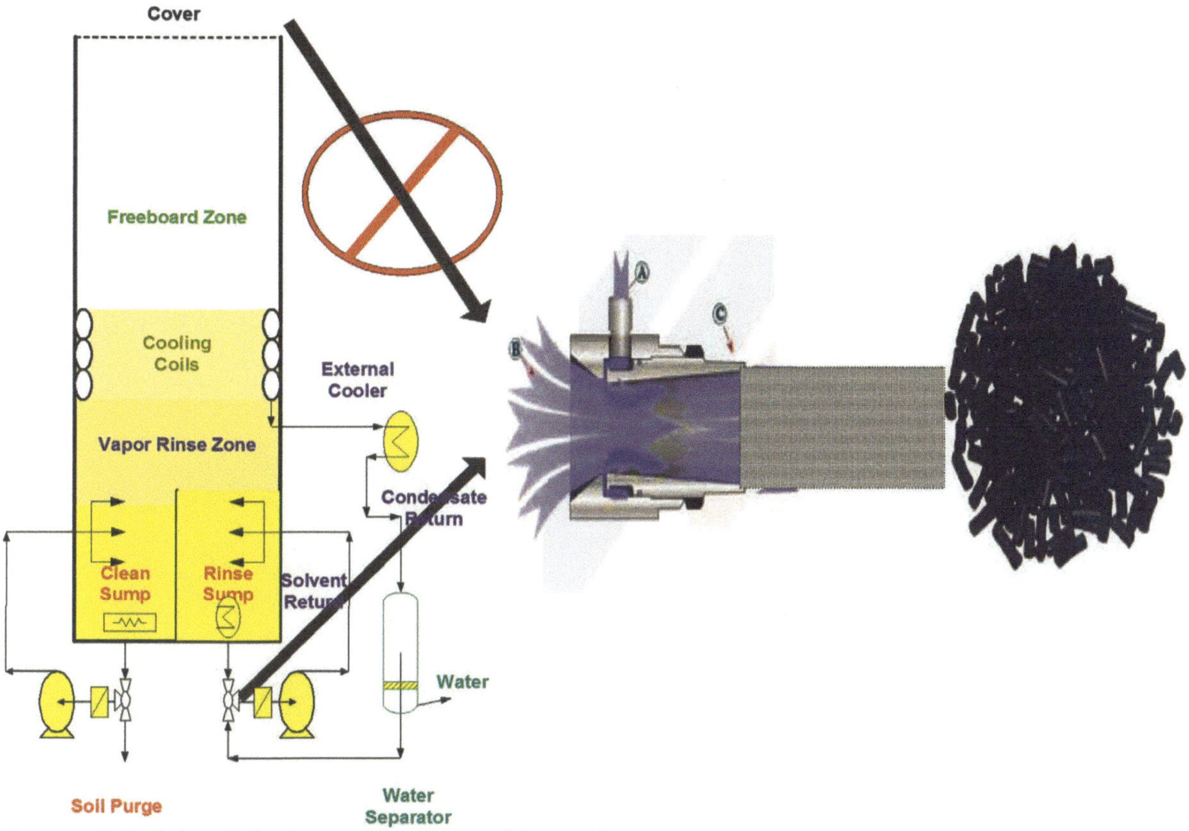

Figure 4.11 **Emission, Collection, and Treatment of Solvent Fumes**

amplifier which dispenses the lean-solvent air stream into a bed of activated carbon. Note again that the direct use of lip vents is discouraged, although solvent does enter the work area by entrainment with the parts and basket as they are removed from the machine.

An air amplifier (Figure 4.11) performs the function of a fan or blower used in suction mode. Air polluted with solvent from all three types of emissions, and room air, is pulled into a common air stream with the air amplifier. The result is twofold: (1) solvent emissions are removed from where workers work and sparks may ignite, and (2) solvent concentrations in air are reduced to below posted exposure limits.

Since cleaning solvents are heavier than air[N]; undisturbed, they don't rise in still air. So one (and *probably more*) suction point(s) for an air amplifier should be located low in the enclosure/pit/area in which solvents are used (Figure 4.12).

Another is where workers access the machine, as is shown in Figure 4.13[O].

Where clean and dirty solvents are stored (in drums or pails) are other areas which are nearly certain to produce fugitive emissions (Figure 4.14[P]).

Figure 4.12

Figure 4.13

Figure 4.14 **Storage of Flammable (Red Can) and Non-Flammable (Black Can) Solvents in Catch Pan**

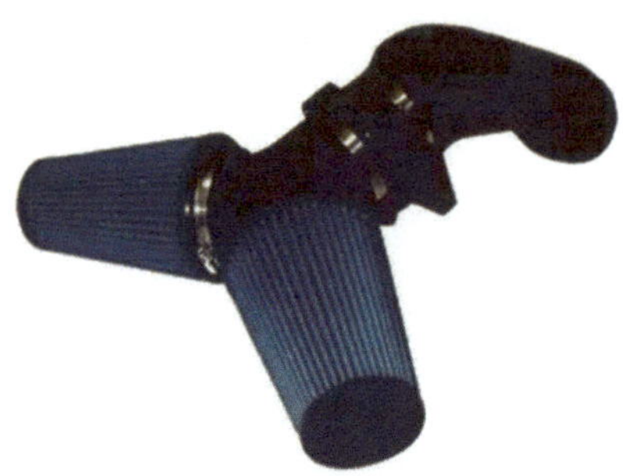

Figure 4.15 **Dual Pickup Air Filter**

Figure 4.16 **Single Pickup Air Filter**

The orange arrows in Figures 4.12, 4.13, and 4.14 point to those very suitable locations for multiple suction points for an air amplifier.

It may be that not all of the locations noted above continuously produce solvent fumes. So to conserve resources (solvent, adsorbent bed life, and user attention), sophisticated users are likely to find it economically attractive to couple the air amplifier to an on-line explosimeter[Q] — so that the air amplifier can only extract air when it is polluted with solvent.

As illustrated in Figures 4.12 to 4.14, there are fugitive emissions emitted from more than one point next to, around, or near the cleaning machine (vapor degreaser or cold cleaner). In other words, the air amplifier may have more than one pickup point (Figure 4.15[R]) , or more than one air amplifier may be needed to keep the work area sanitary, safe, and hospitable.

And the collected air stream should have a filter (Figure 4.16[S]) to remove some airborne particulate which could foul access by air to all areas of the adsorbent bed. If solvent-laden air isn't uniformly distributed through the bed, it will channel through some zones, and not others. So, over time, the full adsorption capacity of the bed won't be used.

Finally, the most important virtue of an air/amplifier[19] is that is offers control of: *where* air is removed from for treatment, *when* it is removed, at *what rate* is removed, and *what air* columns are left undisturbed.

4.6.1 A Critical Flaw in Use of the Air Amplifier

No cleaning solvent is free; they all are obtained for a price. Every g-mole/milligram/cc/gallon of vaporized solvent

[19] The reason an air amplifier is used to collect solvent-contaminated air, instead of a fan or blower, is that one can target just the contaminated volume of air for treatment and not the entire shop volume.

moved by an air amplifier represents a waste of money. The waste is at least the purchase price of the solvent which is being thrown away. It matters not if the solvent is captured and doesn't become a pollutant, the cost of its purchase is wasted (Figure 4.61).

The critical flaw is that every air amplifier (or equivalent ventilation fan) used with solvent vapor degreasers is used (deliberately) to waste money; if no air amplifier is in place, the monetary value of escaping solvent is not being wasted[20].

But, where solvent vapor is emitted into room air for whatever reason in whatever way, it must be collected and removed. That's the reason for use of an air amplifier.

4.6.2 Comparison of Air-Powered Devices

An air amplifier is a sibling of an air eductor (Figure 2.7), and also a step-sibling of an air knife (Figure 3.5). All are powered by compressed air.

The essential difference between the air amplifier and the air eductor is in the diameter of the aperture through which the compressed air is expanded. That aperture is called a venturi. An equation which allows calculation of the performance of either device with a venturi is the Bernoulli Equation[T].

All three devices are powered by compressed air and: have no moving parts, are not driven by electricity (no spark sources), are usually sold by a single firm, are not expensive, start instantaneously, use compressed air at pressures less than 60 to 100 psi (413 to 690 kPa), require negligible maintenance, and have names which are often confused with one another[21]. All three devices are compared in Table 4.2[U].

At least one of the three could or should be part of nearly every installation of an aqueous or solvent vapor degreaser.

4.7 USE OF AN ADSORBENT

The capability of a bed of adsorbent to adsorb solvent vapor depends upon the level of solvent in the air fed to the adsorber.

- Simply, the richer in solvent concentration is the stream fed to the adsorber, the more of it will be retained on the adsorber bed.

This is easily shown in Figure 4.17[22] for four solvents used in cleaning operations.

Adsorption of methylene chloride (brown square symbols in Figure 4.17) is most sensitive to its concentration in the inlet air – capacity for adsorption varies over nearly a decade for a two-decade change in inlet concentration. Yet capacity to adsorb perchloroethylene (green square symbols) is hardly dependent upon inlet concentration.

Data in Figure 4.17 show that the capacity of a bed of activated carbon is not an absolute value (a fixed number of pounds of solvent adsorbed per pound of activated carbon, at a given temperature[23]).

Said another way; bed capacity is a variable[24]!

A bed of activated carbon adsorbent may be filled to "capacity" with solvent vapor molecules, even though little mass of solvent vapor was fed to it; if the waste stream fed to it was sufficiently lean in solvent concentration[25].

That this is not a trivial point is shown within Figure 4.18 (calculated[V] for trichloroethylene), because:

- Emissions purged from airless enclosed vapor degreasers are relatively rich in solvent, and
- Emissions from the lip vents of open-top vapor degreasers are relatively lean in solvent (Chapter 1.21 and Figure 1.39).

Figure 4.19 shows the same calculated outcomes for methylene chloride, which has a substantially higher exposure limit (50 ppm, green solid line), but is not nearly so well adsorbed.

These two different vapor degreasers do use activated carbon adsorbent – but they use it in differently designed

[20] This appears to be an application of Heisenberg's Uncertainty Principle (Ref. 1, Chapter 4.2.3). In the same way that one can't simultaneously measure the position and momentum of a particle, one can't use an air amplifier to collect only the solvent molecules which have escaped beyond the capability of the vapor degreaser to retain them and not pull additional solvent vapor from the throat of the open-top vapor degreaser. Both can't simultaneously be done.

[21] An air eductor is also known as an eductor-jet pump, an aspirator, a water eductor, a vacuum ejector, and a steam-jet ejector (depending upon what fluid is used as the motive fluid). An air amplifier is also erroneously called an air knife or an air eductor.

[22] Note that the ranges of values of both concentration and solvent retention are so broad that both axes are plotted on a logarithmic basis.

[23] Appendix A2 contains detailed information about how to both correlate data and calculate bed capacity as a function of type of solvent, inlet concentration in air, pore size within the adsorbent, and temperature.

[24] For an analogy, consider a balloon. Inflated by an adult with great lung power (or by an air compressor), the final size will be large and perhaps the elastic strength of the material will be exceeded and the balloon will burst. Inflated by a small child with undeveloped lung power, the final size of the same balloon will be disappointingly small. So, there is no fixed volumetric capacity of a balloon – it depends upon how it is filled with air (used).

[25] It is an empirical fact that adsorption of vapor molecules may occur in two places within the carbon pores: (1) the activated sites, and (2) the vapor molecules already adsorbed on the sites. The former is adsorption of a single layer of solvent; the latter is adsorption of multiple layers. Derivation of the BET Equation described in Appendix A2 was based on this fact.

A rationale to explain the calculated results in Figures 4.18 and 4.19 is that at high concentrations of solvent in air, multilayer adsorption dominates because of the higher driving force in concentration. And at lower concentrations of solvent, the reduced concentration driving force only engenders single layer adsorption.

Table 4.2 Comparison of Compressed Air-Driven Devices

	Air Amplifier	Air Knife	Air Eductor
General Use	Collect air from a volume	Blow liquid (usually water) off parts; but NOT solvent (would produce flammable aerosols)	Create vacuum without moving parts
General Effectiveness	Very. About 15 volumes can be entrained by 1 volume of compressed air	Useful. Produces dry to the touch quality (~90% dry)	Helpful. Not capable of producing less than 600 TorrV with a single stage (especially with air as motive fluid)
Operating Principle	Bernoulli's Principle = when a fluid's velocity increases, its pressure declines Flow through a broad aperture produces an increased velocity which entrains air – producing an air current	Simple expansion of compressed air through a thin aperture, supplemented by entrainment of static air by high velocity air	Bernoulli's Principle = when a fluid's velocity increases, its pressure declines Flow through a narrow aperture produces a very high velocity which entrains air – producing a vacuum
Flow Diagram Showing Operation	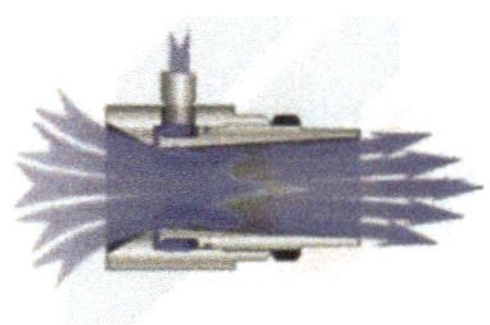	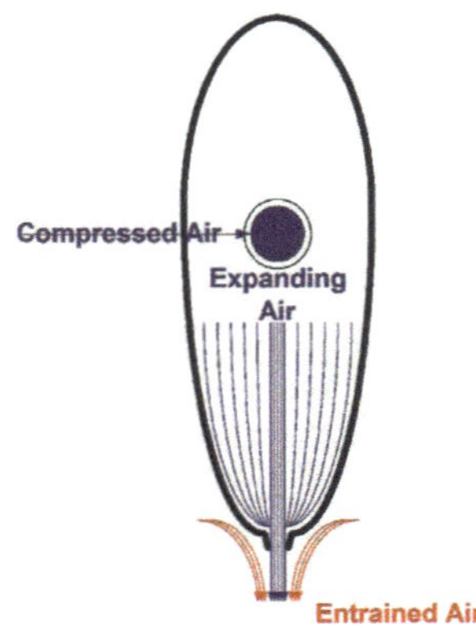	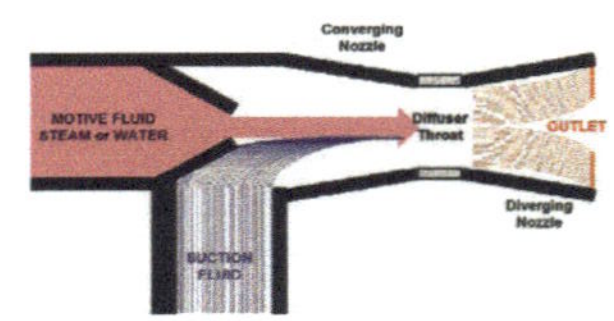
Image of Commercial Product			
Referred to in Chapter	4.6	3.5.2	2.2.2

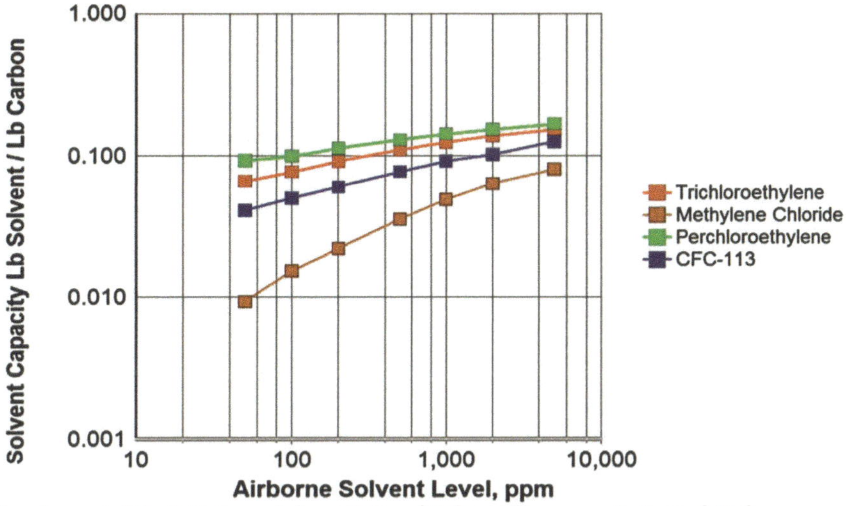

Figure 4.17 **The Fundamental Basic Data — Adsorptivity of Solvent Vapor on Activated Carbon**

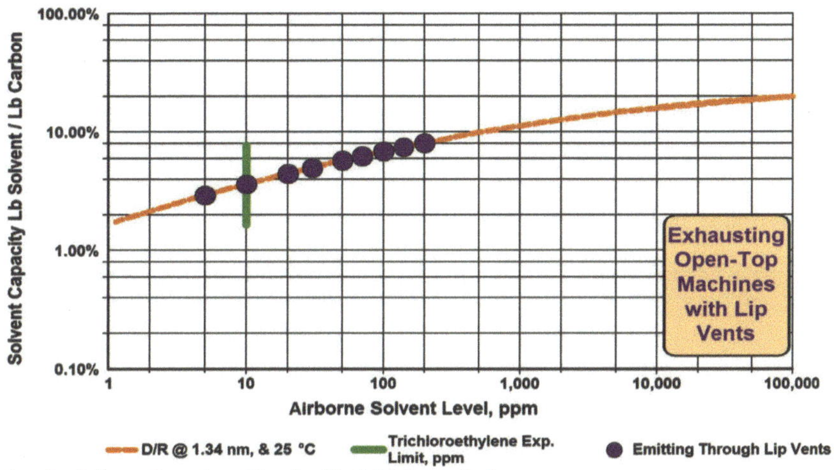

Figure 4.18 **Adsorber Bed Capacity when Used with Trichloroethylene**

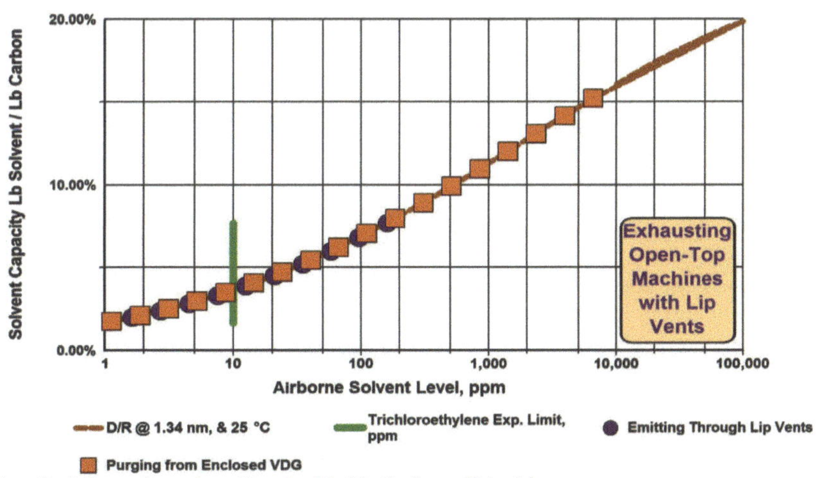

Figure 4.19 **Adsorber Bed Capacity when Used with Methylene Chloride**

Box 4.6 Emission Profile – Two Types of Enclosed Vapor Degreasers

In an airless enclosed vapor degreaser (non-vacuum or vacuum) the starting environment (air) is removed before solvent is added.

- For non-vacuum vapor degreasers, the work chamber contents (air) are displaced by nitrogen prior to addition of solvent. When the cleaning operations are complete, the pressure in the work chamber is 1 atmosphere. So the solvent concentration in the vapor phase is what it would be were the solvent in an open container in open air at the temperature inside the vapor degreaser.
  This concentration, expressed as a fraction, essentially is the solvent vapor pressure divided by 760 Torr (mm Hg). The solvent vapor pressure can be calculated from the Antoine Equation using empirical constants which can be found in Appendix A1, Table A1-1. Calculated values of the solvent concentration and vapor pressure in nitrogen (or air) are tabulated in Endnote W. The assumed temperature is 25°C which is the lowest anticipated temperature for liquid in a vapor degreaser; hence solvent concentrations fed to adsorbers will be even larger than those tabulated in Endnote W.
- For vacuum vapor degreasers, the solvent concentration fed to adsorbers is determined by the level of vacuum in the work chamber at the close of the cleaning operations. The assumed value of vacuum used in the calculations of this book is 750 TorrV (10 mm Hg), producing the same calculated solvent concentration in the feed to adsorbers – 13,158 ppm ([10/760] × 10^6).

facilities, and of course in different ways[26] (Boxes 4.6–4.8) (Chapter 4.8).

The difference in adsorption capacity is significant. An adsorption bed which is perfectly useful to treat the vapor emissions from the lip vents of an open-top vapor degreaser may provide only short-term restricted service when used to treat vapor emissions from an enclosed vapor degreaser. This loss of bed capacity is demonstrated in the non-logarithmic Figure at below.

By analogy, an adsorption bed is a chameleon; its

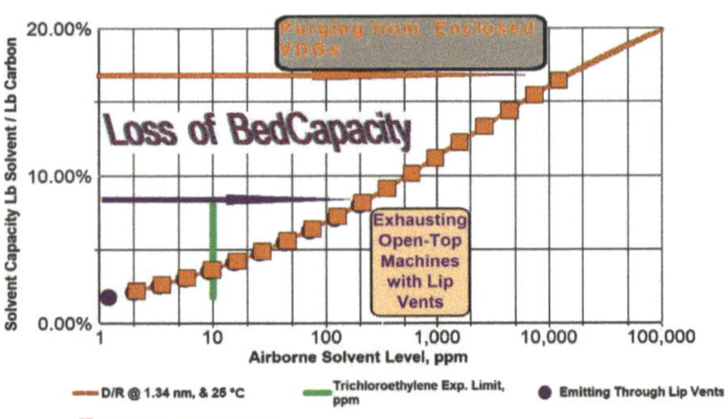

The Adsorber Bed has Two Different Capacities – for Enclosed and for Open-Top Vapor Degreasers

capacity is a function of the solvent concentration in the vapor fed to it. One can't modify this outcome. The reasons for it are the founding basis for the BET approach to description of adsorption – found in Appendix A2.

Box 4.7 Emission Profile – Open-Top Vapor Degreasers

In open-top vapor degreasers, the air flow is chosen to be large (~1500 ACFM; Ref. 5, Endnote H so that solvent fumes in the work area don't accumulate. The flow rate is a compromise – high enough to keep solvent from escaping upward in the air column of the degreaser freeboard and out through the cover of the degreaser causing a concentration in the workplace above the posted exposure limit, and low enough to not waste significant amounts of solvent in the lip vent exhaust. The net effect is that solvent concentrations in this air stream (Figure 1.37) are low – several hundred ppm at most. This author has arbitrarily assumed that this concentration is 20 times the published solvent exposure limit.

Concentrations of solvent in vapor fed from lip vents to an adsorber are often reduced because fresh air is required to supplement the vapor flow to achieve the desired superficial velocity in the adsorber of 150 ft/min.

- *The crucial point is that adsorption beds must be differently designed and used depending upon the nature of the role they are expected to play.*

4.8 DESIGN OF ADSORPTION SYSTEMS

In was noted in Chapter 4.3 that supply of activated carbon for cleanup of emissions from solvent vapor degreasers is a business whose volume is, in 2014, negligible – or less[W]. It's not that such systems don't work or aren't being used, it is that they are only being used where they are specifically required by enforced environmental regulations.

[26] The concentration of solvent in vapor emitted from airless enclosed machines is much higher than the same in vapor emitted from the lip vents of open-top vapor degreasers. The reason for this has to do with the nature of the two vapor degreasers.

Box 4.8 **Comparison of Emissions from Various VDGs**

For non-vacuum enclosed vapor degreasers, the concentration of solvent in the vapor feed to adsorbers can be nearly all solvent – relative to air.

Instead of a few hundred ppm solvent in the vapor feed from the lip vents of an open-top vapor degreaser, or instead of several dozen tens of thousands of ppm solvent in the vapor feed from a vacuum vapor degreaser, the solvent concentration can be hundreds of thousands of ppm in the feed from a non-vacuum enclosed vapor degreaser.

Chamber volumes under vacuum contain less mass of vapor than do those at atmospheric pressure (but that mass may be easier to remove by condensation because it may be colder, see Footnote 46). That's why the concentration of solvent vapor emitted from vacuum enclosed vapor degreasers is substantially less than that emitted from non-vacuum vapor degreasers.

A second concern about managing adsorption beds that treat solvent-rich effluent from enclosed vapor degreasers is that the tiny vapor stream which bypasses the adsorbent granules increases the pollution level of the effluent. Bypass is also normal. It comes about because no packed bed is ever free of small channels through which vapor can exit without effectively contacting adsorbent material. Bypass is typically or more than a few percent of the volumetric flow, and may be less than 1 percent.

The point is that normal bypass is of more concern when the vapor feed is more rich in solvent.

Finally, one must recognize that a multi-purpose adsorber, accepting discontinuously feeds of various solvents at various concentrations from various locations on a site, must be of large capacity – because such a use is probably a most inefficient one.

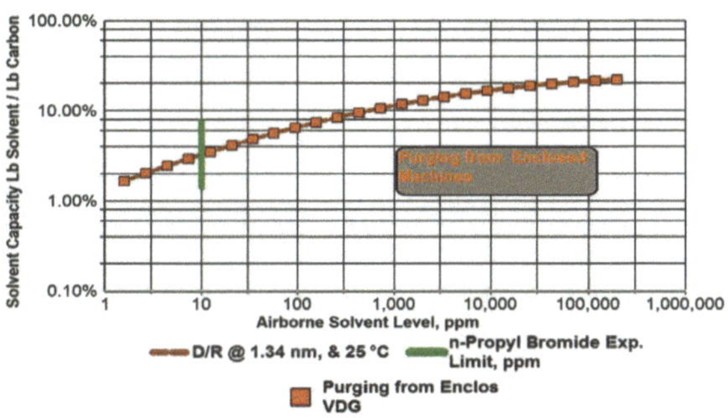

Solvent	Antoine A	Antoine B	Pv, mm Hg	Calculated ppm in air at 25°C
Methylene Chloride	−1,703.0	8.3141	400.1	526,498
Trichloroethylene	−1,752.0	7.7498	74.7	98,344
Perchloroethylene	−1,969.0	7.8811	18.9	24,902
n-Propyl Bromide	−1,619.0	7.5707	138.2	181,859
PCBTF	−2,223.0	8.3327	7.5	9,906
Methyl Formate	−1,641.0	8.2919	613.7	807,505
Methyl Acetate	−1,565.0	7.6320	241.5	317,798
Tertiary Butyl Acetate	−1,739.0	7.5389	50.8	66,904
Dimethyl Carbonate	−2,199.0	8.9550	38.0	49,969
Acetone	−1,598.0	7.7295	234.3	308,296
Isopropanol	−2,260.0	9.2297	44.6	58,723
Ethyl Acetate	−1,815.0	8.0617	94.2	123,979
Toluene	−1,904.0	7.8445	28.7	37,813
HFC-365mfc	−1,345.0	7.0985	386.7	508,785
HFC-43-10 mee	−1,671.0	7.9696	231.8	304,947
HFE-7100	−1,581.0	7.6098	202.8	266,862
CFC-113	−1,362.0	7.2331	462.3	608,297

A high concentration in the feed does not impair adsorption. In fact adsorption capacity is improved (Figure 4.17). Calculated results, based on the D/R method, for n-propyl bromide using the feed concentration in the table above are plotted at above.

However, when the amount of solvent vapor in the adsorbed bed approaches its capacity, "bleedthrough" or "breakthrough" will start to occur – traces of feed vapor will start to appear in the exit stream. This is a normal outcome. But if the feed vapor is very rich in solvent (vs. air), because it is from an enclosed vapor degreaser (non-vacuum), the exit stream may soon become the inlet stream. Prior to this event, conventional on-line gas monitoring equipment (Endnote Q) must have recognized this situation and switched the vapor feed to the regenerated adsorber.

The point is that normal "bleedthrough" is of more concern when the vapor feed is richer in solvent.

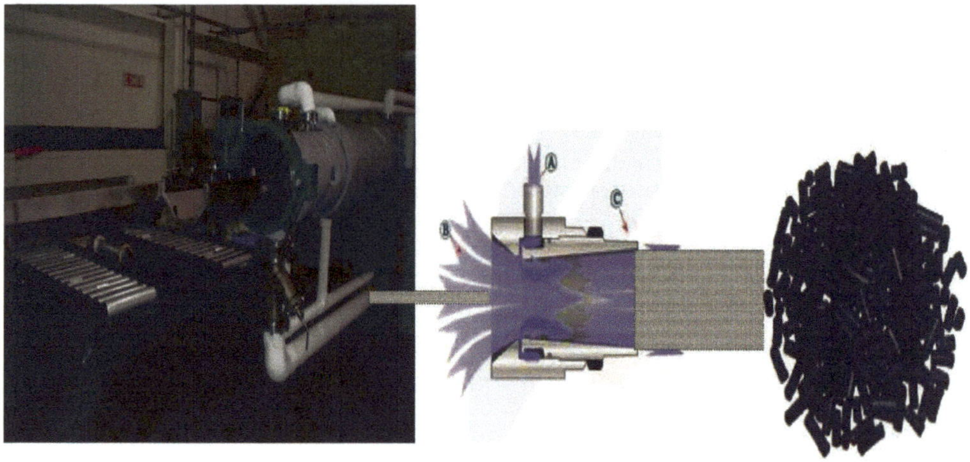

Figure 4.20

Given that the thrust of environmental regulations over the last half-century is toward more and not less restriction of pollution, use of activated carbon to treat exhaust air, and possibly recover the solvent from it, is likely to become much more common – whether the treated solvents are characterized as being halogenated, or VOC.

This means that there will be more "packaged"[27] carbon treatment systems available for use with solvent cleaning systems. As this book is published, the selection of packaged systems is somewhat limited. And none of the available systems are customized for the specific needs of the doing solvent cleaning work.

Consequently, this subchapter provides brief but essential guidance for the engineering department of a site doing cleaning work for them to construct their own "package" system for treatment of solvent-laden air by activated carbon. The necessary engineering (and chemistry) is straightforward – just not well known.

Two design cases will be considered: (1) purging of air at ambient temperature contaminated with solvent vapor from the chamber of an enclosed vapor degreaser just prior to the chamber being opened, and (2) management of the solvent and air discharging from lip vents associated with an open-top vapor degreaser. In both cases, it is necessary to maintain the concentration of solvent vapor in the work area at or beneath the exposure limit.

[27] A packaged system is one that includes all necessary (and possibly some unnecessary) components – supports, piping, vessel(s), instrumentation, process controls, and simplified access to local utilities of electric power, compressed air, steam, and water drains. Such a system is likely purchased (or possibly rented) from the supplier of activated carbon. Some refer to this as a "turn-key" system, in that all the user has to do is "turn the key" and start operation.

4.8.1 Design of Adsorption Systems for Enclosed Vapor Degreasers

The design aim is to purge the cleaning chamber of the last vestiges of solvent material in air, direct the infected air into a suitable adsorption bed of activated carbon, and operate the bed to meet all safety, health, and environmental (SHE) criteria for posted exposure limits and emissions.

4.8.1.1 Management of Emissions from Enclosed Vacuum Vapor Degreasers

First, directly purge solvent vapor from the work chamber into a plenum (duct) which is securely plumbed through an air amplifier into the adsorber bed.

Do follow the visual guidance of Figure 4.20. Don't allow that solvent-rich vapor stream to escape into the work area.

An advantage of enclosed vapor degreasers is that they enable improved use of adsorbent beds (vs. open-top vapor degreasers). This is because the initial volume of exhaust is relatively rich in solvent vapor.

Consider the example where after the cleaning cycle is complete, the final vacuum level in the work chamber is 750 TorrV (10 mm Hg absolute) at 25°C. All remaining vapor molecules are solvent – there is no air present, yet. Suppose the chamber is then filled with clean dry air to raise the chamber pressure to 0 TorrV or 760 mm Hg (as described in Tables 2.6 through 2.10). That action reduces the overall concentration of any solvent in the chamber from 1,000,000 ppm (100 volume %) to about 13,100 ppm ($[10/760] \times 1,000,000$)[28].

[28] If the final vacuum level in the work chamber were 759 TorrV (1 mm Hg), the solvent concentration in the air-filled chamber would be 1310 ppm.

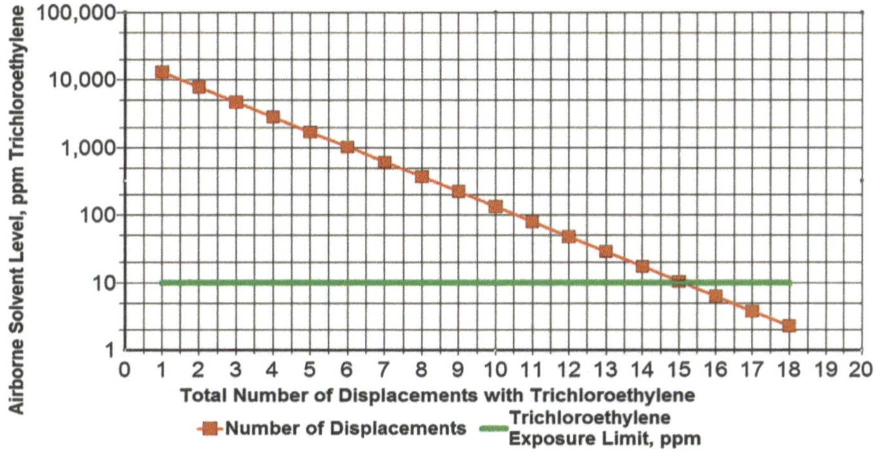

Figure 4.21

At that concentration of trichloroethylene (for example[29]) in air, vs. the posted 10 ppm exposure limit for worker protection, the capacity of the adsorber bed to hold solvent becomes a useful value at 25°C (Figure A2-12) of about 0.15 lb trichloroethylene/lb adsorbent.

4.9 THE IMPERFECT PERFECTLY-MIXED VESSEL

That salubrious situation can't continue.

All of the solvent in the work chamber (at a concentration of 13,100 ppm in air) of an enclosed vapor degreaser isn't removed by displacement when the chamber is continually flushed with clean air. Some solvent always remains in the air within the chamber after displacement with any volume of clean air[X]. This is similar to critical cleaning where it is impossible to remove 100 percent of soil.

Typically, for a chamber whose contents are perfectly mixed in a single phase (homogeneous), it takes three full displacements (not dilutions) of fluid (liquid or gas) to reduce the initial concentration of an impurity to about 5 volume % of the initial concentration (95.02% displacement). The number of well-mixed displacements needed to reduce the solvent concentration in the chamber from 13,100 pm to the low exposure limit of 10 ppm, is not three but seven![30].

The cause of this outcome is that the flow pattern of air through the cleaning chamber is not that of a piston in a cylinder – pushing all ahead with no bypass or recycle. This flow pattern would be called plug flow – causing perfect longitudinal displacement of one immiscible phase with another without mixing. But air (or water) does not flow through a vessel full of air (or water) in plug flow. The air (or water) is a single phase. Preexisting material is mixed with material freshly added. Material exiting the vessel is a mixture of the "old and the new" – causing incomplete displacement of the original contents[Y].

Unfortunately, that's not all. The contents of any cleaning chamber are not perfectly homogeneous! After seven additional volumes of air have passed through the work chamber, the solvent concentration in the air emitted from the work chamber is substantially more than 10 ppm. This is because the original contents work chamber is not perfectly mixed[31] with the freshly added material.

Figure 4.21 shows how, with an assumed level of imperfect mixing, reduction of trichloroethylene concentration from 13,100 ppm to the exposure limit of 10 ppm could take about 15 displacements of the chamber volume with fresh solvent-free air.

Naturally, adsorbent capacity will be high for the initial discharge of solvent-laden air from the cleaning chamber. Solvent contained in subsequent discharges of air won't be as well adsorbed on activated carbon – because the concentration has been reduced (Figure 4.17).

Adsorption capacity of the activated carbon (from Figure 4.17) in the hypothetical case described in Figure 4.21 is plotted in Figure 4.22.

Of more interest is how that capacity is used.

[29] Solvent concentration is a function only of the final level of vacuum and temperature, not the specific solvent.

[30] This represents a Herculean reduction in solvent concentration – 99.92% (= 1 – [10/13,100], or 7 Sigma). Of course, the displacements are sequential in time without interruption.

[31] A cleaning chamber is substantially less well mixed than perfect – at least because the racked parts block fluid flow and cause the flow patterns to be non-uniform. And the chamber shape, while often radially symmetrical, is not longitudinally symmetrical. So, the less than perfectly mixed vessel will retain a fluid component longer than would be expected from the assumption of perfect mixing. Deviation from the well-mixed residence time will be different for every vessel shape and loading of parts, and will be specifically unpredictable (Chapter 2.13.1).

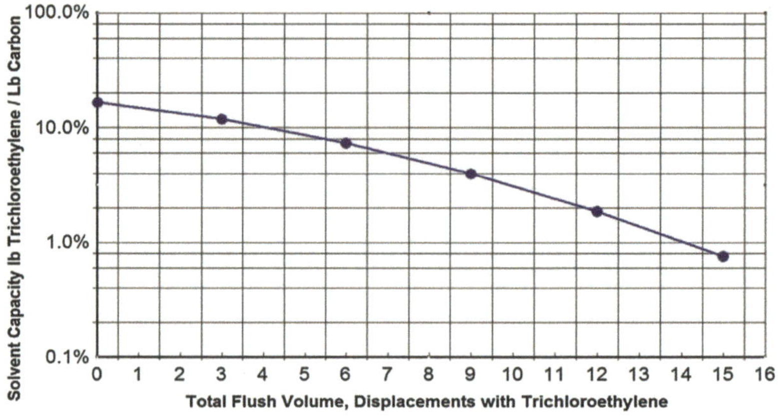

Figure 4.22

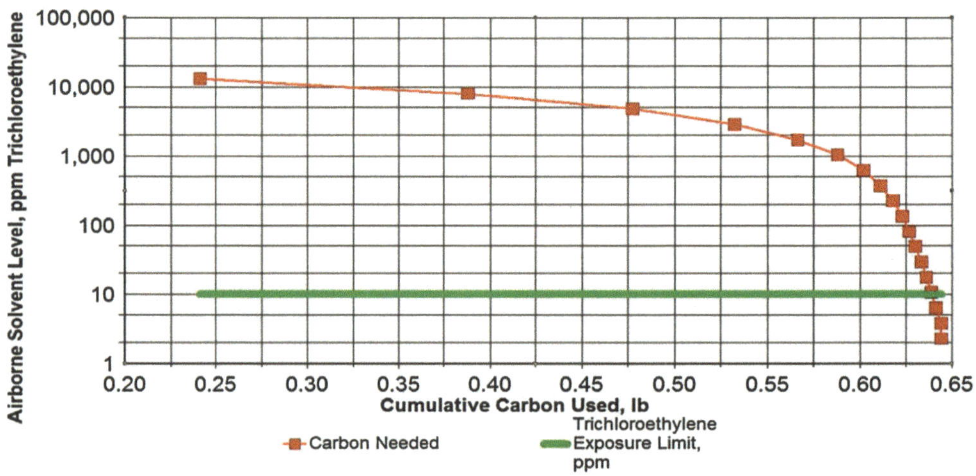

Figure 4.23 **Use of Activated Carbon Adsorbent as Trichloroethylene Vapor is Evacuated from Enclosed Vapor Degreaser**

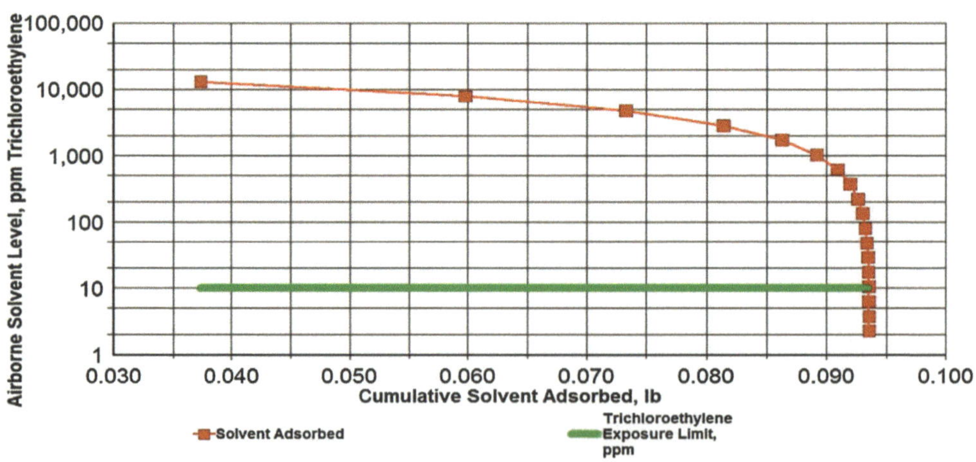

Figure 4.24 **Adsorption of Solvent onto Adsorbent as Trichloroethylene Vapor is Evacuated from Enclosed Vapor Degreaser**

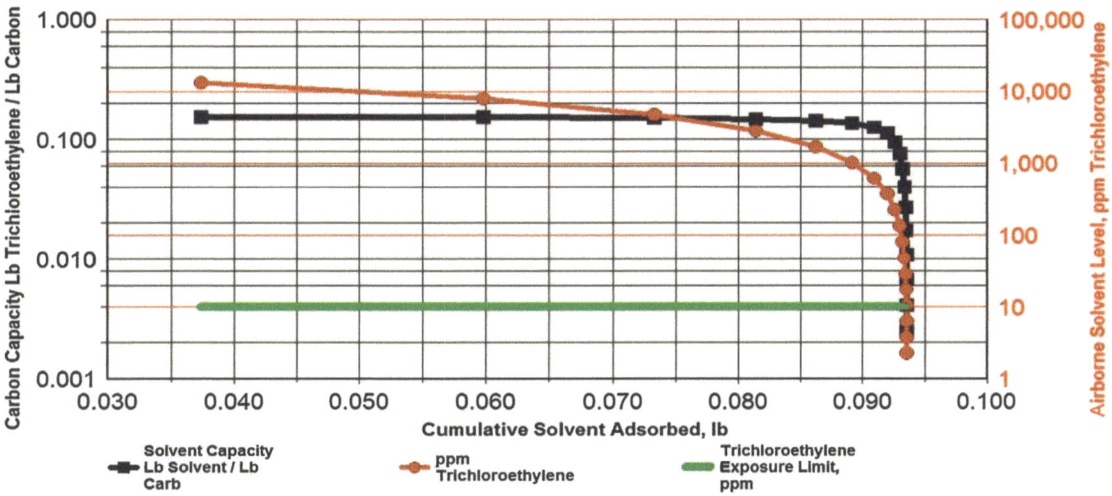

Figure 4.25 **Overall Operation of Adsorber Bed to Remove Trichloroethylene from Air Purged from Enclosed Vapor Degreaser**

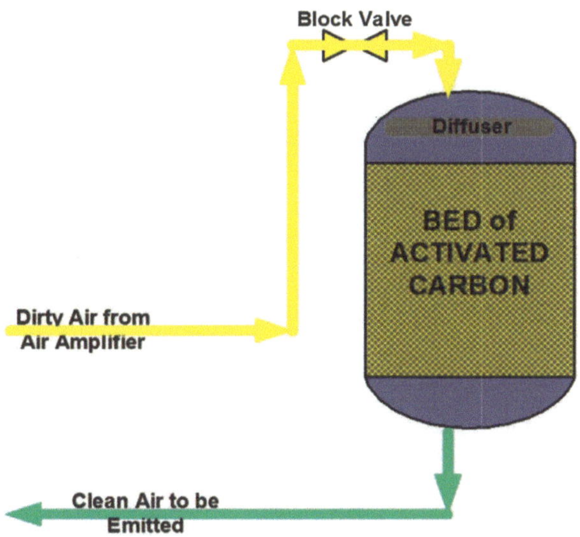

Figure 4.26 **Drawing of Adsorber Vessel and Flows to and from it**

This is plotted in Figure 4.23 – showing that nearly all the activated carbon is loaded with solvent vapor as the concentration of solvent vapor in the work chamber is reduced from 13100 to about 1000 ppm. The remaining reduction from 1000 to 10 ppm, by displacement with more air, requires very little activated carbon.

The reason for this outcome is shown in Figure 4.24 – it is necessary to adsorb only very little mass of solvent vapor to reduce the airborne concentration from 1000 to 10 ppm.

So although the activated carbon sites which adsorb the last-displaced molecules of trichloroethylene aren't efficiently used (Figure 4.22), not many of them are affected (Figure 4.23), because there aren't many solvent molecules remaining to be adsorbed (Figure 4.24).

Both the "instantaneous" capacity of the activated carbon and trichloroethylene concentration are plotted in Figure 4.25.

One can't escape the conclusion that, like cleaning, rinsing, and drying, adsorption will take substantially increased "effort" to obtain the highest quality results.

In the case of adsorption, that's requiring substantial "time" (displacement volumes of clean air) to have the solvent concentration in air be less than the exposure limit. Yet, as shown Figures 4.24–4.26, the full capacity[32] of an adsorption bed is surprisingly well used.

Normally, use of ventilation devices, such an air amplifiers, is discontinuous – at one high rate during a certain phase of the cleaning cycle, at another lower rate during other phases of the cleaning cycle, and at zero ventilation during off hours.

4.10 DESIGN OF THE ADSORPTION SYSTEM FOR USE WITH ENCLOSED VAPOR DEGREASERS

At its heart, such a system is a pressure vessel – a small tank. A simplified drawing of such a tank in use as an adsorber is Figure 4.26.

[32] Estimation of this capacity by the D/R method is described in Appendix A2. There it is noted that adsorption of solvent molecules does occur on the surface of the adsorbent at specific locations called sites. But adsorption of solvent molecules does also occur onto the solvent molecules already adsorbed onto the activated carbon sites. In other words, multilayer adsorption is common and is probably fostered by increased concentrations of solvent in the incoming vapor stream.

Table 4.3 Comparison of Design Specifications

Design Case >>>	"Criteria Design"	"Additional Air"	"Decreased Bed Area"
Issue			
Air Velocity, ft/sec	114	150	
Bed Diameter, inches	16 (L/D = 3)		14 (L/D = 4.5)
Bed Height, inches	50 (L/D = 3)		65 (L/D = 4.5)
Added Air Flow, actual cubic feet per minute (ACFM)	0	54 (beyond the required 160 ACFM)	0
Initial Trichloroethylene Concentration, ppm	13,100	10,000	13,100
Carbon Use per cycle, lb	0.657	0.659	0.640

Solvent-laden air generally enters the adsorber (pumped by the air amplifier) through a port in the top[33] of the tank[34]. A diffuser spreads the volume of flow across the entire area of the adsorber bed.

The adsorber bed is sized to meet the needs[35] of the enclosed vapor degreaser to which it provides pollution abatement service.

- The primary design need of this apparatus is to allow the operator of the enclosed vapor degreaser to open the cover of the work chamber, releasing its contents to the adsorber, and to not be exposed to a solvent concentration greater than the exposure limit when the treated air reaches the work area (or the outside environment[36]).

These needs are defined by three criteria. All are employed in the design of the adsorber vessel, and are that the vessel: (1) has adequate capacity for internal transfer of heat from the adsorbent to the moving air[37], (2) holds up the moving air stream for an adequate time to allow for adsorption to be complete, and (3) contains a sufficient quantity of activated carbon to provide service for whatever long-term period is desired.

- The first two criteria are implemented by specifying that the air velocity through the moving bed is to be between 150 and 200 ft/min (0.75 to about 1 m/sec). In addition, these adsorber beds (Table 4.3) are taller than they are wide, with an L/D (height/diameter) ratio of about 4. Holdup time of the air in the bed is to be at least a few seconds.
- The third criteria is implemented by choosing a mass of activated carbon which will meet the primary design need of performance for a required number of days.

---

[33] Top entry is preferred so that movement of any tramp liquid is aided by the forces of gravity, and liquid is pulled by gravity (drains) downward through the bed.

[34] Though uncommon, some adsorber beds are fed into the center of the vessel with the solvent-laden vapor forced both upward and downward through the adsorber bed.

[35] Consider a very large enclosed vapor degreaser with a cleaning chamber holding about 160 gallons (21.2 CF) of trichloroethylene, operated at a vacuum level of 750 TorrV (10 mm Hg absolute pressure) at 25°C. The cycle time for use of the cleaning chamber is to be 20 minutes and the trichloroethylene is to be only emitted from that chamber during the last 2 minutes of the cleaning cycle. Tables 2.6 to 2.10 show various other schedules of cleaning steps and chamber ventilation.

Displacement by fresh air of all the trichloroethylene vapor in the chamber is required to produce a concentration in emitted air below the posted (by a standard-setting authority or the manufacturer) exposure limit of 10 ppm.

The calculated quantity of trichloroethylene emitted from the cleaning chamber every 20 minutes is about 0.0936 pounds, which is an average rate of: 0.281 pounds per hour, 1.06 actual cubic feet of trichloroethylene vapor per minute (ACFM), or 16.6 gallons of condensed liquid trichloroethylene per month. To reduce this volumetric flow of pure trichloroethylene vapor to a concentration of 10 ppm in air one would need an air flow of about 340 actual cubic feet (ACF), or about 16 times the volume of the work chamber. That's only a tiny flow rate of vapor: (170 ACFM = 340 ACF/ 2 minutes).

If this cleaning chamber were operated at 759 TorrV (1 mm Hg), there would be 90 volume % less trichloroethylene in the cleaning chamber, and so the adsorption vessel would be considerable smaller – or the same charge of activated carbon would last considerably longer.

[36] It may well be that the design limit is not compliance with a posted volumetric exposure limit. The design limit may be a mass amount emitted over time compliant with an operating permit.

[37] Recall from Chapter 3, Footnote 93 that adsorption of water on mole sieves or solvent on activated carbon is exothermic. Heat is given off on adsorption of vapor, and must be supplied to desorb those molecules.

The reason heat is given off is that the adsorbed molecules of solvent vapor have given up some of their freedom of movement when they become adsorbed. After all, when a molecule is pinned (adsorbed) by weak van der Waals (or any forces) to a pore surface, vibrational motion of the molecule is restricted. Intensity of molecular motion requires energy. So when molecules are restricted in their motion they have a surplus of energy, and it is released (exothermic). To restart their level of motion, by separating them from the pore wall (desorption), requires a contribution of energy (endothermic).

There are thermodynamic arguments which support this conclusion (and observation) as well, based on Gibbs free energy, which are outside the scope of this book.

Figure 4.27 **New Adsorption Bed — Ready for Installation**

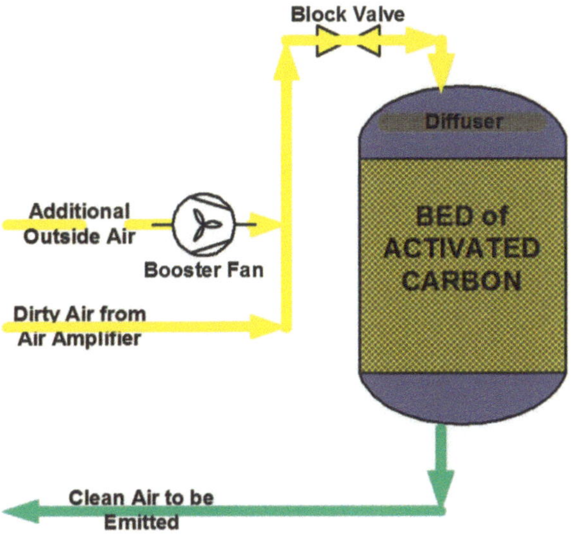

Figure 4.28 **Increased Air Flow**

The specific bed characteristics which fulfill these criteria for the conditions prescribed in Footnote 38 are: a bed working volume of 6.2 CF to hold 140 lbs of activated carbon whose apparent (bulk) density is 0.52 g/cc, whose height is about 65 inches, and whose diameter is about 16 inches[38,39]. An image of an equivalent tank (adsorption bed) is shown in Figure 4.27[Z].

Useful life of the 200 lbs of activated carbon contained within would be about 3 days for 24 hour operation of the enclosed vapor degreaser. After that duration, the adsorbing vessel is regenerated (removing the adsorbed solvent) and put on standby while a second adsorber is put on line.

4.10.1 A Modest Design Concern

This concern pervades designs of adsorber beds used with enclosed vapor degreasers (but not with open-top vapor degreasers). The concern is that the adsorber bed may overheat.

The underlying cause producing that concern is that flushing of the work chamber in an enclosed vapor degreaser doesn't require a substantial volumetric flow of air (as noted in Footnote 38).

- In other words, the adsorption bed designed in the previous subchapter doesn't meet the stated criteria for air velocity.

This concern can be alleviated in one of two ways, by:

- Feeding additional air from a separate blower/fan into the adsorption bed to improve heat transfer, or
- Reducing the diameter of the adsorption bed (while holding the volume constant) to increase the velocity of air flow without increasing the volume flow of air.

The former approach is illustrated by updating Figure 4.26 as Figure 4.28. The latter approach is illustrated in Figure 4.29.

One can examine in Table 4.3 the modest differences in design specifications for the adsorber bed imposed by these two design approaches.

This author always prefers more heat transfer capability rather than less; and a taller bed will have less chance of allowing fluid channeling to occur. Although any of the three design choices could be selected[40], this author prefers a more narrow adsorption bed[41] because addition of added air simply dilutes the efficiency of the activated carbon.

[38] These dimensions would allow the vessel to be fabricated from a piece of available steel pipe (or tubing), and so be relatively inexpensive. Recall that piping is sized by external diameter; tubing is sized by internal diameter (Chapter 2, Endnote FF). Wall thickness of each is identified by a "schedule number." Since the adsorber bed does not have to be an ASME code pressure vessel (because the design pressure and temperature are very low), the wall thickness (calculated by Equations 2.3 and 2.4) can be quite thin. Also, code welding of end caps is not required (Chapter 2, Endnotes V and W). Pipe of the appropriate diameter and any schedule number can be used. Matching the exact numerical value of L/D noted above is not required.

Prior to construction of the adsorber vessel the inside and outside of the welded pipe should be coated with an acid-resistant coating such as a suitable epoxy.

[39] An adsorption vessel of this size would be considered very small by most suppliers of activated carbon.

[40] An explanation is necessary about why the use of carbon is essentially the same for all three design cases in Table 4.3. The explanation can be found in Figure 4.23 or 4.24. It is that the capacity of this particular activated carbon to adsorb trichloroethylene is only slightly dependent upon the feed concentration of trichloroethylene over the range of 13,100 to around 1000 ppm (Figure 4.17).

The weight of carbon given in Table 4.3 is that used in one cycle of cleaning, lasting 20 minutes.

[41] Consequently, the criteria for vessel shape should be revised to implementing a value of L/D in the range of 4 to 7 as long as the chosen diameter matches an available pipe size of carbon steel pipe.

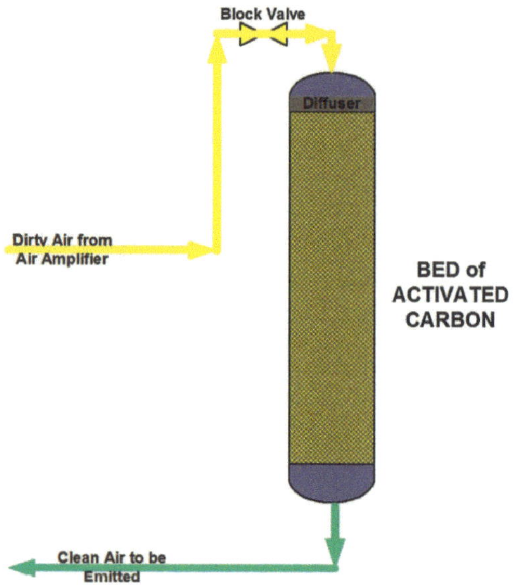

Figure 4.29 Reduced Bed Diameter

4.11 AN INCREMENTAL USE OF VACUUM

One can use the same capabilities of an enclosed vapor degreaser (and not the adsorber) to produce concentrations of solvent in air substantially less than the posted exposure limit (10 ppm for trichloroethylene). One can easily produce, at a modest extra cost, exhaust concentrations of trichloroethylene (for example) of 1 or 2 ppm in air!

As shown in Tables 2.6 through 2.10, this is done with an additional use of the vacuum pump with the work chamber. One simply flushes the contents of the work chamber with clean air into the adsorber for a selected small number of displacements of chamber volume, then uses the vacuum pump to reduce pressure in the work chamber from atmospheric pressure to a selected level of vacuum, and finally displaces the remaining contents of the work chamber into the adsorber with additional clean air at atmospheric pressure.

Consider the example of two schedules of operation in Table 4.4[42].

One can compare Figures 4.30 and 4.23 to observe how the concentration of trichloroethylene is dramatically reduced by scheduled application of vacuum to the work chamber. Perhaps less dramatic is the comparison between Figure 4.31 and Figure 4.24.

This benefit, like all benefits, comes with a price. The price is the energy (and possibly manpower or maintenance) cost of operating the vacuum pump for an incremental cycle, and then condensing the small amount of solvent in the air[43] removed from the work chamber. In most cases, this price is apt to be small, or negligible.

But, the benefit is worthwhile. It is that workers are exposed to lower concentrations of solvent vapor in air when they open the work chamber to remove the parts basket.

Incremental application of vacuum is an excellent and inexpensive way of providing incremental protection to workers from exposure to cleaning solvents.

4.12 DESIGN RESULTS FOR ADSORBERS USED WITH ENCLOSED VAPOR DEGREASERS

In an enclosed vacuum vapor degreaser, operating temperature is an independent variable (Chapter 2.14). Operating pressure is not. It's chosen to adjust the solvent temperature to the desired value[AA]. Should it be necessary to adjust the solvent temperature, because of heat losses or the need to heat (or cool) the thermal masses of parts, during the cleaning cycle temperature can be independently adjusted with heaters or coolers which are part of the apparatus[BB].

After the cleaning cycle is complete, pressure[44] in the work chamber is chosen based on another reason.

[42] The value of 23 ppm trichloroethylene in Table 4.4 is calculated from the concentration of trichloroethylene before vacuum and the level of vacuum which was created [1700 x (10/760) = 23 ppm].

[43] To be clear – two methods of removing solvent from emitted air are involved here. The first is the anticipated adsorption on activated carbon of solvent vapor from air. The second is by condensation of cooler solvent. When vacuum is created, the contents of the work chamber are cooled (Chapter 2.13). When flushed from the work chamber by nitrogen this solvent can be recovered by condensation. However, this action is limited by the interplay between solvent vapor pressure and available cooling temperature which is why adsorption is the final "polishing" treatment to remove solvent from emitted vapor streams.

One may optimize the use of a second cycle of vacuum application in purging solvent-lean air from the work chamber. One does this by using the vacuum pump earlier in the air purge cycle than is shown in Table 4.4. For the hypothetical conditions assumed above, the optimum timing for use of one application of incremental vacuum appears to be after two or three displacements of air from the work chamber, rather than after five displacements as shown in Table 4.4.

However, that's not the best optimization of vacuum application, and time. If a second cycle of incremental vacuum has great effect, so does a third total cycle! Consider the following sequence of operation when the cleaning cycle is complete: (1) fill the work chamber with one volume of air and displace it to the adsorber bed with another volume of air, (2) use the vacuum apparatus to create the same level of vacuum, (3) repeat step 1, (4) repeat step 2, (5) repeat step 1, (6) repeat step 2, and (7) repeat step 1. The calculated concentration of trichloroethylene in air exhaust after three incremental cycles of vacuum and three volume displacements is well below 1 ppm, and probably beyond the resolution of the calculation method! This approach is partially illustrated in Table 2.9.

[44] Actually, it is the vacuum level in the work chamber which is being adjusted.

Table 4.4 Comparison of Steps Toward Air Cleanup

Continuous Displacement by Clean Air[CC]				Intermediate Vacuum Step		
Time min	**Pressure TorrV**	**Step**	**Solvent Conc., ppm**	**Pressure TorrV**	**Solvent Conc., ppm**	**Actions**
1	0	1	13,100	0	13,100	
		2	7,800		7,800	
		3	4,700		4,700	
		4	2,800		2,800	
		5	1,700		1,700	
2		6	1,000	0 to 750 (760 to 10 mm Hg absolute)	Solvent removed from chamber and conveyed to a condenser	Use of vacuum pump to reduce pressure from 1 atmosphere to 750 TorrV
		7	600			
		8	370			
		9	220			
		10	130			
3		11	80	0	23	Increase of pressure to atmospheric pressure, and continuing displacement of chamber contents with clean air
		12	48		13	
		13	29		8	
		14	17		5	
		15	10		3	

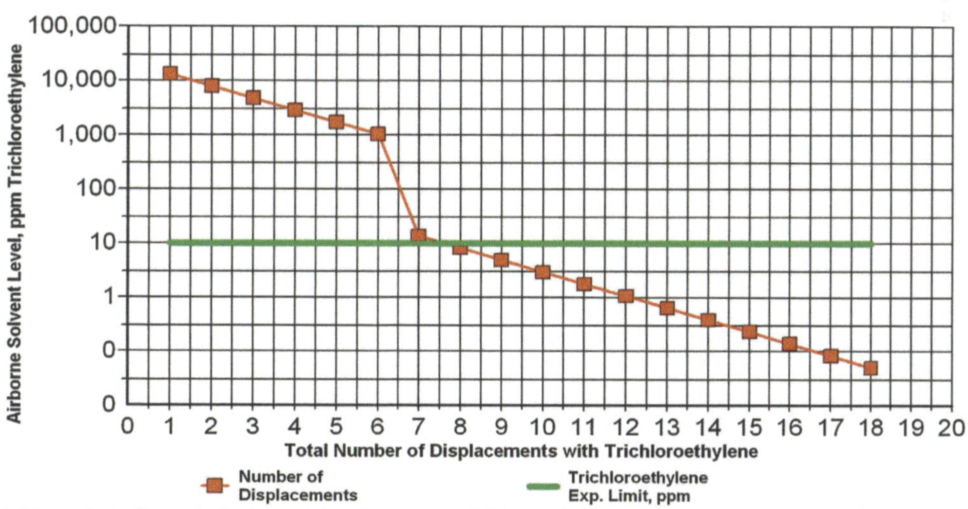

Figure 4.30 **Use of Activated Carbon Adsorbent as Trichloroethylene Vapor is Evacuated from Enclosed Vapor Degreaser — with Additional Vacuum Processing**

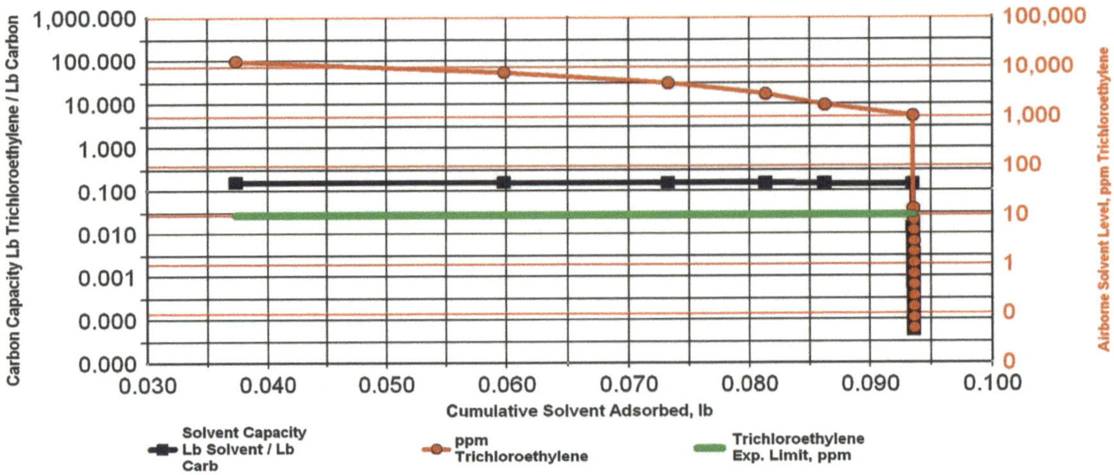

Figure 4.31 Adsorption of Solvent on to Adsorption as Trichloroethylene Vapor is Evacuated from Enclosed Vapor Degreaser – with Additional Vacuum Processing

That is to manage the level of solvent emissions which will be exhausted from the enclosed vapor degreaser.

- Higher levels of absolute pressure in the work chamber mean that more solvent is retained in the work chamber. Higher levels of vacuum mean less.

Solvent retained in the work chamber will either be emitted to the work area when the chamber is opened (so the parts basket can be removed), or pulled from the work chamber by an air amplifier (before the parts basket is removed) and then exhausted into an adsorption bed. There is no third way: emit or treat.

Solvent emitted will come in contact with workers at some concentration.

Solvent treated will be retained within the adsorber, and may later be recovered as wet solvent when the adsorber is regenerated (Chapter 4.16).

Operating staff have a free choice in the level of absolute pressure they choose to maintain in the work chamber before its solvent contents are displaced with clean nitrogen (or air).

- Reduced absolute pressures (increased values of vacuum) mean more solvent is withdrawn from the work chamber and passed into the internal condensation facilities of the system – before the contents of the work chamber are purged with clean gas. In other words, (1) less solvent must be retained within the adsorber and the charge of activated carbon will last longer; and (2) the energy and labor cost to operate the system is increased.
- Higher absolute pressures (reduced values of vacuum) mean better use of the activated carbon adsorbent because the solvent concentration fed to it is increased. In other words, (1) cycle time will be somewhat lengthened because more solvent must be purged from the work chamber; and (2) the energy and labor cost to operate the system is decreased.

These effects are considered for four levels of vacuum (absolute pressure) each with four different common cleaning solvents. The results are tabulated in Table 4.5.

4.13 RECOMMENDATION ABOUT USE OF ADSORBERS WITH ENCLOSED VAPOR DEGREASERS

The focus of this fourth chapter is treatment of emissions being deliberately or unintentionally emitted from enclosed vapor degreasers and open-top vapor degreasers.

Despite that focus, this author believes the use of adsorbers, or any other treatment scheme for solvent emissions, should be avoided – where possible.

- It is always better to "keep the solvent in the tank."[DD] It is always better recycle solvent rather than collect and retain it in adsorber beds, no matter the amount of solvent involved. It is always better to avoid pollution problems rather than to solve them!

Where vacuum-capable vapor degreasers are available (vs. the non-vacuum airless type noted in Chapter 2.1) their capability should be used to the extent possible to retain and recover solvent (Figures 2.31 through 2.33). As noted in Chapter 4.18, the economics of treatment of air emissions to remove solvent by adsorption never favor treating the air vs. keeping the solvent in the tank.

This author recommends following the schedule of operation with intermediate vacuum use described in Table 4.4 – producing the outcomes tabulated in Table 4.5.

Very often, or most often, such as when vacuum technology is not conveniently available, adsorption technology should be applied because it has been proven effective. That's why it was included in this book. But, like

Table 4.5 Summary of Adsorbent Use with Intermediate Use of Vacuum During Chamber Evacuation

Final Level of Vacuum, TorrV	Initial Solvent Concentration, ppm	lb Carbon/ CF Chamber	lb Solvent/ CF Chamber	Days of Use/ CF Chamber	% Carbon Use
Trichloroethylene					
750 (10 mm Hg)	13,100	0.030372	0.004411	0.21	93.8
755 (5 mm Hg)	6,580	0.016265	0.002206	0.39	80.1
759 (1 mm Hg)	1,310	0.004533	0.000441	1.57	69.5
759.5 (.5 mm Hg)	658	0.003108	0.000220	2.36	55.5
Methylene Chloride					
750 (10 mm Hg)	13,100	0.042235	0.002851	0.15	80.1
755 (5 mm Hg)	6,580	0.024290	0.001425	0.25	73.0
759 (1 mm Hg)	1,310	0.009503	0.000284	0.67	49.0
759.5 (.5 mm Hg)	658	0.006146	0.000140	1.05	48.9
Perchloroethylene					
750 (10 mm Hg)	13,100	0.034710	0.005567	0.18	95.2
755 (5 mm Hg)	6,580	0.018300	0.002784	0.35	91.6
759 (1 mm Hg)	1,310	0.004672	0.000566	1.35	75.0
759.5 (.5 mm Hg)	658	0.002790	0.000277	2.36	68.5
n-Propyl Bromide					
750 (10 mm Hg)	13,100	0.029545	0.004129	0.21	91.6
755 (5 mm Hg)	6,580	0.016190	0.002065	0.39	85.4
759 (1 mm Hg)	1,310	0.004944	0.00413	1.35	62.6
759.5 (.5 mm Hg)	658	0.003595	0.000206	1.89	48.7

dieting, paying taxes, and listening to operatic music, avoidance is always better than participation!

4.14 DESIGN OF ADSORPTION SYSTEMS FOR OPEN-TOP VAPOR DEGREASERS

Adsorption beds designed to stanch solvent pollution from open-top vapor degreasers are similar to but different from adsorption beds used to do the same task for pollution from enclosed vapor degreasers. The former hold substantially more activated carbon, use that adsorbent less efficiently, and never treat (fortunately) high concentrations of solvent in air. They are also shaped quite differently!

The key aim is still the same – to have no volumes (or areas) in the workplace contain solvent at concentrations above the posted exposure limit. This is done as with emissions containing higher concentrations of solvent from enclosed vapor degreasers – cost-effective collection of contaminated air and treatment of it by an adsorption bed containing activated carbon. Because of the low concentration of solvent, another and different treatment step is warranted.

4.14.1 Operational Impact of Emissions from Open-Top Vapor Degreasers

One must repeat and amplify the previous guidance about not using lip vents in open-top vapor degreasers to pull

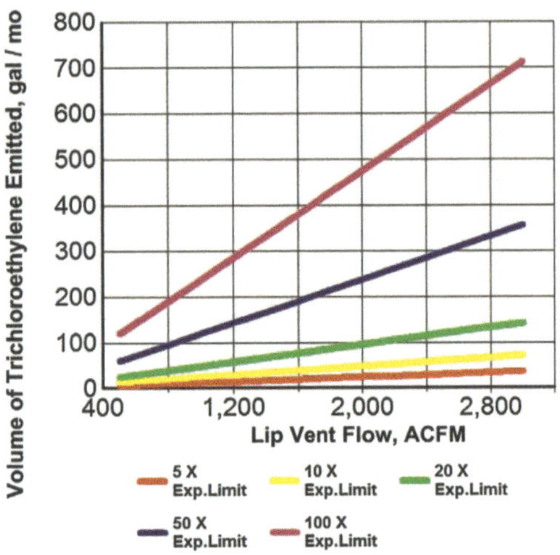

Figure 4.32 **The Potential for Solvent Loss from Lip Vents on Open-Top Vapor Degreasers**

solvent vapor from the top of the vapor degreaser (Chapter 1.21). This action wastes money in two ways:

- By discarding perfectly useful clean[45] solvent. The damage to a proper operating plan can be estimated for lip vents by using the calculations which produced Figure 4.32. The results are striking[46]!
- By forcing the adsorption bed to operate while starved of solvent to be adsorbed. Bed capacity is much less well used (Figure 4.17), and so the budget required to operate it may become starved as well.

This second financial impact is often unexpected and is evaluated in detail in Ref. 5 A3 after a basis for adsorber design is developed and evaluated in this book (Chapter 4.15).

4.14.2 Management of Emissions from Open-Top Vapor Degreasers

Keep solvent in the cleaning tank by both design choices and operating requirements:

- Design with excessive freeboard (Chapter 1.12).
- Design with overmuch freeboard refrigeration (Chapter 1.19).
- Zealot-like attention to prevention of leaks, vents, and spills of solvent (Chapter 4.6).
- Tortoise-like basket vertical speed – to reduce solvent entrainment (Chapter 1.22).
- Magician-like hand (or robot) speed in removing/replacing the washed parts basket (to minimize duration when lip vents may be used (Chapter 1.21).
- Robot-like purging of the washed baskets with nitrogen or air before removal[EE] from the degreaser (Chapter 1.23).

Visually, follow the guidance of Figure 4.11 and Figure 4.12, where the only solvent-infected air which should be fed to a bed of activated carbon is that of unintentional leaks, spills, and vents.

4.14.2.1 A Manager's Choice

A manager who can follow the above traditional and essential guidance can produce a work area that may be infected only with solvent vapor at concentrations probably below and hopefully well below the low posted exposure limit.

The common and routine inability to implement of the above guidance forces a manager to face a choice between two unequally unpleasant options about the rate of air flow through the lip vents and the circulation level of room ventilation:

- Larger flow rates through the lip vents and room circulation levels keep area concentrations of solvent in the workplace low to meet posted exposure limits (more solvent is wasted);
- Smaller flow rates through the lip vents and room circulation levels keep more solvent in the vapor degreaser (less waste of solvent), but workplace concentrations of solvent are higher and worker comfort and safety is more threatened.

Since the workplace must meet all SHE criteria for posted exposure limits and emissions, a prudent manager normally elects to maintain workplace concentrations of solvent low while wasting more solvent – unfortunately and necessarily forcing the feed of solvent-infected air to the adsorber to be starved of solvent.

4.15 DESIGN OF THE ADSORPTION SYSTEMS FOR OPEN-TOP VAPOR DEGREASERS

Exhaust of solvent fumes from lip vents usually exists only during the short period when the parts basket is being loaded into the degreaser freeboard or when it is being removed from it. This is when the cover has been removed. There should be little or no exhaust of solvent fumes from lip vents at the freeboard top when the cover is sealed in place – but there may be exhaust of solvent fumes from leaks and spills at the base of the vapor degreaser (Figures 4.12 to 4.14).

---

45 The solvent found in emissions is perfectly clean – of soil materials. After all, it has been vaporized leaving the low-volatility soil materials behind.

46 Figure 4.32 is based on being able to measure the actual air flow rate through the lip vent, and stipulate the excess by which the solvent concentration therein exceeds the minimum aim (to not exceed the posted exposure limit). For trichloroethylene the loss rate of trichloroethylene can approach 80 gallons per month. This is with a lip vent exhaust rate of 1500 ACFM for a period of 2 minutes when the cover is removed at the end of a 20 minute cleaning cycle, and 20× the posted exposure limit of 10 ppm.

Table 4.6 Comparison of Designs for Beds to Adsorb Trichloroethylene

Treatment of >>>	Exhaust from Lip Vents	Exhaust from Work Chamber
Time of Use	2 minutes exhaust time in 20-minute cycle (solvent vapor is not stored but fed direct to adsorber)	
Volume of Solvent	Unlimited as long as cover open	< 150 gallon chamber capacity
Exhaust Air Flow Rate, ACFM	> 1500	~170
Initial Inlet Solvent Concentration, ppm	~200 ppm trichloroethylene (20 × exposure limit)	13,100 ppm trichloroethylene (750 TorrV in work chamber)
Continuity of Exhaust Fed to Adsorber Bed	Constant solvent composition (probably)	Declines as contents of work chamber are displaced to adsorber bed
Final Inlet Solvent Concentration, ppm	200 ppm trichloroethylene (20×exposure limit)	~10 ppm or so
Number of Treatment Cycles per day	72 (:20 cycles, 24 hrs/day)	
Elapsed Time per Treatment Cycle	2 minutes	
Activated Carbon Use per Treatment Cycle	2.20 lbs	0.64 lbs
Mass of Activated Carbon in Adsorber Bed	~500 lbs	140 lbs
Lifetime of Adsorber Bed Before Regeneration	3 days	3 days
Adsorber Bed Design: Height, Diameter	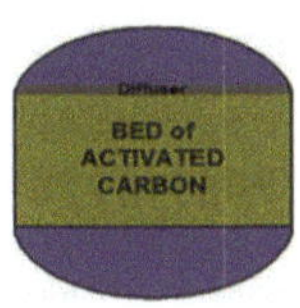 18 inches high × 44 inches diameter (L/D≈0.4)	 65 inches high × 14 inches diameter (L/D ≈ 4.5)
Vapor Velocity (superficial)	150 ft/sec	
Holdup Time (superficial)	0.6 seconds	2.2 seconds
Carbon Use	~0.090 lb trichloroethylene/ lb activated carbon	~0.144 lb trichloroethylene/ lb activated carbon
Efficiency of Removal	Assumed to be 98%	Probably similar
Exit Solvent Concentration	3.8 ppm (estimated)	10 ppm (exposure limit for trichloroethylene)

Typically[FF], air flow from lip vents is 1500[47] to 3000 actual cubic feet per minute (ACFM). These values are those proven necessary to maintain local ground concentrations of halogenated solvent vapor at satisfactory levels, and quell complaints about solvent odor.

At these high flow rates of room air, solvent concentrations are low – relative to those emitted from the work chamber of an enclosed vapor degreaser. Values reported in the references of Endnote GG were measured in the several hundreds of ppm.[48]

Design of an adsorber bed to strip solvent from air exhausted from lip vents is based on the same principles noted in Chapter 4.10 for treatment of exhaust from an enclosed vapor degreaser.

But the outcomes are different, as shown in Table 4.6.

One can use the information in Table 4.6 to scale the size of the adsorption bed should the degreaser be one in which lip vents exhaust to the adsorber bed throughout (Box 4.9)[49] the entire cleaning cycle[50].

Different for each solvent (Figure 4.17) emitted from lip vents, the design outcomes for six common cleaning solvents can be compared in Table 4.7 (Boxes 4.10–4.11)[GG].

If should be clear from the above design outcomes that there is no single design of an adsorption system which should be used with all cleaning solvents, or situations.

The key design specifications are the vapor superficial velocity (to maintain heat transfer rate and holdup time), and the desired length of time the adsorber bed is to last between regeneration cycles (the bed size). As expected, the affinity of the solvent for the activated carbon dominates the outcome (the capacity of the activated carbon, as shown in Figure 4.17).

[47] As noted in Footnote 38 purging of a large enclosed degreaser holding 150 gallons (about 21 CF) of solvent vapor involved about 5 displacements of chamber volume in 1 minute, or about 170 ACFM.

[48] What has changed since the reports noted in Endnote GG were prepared is the development and use of: sealed covers, controlled low-speed hoists, additional coils with lower temperature refrigeration, and superheat devices (Chapters 1.7 to 1.13).

[49] There are three circumstances where lip vents may provide solvent-bearing air to an adsorption bed during the entire cleaning cycle. They are: (1) use of large vapor degreasers ("swimming pools") from which solvent emissions are difficult to contain because freeboard refrigeration is limited in effect beyond about ~36 inches away from the heat transfer surface, (2) careless operation where leaks of liquid solvent and vents of solvent vapor from pump seals, filter or instrument housings, pipe or tubing connections, or drain valves continually contaminate the lower levels of the work area with solvent vapor (Figure 4.12), and (3) use of solvents having very low posted exposure limits (such as trichloroethylene) and/or producing an odor (n-propyl bromide).

[50] In essence, modern vapor degreasers that use lip vents to maintain clean air in and around the opening into the degreaser do so only during the period when the cover is open (2 minutes/cycle in the example designs). Lip vents are "disengaged" when the cover is closed and "engaged" when the cover is opened (18 minutes/cycle in the example designs). "Engagement" means that when the cover is closed little or no solvent escapes from the degreaser freeboard, and the feed to the adsorption bed is nearly solvent-free air. Closure and a shut-off of air flow rate may be activated by electrical contact; but closure may also be managed by human action.

Box 4.9 Beware of Claims Which Overpromise

Suppliers of open-top vapor degreasers often represent their offerings as being capable of maintaining the concentration of specific solvents at or below certain values in the work area around their vapor degreaser. For example, a supplier may represent that their open-top vapor degreaser can meet a proposed 10 ppm exposure limit for n-propyl bromide. One effective method by which this legally binding stricture would be met is a high level of use of lip vents. This management approach distorts the true cost of vapor degreaser ownership by adding the investment for a large adsorber bed to the purchase price, and the cost of its operation to the total cost of use; in addition to the additional cost of the solvent lost via entrainment through the lip vents. Maintaining low posted exposure limits in the work area by "keeping solvent in the tank" through design of refrigeration and freeboard has different consequences than whisking emitted solvent vapors to a bed of adsorbent.

4.16 REGENERATION OF THE ADSORPTION BED

Actually two vessels (beds) are needed in a regeneration system.

- In one vessel, solvent vapors from exhaust air are adsorbed, and
- In a second vessel, solvent vapors are shed (desorbed) from a previously used charge of activated carbon which is solvent-laden.

It may be that adsorption and desorption of different beds are ongoing simultaneously[51], though the time required to complete each operation is not the same.

Desorption of solvent from the activated carbon can be completed in a few hours (or less), while the life of a vessel of activated carbon used for solvent adsorption can be days, weeks, or more (Table 4.7). Visual comparison of the two periods is made in Figure 4.33[52].

When the activated carbon in the adsorption vessel is "full" of solvent, a control system switches flow of solvent-rich exhaust air (or nitrogen) to the other vessel. Then the

[51] It may also be that only a single vessel is needed for both adsorption and desorption if the cleaning work is done during only one shift each work day. Then the desorption may be completed on a later shift when cleaning work is not done.

In this case the bed volume would be considerably smaller than that assumed in the facilities design reported in Tables 4.6 and 4.7. This is because the adsorption bed has to hold only "8 hours' worth" of solvent vs. "72 hours' worth."

[52] Desorption of a solvent-laden bed is often delayed for convenience – as to be done on a specific shift or by a specific crew. However, transition of flow of pollutant-laden air from the solvent-laden bed to the regenerated bed can't be delayed.

Table 4.7 Designs for Adsorber Beds to Treat Various Solvents Emitted from Lip Vents

Solvent	Vessel Capacity lbs Carbon	Days of Use	Vessel Dimensions: Diameter (D), inches	Vessel Dimensions: Height (H), inches	Carbon Use lb Solvent/ lb Carbon	Posted Exposure Limit ppm	Exit Conc. ppm
Methylene Chloride	7,502	3	108	43	0.0187	50	3.1
Trichloroethylene	481		43	17	0.0903	10	3.9
Perchloroethylene	1,120		58	23	0.1225	20	5.6
n-Propyl Bromide	562		46	18	0.0723	10	3.5
Toluene	1,896		67	27	0.1697	100	16.1
CFC-113	63,803		221	88	0.0971	1000	15

Box 4.10 **Design Basis for Operation of Solvent Adsorbers**

Common parameters for all designs in Table 4.6 and Table 4.7 are that:

- The volumes of both vessels are sized to provide the same period of use before regeneration, ~3 days
- Lip vents are used for only 2 minutes per cycle, and 24 hr/day operation
- The superficial vapor velocity is that of vapor passing through an empty adsorber vessel (volumetric vapor flow rate divided by vessel area normal to flow)
- 1,500 ACFM volumetric air flow
- The vapor holdup time is calculated by dividing the adsorber vessel volume by the volumetric superficial vapor velocity
- The adsorber vessel used to treat soiled-air from lip vents is sized as follows:
  (1) the diameter is chosen to provide the same vapor superficial velocity (essentially heat transfer rate) as in the adsorber vessel used to treat exhaust from enclosed vapor degreasers (150 to 200 ft/min [0.75 to about 1 m/sec]), and
  (2) the height is chosen to hold the required volume (mass) of adsorbent.
- The vessel L/D is ~ 0.4
- The adsorption efficiency is 98%
- The influent solvent concentration is 20 X the posted exposure limit (different for each solvent; for CFC-113, this is very high).
- The operating range of airborne solvent concentrations is shown in the plot above. The range is where the line in the plot has been widened. Five different types of cleaning solvents are included.

The images of vessels in Table 4.6 are those holding the same volume of adsorbent. Where necessary to achieve the design goal of 150 ft/min vapor velocity, additional outside air is added. The activated carbon used for all cases is the same – derived from coconut husks and with an internal pore diameter of 1.34 nm (Appendix A2). All of the vessels proposed for use in Table 4.7 must be fabricated from a sheet of carbon steel which has been rolled and seam-welded. Available pipe sizes are too small for use.

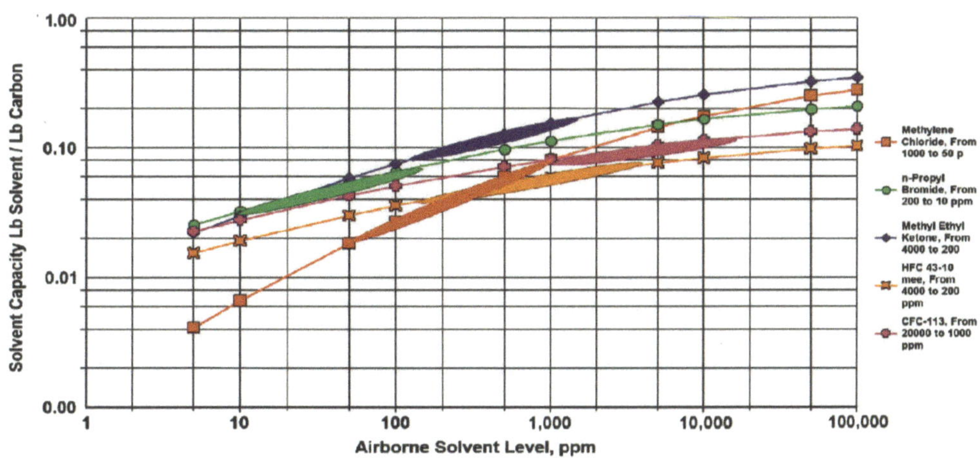

Box 4.11 Adsorber Designs Using Different Parameters

The values of carbon efficiency (lb solvent/lb carbon) in Table 4.15 are calculated for the design and operating conditions specified in Footnotes 50 and 53 of Volume 4. If other design conditions are assumed, such as:

- The adsorber vessel shapes are different than an L/D of 0.4 for adsorber beds designed to treat exhausts from lip vents, and an L/D of 4.5 for adsorber beds designed to treat exhausts from enclosed vapor degreasers; or
- The specification of activated carbon are different than being derived from coconut husks and having an internal pore diameter of 1.34 nm. Or if other operating conditions are assumed, such as:
  - The air velocities are different than 150 ft/min, or
  - The inlet concentrations (especially) of solvent in air are assumed to be 20 times that of the posted exposure limit for each solvent, and exit concentrations of solvent in air required to be less than the posted exposure limit.

Then the values of lb solvent/lb carbon are certain to be similar, but different.

One example is where the adsorber vessel shape factor (L/D) is changed while the superficial gas velocity through it is maintained constant. Calculated results for trichloroethylene solvent emitted from an enclosed vapor degreaser are tabulated in Table 4.8.

In this example, one must first note that at the recommended level of L/D (0.4), a small amount of extra air (45 ACFM – actual cubic feet per minute) must be supplied to maintain the overall superficial gas velocity at the recommended level of 150 ft/min. At L/D values above 0.4, the value of 1500 ACFM is too great to provide a superficial gas velocity of only 150 ft/min because the calculated adsorber diameter is too small.

As is shown in Table 4.10, extra air must be supplied at smaller values of L/D in addition to the normal amount of 1500 ACFM to maintain the superficial velocity at 150 ft/min because the calculated adsorber diameter is too great.

The calculated outcome at an L/D of 0.25 is that the holdup time for solvent-laden vapor within the adsorber is too short to enable the adsorbent to remove solvent as efficiently as if the L/D were 0.4. More activated carbon is required to remove the same amount of solvent.

The second example described above is where the superficial gas velocity is increased beyond 150 ft/min, thereby reducing holdup time for the solvent-laden vapor within the adsorber. Calculated results are tabulated in Table 4.9.

In this second example, one notes that increasing superficial velocity produces, as expected, poorer retention of solvent by activated carbon because the holdup time has been reduced.

Table 4.8 Calculated Effects For Change of Vessel L/D

L/D	Total Gas Velocity ft/min	Extra Air ACFM	Vessel Capacity lb Carbon	Bed Length ft	Bed Diameter ft	Holdup Time, sec	Carbon Capacity lb TCE/lb Carbon
0.40	150.4	45	482	1.45	3.62	0.58	0.0901
0.35	150.4	250	509	1.35	3.85	0.54	0.0854
0.30	150.4	530	545	1.24	4.15	0.50	0.0897
0.25	150.4	940	598	1.14	4.55	0.45	0.0726

Table 4.9 Calculated Effects for Change of Superficial Gas Velocity

L/D	Gas Velocity ft/min	Vessel Capacity lb Carbon	Extra Air ACFM	L ft	D ft	Holdup Time sec	Carbon Capacity lb solvent/lb carbon
0.40	150.4	482	45	1.45	3.62	0.58	0.0901
0.40	175.0	530	415	1.49	3.73	0.51	0.0820
0.40	200.4	586	845	1.54	3.86	0.46	0.0742
0.40	225.7	648	1325	1.60	3.99	0.42	0.0670
0.40	250.2	716	1845	1.65	4.13	0.40	0.0607

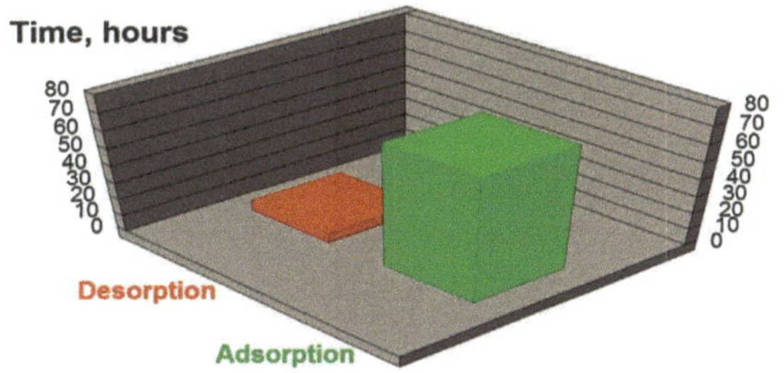

Figure 4.33 **Relative Time Span for Adsorption and Desorption**

Figure 4.34 **Dual Regeneration Column System**

sequence of usage is repeated, and the roles of each vessel are reversed.

- Adsorption processes are normally exothermic (Chapter 4.21 and Footnote 40) so that a reduction in the increase of temperature can be taken to be an indication (or proof) that the adsorption bed is "full" of solvent (Footnote 67).

Such a system, containing two vessels and associated piping is shown in Figure 4.34[HH].

The regeneration (desorption) system for adsorption beds is the same without regard to the circumstances by which any bed was loaded with solvent vapor.

4.16.1 The Desorption Process

There are two core principles which rise above all else:

- Adsorbed molecules are not permanently attached – they can be removed from an activated site. Adsorption of vapor molecules on surfaces is not like the affinity of a fan to their sporting team; it's more like a preference for a restaurant. The affinity of a vapor molecule to an activated site can be swayed by purely local conditions – the concentration of unlike vapor molecules, and temperature.
- That affinity can be quantitatively managed. It's chemistry after all (Appendix A2)!

4.16.2 Chemistry of the Desorption Process

It's all about dihydrogen oxide – steam first, then liquid water; then more solvent. Figure 4.26 is updated as Figure 4.35 showing the use of steam for desorption (regeneration of the bed).

Regeneration of adsorption beds proceeds in two stages:

- First, water vapor "replaces" the molecules of solvent vapor adsorbed on sites of the activated carbon.
- Then the adsorbed water is removed because it doesn't bond as well to activated carbon at higher temperatures.

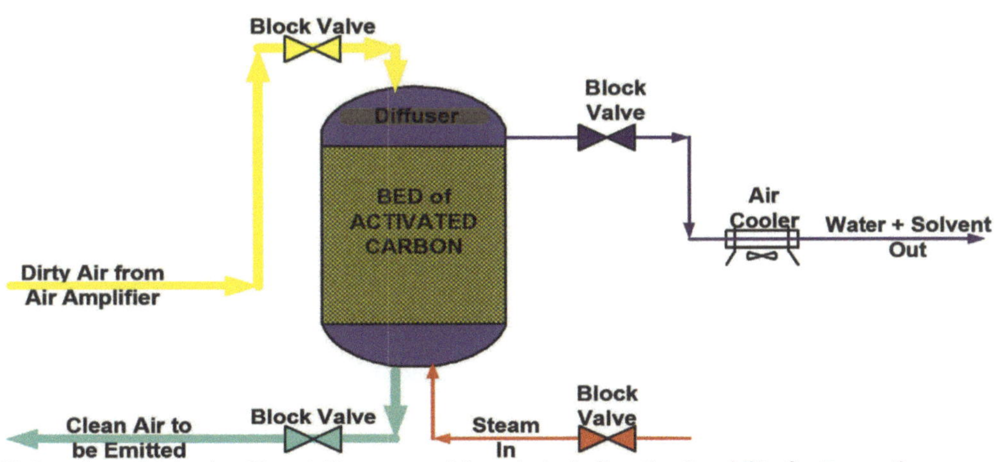

Figure 4.35 **Drawing of Adsorber Vessel, Flows to and from it, Including the Capability for Desorption**

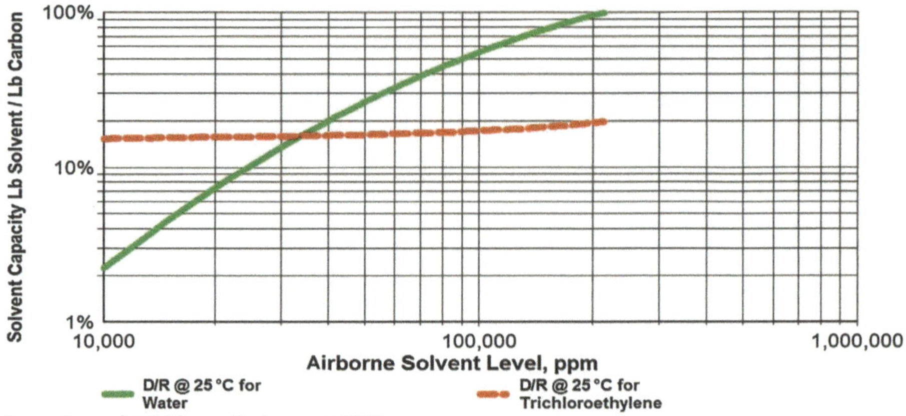

Figure 4.36 **Adsorption of Trichloroethylene at 25°C**

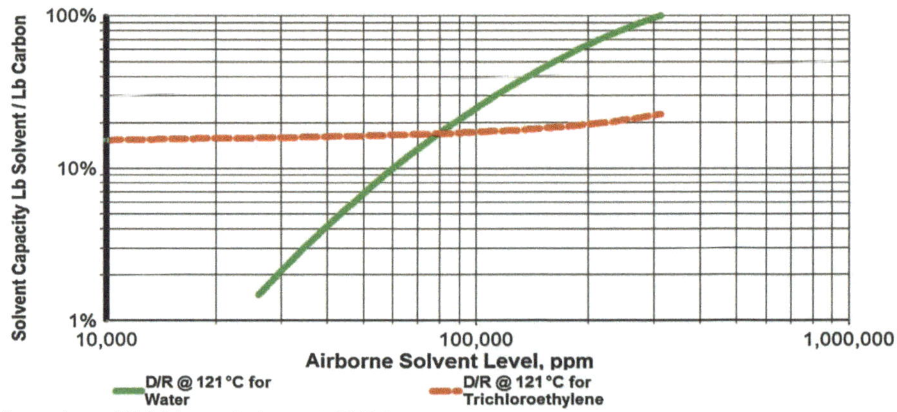

Figure 4.37 **Adsorption of Trichloroethylene at 121°C**

The fate of the solvent is to change those with whom it associates – from being combined in the vapor phase with a large amount of air to being combined in the liquid phase with a large amount of water.

This occurs for two reasons[53,II]:

- Water vapor is preferentially adsorbed over most other solvents at high concentrations in the vapor feed to an adsorber.

A 20% volumetric loading (200,000 ppm) of water in air[54] can fill all sites within a charge of activated carbon at 25°C. This can be recognized (green solid line) in Figure 4.36. Note that 100% of the activated sites can be filled with molecules of water in the vapor phase; substantially less than that can be filled by trichloroethylene.

- Steam is hot[55] (~120°C or 250°C or somewhat higher). Adsorption of any solvent molecule (trichloroethylene, water, methylene chloride, etc.) is not favored at higher temperatures.

This is because at higher temperatures the molecules have a higher level of energy and it is more difficult to bind them to a site on any adsorbent. The calculated situation for steam and trichloroethylene in shown in Figure 4.37[56].

[53] In other words, moisture will have a negative impact on adsorption of cleaning solvents; water will compete with solvent for the available adsorption area, and it will win under some circumstances!

[54] Note that the loading of steam in the vapor fed to the adsorber is 100% steam. There is no air. So all sites should be covered with water molecules.

[55] It is very common to use "waste steam" to desorb solvent from beds of activated carbon. The pressure of waste steam is barely above atmospheric; so its temperature is barely above the boiling point of water. "Waste steam" was generated at a substantially higher temperature and pressure, used for some purpose, and is available at almost no cost because its low temperature doesn't have significant capability to do thermal work, or its pressure significant capacity to do mechanical work.

[56] The basis for Figure 4.37 is the same as for Figure 4.36 – calculations by the D/R method, which is explained in Appendix A2. Notice that for every concentration plotted in Figure 4.37 (121°C) from 100,000 ppm to 200,000 ppm 2 (20 volume %), there is substantially less solvent (trichloroethylene or water) adsorbed on activated carbon then at 25°C (Figure 4.36). At 100,000 ppm in the inlet vapor feed, the values are about 0.55 lb water/lb carbon at 25°C vs. 0.25 lb water/lb carbon at 121°C.

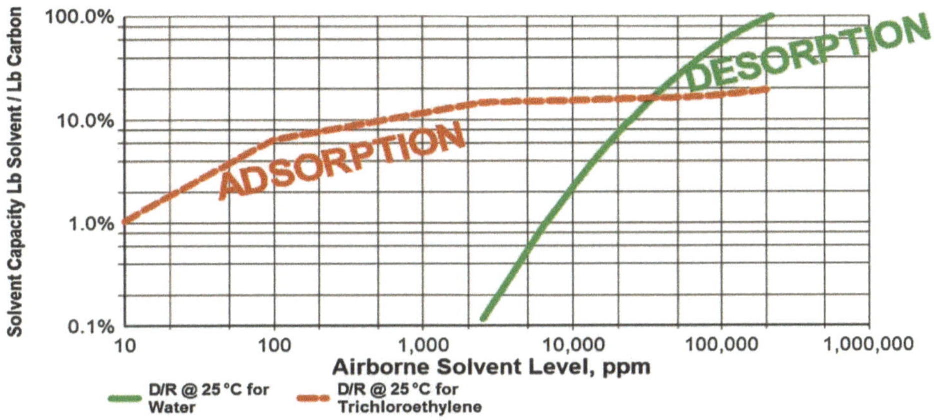

Figure 4.38 **How Solvent Emissions are Managed – Adsorption and Desorption**

The situation reverses itself when the bed is reduced in temperature – trichloroethylene displaces adsorbed water. Trichloroethylene, and all other cleaning solvents, are favored to be adsorbed at ambient temperatures, and especially at the concentrations (20,000 ppm and below) found in air emissions from either type of solvent vapor degreaser[57].

4.16.3 The Difference between Adsorption and Desorption

It's the concentration – of solvent relative to water; of water relative to solvent.

- When little solvent is fed to the adsorption bed, much more of it is adsorbed than when little water vapor is fed (not simultaneously) to the adsorption bed. But –
- When substantial water is fed to the adsorption bed, much more of it is adsorbed than when substantial solvent vapor is fed (not simultaneously) to the adsorption bed.

While temperature does enhance this differentiation (compare Figures 4.37 [25°C] and 4.37 [121°C]), it is not the reason for it.

The reason for the differentiation is illustrated in Figures A2-16 and A2-17.

- Water is a small molecule – relative to trichloroethylene[JJ] or any other cleaning solvent. It takes many water molecules to fill the pore volume in an activated carbon matrix[58].

Said another way, the reason for the differentiation (of concentration) is that vapor molecules only adsorb when they are proximate to pore walls. It takes many tiny water molecules to fill a pore's volume and expose them to activated wall surface. Numerically, few large solvent molecules are required to fill a pore and be adsorbed.

Higher concentrations of water will always impose a greater effect on adsorption than will the same concentrations of solvent – because the water molecule is small and can more easily approach the pore wall. This relationship is the basis for desorption.

Figure 4.38[59] displays the calculated results of using the D/R methodology explained in Appendix A2.

At concentrations of solvent found in air pulled through lip vents (several hundred ppm or somewhat more), and at concentrations of solvent found in air outside enclosable work chambers after being purged with fresh air (several thousand ppm or considerably more), useful adsorption of solvent vapor occurs. But at much higher concentrations of water vapor in air (tens of thousands of ppm or more), water will replace the adsorbed solvent causing it to be desorbed from the bed.

To recall one of the two core principles expressed in Chapter 4.16.1. If the affinity of vapor molecules for activated sites on carbon was permanent, an adsorbent bed couldn't be desorbed.

4.16.4 Heat and Mass Transfer Are Important Too

Steam used for desorption should be saturated at low pressure, though dry or slightly superheated steam works as well. High pressure (temperature) steam is not needed.

[57] Some favor enhancing the rate of removal of water from a solvent-free adsorption bed by passing hot nitrogen through it and removing the water by evaporation. There is another point of view expressed in Footnote 69.

[58] The same is certainly true for solvents. At lower concentrations of solvent in air, fewer solvent molecules move proximate to pore walls, and adsorption of them is poorer (Figure 4.17). This is why adsorption isotherm curves are not flat.

[59] Do not mistake the information in Figure 4.38 for empirical data. It, and that within similar figures based on D/R methodology, are meant to illustrate general effects where specific empirical data is not publicly available to describe them.

It is thought that the following sequence of physical steps[KK] is involved:

- Water condenses from within the flow of steam onto the outer surface of the adsorbent. This happens for two reasons: (1) the steam is hot and the solid contents of the adsorber bed are initially cold, and (2) desorption of solvent vapors requires heat (endothermic, as adsorption was exothermic).
- Condensed water is re-vaporized as the solid contents of the adsorption bed become hot due to transfer of energy from the incoming hot steam.
- Water vapor diffuses into the pores of the adsorbent.
- Smaller numbers of larger molecules of cleaning solvent vapor adsorbed on the activated sites are displaced by a larger number of smaller molecules of water vapor.
- Vapor molecules of cleaning solvent diffuse through the pores of the adsorbent, and into the moving stream of steam.

Management of affairs within an adsorption bed is a flawed concept. The affairs are not managed by human action but by the principles of thermodynamics, chiefly the principle of equilibrium and the role Le Chatelier's Principle (Box 4.12) plays in its attainment.

4.16.5 Managing Steam

The best guidance for managing a desorption (regeneration) process should be available from the supplier of the activated carbon. However, it is likely that for desorption of solvents other than the conventional halogenated cleaning solvents, their experience in recovering cleaning solvents will be limited. It may even be limited for desorption of trichloroethylene, and especially so for n-propyl bromide.

A simple rule for sizing the quantity of steam for desorption is that 3 to 7 pounds of steam should be needed for each pound of solvent in the adsorber bed. If the solvent loading on the adsorbent is not known (as it often is not), use the rule that 0.3 to 0.8 pounds of steam is needed for each pound of carbon in the adsorber bed.

Another simple rule for sizing the time over which steam is fed to the adsorption bed is that desorption should be complete after no more than $^3/_4$ to 2 hrs – without regard to the size of the charge of activated carbon in the adsorption bed[60].

However, the simplest, and most significant rule is to terminate the desorption process at the earliest possible moment. The reasons behind this rule are that: (1) steam is usually expensive, and (2) the solvent-contaminated waste water which it produces is also expensive – to treat as a waste.

Normally, the earliest possible moment to shut off the flow of steam and terminate a desorption process is when one of three events is recognized:

> **Box 4.12 Le Chatlier's Principle Applied to Adsorbers**
>
> Le Chatelier's Principle (LCP) is named for Henri Louis Le Chatelier (1850–1936). He was an influential French chemist of the late 19th and early 20th centuries. The principle synthesized by him and named for him states that if a system is not in equilibrium, the driving force toward equilibrium is greater than the driving force away from equilibrium; i.e. systems have a net driving force toward equilibrium. An adsorption bed provides substantive, dynamic, and interlocking examples of the use of the LCP.
>
> The flow of energy is from steam to adsorbent (producing liquid), and then the reverse (producing vapor). Differences of temperature and pressure, respectively, are the driving forces. They are all in aid of producing an equilibrium distribution of thermal energy.
>
> It's the same with mass flows. Water vapor diffuses from steam through pores onto adsorbent sites, and solvent vapor diffuses from adsorbent sites through pores to be carried away by the steam flow. Both flows are from where there is much of one kind of mass to where there is little, in aid of producing an equilibrium distribution of each material.
>
> It's the same when a desorbed bed is put back into service as a bed of adsorbent. If the flow of steam is stopped, and that of solvent-laden air is begun, diffusion of vapor starts. Molecules of solvent vapor move through pores from where there are many (in the air) to where there are few (the adsorbent), and the reverse for molecules of water – all in aid of producing an equilibrium distribution of each material. Then, with the adsorbent pores enriched with molecules of solvent vapor, and leaned of molecules of water vapor, the situation described in Figure 4.38 can be restated – adsorption of solvent on activated carbon.

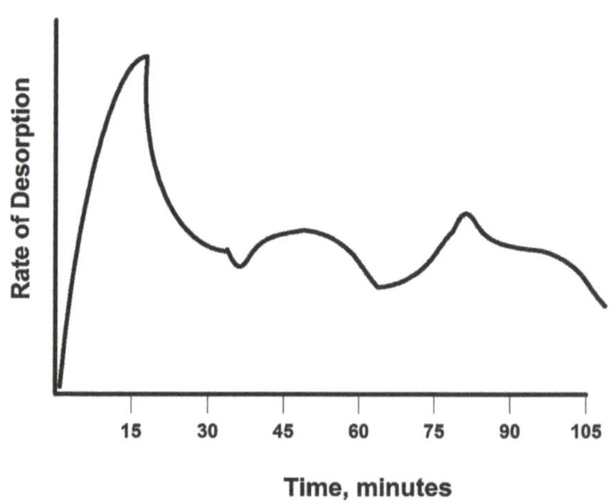

Figure 4.39

[60] The flow rate of steam is sized based on the size of the charge of activated carbon and the chosen desorption time.

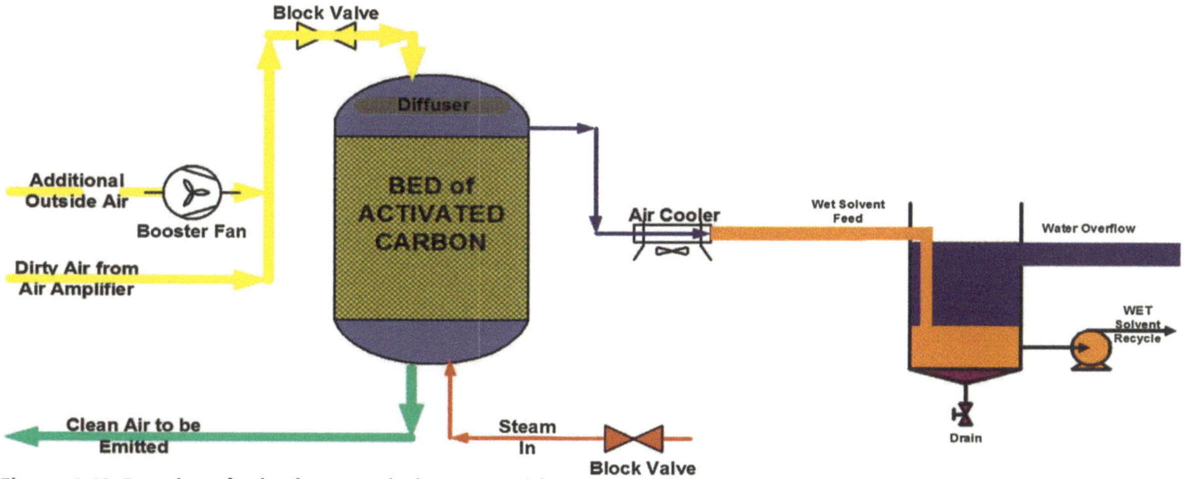

Figure 4.40 **Drawing of Adsorber Vessel, Flows to and from it, Including the Necessary Capabilities to Manage Desorption**

- Since desorption is an endothermic process (Chapter 4.16.1), while it is occurring there is a thermal equilibrium. Addition of thermal energy from steam and removal of thermal energy by the endothermic desorption process are the boundary conditions defining that equilibrium. When desorption is complete, the equilibrium is distorted and that can be recognized by an increase of effluent temperature.
- Desorption is also a mass transfer process – solvent vapor molecules are released from the bed of adsorbent. When the bed is leaned of solvent molecules, their absence in the effluent can be recognized by on-line composition measurements (Footnote 67).
- A fixed period of time has passed.

Actually, most desorption occurs very early in the regeneration cycle – when rates of heat transfer are highest. Figure 4.39 is a generalized plot of this situation, where there is a diminishment of gain with further use of steam[61].

4.16.6 Managing Liquid Water

The appropriate process scheme is shown in Figure 4.40. Here the capabilities shown in two other figures are combined – Figure 4.35 about use of an adsorption bed as a "black hole" to control tailpipe emissions, and Figure 3.14 about the use of a conventional water separator (for removal of insoluble water from cleaning solvent).

Given the rule of thumb in Chapter 4.16.5, one should expect to manage a "flood" of liquid water contaminating the water-immiscible (normally) liquid cleaning solvent after its removal from the adsorbent bed.

That expectation is easily fulfilled; unfortunately.

In a sense, use of activated carbon is simply a neutral intermediate in a pollution control scheme. Activated carbon causes solvent vapor to not be emitted, but to be transferred (in effect) to water. After this transfer, the desorbed, dried, and activated (possibly) carbon pellets are made available for reuse without the prejudice of previous use.

But in another sense, all that has been accomplished is that the solvent vapor has been collected, condensed, and combined with liquid water. The amount of water by mass or volume is significantly larger than the same measure of the amount of condensed solvent. The amount of water is, relatively, a "flood" – as is shown in Table 4.10.

The situation is particularly egregious with methylene chloride – a common cleaning solvent not well adsorbed on activated carbon (Figure 4.17); but especially soluble in water (1.85 % vs 0.0016 % for perchloroethylene). Basing the flow of steam on the amount of activated carbon for this solvent is seen in Table 4.10 to be a poor choice[62].

[61] Operators of adsorption beds should consider experimentation – to learn to what extent the desorption cycle can be shortened in order to reduce energy and especially pollution remediation costs. One can't harm the bed of adsorbent by shortening the desorption cycle by a substantial amount of time (~one-third), and observing if the composition outcomes are satisfactory.

The test of performance should be simple – level of emission of solvent into ambient air (Chapter 4.17). If the adsorption life becomes shortened because the previously adsorbed solvent is still retained, then extend the desorption cycle until it returns to a satisfactory level – considering all costs. This optimization is a perfect problem for study by a short-term intern.

[62] Without unavailable empirical data, an improved basis for selection of the level of applied steam from Table 4.10 might be either the mass of adsorbed solvent, or the moles of adsorbed solvent.

Table 4.10 Operating Outcomes for Desorption of Activated Carbon Used to Contain Solvent from Lip Vents (3 Days of Operation)

Solvent	Use of Activated Carbon @ 1500 ACFM Air, Conc. 20 × Above Exposure Limit (Tables 4.6 and 4.7)			Use of Steam (Water) @ 0.4 Lb Steam/Lb Carbon 90 min Regeneration Time				Mutual Solubilities from Ref. 2, Table 11.2	
	Adsorption Bed Wt., Lbs	Lb Solvent/Lb Carbon	Lb Solvent Captured	Lb Water	Lb Steam/Solvent	Mol Steam/Mol Solvent	Steam Flow pph	% Solvent in Water	% Water in Solvent
Toluene	1,802	0.1690	301.0	721	2.37	12.1	481	0.520%	0.033%
n-Propyl Bromide	562	0.0720	4.0	225	5.53	37.8	150	0.250%	0.500%
Perchloroethylene	1,120	0.1230	137.0	448	3.27	30.1	299	0.0016%	0.001%
Methylene Chloride	7,502	0.0190	84.0	3,001	21.40	100.9	2,001	1.850%	0.240%
Trichloroethylene	481	0.0900	43.0	192	4.43	32.3	128	0.128%	0.330%
CFC-113	63,803	0.0970	6191.0	25,521	4.12	42.9	17,014	0.017%	0.010%

> **Box 4.13 The Path Not Taken**
>
> In a water treatment or chemical plant, those flows may be moderate; in a finishing or manufacturing shop conducting cleaning operations with solvents, those flows are immoderate because there may be only a small use of steam or no supply of steam in the shop. It isn't that the mass amount of steam needed to desorb the bed is large, it's that it is needed only over a period of minutes (while the desorbed bed is used as an adsorber over a period of many thousands of minutes). Simply, those steam flows may not be available without construction of a boiler dedicated to supporting the "black hole" which is supporting the vapor degreaser. Alternately one may elect an uncommonly chosen path — to desorb the adsorption bed with steam over a longer time than ~90 minutes. One might desorb the bed over 900 minutes (15 hours) and reduce the instantaneous flow of steam by a factor of ten. This path is uncommonly chosen because operators of adsorbers in water treatment and chemical plants prefer their standby adsorber bed to be available as soon as possible.

But the most difficult issue may be the instantaneous amount of steam required for desorption. Note in Table 4.10, that flow to be hundreds of pounds per hour, and in some cases thousands of pounds per hour, of low pressure steam[LL] (Box 4.13).

4.17 TRANSITION FROM SIMPLICITY TO COMPLEXITY

Desorption, by using steam, of the solvent attached to the adsorption bed adds complexity to what once was simplicity.

Drawings of a simple tank filled with small black pellets and fed with dirty air were shown in Figure 4.28 and 4.29. It became a dual-use vessel when steam was introduced in countercurrent flow (to remove solvent), and then vented (Figure 4.35). Addition of an air-cooled heat exchanger to condense the vented vapors mixes condensed water and immiscible[63] solvent which can be separated in a decanter. These facilities are shown in the drawing of Figure 4.40.

But the product of the decanter is unusable – the solvent is too wet of water to be used without substantial addition of stabilizer,[MM] or treatment in a distillation column[64]. And the water contains too much soluble organic matter (solvent) to be directly routed to a disposal conduit!

Consequently, the facilities present in Figure 4.40 must be augmented. The expanded facilities are noted in Figure 4.41 as being within the dashed box, and include: (1) the ability to dilute the aqueous stream containing soluble solvent with clean water, and (2) the ability to dry water from the recovered solvent by using another adsorption bed. The second bed has 3 Å mole sieves[NN].

4.18 POLLUTION TRADEOFFS

Projected outcomes for use of the facilities shown in Figure 4.41[65] with six different solvents are found in Table 4.11. Covered are both situations: when the bed of adsorbent is used to treat tailpipe emissions produced by both lip vents, and exhaust from enclosed work chambers.

If one views Figure 4.41 on an overall basis, one can understand that choices have consequences – if one manages fugitive emissions from solvent cleaning equipment by directing them to a "black hole." The main consequence is that solvent pollution of air by solvent is simply and easily exchanged for:

- Air that is free-enough of solvent vapor so that it can be directly emitted.
- Water which is contaminated with soluble solvent, and almost certainly cannot be emitted without treatment[OO].
- Recovered solvent, fully saturated with water[66].
- Reusable activated carbon which will at some future time become unuseful.

So the "black hole" of Figure 4.1 can be seen as an image which camouflages the necessary facilities and actions shown in Figure 4.42, and it does not describe a passive facility.

To summarize: operation of a "black hole" to dispose of solvent emissions in air is convenient, forgiving, common, sometimes cost-effective, and often unavoidable. But, it requires investment, facilities, floor space, attention, and regard for its consequences.

---

[63] Were the recovered solvent to not be immiscible with water, the receiving facility would not be a decanter but a distillation column. Solvents immiscible with water are primarily those bearing oxygen atoms – acetone, isopropanol, methyl ethyl ketone, and the like.

[64] Relative to a decanter, a distillation column is a burden to those using oxygenated cleaning solvents. One might compare the organizational impact of a baby carriage to that of a baby. A distillation column requires additional investment (Ref. 2, Chapter 7.16) as well as energy, floor space, and process management devices/attention, but produces dry solvent ready for use without addition of stabilizer to "titrate" water. Further, for solvents with boiling points sufficiently distant from that of water (such as the three mentioned in Footnote 70), the water product may well be sufficiently free of organic to all direct disposal.

Solvent product from a decanter is not dry nor ready for use.

[65] For solvents more dense than and immiscible with water, such as halogenated and "designer" solvents.

[66] For immiscible solvents only.

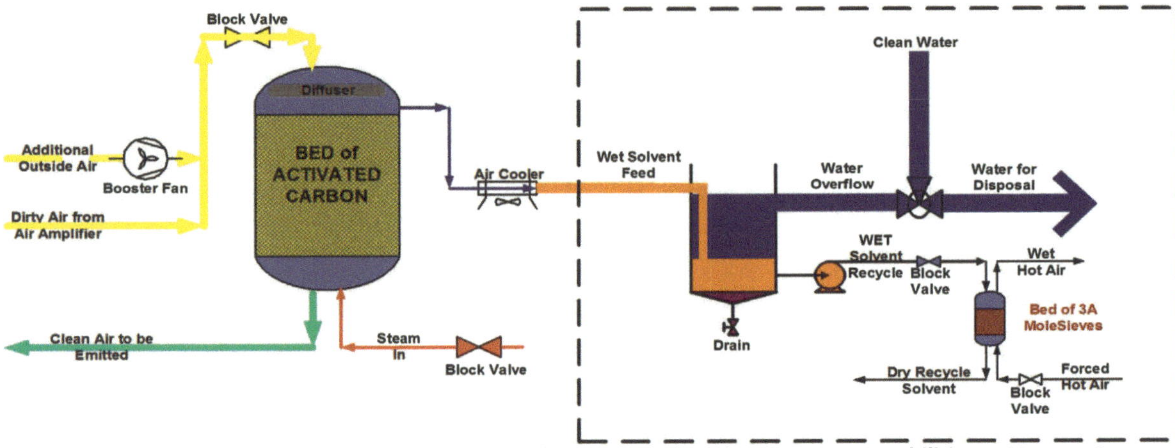

Figure 4.41 **Drawing of Adsorber Vessel, Flows to and from it, Including the Necessary Capabilities to Manage Recovery of Solvent**

4.19 DIFFERENT SOLVENTS ENGENDER DIFFERENT OUTCOMES

The extent of the consequences depends on the character of the fugitive emissions fed to the desorption system – the specific cleaning solvent employed, and the nature of its use (Box 4.14). These consequences are predictable given two types of basic data: (1) the solubility interrelationships between the chosen cleaning solvent and water[PP], and (2) the relationship between adsorption of the solvent vapor on the adsorbent and its inlet concentration in air (Figure 4.17).

- For the common cleaning solvent methylene chloride, it was noted in Chapter 4.16.6 that the consequences are more egregious[67] – because:
  - Methylene chloride is not well adsorbed on activated carbon when present at low concentrations in air; being retained only at ~0.01 lb methylene chloride/lb activated carbon when present at around 100 ppm .
  - Methylene chloride is relatively soluble in water, dissolving at 1.85 wt. %.

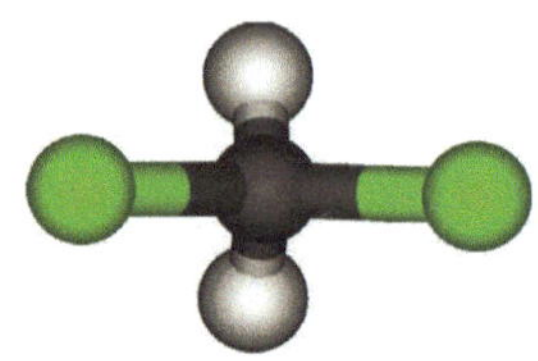

Methylene Chloride

This practical meaning of this information is that larger quantities of activated carbon will be needed – especially to treat air containing low concentrations of methylene chloride vapor such as is found (fortunately) in the exhaust from lip vents. And that outcome spawns three additional negative consequences because larger quantities of: (1) activated carbon will require larger quantities of steam to desorb the solvent, (2) increased steam flows will produce larger volumes of waste water which must be diluted before disposal, and (3) methylene chloride will be dissolved in all that waste water.

- The cleaning solvent whose use is most likely to avoid those three negative outcomes is perchloroethylene. This is because perchloroethylene is well adsorbed onto activated carbon (Figure 4.17) and not compatible with water (Table 4.10).
- Often substituted for one another to delay or resist negative environmental consequences, trichloroethylene and n-propyl bromide have: similar performance as cleaning solvents (Ref. 2, Chapter 2), good adsorption on activated carbon (Table 4.10) similar, compatibility in or with water (Table 4.10), and aggressive reactivity with water to produce a corrosive mineral acid (Ref. 2, Chapter 10). Recovered solvent, of either kind, should be dried in

[67] Consider the following point of view. Use of methylene chloride (BP 40°C) and any similar highly volatile solvents, should not be attempted in open-top vapor degreasers. In addition to the normal losses of highly volatile solvent from an open-top vapor degreaser, the above information reveals that "black box" containment of emitted solvent vapor by activated carbon (with steam desorption) spawns three problems for the one it solves.

Table 4.11 Summary about Abatement of Pollution Using Solvent Adsorption (Basis: Footnotes 55 & 74)

Solvent	Unabated Air		Abated						
	Pollution – Prior to Adsorption		Abatement		Contaminated Streams, per Month				
					Air	Aqueous Waste		Wet Solvent	
	VOC lb/ Month	Solvent lb/ Month	Carbon Bed Size, lb	Steam Use lb/month	Solvent in Air, lb	Volume gal	Total Organic Carbon ppm	Mass lbs	Stabilizer Needed ppm
Emissions from Lip Vent Exhaust									
Toluene		3,047	1,802	10,952	52.1	1,321	4,700	3,031	NA
n-Propyl Bromide		407	562	3,421	11.4	411	730	406	5,000
Perchloroethylene	Exempt	1,372	1,120	6,813	18.0	817	2	1,391	10
Methylene Chloride	Exempt	1,405	7,502	45,637	10.1	5,575	2,600	564	2,400
Trichloroethylene	435		481	2,924	12.6	351	233	438	3,300
CFC-113	Exempt	61,973	63,803	388,135	48.6	46,546	22	62,754	100
Emissions from Work Chamber of an Enclosed Vapor Degreaser									
Toluene		142	83	505	0.32	61	4,700	141	NA
n-Propyl Bromide		189	138	840	0.05	101	730	190	5,000
Perchloroethylene	Exempt	255	162	986	0.18	118	2	258	10
Methylene Chloride	Exempt	131	328	1,995	0.22	244	2,600	95	2,400
Trichloroethylene	202		140	852	0.05	102	230	205	3,300
CFC-113	Exempt	288	236	1,436	8.40	172	22	287	100

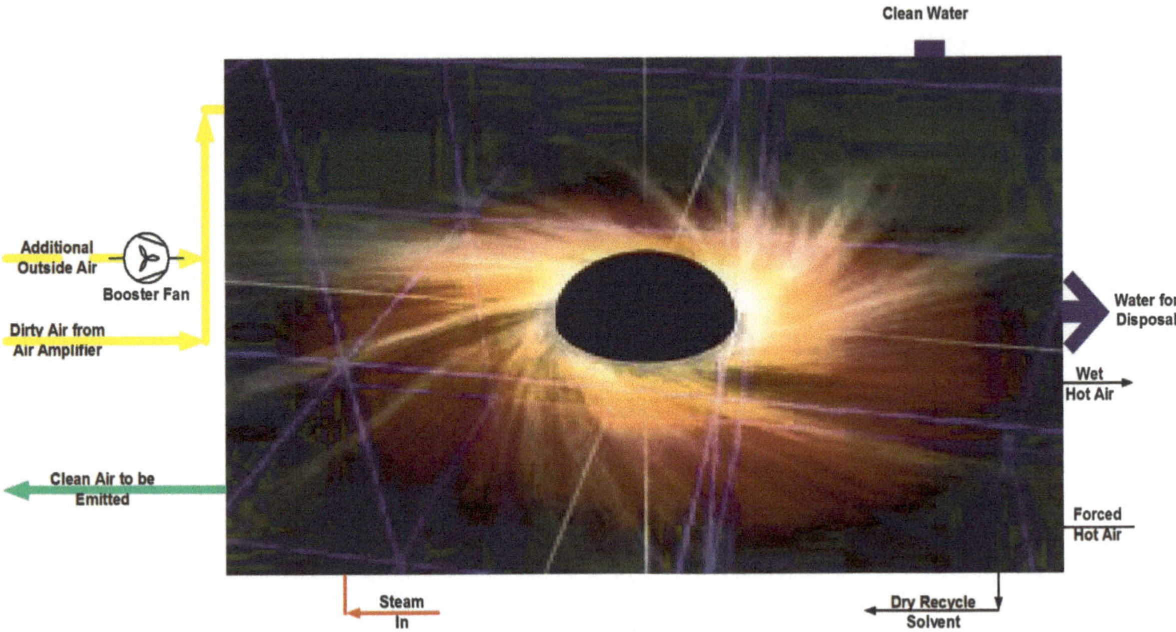

Figure 4.42 **The 'Real' Consequences of Operating with a 'Black Hole'**

Box 4.14 **Basis for Evaluation for Operation of Solvent Adsorbers**

There are two applications in which adsorbent systems are evaluated in this chapter. Both applications have been used to prepare the information collected in Table 4.11. One application is straightforward; the other is made complicated to be relevant.

The straightforward application is purging with clean air the work chamber of a large (21 cubic feet) enclosed vapor degreaser. Conditions in the chamber prior to starting the purge were a vacuum of 750 TorrV (10 mm Hg absolute) with only solvent present. The desired outcome is to have the chamber contents displaced to where the exhaust concentration in ambient air is no more than half of the currently posted exposure limit for the chosen cleaning solvent. These posted exposure limits are: toluene (100 ppm), n-propyl bromide (10 ppm), perchloroethylene (25 ppm), methylene chloride (50 ppm), trichloroethylene (10 ppm), and CFC-113 (1000 ppm).

The complicated application involves use of lip vents attached to an open-top vapor degreaser in which each of the six solvents noted above is being independently used. The assumed feed and exit concentrations are 20× and 1× the exposure limit respectively. Complications arise from the intersection of three factors: (1) the fictitious feed and exit concentrations to and from the adsorber are different for each of the six solvents (because they have different posted exposure limits, Footnote 55), (2) each solvent has a different adsorption vs. concentration response (Figure 4.17), and (3) each solvent has a different solubility (Table 4.10) in and with water (condensed steam). For example, the range of solvent concentrations for n-propyl bromide or trichloroethylene is 200 ppm to 10 ppm, and that for CFC-113 is 1000 ppm to 20,000 ppm. Solubility data of many cleaning solvents in water, and of water in cleaning solvents, are available in Appendix 1, Table A1-5.

a separate bed of 3 Å mole sieves. However, this is seldom done because addition of a stabilizer to consume the reaction product of the water is so cheap[68,QQ].

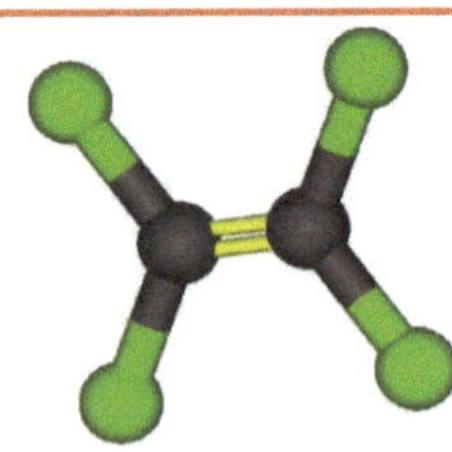

Perchloroethylene

- CFC-113, whose manufacture is banned in "developed countries"(Box 4.15) because it is an

[68] But, it is not always effective – especially with n-propyl bromide which dissolves half again as much water as does trichloroethylene (Table 4.10), and with the bromine atom in n-propyl bromide being much more reactive than are chlorine atoms in trichloroethylene (Ref. 3, Table 22.1).

Box 4.15 **Desiccants – Their Types and Uses**

The Montreal Protocol is a treaty among about 200 states in addition to the European Union. Its collective purpose is to reverse the deterioration of the Earth's ozone layer. Its methodology was to prescribe elimination through gradual phase out of the production and use of substances which cause that deterioration.

The treaty is structured around several groups of halogenated hydrocarbons and two groups of countries.

- The former group includes CFCs (chemical intermediates, refrigerants, reaction media, and cleaning agents), Halons (firefighting agents), carbon tetrachloride (a chemical intermediate), HCFCs (refrigerants and cleaning agents), methyl chloroform (chemical intermediate and cleaning agents), methylbromide (a pesticide), and a few other specialty chemicals.
- The latter group defines developed and developing countries. Although there are various accepted and disputed definitions of the difference between developed and developing countries, the difference devised within the Montreal Protocol was simple. A country capable of replacing the functional performance of the banned halogenated hydrocarbons was considered to be developed; one not so capable was a developing country. Furthermore, developing countries were considered to need the functional performance of some or all of these halogenated hydrocarbons to achieve over time the status of a developed country.
  The parties to the Protocol prescribed distinct phase out schedules for each group of halogenated hydrocarbons within both developed and developing countries. For developed countries, such as France, Germany, Israel, the UK, and the United States, the phase out deadlines produced the most stress in the 1990s. For developing countries, such as Brazil, China, India, Iraq, and North Korea, the phase out deadlines are imminent or ongoing now.
  Developing countries whose annual per capita consumption and production of ozone-depleting halogenated hydrocarbons is less than 0.3 kg are aided by a Multilateral Fund to comply with the control measures of the Protocol.
  The Protocol implemented a triage-based approach. The phase out schedule for specific groups of halogenated hydrocarbons required more immediate and certain action in developed countries that were producing and using halogenated hydrocarbons having more potential to deplete the earth's ozone layer. Because their potential to deplete ozone when emitted is much less than that of CFCs, the phase out of production and consumption of HCFCs was delayed until now.

For readers of this book, one phase out schedule may bring angst; it is for HCFCs, chiefly HCFC-225ca/cb. An excellent replacement for CFC-113 as a cleaning agent, producers and users face the same situation with HCFC-225ca/cb now as they did with CFC-113 in the 1990s. Per the Montreal Protocol, the time to freeze the consumption and production of HCFCs is 2013/2015 in developed countries.

ozone-depleting chemical (ODC), is still used in the US for a few specific applications on an exemption basis. Because of its cost, availability, and concern about atmospheric effects, recovery of emissions is essential. Fortunately it is fairly well adsorbed on activated carbon (Figure 4.17), and not at all compatible in or with water (Table 4.10).

- Oxygenated solvents can be well adsorbed on activated carbon, as shown in Figure A2-20 of Appendix A2. Solvents based on propylene glycol appear to be better adsorbed at the same concentration as is trichloroethylene. Others, which are small molecules, should adsorb poorly at low concentrations as does water. Two such examples from Figure A2-20 are acetone and isopropanol. But, at higher solvent concentrations in air, such as ~2000 ppm or greater, oxygenated solvents and trichloroethylene are all expected to adsorb equally well.

In general, oxygenated solvents are more soluble in and with water than are halogenated solvents. This means that the desorption operation with steam can wreak havoc – the waste water (condensed steam) will contain significant organic pollution; the recovered solvent will be totally wet. For example, acetone, isopropanol, and some propylene glycol ethers are fully compatible with water. So, the decanter shown in Figure 4.41 will do no phase separation. In this case desorption with hot nitrogen is in order per Footnote 62 and Chapter 3.7.7.

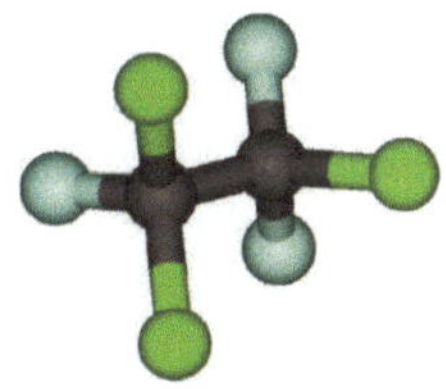

CFC-113

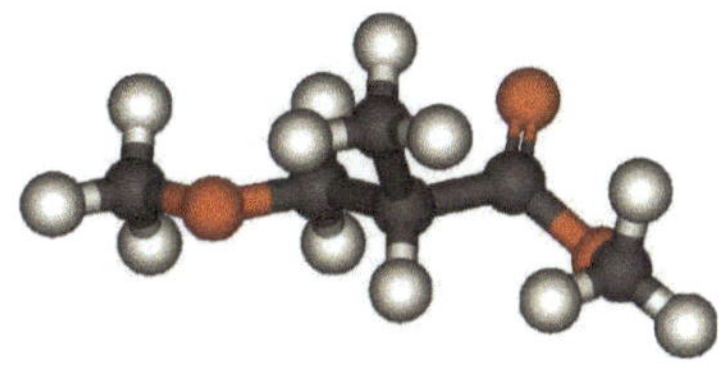

Propylene Glycol Methyl Ether Acetate

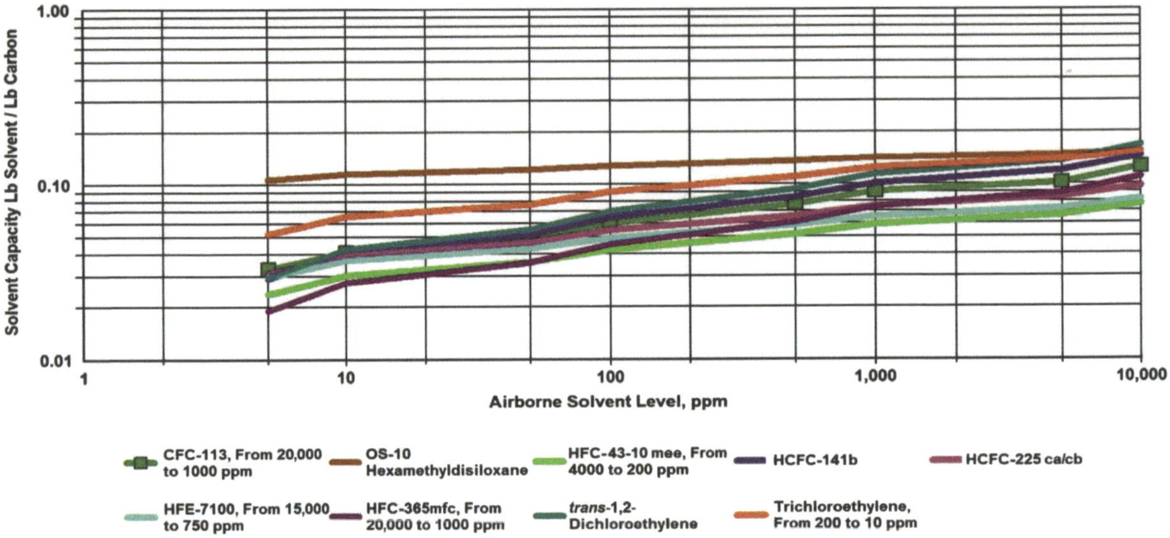

Figure 4.43 **Absorption Isotherms for Various Cleaning Solvents**

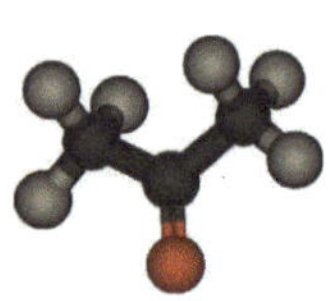

Acetone

- While not an objective of the designers of "designer" solvents[RR], useful but not superior adsorption performance appears by calculation using the methods of Appendix A2 to have been "designed in." Cleaning solvents such as methylsiloxanes (OS-10), HFCs (HFC-43-10mee and HFC-356mfc), HCFCs (HFC-141b and HFC-225ca/cb), and HFEs (HFE-7100) all should be expected to be adsorbed as is their predecessor in use (CFC-113[69]). And their general incompatibility with water is similar to that of CFC-113.

Adsorption outcomes calculated using the D/R methodology of Appendix A2 for a variety of "designer" and other solvents are plotted in Figure 4.43.

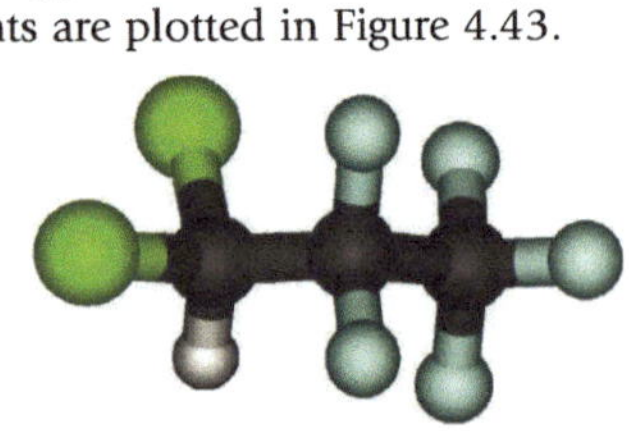

HCFC-225ca

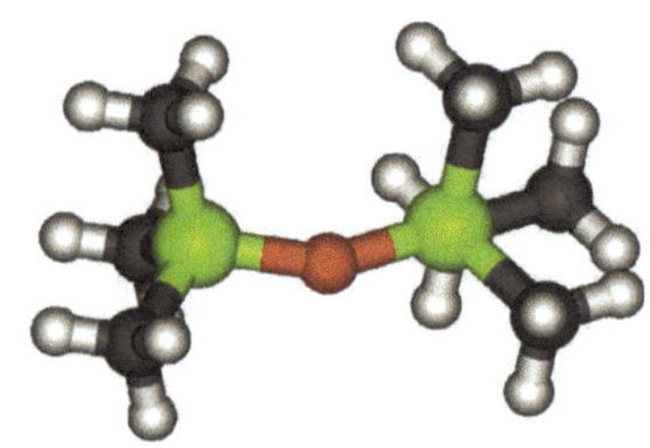

OS-10

In summary, expensive "designer" solvents can be recovered from air emissions by adsorption into activated carbon which can be desorbed with steam. However, the size of the adsorber bed will be fairly large because of the lower level of adsorption on activated carbon.

4.20 LIFE WITHOUT DESORPTION

It may be that the operating site has no wish or ability to conduct the desorption operation[70] (Box 4.16)[SS].

In that case: (1) only one vessel is necessary, and (2) it might be a sealed drum of activated carbon. Many such of various sizes are shown in Figure 4.44.[TT] These drums or canisters are equipped with all necessary fittings for connection to a system of air supply and exhaust.

[69] The square symbols for CFC-113 in Figure 4.43 are of data from Figure 4.17; the lines representing all other solvents are calculated values using the D/R methodology of Appendix A2.

[70] Or, it may be that: (1) a sufficient rate capacity of steam is not available on site, (2) the operating site is not able to manage disposal of the volume of water waste contaminated by soluble solvent, or (3) the recovered wet solvent requires too much stabilizer to be re-purposed for use in boiling service. Or it may be that the site recognizes that a commonly used approach for desorption is incomplete, and bound to fail.

Box 4.16 The Flaw with Desorption Using Hot Nitrogen

That is desorption of solvent by treatment of the adsorbent with hot gas. Nitrogen is commonly, and effectively, used to desorb solvents from activated carbon as well as water from mole sieves (Chapter 3, Footnote 94). Often nitrogen at temperatures of 100°C and above is more effective than steam in removal of solvent from activated carbon. A useful reference is Hwang, K.S., Choi, D.K., Gong, S.Y., and Cho, S.Y., "Adsorption and Thermal Regeneration of Methylene Chloride Vapor on an Activated Carbon Bed," Chemical Engineering Science, Vol. 52, No 7, 1997, pages 1111 to 1123.

What desorption of solvents from activated carbon with hot nitrogen or air accomplishes is to remove the solvent from the adsorption bed. That's all! It does not collect or recover the solvent. The solvent remains vaporized in the nitrogen stream. The net effect of adsorption and this method of desorption is to substitute the vapor in which the solvent vapor is carried — from air, removed from around the open-top vapor degreaser; to nitrogen, from a cylinder or supply pipeline. Contrast this situation with desorption of mole sieves or silica gel with hot nitrogen or air, where there is no interest in recovery of the water.

Figure 4.44 **Drums of Activated Carbon to be Used Intact — Without Opening**

The approach is simple: the user purchases[71] the canister, fills it with solvent vapor which is adsorbed over time onto the contained activated carbon, and then is responsible for disposal of the solvent-laden canister of activated carbon. Two disposal options are conventional: (1) return to the supplier (Chapter 4.20.1), and (2) consignment to a certified landfill (Ref. 2, Chapter 7.8.1.1).

Obviously one needs to retain at least a second canister full of activated carbon in order to continue cleaning operations when the canister adsorbing solvent is full.

The user never opens the sealed canister, nor is able to examine the activated carbon, or is exposed to fumes of the emitted cleaning solvent.

For most situations with a small enclosed vapor degreaser, a drum holding 200 to 500 pounds of pelletized activated carbon should suffice well. Treatment of air emissions from lip vents on an open-top vapor degreaser is less secure. Expected lifetimes for various solvents are calculated[72] and tabulated in Table 4.12. Lifetimes are denominated in cycles of cleaning operation[73].

Table 4.12 Expected Cycles of Use for a 200 Lb Canister of Activated Carbon

	Adsorption of Solvent Fumes from	
Solvent	**Lip Vents**	**Enclosed Vapor Degreasers**
Toluene	8	174
n-Propyl Bromide	26	205
Perchloroethylene	13	89
Methylene Chloride	2	44
Trichloroethylene	20	103
CFC-113	1	62

The information in Table 4.12 reinforces the recommendation previously and repeatedly made in this book, to the effect that use of lip vents with open-top degreasers should be avoided to the extent possible. Especially with the solvent methylene chloride, the outcome of not implementing that recommendation is the near certainty that the desorption operation must be conducted on site.

The lower emission volumes, and higher solvent concentrations, associated with enclosed vapor degreasers

[71] Alternately the user may rent, lease, or toll the canister and return it when saturated with solvent to be replaced with a clean canister.

[72] Duration of a cleaning cycle is only a small number of minutes and includes four steps: immersion of parts in a solvent sump, washing and rinsing in that sump, drying above it, and removal of parts. The common basis for these calculations is the following: 20-minute cleaning cycle (72 cycles/day), 2-minute period within that cycle for solvent emissions, and the solvent retention ratios (lb solvent /lb activated carbon) used in the applications summarized in Table 4.11.

[73] Continuous (24/7) operation of an enclosed vapor degreaser would not be feasible unless an on-site desorption system were present. But, discontinuous operation (day shift only, with a 30-minute cleaning cycle) is very feasible without use of an on-site desorption system.

are certainly compatible with the local decision to avoid desorption of the adsorption bed which had been used as "black hole."

4.20.1 Spent Activated Carbon May Be Hazardous Waste

Use of the canister as an adsorption vessel without local desorption is a "carbon shuttle." It's simply a method of transporting activated carbon from a supplier to a customer, and returning it to the supplier after it has adsorbed solvent fumes to the extent possible. Essentially, it is a system for "toll processing"[74] of the activated carbon, and not the solvent – because an equivalent quantity of solvent-free activated carbon[75] is returned to the customer, but no solvent is recovered or returned to the customer.

This may not be a simple affair where the user simply affixes postage to the canister and delegates return transportation to a shipping company.

When the customer ships the solvent-laden canister to the supplier, they are *possibly* shipping hazardous waste, if the activated carbon contains more adsorbed cleaning solvent than 10 volume[76] percent of:

- Perchloroethylene (tetrachloroethylene), trichloroethylene, methylene chloride, 1,1,1-trichloroethane, carbon tetrachloride, and chlorinated fluorocarbons. Activated carbon containing n-propyl bromide need not be shipped as a hazardous waste – because is it not a Hazardous Air Pollutant (HAP).
- Xylene, acetone, ethyl acetate, ethyl benzene, ethyl ether, methyl isobutyl ketone, n-butyl alcohol, cyclohexanone, and methanol[77].
- Toluene, methyl ethyl ketone, carbon disulfide, isobutanol, pyridine, benzene, and ethylene glycol monoethyl ether (2-ethoxyethanol)[78].

Granules of activated carbon are highly porous. Their bulk density is low, around 0.5 g/cc (Chapter 4.4). That's less than that of all cleaning solvents. In other words, activated carbon containing ~10 weight percent of trichloroethylene (Figure 4.17) will contain less than 10 volume percent of trichloroethylene (a HAP). Users of canisters of activated carbon would do well to manage their use of that product so that canisters of solvent-laden activated carbon do not contain more than 10 volume percent of the solvents noted above so the canisters are not classified as hazardous[UU] from the standpoint of shipping regulations.

When the supplier accepts[79] delivery of the returned canister containing adsorbent saturated with solvent vapor, they must accept legal responsibility for compliant disposal of the adsorbed solvent. Normally the generator of hazardous waste (in the US) is responsible for management of the waste from "cradle to grave." However as part of written business relationships, this responsibility may be transferred.

So, a common outcome when on-site desorption of activated carbon must be avoided, is that one must pay the cost of: return shipping of materials possibly classified as hazardous, providing solvent-free activated carbon, and purchasing of new solvent.

4.20.2 Desorption of Canisters of Activated Carbon

Alternatively, one may seek to avoid these costs and potential consequences by desorbing the small canister of activated carbon on site. In that case, the site must be concerned that:

- The adsorption vessel (drum or canister) is rated for 15 psig internal pressure and 250°F[VV], and constructed of carbon steel[80].
- There is availability in the area of low pressure steam (less than ~20 psig), and suitable disposal for solvent-laden water (polluted steam condensate). This is especially so for some solvents such as methylene chloride which are more water-soluble.
- Adequate trained operational staff are available to desorb the canister when exhaust gases have been switched to the unused adsorption canister. No certification of training is required.

[74] Toll processing (manufacturing) is widely used in the chemical and process industries, where a secondary company carries out a subset of the overall production processes. This is an arrangement in which a company (which has a specialized equipment, such as a chemical or steam furnace for activating carbon) processes raw materials, semi-finished goods, or waste materials for another company. The toll is the negotiated fee to the secondary company (Ref. 2, Chapter 7.15.1).

[75] This activated carbon may be virgin material, having never been used in adsorbent service. But it is more likely to be reprocessed material. Reprocessing is done with a single charge of tens of thousands (or more) pounds of carbon collected from many suppliers. After blending and screening, the granules are reactivated by whatever combination of thermal and chemical treatment is offered by the processor. Specifications of the reprocessed activated carbon are likely to be similar to those of virgin material with only a small diminution in CTC (Chapter 4.4).

[76] This means that more than 10% by volume of the waste is one of the noted solvents, and that it is classified as a generic hazardous waste of type F001. The legal basis for this stricture is Title 40: Protection of Environment, Part 261 – Identification and Listing of Hazardous Waste, Subpart D – Lists of Hazardous Wastes (Ref. 2, Chapter 7.7)

[77] This is the classification as a generic hazardous waste of type F003 (Ref. 2, Table 7.3)

[78] This is the classification as a generic hazardous waste of type F005 (Ref. 2, Table 7.3)

[79] Some suppliers of reprocessed activated carbon will not accept for reprocessing spent material that they did not supply. As of publication, information about reprocessing of activated carbon containing hazardous solvents may be obtained at http://www.carbtrol.com/spent_carbon_reactivation.html.

[80] Stainless steel is not cost-justified for corrosion resistance. However, interior and exterior linings or chemical resistant coatings of the carbon steel are strongly advised. Neither is a plastic vessel advised because of its low mechanical strength at the operating temperature of the steam used in regeneration of the canister's contents. Alternately, if desorption (regeneration) is done off site by a supplier of activated carbon, a vessel used for storage, adsorption, and shipping could well be made of high density polyethylene. An example is shown at top.

- A small and accessible area is available. The footprint of the canister and piping will be similar to that of a large office desk (or possibly larger).
- While a perfectly practical operation, on-site desorption of canister adsorbers is seldom practiced[WW]. If only a small quantity of activated carbon is necessary because of the nature of the degreaser, the overall cost of shipping the canister of possibly-hazardous waste usually doesn't justify the cost of facilities for desorption and of training for their use.
- The price for purchase of a 200 lb canister of virgin coconut-based activated carbon is about $500 to $600 ($2.50 to $3.00 per pound). Acceptance of return of that canister, containing only non-hazardous contents, incurs a modest fee of $75 (plus transportation).
- The price for purchase of a 500 lb canister of virgin coconut-based activated carbon is about $1500 to $2000 ($3.00 to $4.00 per pound[81]). Acceptance of the return of that canister, containing only non-hazardous contents, incurs a less modest fee of only $750 to $1200[82] (plus transportation).

But, should the canister of spent activated carbon contain contents classified by US EPA as hazardous, which is possible (Chapter 4.20.1), the supplier may not accept return of that canister – at any price. Should they accept return of that canister, for higher fees, the canister of activated carbon would not be regenerated but deposited as an unopened container in a landfill, or possibly emptied into a high-temperature incinerator[XX].

4.21 OPERATION OF ADSORPTION SYSTEMS

Adsorption in a bed is a transient process. The amount of material adsorbed within a bed depends both on position within the bed[83], and use time of the bed. Events start at the top of the adsorption bed, and proceed downward, as solvent-laden air is fed from the top of the bed (Figure 4.45).

As solvent-laden air enters the bed, it first comes in contact with the top layers of activated carbon. These layers become a mass transfer zone. Three actions are completed: (1) solvent vapor in air diffuses through pores in the adsorbent, (2) the vaporized solvent is adsorbed, and (3) available adsorption sites are filled.

The mass transfer zone is so named because it is there where some of the mass of solvent vapor diffuses from the main air stream through the network of tiny pores. So it is transferred to be adjacent to the adsorption sites within the activated carbon.

Soon, the activated carbon sites near the top of the bed become saturated with solvent vapor. These sites can retain no more solvent molecules. So the solvent-laden air stream penetrates intact farther into the bed. Thus the mass transfer[84] zone shifts down through the bed as time goes on. This is shown in Figure 4.45.

Detection of when the mass transfer zone has reached the bottom of the adsorption bed can be done in several ways[YY]. That outcome signals the need to take the active adsorption bed off line for desorption, and replace it with one fully regenerated.

- Temperature is an indicator proven effective.[ZZ] And its use can be automated. Since adsorption is an exothermic process, vapor temperature increases in the volumes of activated carbon where solvent vapor has adsorbed to sites. At the base of the bed, when the temperature stops increasing, adsorption has ceased – and the bed is ripe for regeneration. Figure 4.46[85] is a hypothetical plot of this behavior.
- Experience is an excellent reference. One may continually integrate the amount (volume) of solvent vapor entering the adsorber via measurements of flow and concentration. Or one may simply integrate (sum) the elapsed operating time. That value may be compared to a reference value previously shown to represent the capacity of the adsorption bed.

[81] The higher per unit purchase price appears to be a purely temporal phenomenon. It was justified based on the larger container size being non-standard in the 2014 marketplace. This example is offered to make the point that all containers containing hundreds of pounds of activated carbon are non-standard in a market where lots 10,000 pounds in size are common. Packages denominated in ounces and the like for home use are sold by distributors who purchase these large quantities and sell tiny quantities at much larger unit prices. Readers are cautioned that prices quoted in this book, or any book, are intended for reference only and should be expected to change over the useful lifetime of a book.

[82] The actual higher per unit reprocessing price is offered as an example to make the point that treatment of carbon laden with potentially hazardous solvents is not a "mainline" business for suppliers and the service is only offered to those willing to pay a premium price.

[83] Nearly always, one makes the assumption that adsorption is a one-dimensional process. Position refers to length down the bed, and not radial displacement from the centerline of the vessel.

[84] Wherever the mass transfer zone is located within the adsorption bed, the character of it remains the same. At the top of the zone, the concentration of solvent in air is that in the feed stream to the adsorber. At the bottom of the zone, the concentration of solvent in air is nearly zero.

[85] The grey zone in this figure represents the volume of activated carbon at the bottom of the bed. During adsorption, the temperature in this zone will increase at a rapid rate as exothermic adsorption occurs. As sites are filled with solvent molecules and the adsorbent rate declines, so also does the rate of temperature increase. Completion of the bed's use may be recognized by comparing the response of the temperature sensor in the bottom bed volume to a fixed reference value over time, or by use of an integrating transducer to continuously calculate the rate of temperature change with time and compare that to another fixed reference value.

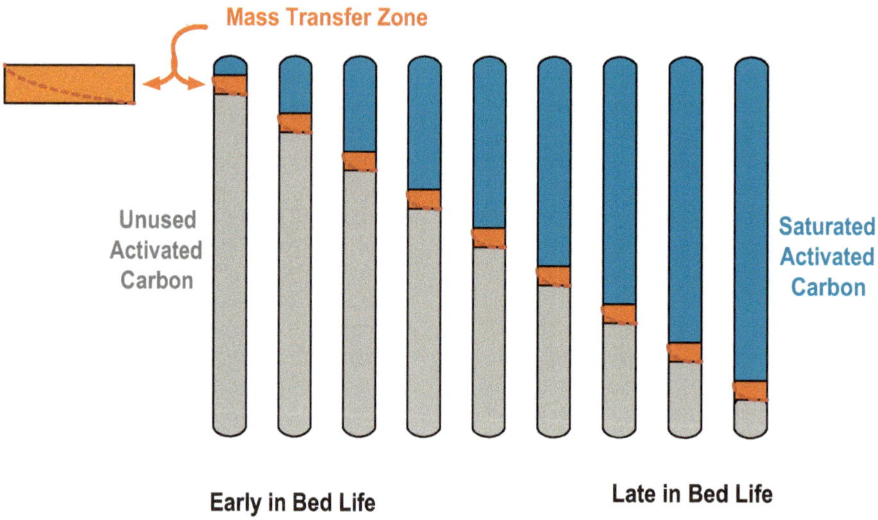

Figure 4.45 Stages of Life in Use of a Bed of Activated Carbon

- The least relevant method is to take a sample of the activated carbon and test it. One might try to learn when the activated carbon has little or no more capability to transfer the mass of solvent vapor through the network of pores and enable it to be adsorbed on activated sites One might complete the carbon tetrachloride (CTC) test (Chapter 4.4) or its more safe replacement the butane adsorption test (Footnote 15). The reason why these tests aren't relevant in this service is that they can't be completed in a few minutes. They probably require off-site participation. And they are not inexpensive to complete.
- The most relevant method for detection of when the mass transfer zone has reached the bottom of the bed is to analyze the exhaust vapor for concentration of the solvent in air. How this is done depends upon the chemical character of the solvent. In some cases it may be done at modest cost with human labor by using personal monitoring badges[86] as disposable direct-reading dosimeters. In many other cases, a sample is taken into a gas bag and analyzed with a gas chromatograph.

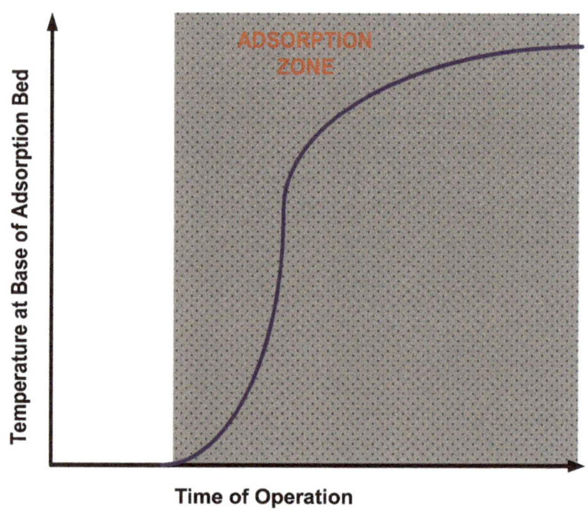

Figure 4.46 Positional Temperature Profile During the Life of Adsorption Bed

The most important guidance about managing the operation of an adsorption bed is to never feed pure solvent vapor to it[87]. Pure solvent is not well adsorbed at low feed concentrations (Figure 4.17), neither is it fully adsorbed at high feed concentrations (Figure 4.38).

[86] A line of colorimetric gas detector tubes is available from many suppliers, under the name of Draeger or Auer tubes. These are colorimetric devices which change color when reacted with the targeted "contaminant." A calibrated 100 ml sample of air is drawn through the tube with the pump. If the targeted "contaminant" chemical is present the reagent in the tube changes color and the length of the color change typically indicates the measured concentration. Purchase price is less than $10 per test.

[87] One can't harm the adsorber bed, but the exhaust from it won't be low in solvent concentration. One avoids this situation by using an air amplifier. This device produces air flows up to 15 times that of a small amount of pressurized air. Both the small volume of solvent-rich air and a large volume of clean dry ambient air are entrained by the compressed air flowing through the air nozzle. The solvent-rich air is thereby diluted by a factor of 15. Design details are covered in Table 4.2.

4.22 ECONOMICS OF ADSORPTION SYSTEMS

A valid cost analysis has several categories of expense. Accumulated, selected categories should represent nearly all significant sums of money spent in the endeavor being analyzed. If an endeavor-significant category is omitted, or not properly evaluated, the analysis is likely to engender flawed decisions.

The following[88] is an analysis of the operation of a "black hole" adsorption system which uses activated carbon. The seven categories of expense below, printed in boldface type, were considered and evaluated for each of six cleaning solvents:

- Whatever **capital investment** is necessary to make the system available on site in working order. Depending upon how site costs are assigned, this might include the activities of: design, purchase, installation, modification (as necessary), and project management. The total of this capital burden will be assigned a useful life, and amortized yearly so that it can be replaced after that use period.
- Cost to purchase the **activated carbon**, including amortization of it so it can be replaced after several years of use.
- **Electric (presumably) power** to provide steam (to desorb the adsorption bed), to operate blowers and air amplifiers, to energize pumps, and for general system operation. This utility cost will be variable as the price of power varies greatly by location and time.
- **Cooling tower water** (CTW) to quench the steam/solvent vapor stream and cool it for recovery. This is a separate utility with a separate cost for electric power.
- **Labor** to operate and maintain the system. For simplicity, this cost will include needed maintenance materials. Further one assumes that a single person provides both the operation and maintenance functions. While an operator may be assigned to manage this system, their presence will be only occasional as the system controls are presumed to be automatic.
- **Waste disposal** of the condensed water, rich with soluble solvent to its solubility limit.

And there is a final category – the positive[89] value of the recovered solvent. This category is one of the two (Box 4.17) positive values provided by an adsorption system.

> **Box 4.17 Alternate Views of Expense**
>
> The other, and more significant, positive category of "expense" is providing "black hole" capability to contain and flexibly manage all the negative issues associated with the Safety, exposure (Health), and pollution (Environmental) (SHE) problems presented by emissions of solvent vapor. This outcome should be a positive addition to the value of an enterprise.
>
> It is difficult to monetize this "expense" as a positive credit in an accounting of cost. Some sites view the cost of the adsorption system as that of an insurance policy. In this view, the expense does not produce any short-term value or benefit – except in a catastrophic event (a direct release). Other sites view the adsorption unit to be an integral component of a modern vapor degreaser. In this view, the adsorption unit serves the same function as does the working mode cover (Chapter 1, Footnote 69) – to keep solvent in the tank.
>
> Not all sites view the adsorption unit as a positive addition to the value of an enterprise. Some sites view it as a distinct drawback to the use of solvent cleaning technology. They ascribe the need for it to the advent of more stringent safety, exposure, and pollution (SHE) regulations.
>
> Readers will have their own views of how to meet all SHE regulations and how operating cost is affected.

These seven categories of cost were evaluated (Box 4.18) for each of two types of operation[90] of the adsorption unit.

- Treatment of emissions when an enclosed vapor degreaser is opened at ambient temperature.
- Treatment of emissions through lip vents when the cover of an open-top vapor degreaser is opened at ambient temperature.

Within each of the two types of operation, a cost evaluation was completed of the adsorption unit fed separately with six different solvents (toluene, n-propyl bromide, perchloroethylene, methylene chloride, trichloroethylene, and CFC-113).

Graphs of the distribution of cost for operation of the adsorption unit are plotted in Table 4.13 (fed with each solvent emitted through lip vents on open-top vapor degreasers), and the same are plotted in Table 4.14 (fed with each solvent emitted from enclosed vapor degreasers).

Plotted outcomes[91] are expressed as overall book-kept (Box 4.19) operating cost per hour of operation. Shown in

---

[88] In Ref. 5 A3 is an evaluation, using the methodology described in this chapter, of the total book-kept operating cost of a solvent cleaning machine. Included is the external cost identified here of operating the adsorption unit.

[89] In the context of a book-kept cost sheet, all categories of expense are considered to be negatives – deductions from whatever value is provided by the operating enterprise. In the same context, some categories of "expense" are positives – increases of value to the operating enterprise.

[90] Vessel designs for adsorbers are found in Table 4.7 and an operation summary of material flows and adsorbent usage is found in Table 4.10.

[91] To scale all six categories of cost for each solvent the vertical axis of each graph in Tables 4.13 and 4.14 has a different range of cost per hour for some solvents. This is especially necessary for methylene chloride and CFC-113.

Box 4.18 **Assumptions Made in Economic Analysis**

The following assumptions were made for each category. All outcomes were converted to an hourly operating charge for 8000 hours yearly operation, with a cleaning cycle time of 20 minutes and an emission time of 2 minutes during that cycle:

$$Investment = \$32,275 \times \left\{ \left(\frac{Carbon\ Capacity}{1200} \right) 0.6 \right\}$$

Capital investment was based on the information provided in the reference of Footnote #12 in Appendix 2, especially pages 340 and 341. That investment amount was $32,275 for an adsorption system based on a charge of 1,200 lbs of activated carbon. Capital investment requirements for the dozen systems evaluated in this chapter were calculated based on a service of 4 years, with an amortization rate of 25%. This capital charge was expec

The cost of **activated carbon** was taken to be $2.25 /lb with a useful service life of 2 years and an amortization rate of 50%.

The cost of **electric power** was estimated as the sum of four terms. The first is the cost to provide the steam flow (in pph) noted in Table 4.11. That value was converted to a requirement for energy as 945 BTU/lb steam, and the conversion factor of 3415 BTU/kilowatt (kW). The second term was the power cost to continually operate the air amplifier, taken to be a 6 HP motor. The third term was the power cost to operate a 3 HP water pump for condensed water and solvent vapor fed to the decanter. The fourth term is a general allocation of 52 kW for miscellaneous and unspecified needs. This value was taken from the reference of Appendix A2, Footnote 12, page 34. Both the third and fourth terms were corrected for the actual capacity of activated carbon using the six-tenth power rule used to estimate capital investment. The sum has the units of kW hours.

The cost of **cooling tower water** was taken to be the sum of two terms. The first is the electrical cost of operating a 10 HP pump for cooling water prorated by the actual capacity of activated carbon using the six-tenth power rule used to estimate capital investment. The second is the thermal equivalent of cooling water necessary to quench the flow of steam, with cooling water priced at $1.30 per thousand gallons.

The price of electric power was assumed to be $10.5 c hour. This is an average for the US in 2013 for commercial operations (residential use is slightly higher).

The total cost of electric power was summed with the cost of providing cooling tower water to be a line item called *utilities* in the Figures 4.46 to 4.59.

The cost of **labor** (maintenance and operating) was taken to be $35/ hour with a 15% utilization applied to the adsorption system since it should require very little attention.

The cost of **waste disposal** (water containing solvent at the solubility limit) was taken to be the cost of disposal of waste water as $0.75/gallon in a treatment facility for non-biodegradable solvents. The volume of waste water was the volume of steam condensate produced over one 20-minute cleaning cycle, expressed on an overall hourly basis.

The cost (actually a negative cost) of **recovery** of condensed and decanted solvent was taken to be the commercial value of the solvent produced per hour from either lip vent exhausts or vented from purging of the work chamber in an enclosed vapor degreaser. This cost was the mass recovered by the activated carbon, expressed on an overall hourly basis, times the following purchase prices: $0.60 /lb (toluene), $3.50 (n-propyl bromide), $0.80 (perchloroethylene), $0.34 (methylene chloride), $0.61 (trichloroethylene), and $25.00 (CFC-113). The latter price is hypothetical.

These assumptions are implemented to produce the detailed cost elements tabulated in the two tables in Appendix 3 (Economic Evaluations of the Operation and Treatment of Emissions from Solvent Vapor Degreasers). It is this detail of cost elements which are used to produce Figures 4.47 through 4.52 and 4.53 through 4.58 above.

the title of each graph is the total cost for each solvent used over a full year.

To repeat, again, the results in Table 4.13 of the cost analysis are agonizingly clear. The use of lip vents with open-top vapor degreasers imposes a total book-kept operating cost which some might describe as horrific!

- Depending upon the cleaning solvent used, the total book-kept operating cost over an operating year of 8000 hrs, can exceed $50/hour. That's a total book-kept cost which can exceed $400,000 in a single year – **just to operate the pollution control device** *in addition to the costs of operating the degreaser*!
- And it costs substantially more than that to manage emissions of one solvent, methylene chloride, because methylene chloride is relatively poorly adsorbed (Figure 4.17) onto activated carbon. The size of the required bed of activated carbon adsorbent is huge (Table 4.10); so the amount of steam required to desorb it at 0.4 lb steam/lb carbon (Chapter 4.16.5) is also huge.

Table 4.13 Cost Analysis for Operation of Adsorbers with Emissions from Lip Vents on Open-Top Vapor Degreasers

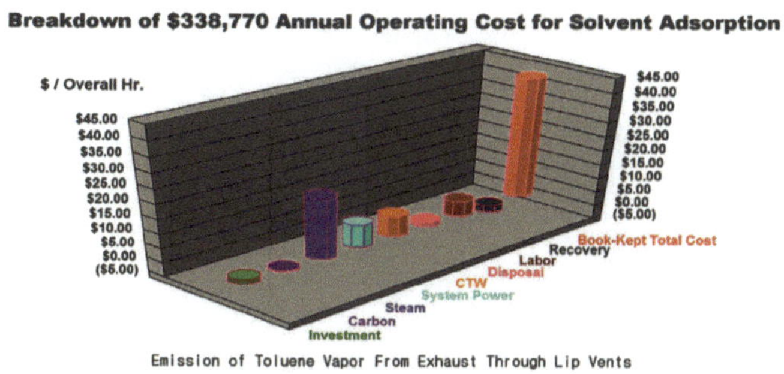

Figure 4.47 Toluene

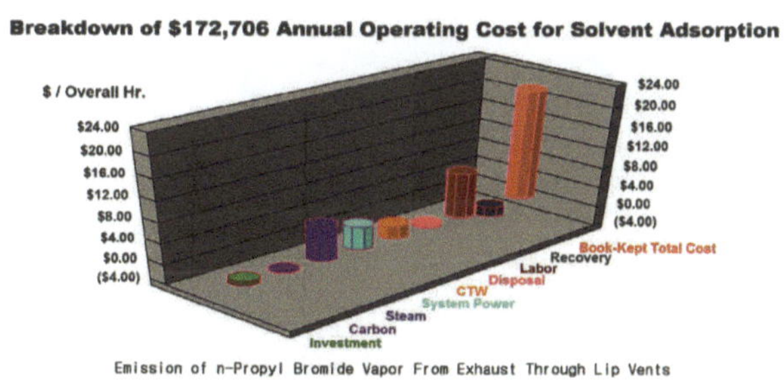

Figure 4.48 n-Propyl Bromide

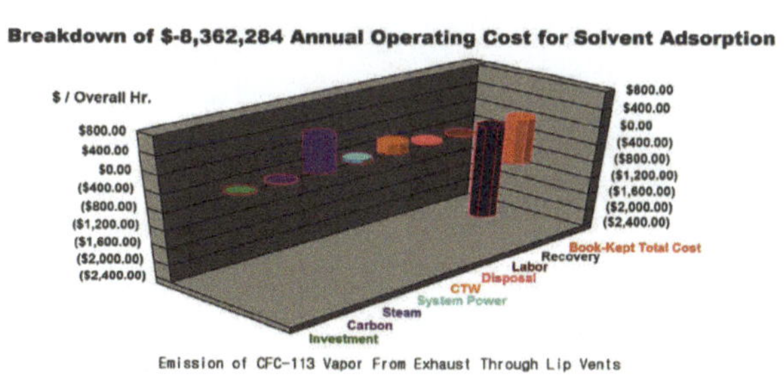

Figure 4.49 CFC-113 — Where the Value of Solvent Recovery Dominates

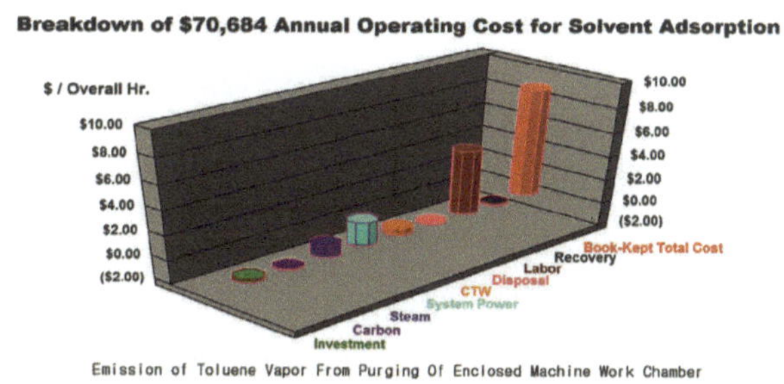

Figure 4.53 Toluene

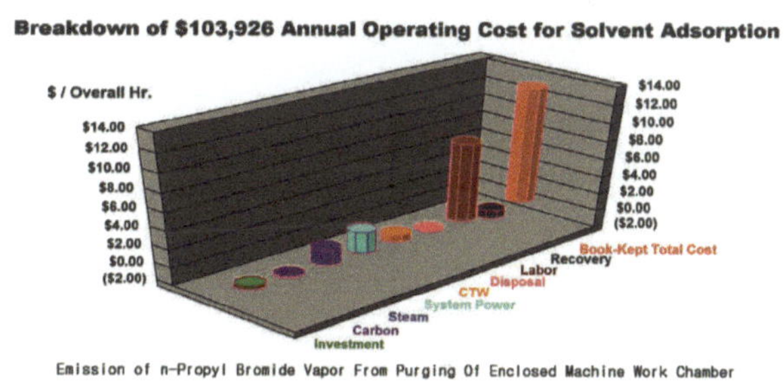

Figure 4.54 n-Propyl Bromide

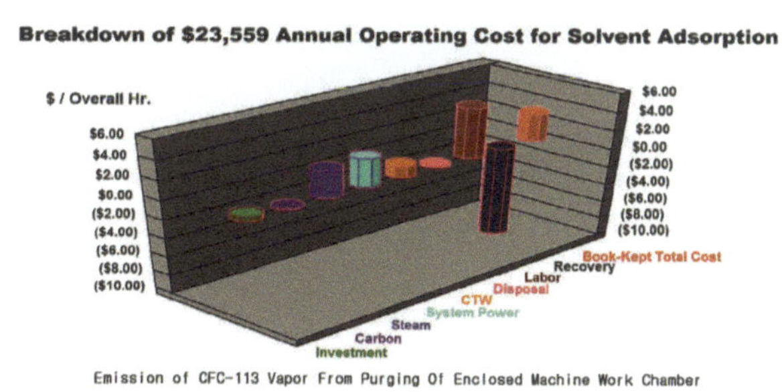

Figure 4.55 CFC-113

Table 4.14 Cost Analysis for Operation of Adsorbers with Emissions from Enclosed Vapor Degreasers

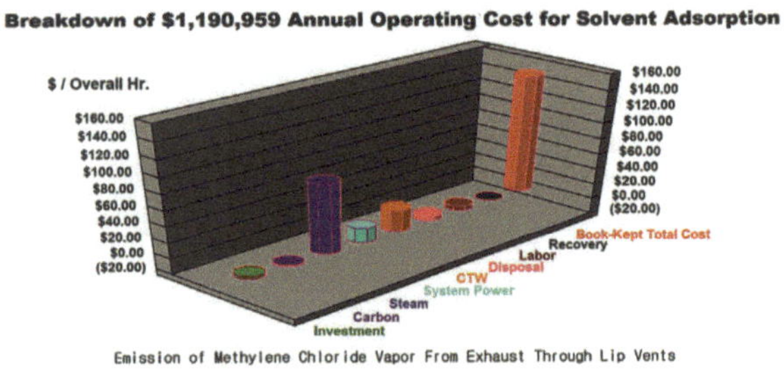

Figure 4.50 Methylene Chloride

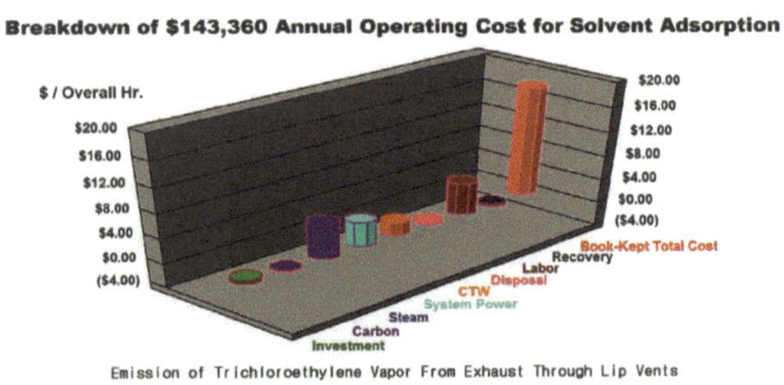

Figure 4.51 Trichloroethylene

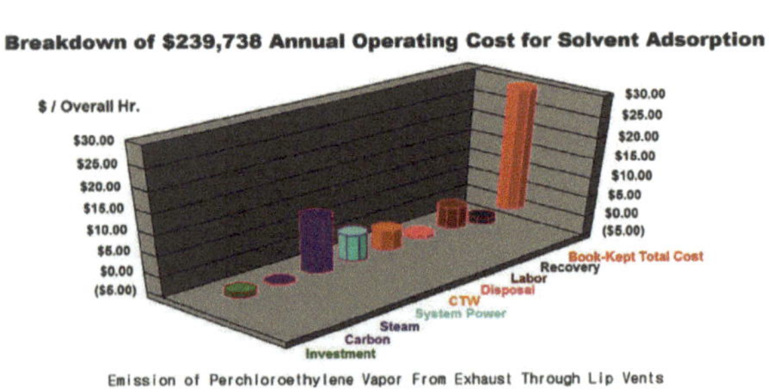

Figure 4.52 Perchloroethylene

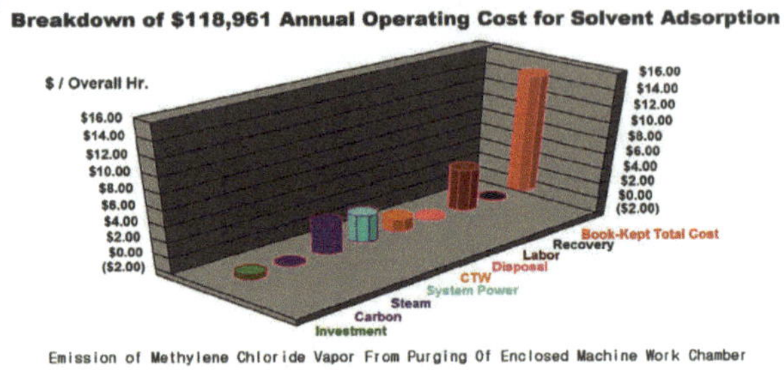

Figure 4.56 Methylene Chloride

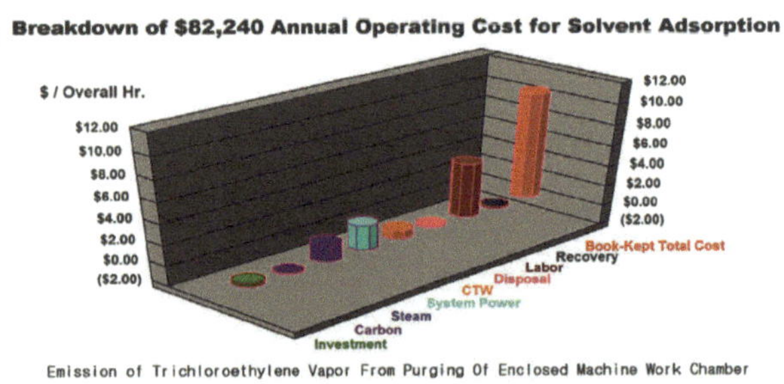

Figure 4.57 Trichloroethylene

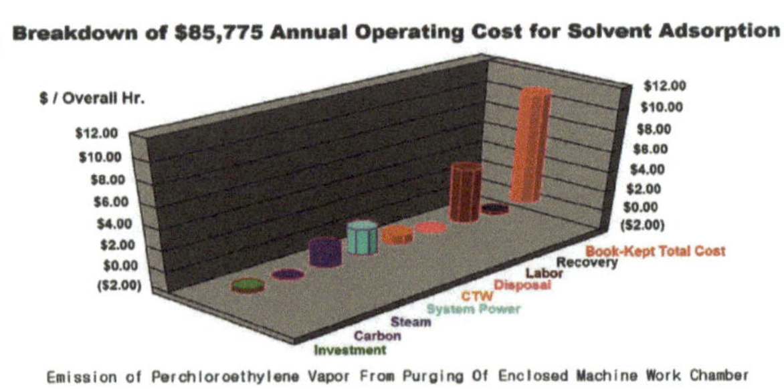

Figure 4.58 Perchloroethylene

Box 4.19 The Meaning of Book-Kept

The phrase "book-kept" has, unfortunately, but little meaning in the industrial cleaning industry. The phrase refers to entries in a ledger made by an accountant. Each entry describes an element of cost to be charged against or paid by a site operation. The elements might include: electric power, natural gas, operating labor, maintenance materials, maintenance labor, etc. The units of each element are the currency associated with the site or business. Each entry by the accountant is about a specific transaction, though general allocation entries (such as legal services, administrative salaries, taxes, insurance, etc. for the entire site) are also included.

One uses the phrase "book-kept" to mean that a manager viewing "book-kept" information has the capability to recognize *all* aspects of cost relating to their operation. The opposite of this phrase would refer to cost information that was not all-inclusive (A3, Ref. 5).

Unfortunately, most managers in the industrial cleaning industry have little idea of the elements of cost associated with their cleaning operation, let alone the magnitude of each. Costs of operating a cleaning unit are typically assigned (at best) to an enterprise or site book-kept cost summary as a general allocation entry.

Yet, the full book-kept cost for a year's operation is substantially less for CFC-113 (Figure 4.49) because of the basis chosen for evaluation[92].

- One simply can't use those two solvents in this application (open-top vapor degreasers) for reasons of cost!
- **The monetary value of a "black hole" is simple: prevention of SHE "mishaps."**

- The value of recovered solvent is negligible relative to the cost of producing it (except for CFC-113[AAA]).
- Utilities, mainly the cost of steam for desorption, are the dominant cost of operating a solvent adsorption system based on activated carbon – except for CFC-113 which should not be used in open-top vapor degreasers because it is so difficult to contain[93].
- Labor can be a significant cost category.
- But, capital payback, purchase of activated carbon, and waste disposal are of little concern in management of a "black hole" to contain emissions from lip vents.

For operation of an enclosed vapor degreaser, the total operating cost over an 8000 hr operating year has been reduced to about \$10-\$15/hr. Yet, recovery of cleaning solvent from the air purged from the work chamber of an enclosed vapor degreaser is still revenue negative (except for CFC-113). And solidly so at around \$80,000 per year!

- Surprisingly, labor is the major component of operating cost[94] for operation of the enclosed vapor degreaser.
- The cost of providing utilities (steam, electric power, and CTW [cooling tower water]) is considerably reduced for treatment of emissions from enclosed vapor degreasers relative to the same from lip vents of open-top vapor degreasers.
- And capital payback, purchase of activated carbon, and waste disposal are still of little concern in cost management of a "black hole" to contain emissions from purging of work chambers.

4.22.1 Evaluation of Book-Kept Economics of Adsorption Systems

The economics of solvent adsorption are not favorable, though it may be necessary to accept them – especially if open-top vapor degreasers are used with lip vents as recommended above.

The economics become more favorable, and more easily accepted, if solvent is better contained and concentrated within the operating vapor degreaser. Sill, the annual cost of operating only the pollution abatement unit is approximated by the purchase price of the enclosed vapor degreaser.

But in "technical" terms, the hourly book-kept cost of using a "black hole" is higher if less solvent is fed to it and fed at a low concentration.

---

92 The economic evaluation above is only of the adsorption process which removes solvent vapor from emitted air. Byproducts of the adsorption process are solvent-laden water, clean air, and recovered wet solvent. In the above economic analysis a modest debit of \$0.75/gallon is made for the necessary off-site treatment of the dirty steam condensate, and no financial credit is given for the value of clean air.

Wet solvent is also produced within this process by regeneration of the adsorbent bed with steam. A financial credit is given for the value of the recovered solvent because it can be reused for cleaning operations. But since no financial charge was made for the solvent when it was a pollutant in the emitted air, the credit for value of the recovered solvent is a windfall.

Consequently, for a few solvents whose purchase price is high (such as CFC-113), the adsorption process appears to produce profit since it is not charged for what will become an asset. But for every cleaning solvent, the actual book-kept operating cost of the adsorption process is understated because credit for the recovered solvent byproduct is given in the cost accounting.

To not give credit would misstate the integrity of the adsorption process; it does produce a valued byproduct – clean air. In any case, this financial credit will be reconciled with a charge to the cleaning process for loss of solvent in an air waste stream when the costs of operating the adsorption process are collected with the costs of operating the degreasing unit (Appendix A3, Ref. 5.).

93 CFC-113 was included in this study because this formerly ubiquitous solvent is still used in some countries, and its recovery value is so high.

94 This numeric outcome may be virtual – real as a spreadsheet output, but not representative of how personnel at a site are assigned. After all, if the site hires no additional staff to operate the adsorption system, is there a real cost of labor to operate it? A book-kept approach to cost accounting requires that such a cost be charged.

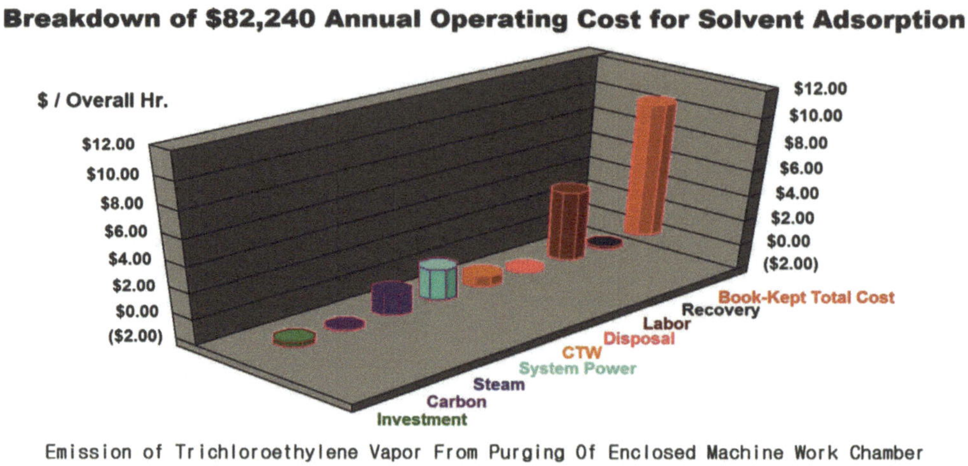

Figure 4.59 **(Also Figure 4.57) Cost Breakdown for an Adsorption Unit Operated as a Stand-Alone Entity**

4.22.2 "Controllable Costs" of Adsorption Systems

Adopt simplifying premises! Possibly, the premises will lead to a more realistic (but not complete) understanding of the costs of operating a "black hole" solvent adsorption system.

The cornerstone premise is that the vapor adsorption system is simply one of many pollution control systems on the site, such as fume scrubbers, biological treatment plants, and landfills. Seldom is a book-kept cost sheet developed for these systems. In that case:

- Site-based costs (all utilities [electric power, steam, and CTW], labor, and investment payback) will not be charged for operation of the adsorption unit – even

Figure 4.61 **Adsorption Unit Operated as a Stand-Alone Unit**

Box 4.20 **False Premises and 'Out of Pocket' Costs**

In cleaning operations this false general premise is most commonly adopted. Many-to-most operators of cleaning systems accept that the only variable cost of their solvent cleaning system is their purchase of virgin solvent and/or their only fixed cost is their purchase of the vapor degreaser (or perhaps not)!

Yet, these two premises may represent reality. It may be that operating labor for the cleaning system is shared with another work area and that the steam used to desorb the adsorber is truly waste steam whose latent heat of vaporization would normally be unused. After all, if the steam is truly a waste and no new employees are hired, should these two items be considered as being associated with the cleaning system? This author's answer to this question is *yes* – which is why Figures 4.47 to 4.58 are published. This answer is justified by two considerations: (1) the site manager may have alternate use for those two assets (utilities and labor), and (2) the operating manager is responsible for and so must truthfully know the complete situation.

An alternate premise is often commonly adopted. It is that utilities supplied to the adsorption unit (and probably also to the vapor degreaser) are not metered. So their consumption is not even known. The justification for this situation, or premise, is that the consumption of electricity, steam, or CTW is not significant relative to that consumed by other operating areas. So the expense of measurement of the consumption of utilities, and correlation of those measurements into the site's overall cost structure, is not spent.

In other words, as earlier noted in Footnote 100, without including site costs the analysis is not of the book-kept type. Some would label this analysis as an incremental one. Better labels are "out-of-pocket costs," and "funds leaving the company."

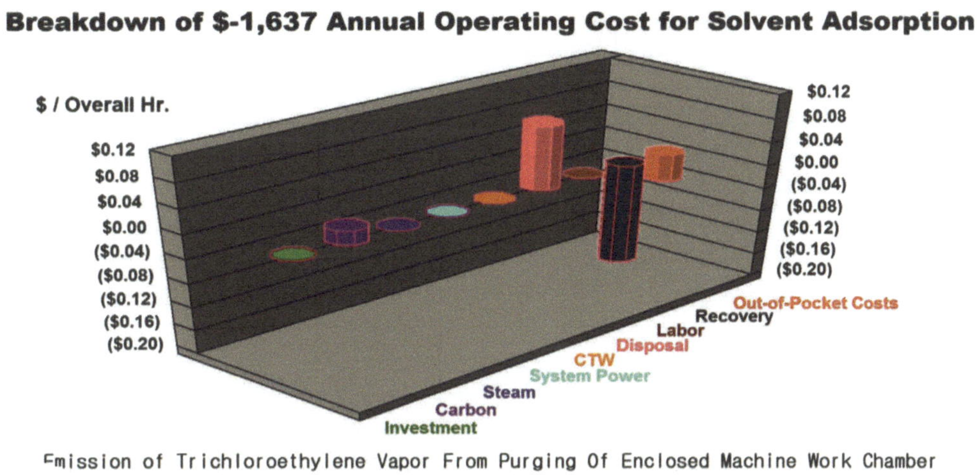

Figure 4.60 Cost Breakdown for an Adsorption Unit Operated as a Unit Integrated into Site Operations

though the unit obviously does consume utilities, employ labor which are provided by the overall site (Box 4.20), and purchase capital goods.

- The specific cost elements which will be recognized as being charged to other emission treatment sites are "out of pocket" costs as: purchase of activated carbon, external waste treatment, and recovery of solvent from desorption treatment.

For use of trichloroethylene in enclosed vapor degreasers, the detailed inventory of all "manager-controllable" costs is re-plotted in Figure 4.59 (also Figure 4.57). When the same plot is re-drawn in Figure 4.60 based on the presumption that the adsorption unit is not an independent operating unit but one integrated into all site operations, the appearance and texture of the situation has changed — at least because the total of "out of pocket" costs is recognized on the plot.

Figure 4.62 Adsorption Unit Operated Within Operating Site

4.23 ALTERNATE VISIONS

It is clear that one can see what they want to see.

If one's view is that the adsorption unit is a stand-alone operating system, where its manager is knowledgeable about and controls *all* costs, then –

- There is a substantial accumulation of many cost elements (Figure 4.59 for trichloroethylene) for which the operating manager is responsible.
- Recovery value of the solvent does not justify the operating cost of adsorption system — elimination of a pollution situation must be the justification.
- Incentives are apparent to reduce cost; at least costs of labor, system power, and steam may be reduced.

For some, this view is absolutely the correct one as it identifies the true "book-kept" cost of operation. With the cost structure known, the operating manager can make choices, about whether or not to invest additional capital in an enclosed vapor degreaser (vs. an open-top unit), whether or not increased automation can be justified based on the current level of labor costs, and if optimization of desorption of the adsorber bed can be done to produce less cost for steam and consequential less production of solvent-laden waste water.

If one's view is that the adsorption unit is one of many included within an operating site, then:

- The total "out-of-pocket" operating cost to capture the solvent emissions from an enclosed vapor degreaser over an 8000 hr operating year has been reduced to a negligible hourly charge.

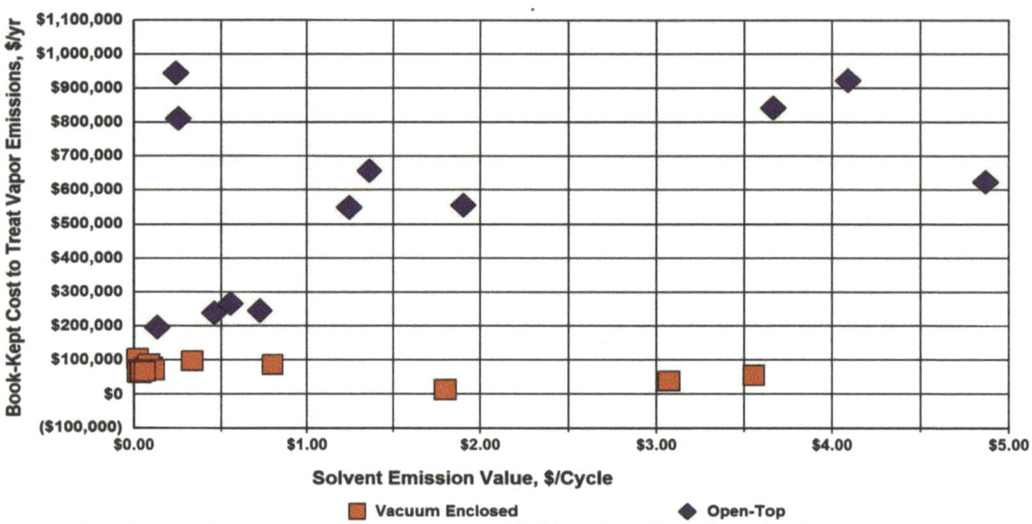

Figure 4.63 **Enclosed Vapor Degreasers are More Cost Justifiable When the Value of Solvent Emissions is Greater**

- In fact, the operation pays for itself on an "out-of-pocket" basis (Figure 4.60 for trichloroethylene) – treatment of the trichloroethylene emissions from an enclosed vapor degreaser does not on the net cause cash to leave the coffers of the enterprise!

For others this is the correct view, as the overall enterprise has the resources to support the adsorption system as a necessary facility to avoid SHE "mishaps."

Here the choice of view is left to the reader (Footnote 98).

This author holds that the "out-of-pocket" view is incorrect even if the adsorption system is both supported by, and used by, other operating units on a site. The "out-of-pocket" view conceals information which could be valuable to support decisions. Yes, the true cost of steam or actual labor use may be difficult to ascertain. But if one doesn't ask or look, one won't find, learn, or be able to take action.

4.24 STRATEGIC CONCLUSIONS

In closing this book, at least three straightforward conclusions must be seriously considered based on consideration of operating costs:

- The cost of treating the emissions from solvent cleaning operations is far more than the purchase price of the replacement solvent, and an amortized charge for purchase of the equipment.
  Disposal/treatment of air-solvent emissions dominates consideration of cost. It is almost always the major element of total operating cost. In managing solvent cleaning operations (Ref. 5, A3), one must always choose the option which most completely "keeps the solvent in the tank."
- New enclosed vapor degreasers can almost always be cost-justified relative to new open-top vapor degreasers – because of the above consideration.
  When one considers only the cost of treatment of emitted solvents from cleaning operations in the total book-kept cost which must be paid by the enterprise, payback of the difference in initial purchase cost can be very short.
  For example consider that the initial purchase price of an enclosed vapor degreaser may be $100,000 and that of an open-top vapor degreaser tasked with the same work may be $40,000. If only the book-kept cost for treatment of emitted solvent is considered, the payback time for six of the seven solvents considered here is less than one year. The values are: ~4, 15, 7, 1, 16, 5, and 0 months respectively for toluene, n-propyl bromide, perchloroethylene, methylene chloride, trichloroethylene, PCBTF, and CFC-113.
- The cleaning operation with the least cost of emission treatment is the one which emits the least monetary value of cleaning agent(s).

This is shown in Figure 4.63. As the value of solvent emitted (emission rate[95] times price), plotted on the horizontal axis increases, so does the total cost of operation for the vapor adsorber system, plotted on the vertical axis[96]. It is found that costs unrelated to direct emissions from the vapor degreaser do not matter in choosing a vapor

[95] A variety of solvent types are included in Figure 4.63: halogenated, oxygenated, aromatic, and "designer." Values of solvent emission rates are taken from Table 4.7, solvent prices are taken from Endnote DDD and total annual operating costs for both the vapor degreaser and the adsorber are taken from the information in Figures 4.47 to 4.58.

[96] The correlation between emitted solvent value and total cost of operation are fairly high—0.98 for vacuum enclosed.

degreaser – investment, utilities, labor, and disposal don't matter relative to consideration of vapor emissions.

All three strategic conclusions from this entire book may be simply summarized by the previously used expression:[BBB] "keep the solvent in the tank."

Endnotes

A. A black hole is a region of space-time from which nothing, not even light, can escape. It called black because it adsorbs all the light that hits its horizon, reflecting nothing. The mass of this stellar object is so great that the gravity field radiating from it is attracts everything – matter *and* energy. Figure 4.1 is a computer-generated image of a "black hole." Black holes can't be seen, but their existence is affirmed by their gravitational effects on surrounding objects. The phrase "black hole" is placed within quotation marks in this text to indicate that an abnormal use of the phrase is intended.

B. Figure 4.3 is courtesy of Wikipedia, http://en.wikipedia.org/wiki/File:Activated-carbon.jpg, and was modified by the author.
Figure 4.4 is courtesy of Mehta Chemicals, http://www.mehtachemicals.com/.
Figure 4.5 is courtesy of Alarco Filters, http://filtra-nerou.alarco.gr/en.
Figure 4.6 is courtesy of the Electron Microscopy Society of Malaysia (EMSM), photo by Jefri Samin, http://ibs.upm.edu.my/~aini/micrograph.htm. The image was magnified 1000X and is of activated carbon from coconut husk fiber.

C. Appendix A2 describes the basis for calculation of specific effects.

D. Intermolecular forces are identified, described, and implemented in Ref. 2, Chapters 2.3.2 (dispersion), 2.3.3 (polar), and 2.3.4 (hydrogen bonding).

E. Activated carbon is available in several physical forms. One common type of physical form is powder (called PAC, powdered activated carbon). PAC is made up of crushed or ground carbon particles, 95–100% of which will pass through a designated mesh sieve, such as a 60 or 80 Tyler mesh sieve (0.251 to 0.178 millimeters [about 0.01 to 0.007 inches] in mesh size). Although PAC has a high level of *external* surface area, it is a poor choice for packed beds due to high pressure drop. PAC is generally added directly to other process units, such as raw water intakes, rapid mix basins, clarifiers, and gravity filters. Hundreds of tons of PAC were added directly to river water to remove benzene from a substantial spill into the Songhua River (China) in 2005. The image of Figure 4.8 is of PAC made from lignin, and it is courtesy of the Chinese firm of Gongyi Songshan (http://www.china-carbon.org/). PAC is also used to decolorize sugar, wine, and fruit juices.

F. The second common type of physical form is granules (GAC, granular activated carbon). Extruded GAC is the type of granule most commonly used in beds for adsorption of solvent vapor from a moving stream of vapor. Common sizes for treatment of moving air streams are delineated as Tyler mesh sizes 4 × 6, 4 × 8, or 4 × 10, meaning that the granules are between 4.75 and 3.35 millimeters, 4.75 and 2.38 millimeters, and 4.75 and 1.68 millimeters, respectively. In the granular form (beads, cylinders or pellets, clusters, etc.) activated carbon can easily be packed into a containment device through which a contaminated vapor stream can be processed for purification. The image of Figure 4.9 is of GAC pellets made from coal and is courtesy of the same image source as in Endnote E.

G. The principle criteria for selection of the Carbonaceous raw material to activate is often cost, or its counterpart – availability. Sources include flax shive (the stems of flax fiber), cotton gin waste, nutshells, peat, pine wood, coir (fiber from the husk of coconut, produced in Sri Lanka and India), the coconut shell itself (from the Philippines), lignite, coal, petroleum pitch, peach or olive pits, eucalyptus, rice husk, apricot stones, grape seeds, cherry stones, and saw dust waste from the manufacture of plywood.
In the US, most activated carbon used is produced from coal and coconut husks. There are three large manufacturers who use coal as a feedstock, and one which uses wood. All GAC pellets are produced in India, Sri Lanka, and the Philippines.

H. An excellent paper on the use of flax shive and cotton gin waste for removal of trichloroethylene from both liquid and gas phases is Klasson, K.T., Wartelle, L.H., Lima, I.M., Marshall, W.E., and Akin, D.E., "Activated Carbons from Flax Shive and Cotton Gin Waste as Environmental Adsorbents for the Chlorinated Hydrocarbon Trichloroethylene," *Bioresource Technology*, Vol. 100, 2009, pages 5045 to 5050.

I. For example, steam activation produces carbon with an alkaline pH (9 to 11). To optimize its pH, steam-activated carbon can be neutralized with phosphoric acid, and thereby maintained at a pH between 4 and 6.

J. There are separate tests for the capacity of activated carbon used in water service to hold impurities. They are known as the Iodine Number test (ASTM D4607-94) and the Molasses Number test (no ASTM test).

K. Given similar adsorber designs and operating parameters, nearly all commercial granular activated carbons can withstand their own weight and the pressure effects induced by changes in fluid flow. So in systems in which the granular activated carbons will be used once or handled very infrequently, hardness characteristics may be of little or no import. This is the exact situation encountered by those doing solvent cleaning who don't regenerate their own activated carbon. In other words, let the supplier of activated carbon be responsible for managing the effects on particle character of thermal excursions by regeneration in place.

L. Chapter 1.21, and Figures 1.38 and 1.39.

M. The image of Figure 4.11 is courtesy of Nexflowair, at http://www.nexflowair.com. It was edited by this author.

N. Because their molecular weights are much greater than that of air (only 28.9 g/g-mole).

O. The image of Figure 4.13 is courtesy of MicroPrecisionCleaners (http://wwwmicroprecisioncleaners.com).

P. The image of Figure 4.14 is courtesy of NewPig.com (http://www.newpig.com).

Q. A continuously operating (on-line) explosimeter (combustible gas detector) may seem to be a example of irrelevance to one doing solvent cleaning but it is not. A basic hand-held device (Figure 3.22) can be purchased for about $250 US. Such a device should be part of the compliment of "tools" available to every operator of any solvent cleaning machine – including aprons, gloves, floor adsorbent, labels, goggles, fluid transfer piping and connectors, a suitable storage cabinet, a first-aid kit, a scale, safety information (MSDS), badges for personnel monitoring of solvent exposure, etc.
As such a tool it is at least a portable device to indicate safe (or not) conditions anywhere about which one wants to inquire (sample). It enables workers to ascertain and manage the safety of their operating environment. It saves solvent by identifying leaks.

But integration of it (figure at above) to a fieldbus (IEC 61158) control system will greatly increase the value of this device, and it's purchase price will not exceed $1000 US. Solvent savings accrue because ventilation to activated carbon is only implemented when area solvent concentrations require it for employee comfort and protection, and not continuously when airborne solvent levels don't justify its use (such as when the vapor degreaser is not operating). As shown in Chapter 4.7 the capacity of an adsorbent bed is substantially diminished when the vapor stream fed to it has only a low concentration of pollutant.

The image above is courtesy of Macro Technology Instrument Co., Ltd. (http://www.asia.ru/en).

R. The image of Figure 4.15 of an air intake plenum for dual pickup is courtesy of http://tanksapperautosport.com.

S. The image of Figure 4.16 is courtesy of Hawks Third Genparts, http://www.hawksthirdgenparts.com.

T. The equation is named for the Dutch-Swiss mathematician Daniel Bernoulli who published this principle in his book *Hydrodynamica* in 1738.

The Bernoulli Equation, at bottom, is written for the common situation where the fluid density is constant (incompressible flow or flow with small pressure changes), the flow is steady, and there are negligible frictional (viscous) forces to consume energy and affect flow.

$$\sum\left\{\left[\frac{P}{\rho\times g}\right]+\left[\frac{v^2}{2\times g}\right]\right\}+H = Constant$$

While the Bernoulli Equation can be derived from several standpoints, it is most simply an energy balance. In words, the equation requires that the sum of the energy changes due to pressure change, velocity change, and elevation change is zero. The nomenclature is P = pressure, ρ = density, g = acceleration of gravity, v = fluid velocity, and H = elevation above a datum.

One can substitute into the Bernoulli Equation what's known as the continuity equation. It relates fluid velocity to the mass flow rate (G) and circular channel diameter (D), and is:

$$G = \left\{\frac{\pi\times D^2}{4}\right\}\times\rho\times v.$$

$$\sum\left\{\left[\frac{P}{\rho\times g}\right]+\left[\frac{8\times G^2}{g\times\pi^2\times\rho^2\times D^4}\right]\right\} = Constant$$

In a device such as an air amplifier or an air eductor, there is no change in elevation (H). So the Bernoulli Equation becomes the equation at above.

Note the dependence of pressure change on circular channel diameter (D), all else being the same. It's to the fourth power! That dependence is how one can design an air amplifier with a large diameter venturi and expect only a small pressure change sufficient to entrain additional air; relative to an eductor with a small diameter venturi which would produce a substantial pressure change – a partial vacuum upstream.

Application of the Bernoulli Equation to the simple apparatus which is an air amplifier implements what is known as the Coanda Effect (named after Romanian aerodynamics pioneer Henri Coanda, who, in 1934, was the first to recognize the practical application of the phenomenon in development of aircraft which would have adequate lift at low air speed). Specifically, in an air amplifier, compressed air expands across a specially configured surface. A high velocity jet is produced. The jet clings to the walls of the surface (the Coanda Effect). A zone of low pressure is created. Available air at a higher pressure is pulled into the zone of lower pressure. The level of low pressure is dependent upon the dimensions of the surface configuration (as predicted by the Bernoulli Equation).

U. The three product images within Table 4.2, from left to right, courtesy of: Nexflowair (http://www.nexflowair.com), Direct Industry ((http://www.directindustry.com), and Humboldt Manufacturing (http://www.humboldtmfg.com).

V. The calculation is made with the D/R methodology which is explained and its use it shown in Appendix A2.

W. However, global business for water purification by removal of chloramines with activated carbon is vibrant.

X. Chapter 2.13 and Footnote 32.

Y. From Sinnott, R.K. and Towler, G., *Chemical Engineering Design, 2nd Edition*, Elsevier Science, 2012, Chapter 15 – Design of Reactors and Mixers, Page 638,

"...In a perfect well-mixed reactor (vessel), the entire contents of the vessel (chamber) are at the same temperature and concentration and there is no spatial variation in any mixture property. The fluid leaving the well-mixed reactor is therefore at the same concentration and temperature as the vessel contents. Material entering the well-mixed reactor is immediately mixed with the contents and comes to much the same conditions. Some of the material that enters can leave immediately, while some leaves much later, so there is a broad distribution in residence time..."..

Z. The image of Figure 4.27 is courtesy of Tigg Corporation, http://www.steel-tank-vessel.com.

AA. Through the locus of pressure values allowed by the Antoine Equation – Chapter 2.14.1.

BB. As illustrated in Figures 2.29 and 2.30 among others

CC. The sequence of continuous displacements in Table 4.4 is the same as plotted in Figure 4.22.

DD. A phrase, and concept, learned from a past employer – Mr. Carl Wolfe.

EE. Rely on the advice of others as reported in the references of Footnotes 65 and 66 of Chapter 1.

FF. Useful, but not current, references about management of solvent losses are:

"Conservation and Recycling Practices for CFC-113 and Methyl Chloroform," *EPA/400/1–91/017*, May 1991, ICOLP Technical Committee, page 56. They report values between 800 and 1600 ACFM.

"Solvent Cleaning and Degreasing – An Assessment of Emission Control Options," *The Center for Emissions Control*, November, 1992, page 20. They report that velocities within the lip openings are between 130 and 160 ft / second. They also report that

"...available data indicate that a lip exhaust, even when properly operated, can double solvent consumption..."

Scheil, G.W. "Source Test Trichloroethylene Adsorber," *EPA Project Report # 76-DEG-1*, February 1976. Lowest reported air flows were 3000 SCFM.

"Degreaser System Pollution Prevention Evaluation," *EPA Report ESL-TR-90-33*, September, 1990.

"...the local exhaust system operated between 2,442 and 2,600 ACFM...". A reduction was achieved to 1,400 ACFM, but "...there, TCA levels exceeded the TCA concentration limits...". See page 23.

GG. So a comparative analysis can be made, the information in the Table 4.15 is provided to supplement that in Table 4.7. Designs are provided for the same solvent when emitted from two very different operating situations: emission of solvent vapor from lip vents during a 2-minute interval when the cover from an open-top vapor degreaser is opened, and emission of solvent vapor in air or nitrogen from the work chamber of an enclosed vapor degreaser when the cleaning cycle is complete. The basis for design is that used to produce the information in Table 4-13.

HH. The image of Figure 4.34 is courtesy of Encotech, http://www.carbonservice.net/. It was edited by this author. The reverse side of the control panel is shown on the right side of the image. Vessel sizes are at or above the largest which might be considered for use with a single enclosed solvent cleaning system.

II. A useful reference (not in English) is: Tsotas, E., "Uber die Warme-und Stoffqbertragung in Durchstrfmten Festbetten," Fortschritte Reihe, Vol. 3, No. 223, VDI-Verlag, Dusseldorf, 1990.

JJ. Trichloroethylene is used as the example cleaning solvent in these calculations. This is because of its commonality of use. The calculations here produce the same general outcomes for methylene chloride, or any solvent.

KK. Gu, J., and Bart, H-J., "Heat and Mass Transfer in Steam Desorption of an Activated Carbon Adsorber," International Communications in Heat and Mass Transfer, Vol. 32, 2005, pages 296 to 304.

Table 4.15 Details of Adsorber Designs

		For Purging of Enclosed Vapor Degreaser Work Chamber						For Lip Vent Exhaust					
				Adsorber Vessel						Adsorber Vessel			
Solvent	Exposure Limit ppm	Chamber Solvent Conc. ppm	Exit Solvent Conc. ppm	Lb Solvent/ Lb Carbon	Capacity Lb Carbon	L ft	D ft	Feed ppm	Exit Solvent Conc. ppm	Lb Solvent/ Lb Carbon	Capacity Lb Carbon	L ft	D Ft
Methylene Chloride	50	13,158	7.8	0.0625	209	5.4	1.2	1,000	3.1	0.0187	7,502	3.6	9.0
Trichloroethylene	10	13,158	9.8	0.1443	140	5.4	1.2	200	3.9	0.0903	481	1.4	3.6
Perchloroethylene	25	13,158	55.5	0.1583	161	5.4	1.2	500	5.6	0.1225	1,120	1.9	4.8
n-Propyl Bromide	10	13,158	9.8	0.1388	136	5.4	1.2	200	3.5	0.0723	562	1.5	3.8
PCBTF	25	13,158	8.5	0.1221	227	5.4	1.2	500	4.4	0.0928	1,609	2.2	5.4
Methyl Formate	100	13,158	7.2	0.0868	107	5.4	1.2	2,000	7.5	0.0330	6,028	3.4	8.4
Methyl Acetate	200	13,158	6.5	0.1425	80	5.4	1.2	4,000	19.8	0.1310	3,740	2.9	7.2
Tertiary Butyl Acetate	200	13,158	5.9	0.1764	101	5.4	1.2	4,000	17.4	0.1695	4,532	3.1	7.6
Dimethyl Carbonate	200	13,158	5.9	0.1358	102	5.4	1.2	4,000	16.4	0.1199	4,968	3.1	7.9
Acetone	750	13,158	15.0	0.1452	62	5.4	1.2	15,000	44.3	0.1760	8,187	3.7	9.3
Isopropanol	400	13,158	18.0	0.1685	55	5.4	1.2	8,000	36.9	0.1904	4,175	3.0	7.4
Ethyl Acetate	400	13,158	21.0	0.1667	81	5.4	1.2	8,000	26.7	0.1715	6,797	3.5	8.7
Toluene	100	13,158	9.8	0.1700	83	5.4	1.2	2,000	35.2	0.1697	1,802	2.2	5.6
HFC-365mfc	1000	13,158	55.5	0.0865	263	5.4	1.2	20,000	15.5	0.0809	60,542	7.2	18.1
HFC-43-10 mee	200	13,158	5.2	0.0677	572	5.4	1.2	4,000	4.1	0.0424	39,308	6.3	15.7
HFE-7100	750	13,158	15.0	0.0715	538	5.4	1.2	15,000	8.3	0.0615	100,796	8.6	21.5
CFC-113	1000	13,158	55.5	0.1034	279	5.4	1.2	20,000	15.00	0.0971	63,803	7.4	18.4

LL. There is significantly more information in the reference of Appendix A2, Footnote 12, pages 339 to 344.

MM. More detail can be found in Ref. 2, Chapter 11.

NN. More information is available in Footnote 94 in Chapter 3.7.7

OO. Halogenated solvents are not biodegradable in conventional aerobic treatment systems. This situation is especially important in remediation of groundwater contaminated by trichloroethylene. In some waste treatment systems, the microorganisms are actually destroyed by halogenated solvents. Yes, halogenated solvents can be bioremediated in special treatment systems such as those described in the reference below. But absent these unusual and non-commercial systems, the only available treatment plan for these water wastes from "black holes" is dilution with clean water. This doesn't change the mass flow of halogenated materials – only their concentration. Damborsky, J., "Tetrachloroethene-dehalogenating Bacteria," Folia Microbiological, Vol. 44, No. 3, 1999; Den, W. "Photo-oxidation and Biotrickling Filtration for Controlling Industrial Emissions of Trichloroethylene and Perchloroethylene, Chemical Engineering Science, Vol. 61, No. 24, 2006.

PP. Ref. 2, Table 11.2.

QQ. A US patent issued in 2006 to Parts Cleaning Technologies (USP 7,611,564) describes a process by which n-propyl bromide is stabilized in the vapor phase with 250 to 300 ppm of an acid scavenger dissolved in a carrier. What's unique is that the stabilizer is continuously added to the solvent-air vapor phase before it flows into the adsorber, and not into the condensed liquid after the adsorbent bed has been desorbed with steam (as is done with trichloroethylene). The acid scavenger is a base – usually triethylamine. Its role is to react with any hydrobromic acid formed by reaction with the solvent and water (from steam). Excess scavenger, the carrier (usually n-propyl alcohol), and reaction products are all soluble in water. All are recovered in the aqueous phase produced in the decanter associated with the adsorber bed (Figure 4.41).

RR. Ref. 3, Chapters 20, 22, 24, and 25.

SS. Chilling either stream will allow recovery of some solvent without contamination by water, but the exhaust vapor (air or nitrogen) will retain too much solvent to either meet an exposure limit or a regulated emission limit. Figures 2.35 through 2.40, including 2.36 (methylene chloride), show that the dew points of these lean-solvent nitrogen streams are very low – less than normally achievable. For example, the calculated dew point of a nitrogen stream with 200 ppm of methylene chloride (exposure limit of 50 ppm) is below $-100°C$.

TT. The image of Figure 4.44 is courtesy of Enviro Equipment (http://www.enviroequipment.com/remediation_new/EEICarbonDrums.html).

UU. See Endnote TT.

VV. At these operating conditions, this vessel would not need to be compliant with the ASME Pressure Vessel Code, Section VIII; 2010 ASME Boiler & Pressure Vessel Code Section VIII Rules for Construction of Unfired Pressure Vessels - Division 1; updated by ASME, July 1, 2011.

WW. USP 5,958,109, *Adsorption Process for Recovery of Solvents*, 1999, describes a process by which solvents can be recovered from activated carbon by management of pressure changes. While this process doesn't involve an additional chemical as a desorbant or produce waste, its complexity is seldom justified for desorption of small volumes of adsorbent.

XX. Personal communication, March 13, 2012, from Ken Schaeffer, President, Carbon Resources.

YY. Michaslis T.B. "Techniques to Detect Failure in Carbon Adsorption Systems," *EPA Report 340/1–80011*, April 1980.

ZZ. USP 8,016,914 claims use of temperature to manage operation of twin adsorption beds – used for recovery of hydrogen from mixed gas systems. The claims include only operation of pressure swing systems for adsorption of hydrogen, and not recovery of solvents from activated carbon.

AAA. Recovery is shown in the images of Figures 4.47 to 4.60 as a positive number because it is more easily seen in these plots. Recognize that the monetary value of recovered solvent is negligible from an overall standpoint (except for CFC-113 because of its unusual purchase price).

BBB. See Endnote EE.

Appendix | A1 |

Basic Data about Cleaning Solvents

This book includes a database of 210 solvents which can, are, and have been used for parts cleaning at various times and in various industries throughout the world. None are perfect choices for many applications, but all have been used successfully for some applications, at some times, in some places.

Data concerning them can be found in this appendix, which contains five tables, each with different data about the same 210 solvents.

The content of the five tables includes general data (Table A I-1), physical properties (Table A I-2), environmental properties/rankings/exposure limits (Table A I-3), flammability data (Table A I-4), and solubility/solvency data (Table A I-5).

The tables are not each sorted on the same basis.

- Table A I-1 (General data) is sorted: first, based on the type of each solvent in alphabetical order (acids, alcohols, aromatics...specials); and then second, within types, based on increasing boiling point. One uses Table A I-1, both as a data resource, and as a tool for identifying specific solvents considered for use as substitutes. In Appendix VIII, one can use Figures A I-4 through A I-22 to identify solvents by their specific type and approximate boiling point, which allows for specific name identification via Table A I-1.
- Table A I-2 (Physical properties) is sorted: first, based on the type of each solvent in alphabetical order (acids, alcohols, aromatics...specials); and then second, based on the name of each solvent in alphabetical order within type (1-hexanol, amyl alcohol, benzyl alcohol...texanol). One uses Table A I-2 to obtain additional physical property information about a solvent, based on its name within type.
- Table A I-3 (Environmental properties/rankings) is sorted: first, based on the type of each solvent in alphabetical order (acids, alcohols, aromatics...specials); second, by whether or not a solvent is a HAP in the U.S. – with those not so being first noted; third, by whether or not a solvent is a VOC exempt in the U.S. – which those being so first noted; then fourth based on the name of each solvent in alphabetical order (1-hexanol, amyl alcohol, benzyl alcohol....texanol). One uses Table A I-3 to select solvents, within type, based on those presenting the least SHE risk first listed alphabetically.
- Table A I-4 (Flammability) is sorted: first based on the type of each solvent in alphabetical order (acids, alcohols, aromatics...specials); second within types based on flash point – with the highest values first; and third based on lower explosive limit (LEL) from highest to lowest. In other words, within type, solvents are sorted based on the ones with the least flammability risk being listed first.
- Table A I-5 (Solubility) is sorted first based on the type[1] of each solvent in alphabetical order (acids, alcohols, aromatics...specials); and then second, based on the name of each solvent in alphabetical order within type (1-hexanol, amyl alcohol, benzyl alcohol...texanol). One uses Table A I-5 to obtain information about solubility parameters and water solubility for and of a solvent based on its name within type.

[1] Note that each table has a common primary sorting parameter – solvent type. The reason for this is found chiefly in Chapter 2, and also throughout this book. It is that the performance of solvents relative to soils to and other solvents depends upon how well the two are matched. And the primary characteristic of a solvent is its type – acid, alcohol, aromatic...or specialty.

Table A1-1

Type	Name	CAS #	EINECS #	DOT #	Density g / cc	Mol. Wt.	Boiling Point °C	Vapor Pressure mm Hg @ 20 °C
Acids	Acetic Acid	64-19-7	200-580-7	2789	1.050	60.05	118.0	11.000
Alcohols	Methanol	67-56-1	200-659-6	1230	0.791	32.04	64.7	97.510
Alcohols	Ethanol	64-17-5	200-578-6	1170	0.790	46.07	78.2	44.430
Alcohols	Isopropanol	67-63-0	200-661-7	1219	0.786	60.10	82.0	31.200
Alcohols	Tertiary Butanol	75-65-0	200-889-7	1120	0.789	74.10	82.2	20.000
Alcohols	Secondary Butanol	78-92-2	201-158-5	1120	0.810	74.12	93.9	18.000
Alcohols	N-Propanol	71-23-8	200-746-9	1274	0.804	60.10	97.2	18.900
Alcohols	Isobutanol	78-83-1	201-148-0	1212	0.802	74.12	108.0	7.230
Alcohols	N-Butanol	71-36-3	200-751-6	1120	0.810	74.12	117.5	4.890
Alcohols	iso-Amyl Alcohol	123-51-3	204-633-5	1105	0.810	88.15	130.0	2.800
Alcohols	Methyl Isobutyl Carbinol	108-11-2	203-551-7	2053	0.808	102.18	133.0	2.180
Alcohols	Amyl Alcohol	71-41-0	200-752-1	1105	0.814	88.15	133.1	2.930
Alcohols	1-Hexanol	111-27-3	203-852-3	2282	0.814	102.18	157.6	0.900
Alcohols	Cyclohexanol	108-93-0	203-630-6	1986	0.963	100.16	160.8	0.980
Alcohols	Furfuryl Alcohol	98-0-0	202-626-1		1.130	98.10	171.0	0.600
Alcohols	Diacetone Alcohol	123-42-2	204-626-7		0.931	116.16	172.2	0.950
Alcohols	Tetrahydrofurfuryl Alcohol	97-99-4			1.054	102.13	178.0	0.000
Alcohols	Benzyl Alcohol	100-51-6	202-859-9		1.044	108.14	205.3	0.077
Alcohols	Texanol	25265-77-4	246-771-9		0.948	216.30	244.0	0.008
Aromatics	Benzene	71-43-2	200-753-7	1114	0.877	78.11	80.0	75.000
Aromatics	Lactol Spirits	64742-49-0		1255	0.725	142.30	98.5	40.000

Aromatics	Toluene	108-88-3	203-625-9	1294	0.867	92.14	110.6	21.890
Aromatics	VM&P Naptha	64742-89-8	232-453-7		0.730	100.00	134.2	53.329
Aromatics	o-Xylene	95-47-6	215-535-7	1307	0.862	106.00	138.0	6.000
Aromatics	Turpentine, oil	8006-64-2	232-350-7	1299	0.861	136.00	150.0	4.000
Aromatics	Stoddard Solvent	8052-41-3			0.780	150.00	159.0	5.000
Aromatics	Mineral Spirits	64741-41-9			0.757	166.00	200.0	0.750
Benzotrifluorides	Benzotrifluoride	98-08-8	202-635-0		1.185	146.11	102.1	20.000
Benzotrifluorides	Parachloro-benzotrifluoride	98-56-6	202-681-1	2234	1.338	180.56	139.0	5.300
Brominateds	n-Propyl Bromide	106-94-5	203-445-0	2344	1.332	123.00	71.1	110.000
CFCs	CFC-113	76-13-1	200-936-1		1.560	187.38	47.8	403.800
Chlorinateds	Methylene Chloride	75-09-2	200-838-9	1593	1.326	84.93	39.8	356.280
Chlorinateds	Trans 1,2 Dichloroethylene	156-60-5		1150	1.257	96.94	48.7	266.849
Chlorinateds	1,1-Dichloroethane	75-34-3	200-863-5	2362	1.174	98.96	57.3	175.000
Chlorinateds	Chloroform	67-66-3	200-663-8	1888	1.489	119.38	60.5	211.980
Chlorinateds	1,1,1-Trichloroethane	71-55-6	200-756-3	2831	1.338	133.41	74.1	99.760
Chlorinateds	Carbon Tetrachloride	56-23-5	200-262-8	1846	1.593	153.82	76.0	90.310
Chlorinateds	Ethylene Dichloride	107-06-2	203-458-1	1184	1.235	98.96	83.5	65.000
Chlorinateds	Trichloroethylene	79-01-6	201-167-4	1710	1.463	131.39	87.0	58.880
Chlorinateds	Perchloroethylene	127-18-4	204-825-9	1897	1.621	165.83	121.0	14.850
Dibasic Acid Esters	DBE-4	106-65-0	203-419-9		1.121	146.00	91.1	0.300
Dibasic Acid Esters	DBE	1119-40-0	214-277-2		1.092	159.00	196.0	0.200
Dibasic Acid Esters	DBE-6	627-93-0	211-020-6		1.064	174.20	227.0	0.050
Esters	Methyl Formate	107-31-3	203-481-7	1243	0.971	60.10	31.7	476.000

(Continued)

Table A1-1 *(Continued)*

Type	Name	CAS #	EINECS #	DOT #	Density g / cc	Mol. Wt.	Boiling Point °C	Vapor Pressure mm Hg @ 20 °C
Esters	Ethyl Formate	109-94-4	203-721-0	1190	0.922	74.08	54.3	195.020
Esters	Methyl Acetate	79-20-9	201-185-2	1231	0.934	74.08	56.9	219.770
Esters	Ethyl Acetate	141-78-6	205-500-4	1173	0.901	88.10	76.5	72.650
Esters	Isopropyl Acetate	108-21-4	203-561-1	1220	0.873	102.10	88.5	45.620
Esters	Tertiary Butyl Acetate	540-88-5	208-760-7		0.867	116.16	98.0	47.000
Esters	n-Propyl Acetate	109-60-4	203-686-1	1276	0.888	102.10	101.6	25.130
Esters	Isobutyl Acetate	110-19-0	203-745-1	1213	0.872	116.16	117.0	15.980
Esters	n-Propyl Propionate	106-36-5	203-389-7		0.881	116.20	123.0	10.700
Esters	n-Propyl Propionate	106-36-5	203-389-7		0.878	116.20	124.0	10.700
Esters	n-Butyl Acetate	123-86-4	204-658-1	1123	0.882	116.16	126.5	9.380
Esters	N-Butyl Propionate	590-01-2	209-669-5		0.874	130.20	144.7	3.000
Esters	Amyl Acetate	628-63-7	211-047-3	1104	0.876	130.19	149.0	4.000
Esters	Amyl Acetate	628-63-7	211-047-3	1104	0.871	130.18	149.0	4.000
Esters	Ethyl-3-Ethoxypropionate	763-69-9	212-112-9		0.951	146.19	169.7	0.700
Esters	Cyclohexyl Acetate	622-45-7	210-736-6		0.966	142.20	172.0	0.730
Esters	Ethylene Glycol Diacetate	111-55-7	203-881-1		1.104	146.15	190.9	0.200
Esters	Propylene Glycol Diacetate	623-84-7	231-765-0		0.966	160.00	191.1	0.233
Esters	Butyl Glycol Acetate	112-07-2	203-933-3		0.972	160.20	192.0	0.233
Esters	2-Ethyl Hexyl Acetate	103-9-3			0.872	172.27	199.0	0.400
Esters	Dimethyl Succinate	106-65-0	203-419-9	1993	1.120	146.14	200.0	0.225

Ethers	Methylal	109-87-5	203-714-2	1234	0.821	76.10	42.0	400.000
Ethers	Methyl t-Butyl Ether	1634-04-4	216-653-1	2398	0.741	88.15	55.2	ERR
Ethers	Methyl Butyl Ether	628-28-4			0.750	88.15	70.0	ERR
Ethers	t-Butyl Ether	6163-66-2		3271	0.766	130.20	101.0	33.000
Ethers	Cyclopentyl Methyl Ether	5614-37-9			0.860	100.20	106.0	27.280
Ethers	Amyl t-Butyl Ether	10100-95-5			0.786	144.25	148.6	ERR
Glycol Ethers	Dipropylene Glycol Dimethyl Ether	111109-77-4	404-640-5	1993	0.902	162.20	175.0	0.550
Glycol Ethers	Dipropylene Glycol n-Propyl Ether	29911-27-1	216-372-4		0.923	176.30	212.0	0.050
Glycol Ether Acetates	Ethylene glycol methyl ether acetate	110-49-6	203-772-9	1189	1.004	118.14	145.0	2.030
Glycol Ether Acetates	Propylene Glycol Methyl Ether Acetate	108-65-6	283-152-2		0.967	132.00	146.5	2.528
Glycol Ether Acetates	Ethylene Glycol Ethyl Ether Acetate	111-15-9	203-839-2		0.974	132.16	156.4	1.100
Glycol Ether Acetates	Propylene Glycol Ethyl Ether Acetate	54839-24-6	259-370-9		0.941	146.00	159.0	1.700
Glycol Ether Acetates	Ethylene Glycol Butyl Ether Acetate	112-07-2	203-933-3		0.941	160.21	186.0	0.300
Glycol Ether Acetates	Dipropylene Glycol Methyl Ether Acetate	88917-22-0			0.976	190.20	200.0	0.050
Glycol Ether Acetates	Diethylene Glycol Monoethyl Ether Acetate	112-15-2			1.008	176.22	214.0	0.050
Glycol Ether Acetates	Diethylene Glycol Monobutyl Ether Acetate	124-17-4	204-685-9		0.980	204.27	231.0	0.006
Glycol Ethers	Ethylene Glycol Diethyl Ether	629-14-1	211-076-1	1153	0.848	118.18	119.4	9.000

(Continued)

Table A1-1 *(Continued)*

Type	Name	CAS #	EINECS #	DOT #	Density g / cc	Mol. Wt.	Boiling Point °C	Vapor Pressure mm Hg @ 20 °C
Glycol Ethers	Propylene Glycol Monomethyl Ether	107-98-2	203-539-1		0.923	90.10	119.6	9.008
Glycol Ethers	Propylene Glycol Monomethyl Ether	107-98-2	203-539-1		0.921	90.10	120.0	9.976
Glycol Ethers	Ethylene Glycol Monomethyl Ether	109-86-4	203-713-7	1188	0.964	76.10	124.0	6.230
Glycol Ethers	Propylene Glycol Monoethyl Ether	1569-02-4	216-374-5		0.903	104.15	132.0	4.000
Glycol Ethers	Propylene Glycol Monoethyl Ether	1569-02-4	216-374-5		0.897	104.00	133.0	7.500
Glycol Ethers	Ethylene Glycol Monoethyl Ether	110-80-5	203-804-1	1171	0.929	90.12	135.6	3.750
Glycol Ethers	Ethylene Glycol Mono Isopropyl Ether	109-59-1	203-685-6		0.911	104.20	142.5	3.780
Glycol Ethers	Propylene Glycol Monopropyl Ether	30136-13-1	250-069-8		0.855	118.20	149.0	1.500
Glycol Ethers	Propylene Glycol n-Propyl Ether	30136-13-1	250-069-8		0.887	118.20	150.0	1.800
Glycol Ethers	Propylene Glycol Mono t-butyl Ether	57018-52-7			0.870	132.20	151.0	0.826
Glycol Ethers	Propylene Glycol Methyl Propyl Ether	34590-94-8	252-104-2		0.951	148.00	155.6	0.750
Glycol Ethers	Propylene Glycol Ethyl Propyl Ether	30025-38-8			0.942	146.20	162.0	0.110
Glycol Ethers	Propylene Glycol Monobutyl Ether	1569-02-4	249-598-7		0.879	132.20	170.2	0.830
Glycol Ethers	Ethylene Glycol Monobutyl Ether	111-76-2	203-905-0	2369	0.902	118.18	171.0	0.660

Glycol Ethers	Butyl Glycol Ether	111-76-2	203-905-0	2369	0.901	118.20	171.0	0.512
Glycol Ethers	3-Methoxy-3-Methyl-1-Butanol	56539-66-3	260-252-4		0.904	118.18	174.0	0.500
Glycol Ethers	Dipropylene Glycol Monomethyl Ether	112-28-7	252-104-2		0.951	148.20	190.0	0.280
Glycol Ethers	Diethylene Glycol Monomethyl Ether	111-77-3	203-906-6		1.035	120.15	193.0	0.190
Glycol Ethers	Dipropylene Glycol Monoethyl Ether	30025-38-8			0.927	162.26	193.9	0.300
Glycol Ethers	Diethylene Glycol Monomethyl Ether	111-77-3	203-906-6		1.025	120.15	196.0	0.300
Glycol Ethers	Diethylene Glycol Monobutyl ether	112-34-5	203-961-6		0.954	162.23	218.0	0.020
Glycol Ethers	Dipropylene Glycol Monobutyl Ether	29911-28-2	249-951-5		0.912	190.30	228.9	0.020
Glycol Ethers	Dipropylene Glycol n-Butyl Ether	29911-28-2	249-951-5		0.912	190.30	230.0	0.015
Glycol Ethers	Dipropylene Glycol Monobutyl Ether	29911-28-2	252-776-7		0.912	190.30	230.0	0.040
Glycol Ethers	Triethylene glycol monomethyl ether	112-35-6	203-962-1		1.026	164.20	250.0	0.008
Glycol Ethers	Tripropylene Glycol Monomethyl Ether	25498-49-1	247-045-4		0.965	206.30	255.4	0.010
Glycol Ethers	Tripropylene Glycol n-Propyl Ether	96077-04-2			0.935	234.33	274.0	0.002
Glycol Ethers	Tripropylene Glycol n-Butyl Ether	55934-93-5	212-222-7		0.934	248.40	276.0	0.010
Glycols	Propylene Glycol	57-55-6	200-338-0		1.035	76.10	188.2	0.150
Glycols	Hexylene Glycol	107-41-5	203-489-0		0.920	118.18	197.0	0.060
Glycols	Ethylene Glycol	107-21-1	203-473-3	1142	1.113	62.07	197.3	0.086
Glycols	Dipropylene Glycol	2396-61-4	246-770-3		1.023	134.07	228.0	0.010
Glycols	1,4-Butane Diol	110-63-4	203-786-5		1.017	90.12	230.0	0.000

(Continued)

Table A1-1 *(Continued)*

Type	Name	CAS #	EINECS #	DOT #	Density g / cc	Mol. Wt.	Boiling Point °C	Vapor Pressure mm Hg @ 20 °C
Glycols	Diethylene Glycol	111-46-6	203-872-2		1.116	106.12	245.8	0.000
Glycols	Triethylene Glycol	112-27-6	203-953-2		1.124	150.17	288.0	0.001
Glymes	Ethylene Glycol Dimethyl Ether (Monoglyme)	110-71-4			0.868	90.12	85.2	54.000
Glymes	Ethylene Glycol Diethyl Ether (Ethyl Glyme)	629-14-1			0.842	118.18	121.0	9.000
Glymes	Diethylene Glycol Dimethyl Ether (Diglyme)	111-96-6	203-924-4		0.945	134.17	162.0	0.000
Glymes	Diethylene Glycol Diethyl Ether (Ethyl Diglyme)	112-36-7			0.908	162.23	189.0	11.999
Glymes	Triethylene Glycol Dimethyl Ether (Triglyme)	112-49-2			0.986	178.22	216.0	0.027
Glymes	Diethylene Glycol Dibutyl Ether (Butyl Diglyme)	112-73-2			0.881	218.34	256.0	0.000
HCFCs	HCFC-141b	1717-00-6			1.230	116.95	32.2	477.200
HCFCs	HCFC-225 ca / cb	422-56-0/507-55-1			1.550	202.94	53.9	290.500
HFCs	HFC-245 fa	460-73-1			1.320	134.05	14.9	915.075
HFCs	HFC-365mfc	405-58-6			1.270	148.07	40.2	356.700
HFCs	HFC 43-10 mee	138495-42-8			1.580	252.00	54.4	176.000
HFEs	HFE-7100	163702-07-6			1.520	250.00	60.0	218.530
HFEs	HFE-7200	163702-06-5			1.430	264.10	76.0	132.290

HFEs	HFE-8200	163702-05-4/ 163702-06-5			1.430	264.00	76.0	109.000
HFEs	HFE-7300	132182-92-4			1.660	350.08	98.0	35.552
Iodateds	Perfluoro n-butyl Iodide	423-39-2			2.010	345.93	67.5	0.300
Ketones	Acetone	67-64-1	200-662-2	1090	0.791	58.08	55.5	185.570
Ketones	Methyl Ethyl Ketone	78-93-3	201-159-0	1193	0.805	72.10	79.6	74.560
Ketones	Diethyl Ketone	96-22-0	202-490-3	1156	0.814	86.13	101.7	26.180
Ketones	Methyl Propyl Ketone	107-87-9	203-528-1	1249	0.809	86.13	102.3	27.830
Ketones	Methyl Isobutyl Ketone	108-10-1	203-550-1	1245	0.801	100.16	117.4	14.840
Ketones	Acetyl Acetone	123-54-6	204-634-0	2310	0.975	100.12	138.3	5.550
Ketones	Dispropyl Ketone	123-19-3		2710	0.800	114.20	145.0	0.840
Ketones	Methyl Isoamyl Ketone	110-12-3	210-844-3	2302	0.810	114.20	145.0	4.500
Ketones	Methyl Amyl Ketone	110-43-0	203-767-1	1110	0.811	114.20	151.5	2.180
Ketones	Cyclohexanone	108-94-1	203-631-1	1915	0.945	98.14	155.7	3.457
Ketones	Diisobutyl Ketone	108-83-8	203-620-1	1157	0.805	142.24	168.0	1.430
Lactates	Methyl Lactate	27871-49-4	547-64-8		1.098	104.00	144.0	2.550
Lactates	Ethyl Lactate	97-64-3	211-694-1		1.033	118.13	154.5	2.025
Lactates	Isopropyl Lactate	63697-0-7	617-51-6		0.986	132.00	160.0	1.275
Lactates	Propyl Lactate	616-09-1	210-464-8		1.000	132.00	170.0	1.650
Lactates	Butyl Lactate	138-22-7	205-316-4	1993	0.984	146.19	187.0	0.300
Lactates	2-Ethylhexyl Lactate	186817-80-1	228-503-2		0.940	202.00	247.0	0.015
Limonene	D-Limonene	5989-27-5	227-813-5		0.841	136.26	176.1	2.100
Limonene	Dipentene (Dl-Limonene)	138-86-3	227-815-6		0.840	136.24	177.8	2.100

(Continued)

Table A1-1 *(Continued)*

Type	Name	CAS #	EINECS #	DOT #	Density g / cc	Mol. Wt.	Boiling Point °C	Vapor Pressure mm Hg @ 20 °C
OSs	OS-10 Hexamethyl-disiloxane	107-46-0	203-492-7		0.764	162.30	100.6	31.320
OSs	OS-20 Octamethyl-trisiloxane	107-51-7	203-497-4		0.820	236.50	152.8	2.560
OSs	OS-30 Decamethyl-trisiloxane	141-62-8	205-491-7		0.854	310.70	193.9	0.270
Paraffins	2-Methyl Pentane	107-83-5	203-523-4	2462	0.653	86.18	16.7	180.000
Paraffins	Pentane	109-66-0	203-692-4	1265	0.626	72.15	36.1	420.780
Paraffins	SBP 1	64742-89-8	265-192-2		0.677	100.00	44.0	170.260
Paraffins	Cyclopentane	287-92-3	206-016-6	1146	0.746	70.13	49.3	318.000
Paraffins	Hexane	110-54-3	203-777-7	1208	0.659	86.18	68.7	121.510
Paraffins	Methyl Cyclopentane	96-37-7	202-503-2	2298	0.749	84.16	71.8	107.780
Paraffins	Cyclohexane	110-82-7	217-077-3	1145	0.779	84.16	80.7	77.260
Paraffins	SBP 80 / 110	64747-49-0	265-151-9		0.710	99.00	87.0	63.760
Paraffins	3-Methyl Hexane	589-34-4	209-643-3	1993	0.687	100.20	91.9	49.500
Paraffins	SBP 5	64742-89-8	265-192-2		0.707	100.00	92.0	45.000
Paraffins	Heptane	142-82-5	205-563-8	1206	0.684	100.20	98.4	35.680
Paraffins	Cypar 7	64742-89-8	265-192-2		0.771	100.00	100.0	ERR
Paraffins	SBP 3	64742-89-8	265-192-2		0.718	100.00	100.0	35.250
Paraffins	Isopar C	90622-56-3	292-458-5		0.695	106.00	100.0	37.062
Paraffins	Methyl Cyclohexane	108-87-2	203-624-3	2296	0.769	98.19	100.9	33.380
Paraffins	SBP 100 / 140	64742-49-0	265-151-9		0.740	112.00	105.0	26.250
Paraffins	2-Methyl Heptane	592-27-8	209-747-9	1262	0.698	114.23	117.6	16.330
Paraffins	Isopar E	90622-56-3	292-458-5		0.720	117.00	118.0	17.690
Paraffins	Ethyl Cyclohexane	1678-91-7	216-835-0	1993	0.788	112.20	131.8	10.950

Paraffins	Octane	111-65-9	203-892-1	1262	0.703	114.23	131.8	10.070
Paraffins	D30	64742-48-9	265-150-3		0.762	133.00	139.0	3.980
Paraffins	DSP 145/160	8002-5-9	232-298-5		0.755	128.00	143.0	2.930
Paraffins	Nonane	111-84-2	203-913-4	1920	0.718	128.26	150.8	3.050
Paraffins	D40	64742-48-9	265-150-3		0.774	143.00	158.0	1.500
Paraffins	Cypar 9	68553-14-0			0.793	128.00	162.0	ERR
Paraffins	Solvesso 100	64742-95-6	265-178-6		0.880	127.00	165.0	3.000
Paraffins	Whitespirit	64742-82-1	265-185-4		0.775	143.00	170.0	2.780
Paraffins	Isododecane	13475-86-6	250-816-8		0.745	170.34	178.0	1.140
Paraffins	SBP 11	64742-88-7	265-191-7		0.770	142.00	179.5	1.500
Paraffins	D60	64742-48-9	265-150-3		0.789	162.00	180.0	0.300
Paraffins	Isopar H	90622-57-4	292-459-0		0.760	156.00	182.0	0.800
Paraffins	Solvesso 150	64742-94-5	265-192-2		0.903	144.00	184.0	0.560
Paraffins	Isopar L	90622-58-5	292-460-6		0.770	164.00	190.0	0.480
Paraffins	D70	64742-47-8	265-149-8		0.796	174.00	195.0	0.450
Paraffins	Isopar M	9622-58-5	265-149-8		0.785	179.00	216.0	0.080
Paraffins	Solvesso 200	64747-94-5	265-192-2		0.999	165.00	244.0	0.050
Specials	Carbon Dioxide	124-38-9			1.162	44.00	-78.4	48300.000
Specials	Tetrahydrofuran	109-99-9	203-726-8	2056	0.889	72.11	65.7	145.000
Specials	1,3 Dioxolane	646-6-0	211-463-5	1166	1.066	74.06	75.6	70.000
Specials	Acetonitrile	75-5-8	200-835-2	1648	0.782	41.05	81.6	72.530
Specials	Dimethyl Carbonate	616-38-6	210-478-4	1161	1.064	90.08	90.0	39.753
Specials	Water	7732-18-5			0.998	18.02	100.0	17.490

(Continued)

Table A1-1 *(Continued)*

Type	Name	CAS #	EINECS #	DOT #	Density g / cc	Mol. Wt.	Boiling Point °C	Vapor Pressure mm Hg @ 20 °C
Specials	1,4-Dioxane	123-91-1	204-661-8	1165	1.034	88.11	101.5	27.000
Specials	4-Bromo-3-Chloro-3,4,4-Trifluoro-1-Butene	374-25-4			1.727	223.40	107.7	184.000
Specials	Pyridine	110-86-1	203-809-9	1282	0.981	79.10	115.2	16.000
Specials	Dimethyl Formamide	68-12-2	200-679-5	2265	0.949	73.10	139.0	2.828
Specials	Cumene	98-82-8	202-704-5		0.862	120.19	152.4	3.500
Specials	Dimethyl Acetamide	127-19-5	204-826-4		0.941	87.12	166.1	1.320
Specials	Dimethyl Sulfoxide	67-68-5	200-664-3		1.101	78.13	189.0	0.420
Specials	N-Methyl Pyrrolidone	872-50-4	212-828-1	1760	1.032	99.00	202.0	0.290
Specials	Gamma Butyrolactone	96-48-0	202-509-5		1.120	86.10	204.0	0.360
Specials	Isophorone	78-59-1	201-126-0	1224	0.921	138.20	210.5	0.310
Specials	Propylene Carbonate	108-32-7	203-572-1		1.205	102.09	242.0	0.030
Specials	Ethylene Carbonate	96-49-1	202-510-0		1.470	88.10	243.0	0.010
Specials	Soygold 1000	67784-80-9	206-102-3		0.882	248.90	333.3	0.040

Type	Name	Autoignition Point °C	Antoine Coefficients	Molecular Formula
Acids	Acetic Acid	801	A = −2105.5, B = 8.2692	CH3COOH
Alcohols	Methanol	470	A = −1951.6, B = 8.6522	CH3OH
Alcohols	Ethanol	365	A = −2146.5, B = 8.9776	CH3CH2OH
Alcohols	Isopropanol	399	A = −2260.1, B = 9.2297	CH3CH(OH)CH3
Alcohols	Tertiary Butanol	478	A = −2524.5, B = 9.9624	(CH3)3COH
Alcohols	Secondary Butanol	405	A = −2432.1, B = 9.4233	CH3CHOHCH2CH3

Alcohols	N-Propanol	413	A = −2307.6, B = 9.1053	CH3CHCH3OH
Alcohols	Isobutanol	427	A = −2543.0, B = 9.5599	CH3CH(CH3)CH2OH
Alcohols	N-Butanol	343	A = −2582.7, B = 9.5003	CH3CH2CH2CH2OH
Alcohols	iso-Amyl Alcohol	350	A = −2731.3, B = 9.6897	(CH3)2CHCH2CH2OH
Alcohols	Methyl Isobutyl Carbinol		A = −2575.8, B = 9.2789	(CH3)2CHCH2CH(OH)CH3
Alcohols	Amyl Alcohol	350	A = −2684.6, B = 9.4473	CH3CH2CH2CH2CH2OH
Alcohols	1-Hexanol	545	A = −2808.9, B = 9.4245	CH2CH2CH2CH2CH2CH2OH
Alcohols	Cyclohexanol	572	A = −2064.2, B = 7.5900	(CH2)6(CH3)
Alcohols	Furfuryl Alcohol	1,680	A = −2702.0, B = 8.9450	O(CH)3C(CH2OH)
Alcohols	Diacetone Alcohol	603	A = −2442.4, B = 8.3952	CH 3(C=O)CH2C(CH 3)2OH,
Alcohols	Tetrahydrofurfuryl Alcohol	1,003	A = −2579.4, B = 8.7203	O(CH2)3CH(CH2OH)
Alcohols	Benzyl Alcohol	817	A = −3041.7, B = 9.2716	C6H6OH
Alcohols	Texanol	120	A = −3319.9, B = 9.2923	C12H24O3
Aromatics	Benzene	928	A = −1704.0, B = 7.6974	C6H6
Aromatics	Lactol Spirits	250	A = −1775.0, B = 7.6566	~C9H18
Aromatics	Toluene	480	A = −1904.0, B = 7.8445	C7H5CH3
Aromatics	VM&P Naptha	550	A = −2632.9, B = 10.0325	~C7H16
Aromatics	o-Xylene	500	A = −2133.3, B = 8.0786	C6H4(CH3)2
Aromatics	Turpentine, oil	123	A = −2144.7, B = 7.8996	~C10H16
Aromatics	Stoddard Solvent	540	A = −1988.7, B = 7.4825	~C10H20
Aromatics	Mineral Spirits	658	A = −2316.3, B = 7.7761	~C10H21
Benzotrifluorides	Benzotrifluoride	1,202	A = −2026.9, B = 8.3073	C6H5(CF3)
Benzotrifluorides	Parachlorobenzotrifluoride	260	A = −2223.9, B = 8.3327	C6H5(CF3)(CL)
Brominateds	n-Propyl Bromide	254	A = −1619.6, B = 7.5707	CH3CH2CH2Br
CFCs	CFC-113		A = −1362.0, B = 7.2331	CCl2FCClF2
Chlorinateds	Methylene Chloride	605	A = −1703.6, B = 8.3141	CH2Cl2
Chlorinateds	Trans 1,2 Dichloroethylene	405	A = −1345.6, B = 7.0985	ClCH=CHCl

(Continued)

Table A1-1 *(Continued)*

Type	Name	Autoignition Point °C	Antoine Coefficients	Molecular Formula
Chlorinateds	1,1-Dichloroethane	856	A = −1548.9, B = 7.5246	CH3CHCl2 (C2H4Cl2)
Chlorinateds	Chloroform	624	A = −1731.3, B = 8.0916	CHCl3
Chlorinateds	1,1,1-Trichloroethane	999	A = −1643.8, B = 7.6091	CH3CCl3
Chlorinateds	Carbon Tetrachloride		A = −1649.2, B = 7.5905	CCl4
Chlorinateds	Ethylene Dichloride	775	A = −1751.6, B = 7.7853	ClCH=CHCl
Chlorinateds	Trichloroethylene		A = −1752.7, B = 7.7498	ClCH=CCl2
Chlorinateds	Perchloroethylene	45	A = −1969.3, B = 7.8811	Cl2C=CCl2
Dibasic Acid Esters	DBE-4	689	A = −5111.4, B = 16.9126	CH3OOC(CH2)2COOCH3
Dibasic Acid Esters	DBE	696	A = −2797.0, B = 8.8426	CH3OOC(CH2)n−COOCH3, n = 2,3 and 4
Dibasic Acid Esters	DBE-6	1,256	A = −2960.6, B = 8.8002	CH3OOC(CH2)4COOCH3
Esters	Methyl Formate	824	A = −1641.0, B = 8.2919	CH3OCHO
Esters	Ethyl Formate	455	A = −1598.9, B = 7.7535	CH3CH2OCHO
Esters	Methyl Acetate	455	A = −1565.9, B = 7.6320	CH3COOCH3
Esters	Ethyl Acetate	425	A = −1815.1, B = 8.0617	CH3COOCH2CH3
Esters	Isopropyl Acetate	460	A = −1920.7, B = 8.2099	CH3COOCH(CH3)2
Esters	Tertiary Butyl Acetate		A = −1739.9, B = 7.5389	CH3COOC(CH3)3
Esters	n-Propyl Acetate	450	A = −1971.1, B = 8.1398	CH3COOCH2CH2CH3
Esters	Isobutyl Acetate	423	A = −2018.7, B = 8.0585	CH3CH(CH3)CH2O(C=O)CH3
Esters	n-Propyl Propionate		A = −2123.4, B = 8.2627	CH3CH2C(O)OCH2CH2CH3
Esters	n-Propyl Propionate		A = −2094.24, B = 8.157036	CH3CH2(C=O)OCH2CH2CH3
Esters	n-Butyl Acetate	370	A = −2109.8, B = 8.1813	CH3CH2CH2CH2O(C=O)CH3
Esters	N-Butyl Propionate	427	A = −2356.3, B = 8.5650	CH3CH2C(=O) OCH2CH2CH2CH3

Esters	Amyl Acetate	182	A = −2222.9, B = 9.1862	CH3COOCH2CH2 CH2CH2CH3
Esters	Amyl Acetate	182	A = −2222.9, B = 9.1862	CH3CH2CH2CH2CH2 OC(=O)CH3
Esters	Ethyl-3-Ethoxypropionate		A = −2632.6, B = 8.8251	CH3–CH2–O–CH2–CH2–C(=O)–O–CH2–CH3
Esters	Cyclohexyl Acetate		A = −2206.0, B = 8.1192	C6H11O(C=O)CH3
Esters	Ethylene Glycol Diacetate	482	A = −2909.9, B = 9.2360	CH3COOCH2CH2OCOCH3
Esters	Propylene Glycol Diacetate	1,486	A = −2928.7, B = 9.1889	CH3CH(=O)0CH2 (CH3)O(=O)CHCH3
Esters	Butyl Glycol Acetate		A = −2793.4, B = 8.9670	C4H9O(CH2)2O(C=O)CH3
Esters	2-Ethyl Hexyl Acetate	958	A = −2513.2, B = 8.3326	CH3CH2CH2CH2CH(CH2CH3) CH2OCH(CH3)(=O)
Esters	Dimethyl Succinate	689	A = −2719.2, B = 8.6277	CH3OOCCH2CH2COOCH3
Ethers	Methylal	455	A = −1495.6, B = 6.7429	CH3OCH2OCH3
Ethers	Methyl t-Butyl Ether	ERR	A = −1549.1, B = 7.6081	CH3CH2OCH2CH3
Ethers	Methyl Butyl Ether	ERR	A = −1699.7, B = 7.7824	CH3CH2CH2OCH2CH2CH3
Ethers	t-Butyl Ether	689	A = −1813.4, B = 7.6474	C6H14O/(CH3)2 CHOCH(CH3)2
Ethers	Cyclopentyl Methyl Ether	356	A = −1870.7, B = 7.8276	C2F5CF(OCH3)CF(CF3)2
Ethers	Amyl t-Butyl Ether	ERR	A = −2168.0, B = 8.0575	CH3CH2CH2CH2 OCH2CH2CH2CH3
Glycol Ethers	Dipropylene Glycol Dimethyl Ether	329	A = −2662.1, B = 8.8208	CH3OCH2CH(CH3) OCH2CH(CH3)oCH3
Glycol Ethers	Dipropylene Glycol n-Propyl Ether		A = −3096.2, B = 9.2626	(CH3CH2CH2O) C3H6OC3H6(OH)
Glycol Ether Acetates	Ethylene glycol methyl ether acetate	380	A = −2425.9, B = 8.7181	CH3(C=O)OCH2CH2OCH3
Glycol Ether Acetates	Propylene Glycol Methyl Ether Acetate		A = −2385.0, B = 8.6061	CH3OCH2CH2CH2O(C=O) CH3

(*Continued*)

Table A1-1 *(Continued)*

Type	Name	Autoignition Point °C	Antoine Coefficients	Molecular Formula
Glycol Ether Acetates	Ethylene Glycol Ethyl Ether Acetate	714	A = −2625.3, B = 9.0794	CH3COOCH2CH2OCH2CH3
Glycol Ether Acetates	Propylene Glycol Ethyl Ether Acetate	325	A = −2415.1, B = 8.4692	CH3CH2OCH2CH2C H2O(C=O)CH3
Glycol Ether Acetates	Ethylene Glycol Butyl Ether Acetate		A = −2760.0, B = 8.8918	CH3(CH2)3OC2H4OOCCH3
Glycol Ether Acetates	Dipropylene Glycol Methyl Ether Acetate		A = −3220.9, B = 9.6881	CH3CH(OCH3) CH2OCH2CH(CH3)O(C=O) CH3
Glycol Ether Acetates	Diethylene Glycol Monoethyl Ether Acetate		A = −3076.9, B = 9.1968	CH3CH2(OC2H4)2OOCCH3
Glycol Ether Acetates	Diethylene Glycol Monobutyl Ether Acetate		A = −3500.0, B = 9.8890	CH3(CH2)3(OC2H4)2OOCCH3
Glycol Ethers	Ethylene Glycol Diethyl Ether	406	A = −1968.7, B = 7.9393	CH3CH2OCH2CH2OCH2CH3
Glycol Ethers	Propylene Glycol Monomethyl Ether	270	A = −2226.6, B = 8.5497	CH3OCH2CH(OH)CH3
Glycol Ethers	Propylene Glycol Monomethyl Ether	270	A = −2104.8, B = 8.2554	CH3OCH2CH2CH2OH
Glycol Ethers	Ethylene Glycol Monomethyl Ether	285	A = −2292.1, B = 8.6573	CH3OCH2CH2OH
Glycol Ethers	Propylene Glycol Monoethyl Ether	491	A = −2179.7, B = 8.2904	CH3CH(OH)CH2OCH2CH3
Glycol Ethers	Propylene Glycol Monoethyl Ether	255	A = −2113.5, B = 8.0843	CH3CH2OCH2CH2CH2OH
Glycol Ethers	Ethylene Glycol Monoethyl Ether	235	A = −2392.4, B = 8.7496	CH3OCH2CH2O(C=O)CH3
Glycol Ethers	Ethylene Glycol Mono Isopropyl Ether		A = −2291.2, B = 8.3929	(CH3)2CHOCH2CH2OH

Glycol Ethers	Propylene Glycol Monopropyl Ether	486	A = −2594.8, B = 9.0274	CH3CH2CH2O CH2CH2CH2OH
Glycol Ethers	Propylene Glycol n-Propyl Ether	486	A = −2650.1, B = 9.1434	CH3(CH2)2OCH2CHOHCH3
Glycol Ethers	Propylene Glycol Mono t-butyl Ether		A = −2813.4, B = 9.5137	CH3OCH2CH(OH)C(CH3)3
Glycol Ethers	Propylene Glycol Methyl Propyl Ether	270	A = −2723.5, B = 8.8608	CH3OC3H6OC3H6OH
Glycol Ethers	Propylene Glycol Ethyl Propyl Ether	190	A = −3467.4, B = 10.8490	CH3CH2OC3H6OC3H6OH
Glycol Ethers	Propylene Glycol Monobutyl Ether		A = −2564.9, B = 8.6657	CH3CH2CH2CH20 CH2CH(OH)CH3
Glycol Ethers	Ethylene Glycol Monobutyl Ether	446	A = −2567.8, B = 8.5821	CH3CH2CH2CH2OCH2 CH2OH (C6H14O2)
Glycol Ethers	Butyl Glycol Ether	245	A = −2675.6, B = 8.9560	C4H9OCH2CH2OH
Glycol Ethers	3-Methoxy-3-Methyl-1-Butanol	1,369	A = −2651.4, B = 8.8200	CH3C(CH3)(OCH3)CH2OH
Glycol Ethers	Dipropylene Glycol Monomethyl Ether		A = −2742.8, B = 8.8027	CH3OCH2CH(CH3) OCH2CH(CH3)OH
Glycol Ethers	Diethylene Glycol Monomethyl Ether	215	A = −2916.1, B = 9.1721	CH3OCH2CH2OCH2CH2OH
Glycol Ethers	Dipropylene Glycol Monoethyl Ether		A = −2680.0, B = 8.6191	CH3CH(OH) CH2OCH2CH(CH3)OCH2CH3
Glycol Ethers	Diethylene Glycol Monomethyl Ether	215	A = −3009.2, B = 9.3353	HOCH2CH2OCH2CH2OCH3
Glycol Ethers	Diethylene Glycol Monobutyl ether		A = −3335.8, B = 9.6927	CH3(CH2) 3OCH2CH2OCH2CH2OH
Glycol Ethers	Dipropylene Glycol Monobutyl Ether		A = −3223.0, B = 9.3006	CH3(CH2)3(OCH2CHCH3)2OH
Glycol Ethers	Dipropylene Glycol n-Butyl Ether	381	A = −3313.8, B = 9.4800	CH3(CH2)3(OCH2CHCH3)2OH
Glycol Ethers	Dipropylene Glycol Monobutyl Ether	381	A = −3015.7, B = 8.8863	CH3CH2CH2CH2OCH(CH3) CH2OCH2CH2CH2OH
Glycol Ethers	Triethylene glycol monomethyl ether	96	A = −3329.4, B = 9.2448	CH3OCH2CH2OCH2 CH2OCH2CH2OH

(Continued)

Table A1-1 *(Continued)*

Type	Name	Autoignition Point °C	Antoine Coefficients	Molecular Formula
Glycol Ethers	Tripropylene Glycol Monomethyl Ether		A = −3219.7, B = 8.9717	OHCH2CH(CH3) OCH2CH(CH3)OCH2CH(CH3)
Glycol Ethers	Tripropylene Glycol n-Propyl Ether	396	A = −3491.4, B = 9.2618	OHCH2CH2(CH3) OCH2CH2(CH3) OCH2CH2(CH3)OH
Glycol Ethers	Tripropylene Glycol n-Butyl Ether	367	A = −3066.7, B = 8.6266	C4H9—O—C3H6—O—C3H6—O—CH2—CHOH—CH3
Glycols	Propylene Glycol	371	A = −3056.1, B = 9.5412	OHCH3CHOHCH3
Glycols	Hexylene Glycol		A = −3204.4, B = 9.7323	(CH3)2COHCH2CHOHCH3
Glycols	Ethylene Glycol	432	A = −3188.5, B = 9.7135	HOCH2CH2OH
Glycols	Dipropylene Glycol	590	A = −3455.3, B = 9.7754	CH3CH(OH) CH2OCH2CH(CH3)OH
Glycols	1,4-Butane Diol	357	A = −3732.9, B = 10.4561	OHCH2CH2CH2CH2 OH(C4H10O2)
Glycols	Diethylene Glycol		A = −3522.3, B = 9.6879	OHCH2CH2OCH2CH2OH
Glycols	Triethylene Glycol	371	A = −3662.4, B = 9.3787	OHCH2CH2OCH2 CHC2OCH2CH2OH
Glymes	Ethylene Glycol Dimethyl Ether (Monoglyme)	396	A = −1802.6, B = 7.9050	CH3OCH2CH2OCH3
Glymes	Ethylene Glycol Diethyl Ether (Ethyl Glyme)	399	A = −2272.4, B = 8.6390	CH3CH3OCH2CH2OCH2CH3
Glymes	Diethylene Glycol Dimethyl Ether (Diglyme)	374	A = −2371.5, B = 8.2853	CH3OCH2CH2 OCH2CH2OCH3
Glymes	Diethylene Glycol Diethyl Ether (Ethyl Diglyme)	383	A = −2765.6, B = 8.8448	CH3CH2OCH2CH2 OCH2CH2OCH2CH3
Glymes	Triethylene Glycol Dimethyl Ether (Triglyme)	374	A = −3360.7, B = 9.7515	CH3OCH2CH2OCH2 CH2OOCH2CH2OCH3

Glymes	Diethylene Glycol Dibutyl Ether (Butyl Diglyme)	154	A = −2941.0, B = 8.4411	CH3CH2CH2CH2 OCH2CH2OCH2CH2 OCH2CH2CH2CH
HCFCs	HCFC-141b		A = −1473.1, B = 7.7038	C2H3Cl2F
HCFCs	HCFC-225 ca / cb		A = −1446.1, B = 7.2983	CF3CF2CHCl2 / CClF2CF2CHClF
HFCs	HFC-245 fa	412	A = −1647.0, B = 8.5623	CF3CH2CHF2
HFCs	HFC-365mfc	405	A = −1345.6, B = 7.0985	CF3CH2CF2CH3
HFCs	HFC 43-10 mee	300	A = −1671.9, B = 7.9696	CF3(CHF)2CFCF3
HFEs	HFE-7100	761	A = −1581.7, B = 7.6098	C(CF3)3OCH3
HFEs	HFE-7200	707	A = −1629.5, B = 7.5551	C(CF3)3OCH2CH3
HFEs	HFE-8200	676	A = −1541.6, B = 7.2960	(CF3)2CFCF2OCH3
HFEs	HFE-7300	ERR	A = −1792.1, B = 7.7165	C2F5CF(OCH3)CF(CF3)2
Iodateds	Perfluoro n-butyl Iodide		A = −1452.2, B = 7.1590	CF2CF2CF2CF2I
Ketones	Acetone	465	A = −1598.3, B = 7.7295	CH3C=OCH3
Ketones	Methyl Ethyl Ketone	404	A = −1738.7, B = 7.8060	CH3C(=O)CH2CH3
Ketones	Diethyl Ketone	452	A = −1939.6, B = 8.0527	CH3CH2C(=O)CH2CH3
Ketones	Methyl Propyl Ketone	452	A = −1902.9, B = 7.9494	CH3CH2CH2C(=O)CH3
Ketones	Methyl Isobutyl Ketone	454	A = −2008.5, B = 8.0402	C6CH3COCH2CH(CH3)2
Ketones	Acetyl Acetone	340	A = −2188.1, B = 8.2435	CH3(C=O)CH2(C=O)CH3
Ketones	Dispropyl Ketone	806	A = −2912.1, B = 9.8687	(CH3CH2CH2)2CO
Ketones	Methyl Isoamyl Ketone		A = −2226.1, B = 8.2217	CH3COCH2CH2CH(CH3)2
Ketones	Methyl Amyl Ketone	393	A = −2384.4, B = 8.5462	CH3(CH2)3CH2C=0CH3
Ketones	Cyclohexanone	420	A = −2206.8, B = 8.0460	C6H10(=O)
Ketones	Diisobutyl Ketone	396	A = −2425.5, B = 8.4141	(CH3)2CHCH2]2CO
Lactates	Methyl Lactate	385	A = −2491.9, B = 8.9396	CH3CH(OH)COOCH3
Lactates	Ethyl Lactate	400	A = −2410.3, B = 8.5383	CH3CH(OH)COOCH2CH3
Lactates	Isopropyl Lactate	380	A = −2537.3, B = 8.7831	CH3CH(OH)COOCH(CH3)2

(Continued)

Table A1-1 *(Continued)*

Type	Name	Autoignition Point °C	Antoine Coefficients	Molecular Formula
Lactates	Propyl Lactate		A = −2301.6, B = 8.0606	CH3CH(OH)COO(CH2)2CH3
Lactates	Butyl Lactate		A = −2757.8, B = 8.9075	CH3CH(OH)COO(CH2)3CH3
Lactates	2-Ethylhexyl Lactate		A = −3159.1, B = 9.0085	CH3CH(OH)COO(CH2) 4CH(CH2CH3)CH3
Limonene	D-Limonene	257	A = −2252.9, B = 7.8903	CH3-C6H8-C(CH3)=CH2
Limonene	Dipentene (Dl-Limonene)	237	A = −2143.7, B = 7.6348	CH3-C6H8-C(CH3)=CH2
OSs	OS-10 Hexamethyldisiloxane	341	A = −1885.7, B = 7.9280	(CH3)3SiOSi(CH3)3
OSs	OS-20 Octamethyltrisiloxane	350	A = −2331.5, B = 8.3642	(CH3)3SiOSi(CH3)2OSi(CH3)3
OSs	OS-30 Decamethyltrisiloxane	350	A = −2713.5, B = 8.7010	(CH3)3SiOSi(CH3)2OSi(CH3) 2OSi(CH3)3
Paraffins	2-Methyl Pentane	583	A = −1677.6, B = 7.9460	CH3CH(CH3)CH2CH2CH3
Paraffins	Pentane	260	A = −1552.3, B = 7.9080	CH3CH2CH2CH2CH3
Paraffins	SBP 1		A = −2517.0, B = 10.8168	~C7H16
Paraffins	Cyclopentane	682	A = −1416.2, B = 7.2858	C5H10
Paraffins	Hexane	223	A = −1756.9, B = 8.0622	CH3CH2CH2CH2CH2CH3
Paraffins	Methyl Cyclopentane	329	A = −1724.2, B = 7.9097	C6H12
Paraffins	Cyclohexane	260	A = −1669.4, B = 7.5914	C6H12
Paraffins	SBP 80 / 110	250	A = −1696.1, B = 7.5902	~C7H15
Paraffins	3-Methyl Hexane	280	A = −1757.6, B = 7.6925	CH3CH2CH(CH3)CH2CH2CH3
Paraffins	SBP 5		A = −1825.2, B = 7.8791	~C7H16
Paraffins	Heptane	204	A = −1843.9, B = 7.8472	CH3(CH2)5CH3
Paraffins	Cypar 7			0 ~C7H16
Paraffins	SBP 3		A = −1823.7, B = 7.7679	~C7H16
Paraffins	Isopar C		A = −1841.8, B = 7.8507	~C8H10
Paraffins	Methyl Cyclohexane	285	A = −1816.7, B = 7.7451	C6H12CH3

Paraffins	SBP 100 / 140	220	A = −1906.4, B = 7.9220	~C8H16
Paraffins	2-Methyl Heptane	420	A = −1963.5, B = 7.9111	CH3CH(CH3) CH2CH2CH2CH2CH3
Paraffins	Isopar E		A = −1965.3, B = 7.9555	~C9H9
Paraffins	Ethyl Cyclohexane	262	A = −1982.4, B = 7.7834	C6H11CH2CH3
Paraffins	Octane	206	A = −2127.5, B = 8.2602	CH3(CH2)6CH3
Paraffins	D30		A = −2316.5, B = 8.5012	C7H14O2
Paraffins	DSP 145/160		A = −2395.0, B = 8.6359	~C9H20
Paraffins	Nonane	205	A = −2282.0, B = 8.2949	CH3(CH2)7CH3
Paraffins	D40	240	A = −2477.3, B = 8.6265	C7H14O2
Paraffins	Cypar 9			0 ~C9H20
Paraffins	Solvesso 100		A = −2129.3, B = 7.7405	~C9H19
Paraffins	Whitespirit	240	A = −5702.4, B = 19.8947	~C11H11
Paraffins	Isododecane	410	A = −2801.5, B = 9.0138	C12H26
Paraffins	SBP 11	464	A = −2250.8, B = 7.8537	~C10H22
Paraffins	D60	217	A = −2826.1, B = 9.1172	C7H14O2
Paraffins	Isopar H		A = −2564.2, B = 8.6498	~C11H24
Paraffins	Solvesso 150		A = −2563.1, B = 8.4875	~C11H12
Paraffins	Isopar L		A = −2578.1, B = 8.4898	~C12H20
Paraffins	D70		A = −2650.1, B = 8.5416	C7H14O2
Paraffins	Isopar M		A = −2900.5, B = 8.8103	~C13H23
Paraffins	Solvesso 200		A = −2816.0, B = 8.3260	~C12H21
Specials	Carbon Dioxide		A = 7.5859, B= -865.2	O=C=O
Specials	Tetrahydrofuran	321	A = −1600.1, B = 7.5985	C4H8O
Specials	1,3 Dioxolane	525	A = −1815.2, B = 8.0939	C3H6O2 (C3H6O2)
Specials	Acetonitrile	524	A = −1732.4, B = 7.7646	CH3C==N
Specials	Dimethyl Carbonate	856	A = −2199.6, B = 8.9550	CH3O(C=O)OCH3
Specials	Water		A = −2225.3, B = 8.8423	HOH

(Continued)

Table A1-1 *(Continued)*

Type	Name	Autoignition Point °C	Antoine Coefficients	Molecular Formula
Specials	1,4-Dioxane	356	A = −1929.2, B = 8.0317	OHCH=CHOH
Specials	4-Bromo-3-Chloro-3,4,4-Trifluoro-1-Butene		A = −1354.0, B = 6.4291	CH3CON(CH3)2
Specials	Pyridine	900	A = −2002.4, B = 8.0405	C5H5N
Specials	Dimethyl Formamide		A = −2466.7, B = 8.8180	HCON(CH3)2
Specials	Cumene	795	A = −2217.5, B = 8.1170	C6H6(CH(CH3)2)
Specials	Dimethyl Acetamide		A = −2711.2, B = 9.1234	CH3CON(CH3)2
Specials	Dimethyl Sulfoxide	572	A = −2614.9, B = 8.5797	C2H6OS
Specials	N-Methyl Pyrrolidone		A = −2622.3, B = 8.4070	CH2CH2CH2CH(=O)N(CH3)
Specials	Gamma Butyrolactone		A = −2532.1, B = 8.2002	C4H6O2
Specials	Isophorone	462	A = −2516.9, B = 8.0904	C6H5(=O)(CH3)3
Specials	Propylene Carbonate	815	A = −2903.6, B = 8.5726	(CH3)CHCH2OOC(=O)
Specials	Ethylene Carbonate	869	A = −3319.3, B = 9.3116	CHCH2OOC(=O)
Specials	Soygold 1000	637	A = −2429.5, B = 6.8866	CH3CH2CH=CHCH2 CH=CHCH2CH=CHCH2(

Table A1-2

Type	Name	CAS #	Surface Tension dynes / cm	Viscosity cps	Heat of Vaporization BTU / lb	Heat of Combustion BTU / lb	Dielectric Constant	Dipole Moment
Acids	Acetic Acid	64-19-7	28.8	1.060	319.0	22,538.1	6.10	1.500
Alcohols	1-Hexanol	111-27-3	25.8	4.578	579.8		13.03	1.65 @ 20 °C
Alcohols	Amyl Alcohol	71-41-0	25.4	3.700	671.8	16,246.8	15.13	
Alcohols	Benzyl Alcohol	100-51-6	34.1	8.000	276.2	14,858.1	5.47	
Alcohols	Cyclohexanol	108-93-0	32.9	32.400	1,016.8		16.40	
Alcohols	Diacetone Alcohol	123-42-2	31.0	2.798	857.2		18.20	2.860
Alcohols	Ethanol	64-17-5	22.0	1.263	976.6	12,755.8	25.30	1.660
Alcohols	Furfuryl Alcohol	98-0-0	38.0	4.400	903.2		16.85	
Alcohols	iso-Amyl Alcohol	123-51-3	23.7	4.300	689.3		15.63	
Alcohols	Isobutanol	78-83-1	23.0	4.113	749.5	15,477.2	17.93	
Alcohols	Isopropanol	67-63-0	20.9	2.457	805.2	14,350.8	20.18	2.660
Alcohols	Methanol	67-56-1	22.1	0.590	1,161.6	9,742.0	33.00	2.870
Alcohols	Methyl Isobutyl Carbinol	108-11-2	22.6	5.100	179.9	16,693.9	4.07	
Alcohols	N-Butanol	71-36-3	24.9	3.006	772.5	15,520.2	17.84	1.750
Alcohols	N-Propanol	71-23-8	23.3	2.250	859.0	14,458.3	20.80	3.090
Alcohols	Secondary Butanol	78-92-2	22.5	3.100	704.2		17.26	
Alcohols	Tertiary Butanol	75-65-0	20.0		487.6		12.47	
Alcohols	Tetrahydrofurfuryl Alcohol	97-99-4	37.0	4.010	609.3		13.48	
Alcohols	Texanol	25265-77-4	28.9	13.500				
Aromatics	Benzene	71-43-2	28.2	0.604	70.0		2.28	
Aromatics	Lactol Spirits	64742-49-0		3.700				

(Continued)

Table A1-2 *(Continued)*

Type	Name	CAS #	Surface Tension dynes / cm	Viscosity cps	Heat of Vaporization BTU / lb	Heat of Combustion BTU / lb	Dielectric Constant	Dipole Moment
Aromatics	Mineral Spirits	64741-41-9		0.740				
Aromatics	o-Xylene	95-47-6	28.5	0.700		18,469.5	2.36	
Aromatics	Stoddard Solvent	8052-41-3		0.740				
Aromatics	Toluene	108-88-3	27.9	0.587	79.0	18,245.9	2.38	
Aromatics	Turpentine, oil	8006-64-2	29.0	1.375				
Aromatics	VM&P Naptha	64742-89-8		0.970				
Benzotrifluorides	Benzotrifluoride	98-08-8	23.4	0.574	300.6		9.22	
Benzotrifluorides	Parachloroben-zotrifluoride	98-56-6	25.0	0.790		7,738.6		
Brominateds	n-Propyl Bromide	106-94-5	25.3	0.490	241.6		8.09	2.020
CFCs	CFC-113	76-13-1	17.0	0.650	68.6		2.60	
Chlorinateds	1,1,1-Trichloroethane	71-55-6	25.2	0.868	236.3	3,572.7	7.27	
Chlorinateds	1,1-Dichloroethane	75-34-3	24.8	0.490	287.2		10.10	
Chlorinateds	Carbon Tetrachloride	56-23-5	26.4	1.049	66.8	11,835.8	2.24	
Chlorinateds	Chloroform	67-66-3	26.7	0.581	140.5		4.81	1.105
Chlorinateds	Ethylene Dichloride	107-06-2	31.9	0.779	333.4		10.42	
Chlorinateds	Methylene Chloride	75-09-2	27.2	0.437	250.9	2,824.6	8.93	1.140
Chlorinateds	Perchloroethylene	127-18-4	31.3	0.903	78.8		2.27	
Chlorinateds	Trans 1,2 Dichloroethylene	156-60-5	25.0	0.317	61.8		2.14	
Chlorinateds	Trichloroethylene	79-01-6	28.8	0.592	106.4		3.39	
Dibasic Acid Esters	DBE	1119-40-0	35.2	2.400		8,981.1		
Dibasic Acid Esters	DBE-4	106-65-0	35.2	2.500				

Dibasic Acid Esters	DBE-6	627-93-0	35.6	3.290				
Esters	2-Ethyl Hexyl Acetate	103-9-3		1.500				
Esters	Amyl Acetate	628-63-7	25.2	0.867	229.9	13,877.9	4.79	
Esters	Amyl Acetate	628-63-7	25.2	0.862	229.9	13,877.9	4.79	
Esters	Butyl Glycol Acetate	112-07-2	26.0	1.808		12,115.2		
Esters	Cyclohexyl Acetate	622-45-7	31.2	2.270				
Esters	Dimethyl Succinate	106-65-0	34.1	2.820		8,666.5		
Esters	Ethyl Acetate	141-78-6	23.4	0.459	194.0	10,920.0	6.08	
Esters	Ethyl Formate	109-94-4	23.2	0.412	256.2		8.57	
Esters	Ethyl-3-Ethoxypropionate	763-69-9	28.1	1.300				
Esters	Ethylene Glycol Diacetate	111-55-7	20.0	2.900	350.4		7.70	
Esters	Isobutyl Acetate	110-19-0	23.1	0.709	182.0	13,104.0	5.07	
Esters	Isopropyl Acetate	108-21-4	21.8	0.525		12,119.5		
Esters	Methyl Acetate	79-20-9	24.7	0.385	214.2	9,239.0	7.07	1.74 @ 40 °C
Esters	Methyl Formate	107-31-3	24.6	0.351		7,225.0		
Esters	n-Butyl Acetate	123-86-4	24.9	0.725	184.0	13,125.5	5.07	
Esters	N-Butyl Propionate	590-01-2	25.9	0.780		12,876.2		
Esters	n-Propyl Acetate	109-60-4	23.8	0.585	190.5	12,184.0	5.62	1.780
Esters	n-Propyl Propionate	106-36-5	24.7	0.700				
Esters	n-Propyl Propionate	106-36-5	24.7	0.700				
Esters	Propylene Glycol Diacetate	623-84-7	32.6	2.590				
Esters	Tertiary Butyl Acetate	540-88-5	22.4	1.200	74.4		1.94	
Ethers	Cyclopentyl Methyl Ether	5614-37-9	25.2	0.550	138.2	37,607.3	4.76	1.270

(Continued)

Table A1-2 *(Continued)*

Type	Name	CAS #	Surface Tension dynes / cm	Viscosity cps	Heat of Vaporization BTU / lb	Heat of Combustion BTU / lb	Dielectric Constant	Dipole Moment
Ethers	Methyl Butyl Ether	628-28-4	20.8	1.900				
Ethers	Methyl t-Butyl Ether	1634-04-4	19.4	0.340		16,392.2		1.230
Ethers	Methylal	109-87-5	21.1	0.330		10,937.3		
Ethers	t-Butyl Ether	6163-66-2	21.6			17,573.3		
Ethers	Amyl t-Butyl Ether	10100-95-5	23.7	5.100				
Glycol Ethers	Dipropylene Glycol Dimethyl Ether	111109-77-4	26.3	1.120		12,424.8		
Glycol Ethers	Dipropylene Glycol n-Propyl Ether	29911-27-1	26.0	4.400		12,768.7		
Glycol Ether Acetates	Diethylene Glycol Monobutyl Ether Acetate	124-17-4	30.0	3.200				
Glycol Ether Acetates	Diethylene Glycol Monoethyl Ether Acetate	112-15-2	31.7	2.800				
Glycol Ether Acetates	Dipropylene Glycol Methyl Ether Acetate	88917-22-0	28.3	2.100				
Glycol Ether Acetates	Ethylene Glycol Butyl Ether Acetate	112-07-2	30.3	1.800				
Glycol Ether Acetates	Ethylene Glycol Ethyl Ether Acetate	111-15-9	31.8	1.020	302.5	10,699.8	7.57	
Glycol Ether Acetates	Ethylene glycol methyl ether acetate	110-49-6		1.100	362.2	9,707.7	8.25	
Glycol Ether Acetates	Propylene Glycol Ethyl Ether Acetate	54839-24-6		1.328				
Glycol Ether Acetates	Propylene Glycol Methyl Ether Acetate	108-65-6	28.2	1.121		10,404.1		

Glycol Ethers	3-Methoxy-3-Methyl-1-Butanol	56539-66-3	29.9	7.350				
Glycol Ethers	Butyl Glycol Ether	111-76-2	26.1	3.282	404.5	13,129.8	9.30	
Glycol Ethers	Diethylene Glycol Monobutyl ether	112-34-5	30.0	4.900				
Glycol Ethers	Diethylene Glycol Monomethyl Ether	111-77-3	28.5	3.480				
Glycol Ethers	Diethylene Glycol Monomethyl Ether	111-77-3	29.5	3.450				
Glycol Ethers	Dipropylene Glycol Monobutyl Ether	29911-28-2	28.4	5.720				
Glycol Ethers	Dipropylene Glycol Monobutyl Ether	29911-28-2	28.8	4.400				
Glycol Ethers	Dipropylene Glycol Monoethyl Ether	30025-38-8						
Glycol Ethers	Dipropylene Glycol Monomethyl Ether	112-28-7	29.0	3.400				
Glycol Ethers	Dipropylene Glycol n-Butyl Ether	29911-28-2	28.8	4.400				
Glycol Ethers	Ethylene Glycol Diethyl Ether	629-14-1	26.0	0.700	141.6		3.90	
Glycol Ethers	Ethylene Glycol Mono Isopropyl Ether	109-59-1		1.600		12,360.3		
Glycol Ethers	Ethylene Glycol Monobutyl Ether	111-76-2	26.1	2.900	438.0	15,144.4	9.30	
Glycol Ethers	Ethylene Glycol Monoethyl Ether	110-80-5	28.4	2.100	524.5	11,350.0	13.38	2.040
Glycol Ethers	Ethylene Glycol Monomethyl Ether	109-86-4	30.8	1.800	645.0	10,051.6	17.20	2.040
Glycol Ethers	Propylene Glycol Ethyl Propyl Ether	30025-38-8		4.710				
Glycol Ethers	Propylene Glycol Methyl Propyl Ether	34590-94-8	30.7	3.900		11,835.8		

(Continued)

Table A1-2 *(Continued)*

Type	Name	CAS #	Surface Tension dynes / cm	Viscosity cps	Heat of Vaporization BTU / lb	Heat of Combustion BTU / lb	Dielectric Constant	Dipole Moment
Glycol Ethers	Propylene Glycol Mono T-Butyl Ether	57018-52-7	24.4	3.400				
Glycol Ethers	Propylene Glycol Monobutyl Ether	1569-02-4	27.5	3.630				
Glycol Ethers	Propylene Glycol Monoethyl Ether	1569-02-4		2.215		12,313.0		
Glycol Ethers	Propylene Glycol Monoethyl Ether	1569-02-4		2.090		12,313.0		
Glycol Ethers	Propylene Glycol Monomethyl Ether	107-98-2	27.0	1.670		11,089.2		
Glycol Ethers	Propylene Glycol Monomethyl Ether	107-98-2	27.7	1.906		11,264.0		
Glycol Ethers	Propylene Glycol Monopropyl Ether	30136-13-1	25.4	2.400		12,897.7		
Glycol Ethers	Propylene Glycol n-Propyl Ether	30136-13-1	27.0	2.300		12,897.7		
Glycol Ethers	Triethylene glycol monomethyl ether	112-35-6	37.8	7.250				
Glycol Ethers	Tripropylene Glycol Monomethyl Ether	25498-49-1	29.0	5.600				
Glycol Ethers	Tripropylene Glycol n-Butyl Ether	55934-93-5	29.7	8.700		13,069.6		
Glycol Ethers	Tripropylene Glycol n-Propyl Ether	96077-04-2	29.3	7.000				
Glycols	1,4-Butane Diol	110-63-4	44.6	89.240	2,529.7	11,926.1	31.90	
Glycols	Diethylene Glycol	111-46-6	44.8	38.000		9,621.7	31.82	5.500
Glycols	Dipropylene Glycol	2396-61-4		107.000				
Glycols	Ethylene Glycol	107-21-1	48.0	21.000	813.1	8,237.3	16.10	3.505

Glycols	Hexylene Glycol	107-41-5	33.1	38.900	166.2		2.90	2.900
Glycols	Propylene Glycol	57-55-6	36.5	54.650	1,441.0	10,361.1	27.50	
Glycols	Triethylene Glycol	112-27-6	45.2	49.000	1,691.5	10,184.9	23.69	
Glymes	Diethylene Glycol Dibutyl Ether (Butyl Diglyme)	112-73-2	27.0	2.400				
Glymes	Diethylene Glycol Diethyl Ether (Ethyl Diglyme)	112-36-7	27.2	1.400				
Glymes	Diethylene Glycol Dimethyl Ether (Diglyme)	111-96-6	27.0	2.000				
Glymes	Ethylene Glycol Diethyl Ether (Ethyl Glyme)	629-14-1		0.700				
Glymes	Ethylene Glycol Dimethyl Ether (Monoglyme)	110-71-4	22.9	1.100	204.8		7.30	
Glymes	Triethylene Glycol Dimethyl Ether (Triglyme)	112-49-2	29.4	3.800				
HCFCs	HCFC-141b	1717-00-6	21.0	0.430				
HCFCs	HCFC-225 ca / cb	422-56-0/ 507-55-1	16.2	0.590				
HFCs	HFC 43-10 mee	138495-42-8	14.1	0.670	277.5		8.50	
HFCs	HFC-245 fa	460-73-1	13.6	0.974				
HFCs	HFC-365mfc	405-58-6	15.0	0.421				
HFEs	HFE-7100	163702-07-6	13.6	0.610	284.0		7.40	
HFEs	HFE-7200	163702-06-5	13.6	0.610	242.1		7.30	
HFEs	HFE-7300	132182-92-4	15.0	1.180		131,392.9		

(Continued)

Table A1-2 *(Continued)*

Type	Name	CAS #	Surface Tension dynes / cm	Viscosity cps	Heat of Vaporization BTU / lb	Heat of Combustion BTU / lb	Dielectric Constant	Dipole Moment
HFEs	HFE-8200	163702-05-4/ 163702-06-5	13.6	0.610				
Iodateds	Perfluoro n-butyl Iodide	423-39-2		0.640				
Ketones	Acetone	67-64-1	23.5	0.330	611.4	13,250.2	21.01	2.690
Ketones	Acetyl Acetone	123-54-6	32.1	0.700	909.6	11,530.5	26.52	
Ketones	Cyclohexanone	108-94-1	34.6	2.250	115.7	15,417.0	2.87	3.080
Ketones	Diethyl Ketone	96-22-0	24.7	0.470	569.5	15,472.9	17.00	
Ketones	Diisobutyl Ketone	108-83-8	25.5	0.950	107.5	17,244.2	2.70	
Ketones	Dispropyl Ketone	123-19-3	24.9	0.730	480.2		12.60	
Ketones	Methyl Amyl Ketone	110-43-0	26.1	0.810	457.7	15,627.7	11.95	
Ketones	Methyl Ethyl Ketone	78-93-3	24.0	0.426	579.3	14,574.4	18.56	
Ketones	Methyl Isoamyl Ketone	110-12-3		0.740		15,434.2	13.53	
Ketones	Methyl Isobutyl Ketone	108-10-1	23.6	0.585	452.3	13,211.5	13.11	2.290
Ketones	Methyl Propyl Ketone	107-87-9	23.3	0.480	90.2	15,472.9	2.70	
Lactates	2-Ethylhexyl Lactate	186817-80-1	29.8	7.700				
Lactates	Butyl Lactate	138-22-7	28.8	3.220				
Lactates	Ethyl Lactate	97-64-3	28.9	2.610	702.2		15.40	
Lactates	Isopropyl Lactate	63697-0-7	28.0	2.900				
Lactates	Methyl Lactate	27871-49-4	34.2	2.940				
Lactates	Propyl Lactate	616-09-1	30.4	3.300				
Limonene	D-Limonene	5989-27-5	26.7	0.900	93.6		2.37	

Limonene	Dipentene (Dl-Limonene)	138-86-3		0.900				
OSs	OS-10 Hexamethyldisiloxane	107-46-0	15.2	0.496	75.9		2.18	
OSs	OS-20 Octamethyltrisiloxane	107-51-7	16.5	0.574	81.1	800.2	2.29	
OSs	OS-30 Decamethyltrisiloxane	141-62-8	17.3	1.280	113.5	1,280.2	2.36	
Paraffins	2-Methyl Heptane	592-27-8	20.8	0.670	64.9	20,571.8	1.95	
Paraffins	2-Methyl Pentane	107-83-5	19.4	0.280		20,802.7		0.022
Paraffins	3-Methyl Hexane	589-34-4	19.6	0.380		20,657.8		
Paraffins	Cyclohexane	110-82-7	24.7	0.980	60.6	20,021.5	2.02	
Paraffins	Cyclopentane	287-92-3	21.9	0.413	53.8		1.97	
Paraffins	Cypar 7	64742-89-8						
Paraffins	Cypar 9	68553-14-0						
Paraffins	D30	64742-48-9		0.750				
Paraffins	D40	64742-48-9		0.960		19,561.5		
Paraffins	D60	64742-48-9		1.330				
Paraffins	D70	64742-47-8		1.460				
Paraffins	DSP 145 / 160	8002-5-9		0.750				
Paraffins	Ethyl Cyclohexane	1678-91-7	25.2	0.784	69.7	20,012.9	2.05	
Paraffins	Heptane	142-82-5	19.7	0.418	61.1	20,666.4	1.92	
Paraffins	Hexane	110-54-3	17.9	0.313	54.6	20,769.6	1.89	
Paraffins	Isododecane	13475-86-6		1.345				
Paraffins	Isopar C	90622-56-3	20.3	1.000				
Paraffins	Isopar E	90622-56-3	22.1	1.000				
Paraffins	Isopar H	90622-57-4	24.1	1.000				

(Continued)

Table A1-2 *(Continued)*

Type	Name	CAS #	Surface Tension dynes / cm	Viscosity cps	Heat of Vaporization BTU / lb	Heat of Combustion BTU / lb	Dielectric Constant	Dipole Moment
Paraffins	Isopar L	90622-58-5	25.1	1.000				
Paraffins	Isopar M	9622-58-5	26.4	0.520				
Paraffins	Methyl Cyclohexane	108-87-2	23.3	0.670	63.2	19,987.1	2.02	
Paraffins	Methyl Cyclopentane	96-37-7	21.7	0.520	57.9	20,374.0	1.99	
Paraffins	Nonane	111-84-2	22.4	0.716	72.7	20,533.1	1.97	
Paraffins	Octane	111-65-9	21.1	3.450	67.0	20,589.0	1.95	
Paraffins	Pentane	109-66-0	15.5	0.235	47.5	20,907.1	1.84	
Paraffins	SBP 1	64742-89-8		0.340				
Paraffins	SBP 100 / 140	64742-49-0		0.800		19,080.0		
Paraffins	SBP 11	64742-88-7		0.830		18,701.6		
Paraffins	SBP 3	64742-89-8		0.530		19,131.6		
Paraffins	SBP 5	64742-89-8		0.450		19,776.4		
Paraffins	SBP 80 / 110	64747-49-0		0.600		19,080.0		
Paraffins	Solvesso 100	64742-95-6		0.810		17,639.7		
Paraffins	Solvesso 150	64742-94-5		1.150		17,661.2		
Paraffins	Solvesso 200	64747-94-5		2.700				
Paraffins	Whitespirit	64742-82-1		0.840				
Specials	1,3 Dioxolane	646-6-0	34.1	0.589	259.5		7.34	
Specials	1,4-Dioxane	123-91-1	32.8	1.180	75.9		2.22	0.450
Specials	4-Bromo-3-Chloro-3, 4,4-Trifluoro-1-Butene	374-25-4	24.1	2.800				
Specials	Acetonitrile	75-5-8	28.7	0.358	1,091.9	13,061.0	36.64	

Specials	Carbon Dioxide	124-38-9	16.2					
Specials	Cumene	98-82-8	27.7	0.737	89.3		2.38	
Specials	Dimethyl Acetamide	127-19-5	32.4	0.920	1,469.1		33.85	3.700
Specials	Dimethyl Carbonate	616-38-6	28.5	0.590		6,820.2		
Specials	Dimethyl Formamide	68-12-2		0.924	1,468.8	11,418.7	38.25	3.240
Specials	Dimethyl Sulfoxide	67-68-5	42.9	1.990	2,036.0	66,430.7	47.24	
Specials	Ethylene Carbonate	96-49-1	54.6	2.547				
Specials	Gamma Butyrolactone	96-48-0	43.4	1.700	2,035.8	9,501.3	39.00	
Specials	Isophorone	78-59-1	31.0	2.632		15,632.0		
Specials	N-Methyl Pyrrolidone	872-50-4	40.7	2.222	1,628.3	12,992.3	32.20	4.090
Specials	Propylene Carbonate	108-32-7	41.4	1.600	3,124.5		66.14	
Specials	Pyridine	110-86-1	36.6	0.879	465.4		13.26	2.370
Specials	Soygold 1000	67784-80-9	31.6	4.500		80,352.2	42.20	
Specials	Tetrahydrofuran	109-99-9	26.4	0.575	215.7	14,914.0	7.52	1.750
Specials	Water	7732-18-5	73.0	1.002	3,188.7	13,061.0	78.54	

Type	Name	Refractive Index	Critical Point °C	Critical Pressure Psi	Critical Volume cc/mol	Relative Evaporation Rate RER	Ohnesorge Number
Acids	Acetic Acid	1.372	321.6	839.2	171.0	0.970	0.00542
Alcohols	1-Hexanol	1.418	337.2	495.6	381.0	0.050	0.02807
Alcohols	Amyl Alcohol	1.401	315.0	564.1	326.0	0.237	0.02288
Alcohols	Benzyl Alcohol	1.540	442.0	623.7		0.006	0.03768
Alcohols	Cyclohexanol	1.464	376.9	617.9	322.0	0.080	0.16174
Alcohols	Diacetone Alcohol	1.422	334.0	558.4		0.082	0.01464
Alcohols	Ethanol	1.362	240.8	890.1	167.0	3.300	0.00852
Alcohols	Furfuryl Alcohol	1.487				0.040	0.01887

(Continued)

Table A1-2 *(Continued)*

Type	Name	Refractive Index	Critical Point °C	Critical Pressure Psi	Critical Volume cc/mol	Relative Evaporation Rate RER	Ohnesorge Number
Alcohols	iso-Amyl Alcohol	1.405	304.1	570.0	327.0	0.200	0.02757
Alcohols	Isobutanol	1.396	274.6	622.9	273.0	0.670	0.02692
Alcohols	Isopropanol	1.378	235.2	691.0	220.0	2.830	0.01702
Alcohols	Methanol	1.329	239.4	1138.5	118.0	5.900	0.00397
Alcohols	Methyl Isobutyl Carbinol	1.410	301.3			0.300	0.03352
Alcohols	N-Butanol	1.400	289.9	640.2	275.0	2.797	0.01880
Alcohols	N-Propanol	1.384	263.7	749.7	219.0	1.300	0.01460
Alcohols	Secondary Butanol	1.398	263.1	609.4	269.0	1.300	0.02039
Alcohols	Tertiary Butanol	1.388	233.1	577.5	275.0	0.030	0.00000
Alcohols	Tetrahydrofurfuryl Alcohol	1.452					0.01804
Alcohols	Texanol	1.442				0.002	0.07247
Aromatics	Benzene	1.501	288.9	710.0	255.0	0.160	0.00341
Aromatics	Lactol Spirits	1.413				2.400	0.00000
Aromatics	Mineral Spirits					0.100	0.00000
Aromatics	o-Xylene		343.9	513.6	375.0	0.896	0.00397
Aromatics	Stoddard Solvent					0.150	0.00000
Aromatics	Toluene	1.497	318.6	595.8	316.0	1.984	0.00335
Aromatics	Turpentine, oil					0.860	0.00773
Aromatics	VM&P Naptha					7.563	0.00000
Benzotrifluorides	Benzotrifluoride	1.415				2.800	0.00306
Benzotrifluorides	Parachlorobenzotrifluoride	1.444				0.900	0.00384
Brominateds	n-Propyl Bromide	1.434	262.0			4.500	0.00237
CFCs	CFC-113	1.354	214.1	491.7		7.729	0.00355

Chlorinateds	1,1,1-Trichloroethane	1.438	272.0	623.7		6.000	0.00420
Chlorinateds	1,1-Dichloroethane	1.417	261.5	1682.4			0.00255
Chlorinateds	Carbon Tetrachloride	1.421	283.5	655.0	276.0	7.500	0.00454
Chlorinateds	Chloroform	1.446	263.3	793.4	239.0	10.200	0.00259
Chlorinateds	Ethylene Dichloride	1.445	288.0	783.2	225.0	6.100	0.00349
Chlorinateds	Methylene Chloride	1.424	237.0	884.7		14.500	0.00205
Chlorinateds	Perchloroethylene	1.438	347.1			2.600	0.00356
Chlorinateds	Trans 1,2 Dichloroethylene	1.445	243.3	799.2		7.434	0.00159
Chlorinateds	Trichloroethylene	1.477	271.1	728.1		6.400	0.00256
Dibasic Acid Esters	DBE	1.420					0.01088
Dibasic Acid Esters	DBE-4	1.420					0.01118
Dibasic Acid Esters	DBE-6	1.427					0.01502
Esters	2-Ethyl Hexyl Acetate	1.420				0.030	0.00000
Esters	Amyl Acetate	1.401	325.0	389.4		0.420	0.00520
Esters	Amyl Acetate	1.402	325.0	389.4		0.420	0.00516
Esters	Butyl Glycol Acetate	1.414				0.033	0.01011
Esters	Cyclohexyl Acetate	1.440				0.157	0.01162
Esters	Dimethyl Succinate	1.420					0.01282
Esters	Ethyl Acetate	1.373	250.2	563.0	286.0	4.172	0.00281
Esters	Ethyl Formate	1.360	235.4	687.5	229.0	6.930	0.00250
Esters	Ethyl-3-Ethoxypropionate					0.120	0.00707
Esters	Ethylene Glycol Diacetate	1.416				0.020	0.01734
Esters	Isobutyl Acetate	1.391	288.0	458.3		1.700	0.00444
Esters	Isopropyl Acetate	1.378	258.0	476.1		2.881	0.00339
Esters	Methyl Acetate	1.361	233.4	688.9	228.0	5.500	0.00225

(Continued)

Table A1-2 *(Continued)*

Type	Name	Refractive Index	Critical Point °C	Critical Pressure Psi	Critical Volume cc/mol	Relative Evaporation Rate RER	Ohnesorge Number
Esters	Methyl Formate	1.344	214.0	870.7	173.0		0.00202
Esters	n-Butyl Acetate	1.395	306.0		389.0	1.000	0.00435
Esters	N-Butyl Propionate	1.401	321.0			0.450	0.00460
Esters	n-Propyl Acetate	1.384	276.6	487.3	345.0	1.494	0.00357
Esters	n-Propyl Propionate					1.200	0.00422
Esters	n-Propyl Propionate					1.200	0.00422
Esters	Propylene Glycol Diacetate					0.036	0.01297
Esters	Tertiary Butyl Acetate	1.395	266.8	459.8	389.0	2.800	0.00765
Ethers	Cyclopentyl Methyl Ether	1.419	325.4	522.1	343.7	3.500	0.00332
Ethers	Methyl Butyl Ether	1.376	229.7	479.6		7.201	0.01352
Ethers	Methyl t-Butyl Ether	1.377	223.9	492.7		19.953	0.00252
Ethers	Methylal	1.359	207.4	573.8		0.880	0.00223
Ethers	t-Butyl Ether	1.433	282.0	3470.8		3.094	0.00000
Ethers	Amyl t-Butyl Ether	1.410	311.6	332.3		0.691	0.03320
Glycol Ethers	Dipropylene Glycol Dimethyl Ether					0.130	0.00646
Glycol Ethers	Dipropylene Glycol n-Propyl Ether	1.422				0.014	0.02524
Glycol Ether Acetates	Diethylene Glycol Monobutyl Ether Acetate						0.01658
Glycol Ether Acetates	Diethylene Glycol Monoethyl Ether Acetate	1.422				0.080	0.01392
Glycol Ether Acetates	Dipropylene Glycol Methyl Ether Acetate					0.015	0.01123

Glycol Ether Acetates	Ethylene Glycol Butyl Ether Acetate					0.064	0.00947
Glycol Ether Acetates	Ethylene Glycol Ethyl Ether Acetate	1.405	334.2	459.2	443.0	0.233	0.00515
Glycol Ether Acetates	Ethylene glycol methyl ether acetate	1.403				0.403	0.00000
Glycol Ether Acetates	Propylene Glycol Ethyl Ether Acetate	1.404				0.210	0.00000
Glycol Ether Acetates	Propylene Glycol Methyl Ether Acetate	1.402				0.356	0.00603
Glycol Ethers	3-Methoxy-3-Methyl-1-Butanol	1.428				0.070	0.03973
Glycol Ethers	Butyl Glycol Ether	1.419	370.0	565.8	424.0	0.065	0.01900
Glycol Ethers	Diethylene Glycol Monobutyl ether					0.004	0.02574
Glycol Ethers	Diethylene Glycol Monomethyl Ether	1.426	399.0	532.3		0.019	0.01801
Glycol Ethers	Diethylene Glycol Monomethyl Ether	1.430	399.0	532.3		0.013	0.01762
Glycol Ethers	Dipropylene Glycol Monobutyl Ether	1.425				0.006	0.03158
Glycol Ethers	Dipropylene Glycol Monobutyl Ether					0.4	0.02412
Glycol Ethers	Dipropylene Glycol Monoethyl Ether						0.00000
Glycol Ethers	Dipropylene Glycol Monomethyl Ether	1.420				0.035	0.01819
Glycol Ethers	Dipropylene Glycol n-Butyl Ether	1.425				0.006	0.02412
Glycol Ethers	Ethylene Glycol Diethyl Ether	1.386	269.0				0.00419

(Continued)

Table A1-2 *(Continued)*

Type	Name	Refractive Index	Critical Point °C	Critical Pressure Psi	Critical Volume cc/mol	Relative Evaporation Rate RER	Ohnesorge Number
Glycol Ethers	Ethylene Glycol Mono Isopropyl Ether	1.410				0.233	0.00000
Glycol Ethers	Ethylene Glycol Monobutyl Ether	1.420	360.8	565.8	424.0	0.074	0.01678
Glycol Ethers	Ethylene Glycol Monoethyl Ether	1.408	312.0	630.9		0.281	0.01150
Glycol Ethers	Ethylene Glycol Monomethyl Ether	1.402	324.0	756.5		0.356	0.00928
Glycol Ethers	Propylene Glycol Ethyl Propyl Ether	1.422				0.028	0.00000
Glycol Ethers	Propylene Glycol Methyl Propyl Ether	1.421				0.034	0.02028
Glycol Ethers	Propylene Glycol Mono T-Butyl Ether					0.250	0.02074
Glycol Ethers	Propylene Glycol Monobutyl Ether	1.416				0.093	0.02075
Glycol Ethers	Propylene Glycol Monoethyl Ether	1.408	299.0	548.2		0.410	0.00000
Glycol Ethers	Propylene Glycol Monoethyl Ether	1.409	299.0	548.2		0.540	0.00000
Glycol Ethers	Propylene Glycol Monomethyl Ether	1.403	307.0	596.5		0.660	0.00940
Glycol Ethers	Propylene Glycol Monomethyl Ether	1.404	307.0	596.5		0.620	0.01060
Glycol Ethers	Propylene Glycol Monopropyl Ether	1.410				0.210	0.01447
Glycol Ethers	Propylene Glycol n-Propyl Ether	1.410				0.210	0.01321

Glycol Ethers	Triethylene glycol monomethyl ether	1.439				0.010	0.03271
Glycol Ethers	Tripropylene Glycol Monomethyl Ether	1.430				0.003	0.02975
Glycol Ethers	Tripropylene Glycol n-Butyl Ether	1.433				0.010	0.04642
Glycol Ethers	Tripropylene Glycol n-Propyl Ether	1.430				0.001	0.03758
Glycols	1,4-Butane Diol	1.446	454.0	902.2		0.001	0.37234
Glycols	Diethylene Glycol	1.448	480.0	691.9		0.001	0.15104
Glycols	Dipropylene Glycol	1.439					0.00000
Glycols	Ethylene Glycol	1.432	445.0	1189.3	191.0	0.010	0.08074
Glycols	Hexylene Glycol	1.427	347.8	581.6		0.003	0.19808
Glycols	Propylene Glycol	1.430	403.0	861.6	239.0	0.010	0.24982
Glycols	Triethylene Glycol	1.456	524.0	478.6		0.001	0.19322
Glymes	Diethylene Glycol Dibutyl Ether (Butyl Diglyme)	1.424					0.01382
Glymes	Diethylene Glycol Diethyl Ether (Ethyl Diglyme)	1.408				0.040	0.00792
Glymes	Diethylene Glycol Dimethyl Ether (Diglyme)	1.408				0.360	0.01113
Glymes	Ethylene Glycol Diethyl Ether (Ethyl Glyme)	1.392					0.00000
Glymes	Ethylene Glycol Dimethyl Ether (Monoglyme)	1.380	263.0	561.3	271.0	4.900	0.00693
Glymes	Triethylene Glycol Dimethyl Ether (Triglyme)	1.412				0.001	0.01983
HCFCs	HCFC-141b	1.369	204.2	616.4	270.1	13.320	0.00238
HCFCs	HCFC-225 ca / cb	1.321	202.0	429.5	365.7	6.732	0.00331

(Continued)

Table A1-2 *(Continued)*

Type	Name	Refractive Index	Critical Point °C	Critical Pressure Psi	Critical Volume cc/mol	Relative Evaporation Rate RER	Ohnesorge Number
HFCs	HFC 43-10 mee	1.253	181.0	331.8	433.0	29.000	0.00399
HFCs	HFC-245 fa	1.295	154.1	642.5		104.000	0.00647
HFCs	HFC-365mfc	1.280					0.00271
HFEs	HFE-7100	1.269	194.8	323.4	450.5	49.000	0.00377
HFEs	HFE-7200	1.281	208.8	286.6	509.7	33.000	0.00389
HFEs	HFE-7300	1.263	231.8	232.2	608.3	478.000	0.00664
HFEs	HFE-8200					33.000	0.00389
Iodateds	Perfluoro n-butyl Iodide	1.328				14.500	0.00000
Ketones	Acetone	1.360	235.0	681.7	209.0	5.762	0.00215
Ketones	Acetyl Acetone	1.449				0.328	0.00351
Ketones	Cyclohexanone	1.451	379.9	580.2	311.0	0.303	0.01106
Ketones	Diethyl Ketone	1.392	288.3	540.8	336.0	2.300	0.00294
Ketones	Diisobutyl Ketone	1.412				0.200	0.00589
Ketones	Dispropyl Ketone						0.00460
Ketones	Methyl Amyl Ketone	1.409	338.4	498.3		0.340	0.00495
Ketones	Methyl Ethyl Ketone	1.379	263.6	610.2	267.0	4.481	0.00272
Ketones	Methyl Isoamyl Ketone	1.408				0.460	0.00000
Ketones	Methyl Isobutyl Ketone	1.396	298.0	474.3	369.0	1.210	0.00378
Ketones	Methyl Propyl Ketone	1.390	287.9	535.8	301.0	2.400	0.00311
Lactates	2-Ethylhexyl Lactate	1.435				0.003	0.04088
Lactates	Butyl Lactate	1.421	341.4	455.1	466.5	0.036	0.01700
Lactates	Ethyl Lactate	1.412	330.0	571.6	354.5	0.180	0.01342
Lactates	Isopropyl Lactate	1.408	311.8	513.3	404.5	0.180	0.01551

Lactates	Methyl Lactate	1.414				0.260	0.01348
Lactates	Propyl Lactate	1.417				0.110	0.01682
Limonene	D-Limonene	1.475				0.200	0.00534
Limonene	Dipentene (Dl-Limonene)	1.475				0.200	0.00000
OSs	OS-10 Hexamethyl-disiloxane	1.375	276.4	274.1		3.800	0.00409
OSs	OS-20 Octamethyl-trisiloxane	1.383	402.7			1.000	0.00438
OSs	OS-30 Decamethyl-trisiloxane	1.387				0.150	0.00936
Paraffins	2-Methyl Heptane	1.395	286.6	360.3	488.0	0.157	0.00494
Paraffins	2-Methyl Pentane	1.372	224.8	438.0	366.0		0.00221
Paraffins	3-Methyl Hexane	1.389	262.4	413.0		4.750	0.00291
Paraffins	Cyclohexane	1.427	280.4	590.3	308.0	4.840	0.00628
Paraffins	Cyclopentane	1.407	238.6	653.8	260.0		0.00287
Paraffins	Cypar 7					3.100	0.00000
Paraffins	Cypar 9					0.680	0.00000
Paraffins	D30	1.419				0.400	0.00000
Paraffins	D40	1.427				0.120	0.00000
Paraffins	D60	1.434				0.040	0.00000
Paraffins	D70	1.441				0.020	0.00000
Paraffins	DSP 145 / 160	1.420				0.600	0.00000
Paraffins	Ethyl Cyclohexane	1.433			430.0	1.070	0.00495
Paraffins	Heptane	1.388	267.1	397.4	428.0	3.180	0.00321
Paraffins	Hexane	1.375	234.5	438.7	368.0	6.820	0.00256
Paraffins	Isododecane					0.140	0.00000
Paraffins	Isopar C	1.393				5.450	0.00748
Paraffins	Isopar E	1.406				1.550	0.00704

(Continued)

Table A1-2 *(Continued)*

Type	Name	Refractive Index	Critical Point °C	Critical Pressure Psi	Critical Volume cc/mol	Relative Evaporation Rate RER	Ohnesorge Number
Paraffins	Isopar H	1.426				0.090	0.00657
Paraffins	Isopar L	1.429				0.040	0.00639
Paraffins	Isopar M	1.438				0.010	0.00321
Paraffins	Methyl Cyclohexane	1.423	299.1	503.4	368.0	2.990	0.00445
Paraffins	Methyl Cyclopentane	1.410	259.6	548.8	319.0	5.930	0.00362
Paraffins	Nonane		321.5	332.1	555.0	0.410	0.00502
Paraffins	Octane	1.397	295.6	361.1	492.0	1.230	0.02515
Paraffins	Pentane	1.357	196.6	488.8	311.0	10.460	0.00212
Paraffins	SBP 1	1.389				5.600	0.00000
Paraffins	SBP 100 / 140	1.405				1.900	0.00000
Paraffins	SBP 11	1.434				0.290	0.00000
Paraffins	SBP 3	1.403				2.700	0.00000
Paraffins	SBP 5	1.397				3.900	0.00000
Paraffins	SBP 80 / 110	1.397				4.500	0.00000
Paraffins	Solvesso 100	1.502				0.160	0.00000
Paraffins	Solvesso 150	1.517				0.050	0.00000
Paraffins	Solvesso 200	1.592				0.001	0.00000

Paraffins	Whitespirit	1.437				0.242	0.00000
Specials	1,3 Dioxolane	1.397				3.500	0.00275
Specials	1,4-Dioxane	1.422	314.0	755.6	238.0	2.200	0.00570
Specials	4-Bromo-3-Chloro-3,4, 4-Trifluoro-1-Butene	1.429	296.7	470.1	394.8	244.000	0.01220
Specials	Acetonitrile	1.344	272.4	703.4	173.0	2.300	0.00212
Specials	Carbon Dioxide	1.120	31.0	1070.0	95.0		0.00000
Specials	Cumene	1.492	357.9	465.4	429.0		0.00424
Specials	Dimethyl Acetamide	1.437	385.0	583.4		0.138	0.00468
Specials	Dimethyl Carbonate	1.369	284.0	696.1		3.220	0.00301
Specials	Dimethyl Formamide	1.430	376.5		262.0	0.200	0.00000
Specials	Dimethyl Sulfoxide	1.417	447.0				0.00813
Specials	Ethylene Carbonate	1.415	533.0	968.9			0.00799
Specials	Gamma Butyrolactone	1.434				0.030	0.00685
Specials	Isophorone	1.477				0.025	0.01384
Specials	N-Methyl Pyrrolidone	1.469	448.7	684.6	311.0	0.060	0.00963
Specials	Propylene Carbonate	1.419	502.0	784.5			0.00637
Specials	Pyridine	1.510	346.9	822.4	243.0		0.00412
Specials	Soygold 1000					0.005	0.02395
Specials	Tetrahydrofuran	1.407	267.0	752.7	224.0	6.300	0.00333
Specials	Water	1.333	373.8	3205.3	56.0	0.360	0.00330

Table A1-3

Type	Name	CAS #	VOC in US	HAP in US	VOC in Europe	Ozone Reactivity MIR	Exposure Limit in US
Acids	Acetic Acid	64-19-7	YES	NO	YES	0.685	10(OSHA)
Alcohols	1-Hexanol	111-27-3	YES	NO	YES	2.691	50 (NIOSH)
Alcohols	Amyl Alcohol	71-41-0	YES	NO	YES	2.827	100 (PEL)
Alcohols	Benzyl Alcohol	100-51-6	YES	NO	YES	5.114	10 (AI HA)
Alcohols	Cyclohexanol	108-93-0	YES	NO	YES	1.947	50(OSHA)
Alcohols	Diacetone Alcohol	123-42-2	YES	NO	YES	0.595	50 (ACGIH)
Alcohols	Ethanol	64-17-5	YES	NO	YES	1.526	1000 (OSHA)
Alcohols	Furfuryl Alcohol	98-0-0	YES	NO	YES		50 (PEL)
Alcohols	iso-Amyl Alcohol	123-51-3	YES	NO	YES	3.162	100 (ACGIH)
Alcohols	Isobutanol	78-83-1	YES	NO	YES	2.509	100(OSHA)
Alcohols	Isopropanol	67-63-0	YES	NO	YES	0.614	400 (ACGIH)
Alcohols	Methyl Isobutyl Carbinol	108-11-2	YES	NO	YES	2.637	25(OSHA)
Alcohols	N-Butanol	71-36-3	YES	NO	YES	2.880	50 (ACGIH)
Alcohols	N-Propanol	71-23-8	YES	NO	YES	2.497	200 (ACGIH)
Alcohols	Secondary Butanol	78-92-2	YES	NO	YES	1.360	100 (ACGIH)
Alcohols	Tertiary Butanol	75-65-0	YES	NO	YES	0.406	300 (PEL)
Alcohols	Tetrahydrofurfuryl Alcohol	97-99-4	YES	NO	YES	3.311	2 (AIHA)
Alcohols	Texanol	25265-77-4	YES	NO	YES	0.809	
Alcohols	Methanol	67-56-1	YES	YES	YES	0.672	200 (ACGIH)
Aromatics	Lactol Spirits	64742-49-0	YES	NO	YES	1.410	200 (OSHA)
Aromatics	Mineral Spirits	64741-41-9	YES	NO	YES		100 (AEL)
Aromatics	Stoddard Solvent	8052-41-3	YES	NO	YES		100 (ACGIH))
Aromatics	Turpentine, oil	8006-64-2	YES	NO	YES	4.279	100(OSHA)

Aromatics	VM&P Naptha	64742-89-8	YES	NO	YES		300 (ACGIH)
Aromatics	Benzene	71-43-2	YES	YES	YES	0.721	10 (ACGIH)
Aromatics	o-Xylene	95-47-6	YES	YES	YES	7.640	100 (ACGIH)
Aromatics	Toluene	108-88-3	YES	YES	YES	4.005	100 (ACGIH)
Benzotrifluorides	Parachlorobenzotrifluoride	98-56-6	NO	NO	YES	0.126	25 (PEL)
Benzotrifluorides	Benzotrifluoride	98-08-8	YES	NO	YES		100 (PEL)
Brominateds	n-Propyl Bromide	106-94-5	YES	NO	YES	0.424	10 (ACGIH)
CFCs	CFC-113	76-13-1	NO	NO	YES		1000 (PEL)
Chlorinateds	Trans 1,2 Dichloroethylene	156-60-5	YES	NO	YES	1.704	200 (ACGIH)
Chlorinateds	1,1,1-Trichloroethane	71-55-6	NO	YES	YES	0.005	350 (ACGIH)
Chlorinateds	Methylene Chloride	75-09-2	NO	YES	YES		50 (ACGIH)
Chlorinateds	Perchloroethylene	127-18-4	NO	YES	YES	0.031	25 (ACGIH)
Chlorinateds	1,1-Dichloroethane	75-34-3	YES	YES	YES	0.069	100(OSHA)
Chlorinateds	Carbon Tetrachloride	56-23-5	YES	YES	YES		10(OSHA)
Chlorinateds	Chloroform	67-66-3	YES	YES	YES	0.022	10 (ACGIH)
Chlorinateds	Ethylene Dichloride	107-06-2	YES	YES	YES		1 (NIOSH)
Chlorinateds	Trichloroethylene	79-01-6	YES	YES	YES		10 (ACGIH)
Dibasic Acid Esters	DBE	1119-40-0	YES	NO	YES	0.419	1.5 (PEL)
Dibasic Acid Esters	DBE-4	106-65-0	YES	NO	YES	0.225	1.5 (PEL)
Dibasic Acid Esters	DBE-6	627-93-0	YES	NO	YES	1.797	1.5 (PEL)
Esters	Methyl Acetate	79-20-9	NO	NO	YES	0.072	200 (ACGIH)
Esters	Methyl Formate	107-31-3	NO	NO	YES	0.057	100(OSHA)
Esters	Tertiary Butyl Acetate	540-88-5	NO	NO	YES	0.180	200 (OSHA)
Esters	2-Ethyl Hexyl Acetate	103-9-3	YES	NO	YES	0.656	
Esters	Amyl Acetate	628-63-7	YES	NO	YES	0.839	100 (ACHIG)

(Continued)

Table A1-3 *(Continued)*

Type	Name	CAS #	VOC in US	HAP in US	VOC in Europe	Ozone Reactivity MIR	Exposure Limit in US
Esters	Amyl Acetate	628-63-7	YES	NO	YES	0.839	100 (ACGIH)
Esters	Butyl Glycol Acetate	112-07-2	YES	NO	YES		5 (NIOSH)
Esters	Cyclohexyl Acetate	622-45-7	YES	NO	YES		
Esters	Dimethyl Succinate	106-65-0	YES	NO	YES	0.225	1.5 (OSHA)
Esters	Ethyl Acetate	141-78-6	YES	NO	YES	0.626	400 (ACGIH)
Esters	Ethyl Formate	109-94-4	YES	NO	YES	0.475	100(OSHA)
Esters	Ethyl-3-Ethoxypropionate	763-69-9	YES	NO	YES	3.582	
Esters	Ethylene Glycol Diacetate	111-55-7	YES	NO	YES	0.665	
Esters	Isobutyl Acetate	110-19-0	YES	NO	YES	0.623	150 (ACGIH)
Esters	Isopropyl Acetate	108-21-4	YES	NO	YES	1.073	250 (OSHA)
Esters	n-Butyl Acetate	123-86-4	YES	NO	YES	0.835	150 (ACGIH)
Esters	N-Butyl Propionate	590-01-2	YES	NO	YES		
Esters	n-Propyl Acetate	109-60-4	YES	NO	YES	0.779	200 (OSHA)
Esters	n-Propyl Propionate	106-36-5	YES	NO	YES	0.843	
Esters	n-Propyl Propionate	106-36-5	YES	NO	YES	0.843	
Esters	Propylene Glycol Diacetate	623-84-7	YES	NO	YES	0.613	1 (NIOSH)
Ethers	Cyclopentyl Methyl Ether	5614-37-9	YES	NO	YES		
Ethers	Methyl Butyl Ether	628-28-4	YES	NO	YES	3.150	
Ethers	Methyl t-Butyl Ether	1634-04-4	YES	NO	YES		50 (ACGIH)
Ethers	Methylal	109-87-5	YES	NO	YES	0.944	1000 (ACGIH)
Ethers	t-Butyl Ether	6163-66-2	YES	NO	YES		
Ethers	Amyl t-Butyl Ether	10100-95-5	YES	NO	YES		
Glycol Ethers	Dipropylene Glycol Dimethyl Ether	111109-77-4	YES	NO	YES		
Glycol Ethers	Dipropylene Glycol n-Propyl Ether	29911-27-1	YES	NO	YES		

Glycol Ether Acetates	Diethylene Glycol Monobutyl Ether Acetate	124-17-4	YES	NO	YES	1.380	
Glycol Ether Acetates	Diethylene Glycol Monoethyl Ether Acetate	112-15-2	YES	NO	YES	1.479	5 (AEL)
Glycol Ether Acetates	Dipropylene Glycol Methyl Ether Acetate	88917-22-0	YES	NO	YES	1.452	
Glycol Ether Acetates	Ethylene Glycol Butyl Ether Acetate	112-07-2	YES	NO	YES		5 (NIOSH)
Glycol Ether Acetates	Ethylene Glycol Ethyl Ether Acetate	111-15-9	YES	NO	YES	1.838	5 (PEL)
Glycol Ether Acetates	Ethylene Glycol Methyl Ether Acetate	110-49-6	YES	NO	YES	1.148	25(OSHA)
Glycol Ether Acetates	Propylene Glycol Ethyl Ether Acetate	54839-24-6	YES	NO	YES		
Glycol Ether Acetates	Propylene Glycol Methyl Ether Acetate	108-65-6	YES	NO	YES	1.705	50 (ACGIH)
Glycol Ethers	3-Methoxy-3-Methyl-1-Butanol	56539-66-3	YES	NO	YES	2.881	
Glycol Ethers	Butyl Glycol Ether	111-76-2	YES	NO	YES	2.895	50(OSHA)
Glycol Ethers	Diethylene Glycol Monobutyl Ether	112-34-5	YES	NO	YES	2.387	5 (AEL)
Glycol Ethers	Diethylene Glycol Monomethyl Ether	111-77-3	YES	NO	YES	2.660	
Glycol Ethers	Diethylene Glycol Monomethyl Ether	111-77-3	YES	NO	YES	2.660	
Glycol Ethers	Dipropylene Glycol Monobutyl Ether	29911-28-2	YES	NO	YES	1.828	
Glycol Ethers	Dipropylene Glycol Monobutyl Ether	29911-28-2	YES	NO	YES	1.828	
Glycol Ethers	Dipropylene Glycol Monoethyl Ether	30025-38-8	YES	NO	YES		
Glycol Ethers	Dipropylene Glycol Monomethyl Ether	112-28-7	YES	NO	YES		100 (PEL)

(*Continued*)

Table A1-3 *(Continued)*

Type	Name	CAS #	VOC in US	HAP in US	VOC in Europe	Ozone Reactivity MIR	Exposure Limit in US
Glycol Ethers	Dipropylene Glycol n-Butyl Ether	29911-28-2	YES	NO	YES	1.828	
Glycol Ethers	Ethylene Glycol Diethyl Ether	629-14-1	YES	NO	YES	2.954	
Glycol Ethers	Ethylene Glycol Mono Isopropyl Ether	109-59-1	YES	NO	YES		25 (NIOSH)
Glycol Ethers	Ethylene Glycol Monobutyl Ether	111-76-2	YES	NO	YES	2.895	50(OSHA)
Glycol Ethers	Ethylene Glycol Monoethyl Ether	110-80-5	YES	NO	YES	3.710	200 (OSHA)
Glycol Ethers	Ethylene Glycol Monomethyl Ether	109-86-4	YES	NO	YES	2.929	25(OSHA)
Glycol Ethers	Propylene Glycol Ethyl Propyl Ether	30025-38-8	YES	NO	YES		
Glycol Ethers	Propylene Glycol Methyl Propyl Ether	34590-94-8	YES	NO	YES		100(OSHA)
Glycol Ethers	Propylene Glycol Mono t-butyl Ether	57018-52-7	YES	NO	YES	1.610	
Glycol Ethers	Propylene Glycol Monobutyl Ether	1569-02-4	YES	NO	YES		
Glycol Ethers	Propylene Glycol Monoethyl Ether	1569-02-4	YES	NO	YES		
Glycol Ethers	Propylene Glycol Monoethyl Ether	1569-02-4	YES	NO	YES		
Glycol Ethers	Propylene Glycol Monomethyl Ether	107-98-2	YES	NO	YES	2.443	100 (NIOSH)
Glycol Ethers	Propylene Glycol Monomethyl Ether	107-98-2	YES	NO	YES	2.443	100 (ACGIH)
Glycol Ethers	Propylene Glycol Monopropyl Ether	30136-13-1	YES	NO	YES	3.767	

Glycol Ethers	Propylene Glycol n-Propyl Ether	30136-13-1	YES	NO	YES	3.767	
Glycol Ethers	Triethylene glycol monomethyl ether	112-35-6	YES	NO	YES	2.575	
Glycol Ethers	Tripropylene Glycol Monomethyl Ether	25498-49-1	YES	NO	YES	1.920	100 (AEL)
Glycol Ethers	Tripropylene Glycol n-Butyl Ether	55934-93-5	YES	NO	YES	1.642	
Glycol Ethers	Tripropylene Glycol n-Propyl Ether	96077-04-2	YES	NO	YES		
Glycols	1,4-Butane Diol	110-63-4	YES	NO	YES	2.718	
Glycols	Diethylene Glycol	111-46-6	YES	NO	YES	3.355	
Glycols	Dipropylene Glycol	2396-61-4	YES	NO	YES		
Glycols	Hexylene Glycol	107-41-5	YES	NO	YES	1.454	25 (NIOSH)
Glycols	Propylene Glycol	57-55-6	YES	NO	YES	2.578	50 (AI HA)
Glycols	Triethylene Glycol	112-27-6	YES	NO	YES	3.253	
Glycols	Ethylene Glycol	107-21-1	YES	YES	YES	3.133	50 (ACGIH)
Glymes	Diethylene Glycol Dibutyl Ether (Butyl Diglyme)	112-73-2	YES	NO	YES		
Glymes	Diethylene Glycol Diethyl Ether (Ethyl Diglyme)	112-36-7	YES	NO	YES		
Glymes	Diethylene Glycol Dimethyl Ether (Diglyme)	111-96-6	YES	NO	YES		
Glymes	Ethylene Glycol Diethyl Ether (Ethyl Glyme)	629-14-1	YES	NO	YES	2.954	
Glymes	Ethylene Glycol Dimethyl Ether (Monoglyme)	110-71-4	YES	NO	YES		
Glymes	Triethylene Glycol Dimethyl Ether (Triglyme)	112-49-2	YES	NO	YES		
HCFCs	HCFC-141b	1717-00-6	NO	NO	YES		500 (PEL)

(Continued)

Table A1-3 *(Continued)*

Type	Name	CAS #	VOC in US	HAP in US	VOC in Europe	Ozone Reactivity MIR	Exposure Limit in US
HCFCs	HCFC-225 ca / cb	422-56-0/507-55-1	NO	NO	YES		100 (PEL)
HFCs	HFC 43-10 mee	138495-42-8	NO	NO	YES	0.000	200 (CEL)
HFCs	HFC-245 fa	460-73-1	NO	NO	YES	0.001	300 (PEL)
HFCs	HFC-365mfc	405-58-6	NO	NO	YES		1000 (CEL)
HFEs	HFE-7100	163702-07-6	NO	NO	YES		750 (AI HA)
HFEs	HFE-7200	163702-06-5	NO	NO	YES		200 (CEL)
HFEs	HFE-7300	132182-92-4	NO	NO	YES		
HFEs	HFE-8200	163702-05-4/ 163702-06-5	NO	NO	YES		200 (PEL)
Iodateds	Perfluoro n-butyl Iodide	423-39-2	YES	NO	YES		
Ketones	Acetone	67-64-1	NO	NO	YES	0.356	750 (ACGIH)
Ketones	Acetyl Acetone	123-54-6	YES	NO	YES	1.006	
Ketones	Cyclohexanone	108-94-1	YES	NO	YES	1.345	50(OSHA)
Ketones	Diethyl Ketone	96-22-0	YES	NO	YES	1.236	200 (NIOSH)
Ketones	Diisobutyl Ketone	108-83-8	YES	NO	YES	2.681	25 (ACGIH)
Ketones	Dispropyl Ketone	123-19-3	YES	NO	YES		50 (ACGIH)
Ketones	Methyl Amyl Ketone	110-43-0	YES	NO	YES	2.360	50 (ACGIH)
Ketones	Methyl Isoamyl Ketone	110-12-3	YES	NO	YES	2.408	100(OSHA)
Ketones	Methyl Propyl Ketone	107-87-9	YES	NO	YES	2.812	200 (ACGIH)
Ketones	Methyl Ethyl Ketone	78-93-3	YES	YES	YES	1.481	200 (ACGIH)
Ketones	Methyl Isobutyl Ketone	108-10-1	YES	YES	YES	3.881	50 (ACGIH)
Lactates	2-Ethylhexyl Lactate	186817-80-1	YES	NO	YES		0.48 (PEL)
Lactates	Butyl Lactate	138-22-7	YES	NO	YES		5 (NIOSH)
Lactates	Ethyl Lactate	97-64-3	YES	NO	YES	2.480	4.08 (PEL)

Lactates	Isopropyl Lactate	63697-0-7	YES	NO	YES		3.60 (PEL)
Lactates	Methyl Lactate	27871-49-4	YES	NO	YES		4.08 (PEL)
Lactates	Propyl Lactate	616-09-1	YES	NO	YES		3.64 (PEL)
Limonene	D-Limonene	5989-27-5	YES	NO	YES	4.552	30 (AI HA)
Limonene	Dipentene (Dl-Limonene)	138-86-3	YES	NO	YES		
OSs	OS-10 Hexamethyldisiloxane	107-46-0	NO	NO	YES	−0.025	200 (CEL)
OSs	OS-20 Octamethyltrisiloxane	107-51-7	NO	NO	YES		200 (CEL)
OSs	OS-30 Decamethyltrisiloxane	141-62-8	NO	NO	YES		200 (CEL)
Paraffins	2-Methyl Heptane	592-27-8	YES	NO	YES	1.073	
Paraffins	2-Methyl Pentane	107-83-5	YES	NO	YES	1.502	500 (ACGIH)
Paraffins	3-Methyl Hexane	589-34-4	YES	NO	YES	1.614	
Paraffins	Cyclohexane	110-82-7	YES	NO	YES	1.250	300 (ACGIH)
Paraffins	Cyclopentane	287-92-3	YES	NO	YES	2.392	600 (NIOSH)
Paraffins	Cypar 7	64742-89-8	YES	NO	ERR		400 (ACGIH)
Paraffins	Cypar 9	68553-14-0	YES	NO	ERR		
Paraffins	D30	64742-48-9	YES	NO	YES		
Paraffins	D40	64742-48-9	YES	NO	YES		
Paraffins	D60	64742-48-9	YES	NO	YES		
Paraffins	D70	64742-47-8	YES	NO	YES		300 (ACGIH)
Paraffins	DSP 145/160	8002-5-9	YES	NO	YES		500 (PEL)
Paraffins	Ethyl Cyclohexane	1678-91-7	YES	NO	YES	1.474	
Paraffins	Heptane	142-82-5	YES	NO	YES	1.074	400 (ACGIH)
Paraffins	Isododecane	13475-86-6	YES	NO	YES		
Paraffins	Isopar C	90622-56-3	YES	NO	YES		300 (PEL)
Paraffins	Isopar E	90622-56-3	YES	NO	YES		280 (PEL)
Paraffins	Isopar H	90622-57-4	YES	NO	YES		177 (PEL)

(Continued)

Table A1-3 *(Continued)*

Type	Name	CAS #	VOC in US	HAP in US	VOC in Europe	Ozone Reactivity MIR	Exposure Limit in US
Paraffins	Isopar L	90622-58-5	YES	NO	YES		171 (PEL)
Paraffins	Isopar M	9622-58-5	YES	NO	YES		150 (PEL)
Paraffins	Methyl Cyclohexane	108-87-2	YES	NO	YES	1.698	400 (ACGIH)
Paraffins	Methyl Cyclopentane	96-37-7	YES	NO	YES	2.191	
Paraffins	Nonane	111-84-2	YES	NO	YES	0.781	200 (NIOSH)
Paraffins	Octane	111-65-9	YES	NO	YES	0.899	500(OSHA)
Paraffins	Pentane	109-66-0	YES	NO	YES	1.313	1000 (OSHA)
Paraffins	SBP 1	64742-89-8	YES	NO	YES		400 (ACGIH)
Paraffins	SBP 100/140	64742-49-0	YES	NO	YES		
Paraffins	SBP 11	64742-88-7	YES	NO	YES		100 (ACGIH)
Paraffins	SBP 3	64742-89-8	YES	NO	YES		400 (ACGIH)
Paraffins	SBP 5	64742-89-8	YES	NO	YES		400 (ACGIH)
Paraffins	SBP 80 / 110	64747-49-0	YES	NO	YES		
Paraffins	Solvesso 100	64742-95-6	YES	NO	YES		50 (ACGIH)
Paraffins	Solvesso 150	64742-94-5	YES	NO	YES		50 (ACGIH)
Paraffins	Solvesso 200	64747-94-5	YES	NO	YES		
Paraffins	Whitespirit	64742-82-1	YES	NO	YES		100 (ACGIH)
Paraffins	Hexane	110-54-3	YES	YES	YES	1.244	50 (ACGIH)
Specials	Carbon Dioxide	124-38-9	NO	NO	ERR		5000 (OSHA)
Specials	Dimethyl Carbonate	616-38-6	NO	NO	YES	0.059	200 (CEL)
Specials	Water	7732-18-5	NO	NO	YES		NONE
Specials	1,3 Dioxolane	646-6-0	YES	NO	YES	4.959	20 (ACGIH)
Specials	4-Bromo-3-Chloro-3, 4,4-Trifluoro-1-Butene	374-25-4	YES	NO	YES		
Specials	Dimethyl Acetamide	127-19-5	YES	NO	YES		10 (NIOSH)

Specials	Dimethyl Sulfoxide	67-68-5	YES	NO	YES	6.676	10 (AEL)
Specials	Ethylene Carbonate	96-49-1	YES	NO	YES		
Specials	Gamma Butyrolactone	96-48-0	YES	NO	YES	0.956	
Specials	N-Methyl Pyrrolidone	872-50-4	YES	NO	YES	2.411	10 (AI HA)
Specials	Propylene Carbonate	108-32-7	YES	NO	YES	0.280	20 (AEL)
Specials	Pyridine	110-86-1	YES	NO	YES		5 (ACGIH)
Specials	Soygold 1000	67784-80-9	YES	NO	YES		
Specials	Tetrahydrofuran	109-99-9	YES	NO	YES	4.307	200 (ACGIH)
Specials	1,4-Dioxane	123-91-1	YES	YES	YES	2.622	100(OSHA)
Specials	Acetonitrile	75-5-8	YES	YES	YES		40(OSHA)
Specials	Cumene	98-82-8	YES	YES	YES	2.516	50(OSHA)
Specials	Dimethyl Formamide	68-12-2	YES	YES	YES		10(OSHA)
Specials	Isophorone	78-59-1	YES	YES	YES	4.625	25(OSHA)
Acids	Acetic Acid			3	2	0	29.8
Alcohols	1-Hexanol			1	2	0	Not Included
Alcohols	Amyl Alcohol			1	3	0	Not Included
Alcohols	Benzyl Alcohol			2	1	0	13.0
Alcohols	Cyclohexanol			1	2	0	11.6
Alcohols	Diacetone Alcohol			1	2	0	17.8
Alcohols	Ethanol			0	3	0	13.2
Alcohols	Furfuryl Alcohol			1	2	1	Not Included
Alcohols	iso-Amyl Alcohol			1	2	0	Not Included
Alcohols	Isobutanol			2	3	0	22.9
Alcohols	Isopropanol			1	3	0	14.2
Alcohols	Methyl Isobutyl Carbinol			2	3	0	Not Included
Alcohols	N-Butanol			2	3	0	19.4
Alcohols	N-Propanol			1	3	0	14.0

(Continued)

Table A1-3 *(Continued)*

Type	Name	CAS #	VOC in US	HAP in US	VOC in Europe	Ozone Reactivity MIR	Exposure Limit in US
Alcohols	Secondary Butanol						12.4
Alcohols	Tertiary Butanol			2	3	0	13.1
Alcohols	Tetrahydrofurfuryl Alcohol			2	2	0	Not Included
Alcohols	Texanol			1	1	0	6.4
Alcohols	Methanol			1	3	0	24.7
Aromatics	Lactol Spirits			1	3	0	Not Included
Aromatics	Mineral Spirits			0	2	0	11.6
Aromatics	Stoddard Solvent			1	2	0	11.6
Aromatics	Turpentine, oil			1	3	0	Not Included
Aromatics	VM&P Naptha			1	3	0	13.3
Aromatics	Benzene			1	3	0	48.0
Aromatics	o-Xylene			2	3	0	26.1
Aromatics	Toluene			2	3	0	29.1
Benzotrifluorides	Parachloroben-zotrifluoride			1	2	1	28.1
Benzotrifluorides	Benzotrifluoride			3	3	1	Not Included
Brominateds	n-Propyl Bromide			2	3	1	9.2
CFCs	CFC-113	1.000	14190	1	0	1	14.5
Chlorinateds	Trans 1,2 Dichloroethylene			2	3	2	28.6
Chlorinateds	1,1,1-Trichloroethane	0.12 (Class I)	144	2	1	0	36.6
Chlorinateds	Methylene Chloride	0.007	Negligible	2	1	0	Not Included
Chlorinateds	Perchloroethylene	0.005		2	0	0	37.5
Chlorinateds	1,1-Dichloroethane			1	3	0	23.7
Chlorinateds	Carbon Tetrachloride	1.200	2540	3	0	0	41.6

Chlorinateds	Chloroform		2	0	0	36.1
Chlorinateds	Ethylene Dichloride		1	3	0	Not Included
Chlorinateds	Trichloroethylene	0.005	2	1	0	41.3
Dibasic Acid Esters	DBE		1	1	0	Not Included
Dibasic Acid Esters	DBE-4		1	1	0	Not Included
Dibasic Acid Esters	DBE-6		1	1	0	
Esters	Methyl Acetate		2	3	0	Not Included
Esters	Methyl Formate		2	4	0	19.3
Esters	Tertiary Butyl Acetate		1	4	0	11.3
Esters	2-Ethyl Hexyl Acetate		1	2	0	Not Included
Esters	Amyl Acetate		2	3	0	Not Included
Esters	Amyl Acetate		2	3	0	16.5
Esters	Butyl Glycol Acetate		1	2	0	Not Included
Esters	Cyclohexyl Acetate		1	2	0	Not Included
Esters	Dimethyl Succinate		1	2	0	
Esters	Ethyl Acetate		1	3	0	30.7
Esters	Ethyl Formate		2	3	0	Not Included
Esters	Ethyl-3-Ethoxypropionate					13.4
Esters	Ethylene Glycol Diacetate		1	2	0	Not Included
Esters	Isobutyl Acetate		1	3	0	16.8
Esters	Isopropyl Acetate		1	3	0	12.5
Esters	n-Butyl Acetate		2	3	0	18.4
Esters	N-Butyl Propionate		2	3	0	Not Included
Esters	n-Propyl Acetate		1	3	0	Not Included
Esters	n-Propyl Propionate		2	3	0	

(Continued)

Table A1-3 *(Continued)*

Type	Name	CAS #	VOC in US	HAP in US	VOC in Europe	Ozone Reactivity MIR	Exposure Limit in US
Esters	n-Propyl Propionate			2	3	0	
Esters	Propylene Glycol Diacetate			1	2	0	24.2
Ethers	Cyclopentyl Methyl Ether	NO		0	0	0	
Ethers	Methyl Butyl Ether			0	0	0	Not Included
Ethers	Methyl t-Butyl Ether			2	3	0	
Ethers	Methylal			1	3	1	
Ethers	t-Butyl Ether			0	0	0	Not Included
Ethers	Amyl t-Butyl Ether			0	0	0	
Glycol Ethers	Dipropylene Glycol Dimethyl Ether			0	0	0	
Glycol Ethers	Dipropylene Glycol n-Propyl Ether			0	0	0	13.0
Glycol Ether Acetates	Diethylene Glycol Monobutyl Ether Acetate			1	1	0	Not Included
Glycol Ether Acetates	Diethylene Glycol Monoethyl Ether Acetate			1	1	0	Not Included
Glycol Ether Acetates	Dipropylene Glycol Methyl Ether Acetate						Not Included
Glycol Ether Acetates	Ethylene Glycol Butyl Ether Acetate			1	2	0	Not Included
Glycol Ether Acetates	Ethylene Glycol Ethyl Ether Acetate			1	2	1	
Glycol Ether Acetates	Ethylene glycol methyl ether acetate			1	2	0	17.6
Glycol Ether Acetates	Propylene Glycol Ethyl Ether Acetate						Not Included
Glycol Ether Acetates	Propylene Glycol Methyl Ether Acetate			0	2	0	10.7

Glycol Ethers	3-Methoxy-3-Methyl-1-Butanol	1	2	0	5.4
Glycol Ethers	Butyl Glycol Ether	3	2	0	20.5
Glycol Ethers	Diethylene Glycol Monobutyl ether	1	2	0	15.5
Glycol Ethers	Diethylene Glycol Monomethyl Ether	1	2	0	Not Included
Glycol Ethers	Diethylene Glycol Monomethyl Ether	1	2	0	17.5
Glycol Ethers	Dipropylene Glycol Monobutyl Ether				Not Included
Glycol Ethers	Dipropylene Glycol Monobutyl Ether	2	1	0	Not Included
Glycol Ethers	Dipropylene Glycol Monoethyl Ether				Not Included
Glycol Ethers	Dipropylene Glycol Monomethyl Ether	0	2	0	13.4
Glycol Ethers	Dipropylene Glycol n-Butyl Ether	2	1	0	Not Included
Glycol Ethers	Ethylene Glycol Diethyl Ether	2	3	0	Not Included
Glycol Ethers	Ethylene Glycol Mono Isopropyl Ether	3	2	1	Not Included
Glycol Ethers	Ethylene Glycol Monobutyl Ether	3	2	0	20.5
Glycol Ethers	Ethylene Glycol Monoethyl Ether	1	2	0	33.6
Glycol Ethers	Ethylene Glycol Monomethyl Ether	1	2	1	26.0
Glycol Ethers	Propylene Glycol Ethyl Propyl Ether				Not Included
Glycol Ethers	Propylene Glycol Methyl Propyl Ether				13.4

(Continued)

Table A1-3 *(Continued)*

Type	Name	CAS #	VOC in US	HAP in US	VOC in Europe	Ozone Reactivity MIR	Exposure Limit in US
Glycol Ethers	Propylene Glycol Mono t-butyl Ether						10.3
Glycol Ethers	Propylene Glycol Monobutyl Ether				3		Not Included
Glycol Ethers	Propylene Glycol Monoethyl Ether			0	3	0	Not Included
Glycol Ethers	Propylene Glycol Monoethyl Ether			0	3	0	Not Included
Glycol Ethers	Propylene Glycol Monomethyl Ether			0	3	0	18.5
Glycol Ethers	Propylene Glycol Monomethyl Ether			0	3	0	18.5
Glycol Ethers	Propylene Glycol Monopropyl Ether			0	1	0	Not Included
Glycol Ethers	Propylene Glycol n-Propyl Ether			0	1	0	Not Included
Glycol Ethers	Triethylene glycol monomethyl ether			0	0	0	Not Included
Glycol Ethers	Tripropylene Glycol Monomethyl Ether			0	1	0	Not Included
Glycol Ethers	Tripropylene Glycol n-Butyl Ether			0	0	0	8.6
Glycol Ethers	Tripropylene Glycol n-Propyl Ether			0	0	0	
Glycols	1,4-Butane Diol			1	1	1	Not Included
Glycols	Diethylene Glycol			1	1	0	14.7
Glycols	Dipropylene Glycol			1	1	0	11.9
Glycols	Hexylene Glycol			0	1	0	14.4
Glycols	Propylene Glycol			0	1	0	14.2
Glycols	Triethylene Glycol			1	1	0	Not Included
Glycols	Ethylene Glycol			1	1	0	16.0

Glymes	Diethylene Glycol Dibutyl Ether (Butyl Diglyme)			1	1	0	Not Included
Glymes	Diethylene Glycol Diethyl Ether (Ethyl Diglyme)			1	2	0	Not Included
Glymes	Diethylene Glycol Dimethyl Ether (Diglyme)			1	2	1	Not Included
Glymes	Ethylene Glycol Diethyl Ether (Ethyl Glyme)			1	3	0	Not Included
Glymes	Ethylene Glycol Dimethyl Ether (Monoglyme)			2	2	0	Not Included
Glymes	Triethylene Glycol Dimethyl Ether (Triglyme)			1	2	1	Not Included
HCFCs	HCFC-141b	0.120	713	2 NCPA HMIS	1 NCPA HMIS	0 NCPA HMIS	Not Included
HCFCs	HCFC-225 ca / cb		353	2 NCPA HMIS	0 NCPA HMIS	0 NCPA HMIS	Not Included
HFCs	HFC 43-10 mee		1610	1 NCPA HMIS	0 NCPA HMIS	1 NCPA HMIS	12.7
HFCs	HFC-245 fa		1020				Not Included
HFCs	HFC-365mfc		782	1	0	1	Not Included
HFEs	HFE-7100		397	3	1	0	Not Included
HFEs	HFE-7200		56	3	1	0	Not Included
HFEs	HFE-7300		200	0	0	0	
HFEs	HFE-8200		55	3	1	0	
Iodateds	Perfluoro n-butyl Iodide						Not Included
Ketones	Acetone			1	3	0	15.9
Ketones	Acetyl Acetone			2	3	0	Not Included
Ketones	Cyclohexanone			2	3	0	32.2
Ketones	Diethyl Ketone			1	3	0	Not Included
Ketones	Diisobutyl Ketone			1	2	0	Not Included

(*Continued*)

Table A1-3 *(Continued)*

Type	Name	CAS #	VOC in US	HAP in US	VOC in Europe	Ozone Reactivity MIR	Exposure Limit in US
Ketones	Dispropyl Ketone						Not Included
Ketones	Methyl Amyl Ketone			1	2	0	12.9
Ketones	Methyl Isoamyl Ketone			1	2	0	Not Included
Ketones	Methyl Propyl Ketone				3	0	13.7
Ketones	Methyl Ethyl Ketone			1	3	0	27.9
Ketones	Methyl Isobutyl Ketone				3	1	31.3
Lactates	2-Ethylhexyl Lactate						Not Included
Lactates	Butyl Lactate			1	2	0	Not Included
Lactates	Ethyl Lactate				2	0	Not Included
Lactates	Isopropyl Lactate			1	2	0	Not Included
Lactates	Methyl Lactate			1	2	0	Not Included
Lactates	Propyl Lactate						Not Included
Limonene	D-Limonene				2	0	7.8
Limonene	Dipentene (Dl-Limonene)			1	3	0	Not Included
OSs	OS-10 Hexamethyldisiloxane			1	3	0	Not Included
OSs	OS-20 Octamethyltrisiloxane			1	3	0	Not Included
OSs	OS-30 Decamethyltrisiloxane			1	2	0	Not Included
Paraffins	2-Methyl Heptane				3	0	Not Included
Paraffins	2-Methyl Pentane			1	3	0	
Paraffins	3-Methyl Hexane				3	0	Not Included
Paraffins	Cyclohexane			1	3	0	23.5
Paraffins	Cyclopentane			1	3	0	Not Included
Paraffins	Cypar 7				3	0	13.3
Paraffins	Cypar 9						Not Included
Paraffins	D30			1	3	0	Not Included

Paraffins	D40	1	3	0	Not Included
Paraffins	D60	1	3	0	Not Included
Paraffins	D70		3		Not Included
Paraffins	DSP 145/160	1	3	1	Not Included
Paraffins	Ethyl Cyclohexane	1	3	0	Not Included
Paraffins	Heptane	1	3	0	14.7
Paraffins	Isododecane	0	2	0	Not Included
Paraffins	Isopar C	1	3	0	Not Included
Paraffins	Isopar E	1	3	0	Not Included
Paraffins	Isopar H	1	2	0	Not Included
Paraffins	Isopar L	1	2	0	Not Included
Paraffins	Isopar M	1	2	0	Not Included
Paraffins	Methyl Cyclohexane	1	3	1	Not Included
Paraffins	Methyl Cyclopentane	1	3	0	Not Included
Paraffins	Nonane	1	3	0	Not Included
Paraffins	Octane	0	3	0	Not Included
Paraffins	Pentane	1	4	0	16.3
Paraffins	SBP 1	2	3	0	13.3
Paraffins	SBP 100/140	1	3	0	Not Included
Paraffins	SBP 11				16.6
Paraffins	SBP 3	2	3	0	13.3
Paraffins	SBP 5	2	3	0	13.3
Paraffins	SBP 80 / 110	1	3	0	Not Included
Paraffins	Solvesso 100	1	3	0	14.2
Paraffins	Solvesso 150	2	3	0	Not Included
Paraffins	Solvesso 200	2	3	0	Not Included
Paraffins	Whitespirit	1	3	0	Not Included

(Continued)

Table A1-3 *(Continued)*

Type	Name	CAS #	VOC in US	HAP in US	VOC in Europe	Ozone Reactivity MIR	Exposure Limit in US
Paraffins	Hexane			1	3	0	31.0
Specials	Carbon Dioxide			0	0	0	4.1
Specials	Dimethyl Carbonate			1	3	1	
Specials	Water			0	0	0	
Specials	1,3 Dioxolane			1	3	2	Not Included
Specials	4-Bromo-3-Chloro-3, 4,4-Trifluoro-1-Butene	0.010		0	0	0	
Specials	Dimethyl Acetamide			2	2	0	Not Included
Specials	Dimethyl Sulfoxide			2	2	0	Not Included
Specials	Ethylene Carbonate			2	1	1	
Specials	Gamma Butyrolactone			1	1	0	Not Included
Specials	N-Methyl Pyrrolidone			2	2	0	7.1
Specials	Propylene Carbonate			1	1	0	Not Included
Specials	Pyridine			3	3	0	27.2
Specials	Soygold 1000			0	1	0	Not Included
Specials	Tetrahydrofuran			2	3	1	31.3
Specials	1,4-Dioxane			2	3	1	33.8
Specials	Acetonitrile			2	3	0	Not Included
Specials	Cumene			2	3	1	31.7
Specials	Dimethyl Formamide			2	2	0	25.4
Specials	Isophorone			2	2	0	29.9

Table A1-4

Type	Name	CAS #	Flash Point °F	NFPA Flammability Classification	LEL Vol %	UEL Vol %
Acids	Acetic Acid	64-19-7	104.0	Combustible Class II	4.00%	16.00%
Alcohols	Texanol	25265-77-4	248.0	Combustible Class IIIB or NOS	0.62%	4.24%
Alcohols	Methyl Isobutyl Carbinol	108-11-2	102.9	Combustible Class II	1.00%	5.50%
Alcohols	Amyl Alcohol	71-41-0	107.6	Combustible Class II	1.20%	10.00%
Alcohols	Diacetone Alcohol	123-42-2	113.0	Combustible Class II	1.80%	6.90%
Alcohols	1-Hexanol	111-27-3	145.4	Combustible Class IIIA	1.20%	7.70%
Alcohols	Cyclohexanol	108-93-0	154.4	Combustible Class IIIA	1.00%	9.00%
Alcohols	Furfuryl Alcohol	98-0-0	167.0	Combustible Class IIIA	1.80%	16.30%
Alcohols	Tetrahydrofurfuryl Alcohol	97-99-4	167.0	Combustible Class IIIA	1.50%	9.70%
Alcohols	Benzyl Alcohol	100-51-6	199.4	Combustible Class IIIA	1.30%	13.00%
Alcohols	iso-Amyl Alcohol	123-51-3	113.0	Combustible Class II	1.20%	9.00%
Alcohols	Tertiary Butanol	75-65-0	52.0	Flammable Class IB	2.40%	8.00%
Alcohols	Methanol	67-56-1	53.6	Flammable Class IB	6.00%	36.00%
Alcohols	Ethanol	64-17-5	55.4	Flammable Class IB	3.30%	19.00%
Alcohols	Isopropanol	67-63-0	57.2	Flammable Class IB	2.00%	12.70%
Alcohols	N-Propanol	71-23-8	59.0	Flammable Class IB	2.20%	13.70%
Alcohols	Secondary Butanol	78-92-2	75.0	Flammable Class IC	1.70%	9.80%
Alcohols	Isobutanol	78-83-1	80.6	Flammable Class IC	1.70%	10.60%
Alcohols	N-Butanol	71-36-3	80.6	Flammable Class IC	1.40%	11.20%
Aromatics	Stoddard Solvent	8052-41-3	102.2	Combustible Class II	0.90%	7.00%
Aromatics	Mineral Spirits	64741-41-9	113.0	Combustible Class II	1.00%	7.00%
Aromatics	Benzene	71-43-2	12.2	Flammable Class IB	1.20%	7.80%
Aromatics	Lactol Spirits	64742-49-0	17.6	Flammable Class IB	1.00%	7.00%

(Continued)

Table A1-4 *(Continued)*

Type	Name	CAS #	Flash Point °F	NFPA Flammability Classification	LEL Vol %	UEL Vol %
Aromatics	Toluene	108-88-3	39.9	Flammable Class IB	1.10%	7.10%
Aromatics	VM&P Naptha	64742-89-8	40.3	Flammable Class IB	1.10%	5.90%
Aromatics	o-Xylene	95-47-6	75.2	Flammable Class IC	1.10%	7.00%
Aromatics	Turpentine, oil	8006-64-2	95.0	Flammable Class IC	0.80%	6.00%
Benzotrifluorides	Parachlorobenzotrifluoride	98-56-6	109.4	Combustible Class II	0.90%	10.50%
Benzotrifluorides	Benzotrifluoride	98-08-8	53.8	Flammable Class IB	3.10%	9.30%
Brominateds	n-Propyl Bromide	106-94-5	75.2	Flammable Class IC	4.60%	8.50%
CFCs	CFC-113	76-13-1	None	NOS		
Chlorinateds	Perchloroethylene	127-18-4	113.0	Combustible Class II	13.00%	23.00%
Chlorinateds	Carbon Tetrachloride	56-23-5	None	NOS		
Chlorinateds	Chloroform	67-66-3	none	NOS		
Chlorinateds	1,1-Dichloroethane	75-34-3	10.4	Flammable Class IB	5.40%	11.40%
Chlorinateds	Ethylene Dichloride	107-06-2	15.8	Flammable Class IB	6.20%	16.00%
Chlorinateds	Methylene Chloride	75-09-2	24.8	Flammable Class IB	12.00%	19.00%
Chlorinateds	1,1,1-Trichloroethane	71-55-6	30.2	Flammable Class IB	7.50%	12.50%
Chlorinateds	Trans 1,2 Dichloroethylene	156-60-5	35.6	Flammable Class IB	6.00%	13.00%
Chlorinateds	Trichloroethylene	79-01-6	89.6	Flammable Class IC	7.90%	11.00%
Dibasic Acid Esters	DBE	1119-40-0	212.0	Combustible Class IIIB or NOS	1.40%	14.30%
Dibasic Acid Esters	DBE-4	106-65-0	200.0	Combustible Class IIIB or NOS		
Dibasic Acid Esters	DBE-6	627-93-0	235.4	Combustible Class IIIB or NOS	0.81%	8.10%
Esters	Cyclohexyl Acetate	622-45-7	134.6	Combustible Class II		
Esters	Ethyl-3-Ethoxypropionate	763-69-9	138.0	Combustible Class II	1.10%	8.70%
Esters	Methyl Acetate	79-20-9	14.0	Flammable Class IB	3.10%	16.00%

Esters	Butyl Glycol Acetate	112-07-2	183.2	Combustible Class IIIA	1.70%	8.40%
Esters	Dimethyl Succinate	106-65-0	185.0	Combustible Class IIIA	1.00%	8.50%
Esters	Propylene Glycol Diacetate	623-84-7	187.0	Combustible Class IIIA	2.80%	12.70%
Esters	2-Ethyl Hexyl Acetate	103-9-3	159.8	Combustible Class IIIA	0.76%	8.14%
Esters	Methyl Formate	107-31-3	-2.0	Flammable Class IA	4.50%	23.00%
Esters	Ethylene Glycol Diacetate	111-55-7	210.2	Combustible Class IIIB or NOS	1.60%	8.40%
Esters	Ethyl Acetate	141-78-6	26.6	Flammable Class IB	2.00%	11.50%
Esters	Tertiary Butyl Acetate	540-88-5	33.8	Flammable Class IB	1.30%	7.30%
Esters	Ethyl Formate	109-94-4	-4.0	Flammable Class IB	2.80%	16.00%
Esters	Isopropyl Acetate	108-21-4	41.0	Flammable Class IB	1.80%	8.00%
Esters	n-Propyl Acetate	109-60-4	57.2	Flammable Class IB	1.70%	8.00%
Esters	n-Propyl Propionate	106-36-5	66.2	Flammable Class IB		
Esters	n-Propyl Propionate	106-36-5	66.9	Flammable Class IB		
Esters	Isobutyl Acetate	110-19-0	69.8	Flammable Class IB	1.30%	10.50%
Esters	Amyl Acetate	628-63-7	71.6	Flammable Class IB	1.10%	7.50%
Esters	Amyl Acetate	628-63-7	73.4	Flammable Class IC	1.10%	7.50%
Esters	n-Butyl Acetate	123-86-4	80.6	Flammable Class IC	1.70%	7.60%
Esters	N-Butyl Propionate	590-01-2	89.6	Flammable Class IC	1.10%	6.80%
Ethers	Methylal	109-87-5	-0.4	FlammableClass IB	3.60%	12.60%
Ethers	Methyl Butyl Ether	628-28-4	14.0	Flammable Class IB		
Ethers	Methyl t-Butyl Ether	1634-04-4	-18.4	FlammableClass IB	1.60%	8.40%
Ethers	t-Butyl Ether	6163-66-2	26.6	Flammable Class IB	0.90%	6.80%
Ethers	Cyclopentyl Methyl Ether	5614-37-9	31.8	Flammable Class IB	1.10%	9.90%
Ethers	Amyl t-Butyl Ether	10100-95-5	89.9	Flammable Class IC		
Glycol Ethers	Dipropylene Glycol Dimethyl Ether	111109-77-4	149.0	Combustible Class IIIA	0.77%	5.57%

(Continued)

Table A1-4 *(Continued)*

Type	Name	CAS #	Flash Point °F	NFPA Flammability Classification	LEL Vol %	UEL Vol %
Glycol Ethers	Dipropylene Glycol n-Propyl Ether	29911-27-1	190.4	Combustible Class IIIA	0.68%	
Glycol Ether Acetates	Ethylene glycol methyl ether acetate	110-49-6	114.8	Combustible Class II	1.50%	12.30%
Glycol Ether Acetates	Propylene Glycol Methyl Ether Acetate	108-65-6	122.0	Combustible Class II		
Glycol Ether Acetates	Propylene Glycol Ethyl Ether Acetate	54839-24-6	129.2	Combustible Class II		
Glycol Ether Acetates	Ethylene Glycol Ethyl Ether Acetate	111-15-9	132.8	Combustible Class II	1.30%	7.20%
Glycol Ether Acetates	Ethylene Glycol Butyl Ether Acetate	112-07-2	160.0	Combustible Class IIIA	0.88%	8.54%
Glycol Ether Acetates	Dipropylene Glycol Methyl Ether Acetate	88917-22-0	187.0	Combustible Class IIIA		
Glycol Ether Acetates	Diethylene Glycol Monoethyl Ether Acetate	112-15-2	225.0	Combustible Class IIIB or NOS	1.50%	19.40%
Glycol Ether Acetates	Diethylene Glycol Monobutyl Ether Acetate	124-17-4	240.0	Combustible Class IIIB or NOS		
Glycol Ethers	Ethylene Glycol Monomethyl Ether	109-86-4	104.0	Combustible Class II	1.80%	14.00%
Glycol Ethers	Propylene Glycol Monoethyl Ether	1569-02-4	104.0	Combustible Class II	1.30%	12.00%
Glycol Ethers	Propylene Glycol Monoethyl Ether	1569-02-4	109.4	Combustible Class II	1.30%	12.00%
Glycol Ethers	Ethylene Glycol Monoethyl Ether	110-80-5	111.9	Combustible Class II	3.00%	18.00%
Glycol Ethers	Propylene Glycol Mono t-butyl Ether	57018-52-7	113.0	Combustible Class II		

Glycol Ethers	Propylene Glycol Monopropyl Ether	30136-13-1	118.4	Combustible Class II		
Glycol Ethers	Propylene Glycol n-Propyl Ether	30136-13-1	118.4	Combustible Class II		
Glycol Ethers	Ethylene Glycol Mono Isopropyl Ether	109-59-1	129.2	Combustible Class II	1.60%	13.00%
Glycol Ethers	Propylene Glycol Monobutyl Ether	1569-02-4	145.4	Combustble Class IIIA		
Glycol Ethers	Ethylene Glycol Monobutyl Ether	111-76-2	146.3	Combustble Class IIIA	1.10%	12.70%
Glycol Ethers	3-Methoxy-3-Methyl-1-Butanol	56539-66-3	154.4	Combustble Class IIIA	1.30%	13.10%
Glycol Ethers	Butyl Glycol Ether	111-76-2	154.4	Combustble Class IIIA	1.10%	12.70%
Glycol Ethers	Dipropylene Glycol Monomethyl Ether	112-28-7	167.0	Combustble Class IIIA	1.10%	3.00%
Glycol Ethers	Diethylene Glycol Monobutyl ether	112-34-5	172.0	Combustble Class IIIA	0.85%	24.60%
Glycol Ethers	Propylene Glycol Methyl Propyl Ether	34590-94-8	174.2	Combustble Class IIIA		
Glycol Ethers	Propylene Glycol Ethyl Propyl Ether	30025-38-8	187.7	Combustble Class IIIA		
Glycol Ethers	Dipropylene Glycol Monoethyl Ether	30025-38-8	195.0	Combustble Class IIIA		
Glycol Ethers	Diethylene Glycol Monomethyl Ether	111-77-3	204.8	Combustble Class IIIB or NOS	1.60%	18.10%
Glycol Ethers	Diethylene Glycol Monomethyl Ether	111-77-3	204.8	Combustble Class IIIB or NOS	1.40%	22.70%
Glycol Ethers	Dipropylene Glycol n-Butyl Ether	29911-28-2	212.9	Combustble Class IIIB or NOS	0.60%	20.40%
Glycol Ethers	Dipropylene Glycol Monobutyl Ether	29911-28-2	212.9	Combustble Class IIIB or NOS	0.60%	20.40%
Glycol Ethers	Dipropylene Glycol Monobutyl Ether	29911-28-2	212.0	Combustble Class IIIB or NOS		

(Continued)

Table A1-4 *(Continued)*

Type	Name	CAS #	Flash Point °F	NFPA Flammability Classification	LEL Vol %	UEL Vol %
Glycol Ethers	Tripropylene Glycol Monomethyl Ether	25498-49-1	249.8	Combustble Class IIIB or NOS		
Glycol Ethers	Tripropylene Glycol n-Propyl Ether	96077-04-2	259.0	Combustble Class IIIB or NOS		
Glycol Ethers	Triethylene glycol monomethyl ether	112-35-6	275.0	Combustble Class IIIB or NOS		
Glycol Ethers	Tripropylene Glycol n-Butyl Ether	55934-93-5	491.0	Combustble Class IIIB or NOS		
Glycol Ethers	Propylene Glycol Monomethyl Ether	107-98-2	87.8	Flammable Class IC	1.60%	13.80%
Glycol Ethers	Propylene Glycol Monomethyl Ether	107-98-2	89.6	Flammable Class IC	1.60%	13.80%
Glycol Ethers	Ethylene Glycol Diethyl Ether	629-14-1	95.0	Flammable Class IC		
Glycols	Hexylene Glycol	107-41-5	199.4	Combustible Class IIIA	1.00%	9.00%
Glycols	Propylene Glycol	57-55-6	219.2	Combustible Class IIIB or NOS	2.40%	17.40%
Glycols	Ethylene Glycol	107-21-1	240.8	Combustible Class IIIB or NOS	3.20%	22.00%
Glycols	Diethylene Glycol	111-46-6	255.2	Combustible Class IIIB or NOS	2.00%	17.00%
Glycols	1,4-Butane Diol	110-63-4	273.2	Combustible Class IIIB or NOS	2.40%	15.30%
Glycols	Dipropylene Glycol	2396-61-4	280.4	Combustible Class IIIB or NOS	2.90%	12.60%
Glycols	Triethylene Glycol	112-27-6	309.0	Combustible Class IIIB or NOS	0.90%	9.20%
Glymes	Diethylene Glycol Dimethyl Ether (Diglyme)	111-96-6	134.6	Combustible Class II	1.40%	17.40%

Glymes	Diethylene Glycol Diethyl Ether (Ethyl Diglyme)	112-36-7	194.0	Combustible Class IIIA	0.90%	10.10%
Glymes	Ethylene Glycol Dimethyl Ether (Monoglyme)	110-71-4	21.2	Flammable Class IB		
Glymes	Triethylene Glycol Dimethyl Ether (Triglyme)	112-49-2	235.4	Combustible Class IIIB or NOS	1.40%	17.40%
Glymes	Diethylene Glycol Dibutyl Ether (Butyl Diglyme)	112-73-2	246.2	Combustible Class IIIB or NOS		
Glymes	Ethylene Glycol Diethyl Ether (Ethyl Glyme)	629-14-1	80.6	Flammable Class IC		
HCFCs	HCFC-225 ca / cb	422-56-0/507-55-1	None	NOS		
HCFCs	HCFC-141b	1717-00-6	None	NOS	10.80%	14.80%
HFCs	HFC-245 fa	460-73-1	None	NOS		
HFCs	HFC 43-10 mee	138495-42-8	None	NOS		
HFCs	HFC-365mfc	405-58-6	-13.0	Flammable Class IB	7.00%	9.00%
HFEs	HFE-7300	132182-92-4	None	NOS		
HFEs	HFE-7100	163702-07-6	None	NOS		
HFEs	HFE-8200	163702-05-4/163702-06-5	None	NOS	2.40%	12.40%
HFEs	HFE-7200	163702-06-5	None	NOS	2.40%	12.40%
Iodateds	Perfluoro n-butyl Iodide	423-39-2	0.0	Flammable Class IB		
Ketones	Acetone	67-64-1	1.4	Flammable Class IB	2.60%	12.80%
Ketones	Methyl Amyl Ketone	110-43-0	102.2	Combustible Class II	1.10%	7.90%
Ketones	Cyclohexanone	108-94-1	111.2	Combustible Class II	1.10%	9.40%
Ketones	Diisobutyl Ketone	108-83-8	120.2	Combustible Class II	0.80%	7.10%
Ketones	Dispropyl Ketone	123-19-3	120.2	Combustible Class II		
Ketones	Methyl Ethyl Ketone	78-93-3	23.0	Flammable Class IB	1.40%	11.40%

(Continued)

Table A1-4 *(Continued)*

Type	Name	CAS #	Flash Point °F	NFPA Flammability Classification	LEL Vol %	UEL Vol %
Ketones	Methyl Propyl Ketone	107-87-9	44.6	Flammable Class IB	1.50%	8.20%
Ketones	Diethyl Ketone	96-22-0	55.4	Flammable Class IB	1.60%	7.70%
Ketones	Methyl Isobutyl Ketone	108-10-1	60.8	Flammable Class IB	1.20%	8.00%
Ketones	Acetyl Acetone	123-54-6	93.2	Flammable Class IC	2.40%	11.60%
Ketones	Methyl Isoamyl Ketone	110-12-3	96.8	Flammable Class IC	1.00%	8.20%
Lactates	Ethyl Lactate	97-64-3	116.6	Combustible Class II	1.50%	11.40%
Lactates	Methyl Lactate	27871-49-4	136.4	Combustible Class II	2.20%	
Lactates	Isopropyl Lactate	63697-0-7	140.0	Combustible Class IIIA		
Lactates	Propyl Lactate	616-09-1	156.2	Combustible Class IIIA		
Lactates	Butyl Lactate	138-22-7	174.2	Combustible Class IIIA	1.15%	7.90%
Lactates	2-Ethylhexyl Lactate	186817-80-1	235.4	Combustible Class IIIB or NOS		
Limonene	Dipentene (Dl-Limonene)	138-86-3	118.0	Combustible Class II	0.70%	6.10%
Limonene	D-Limonene	5989-27-5	118.0	Combustible Class II	0.70%	6.10%
OSs	OS-10 Hexamethyldisiloxane	107-46-0	30.0	Flammable Class IB	1.25%	18.60%
OSs	OS-30 Decamethyltrisiloxane	141-62-8	350.0	Combustible Class IIIB or NOS	0.90%	
OSs	OS-20 Octamethyltrisiloxane	107-51-7	84.0	Flammable Class IC	0.90%	13.80%
Paraffins	Cyclohexane	110-82-7	-0.4	Flammable Class IB	1.30%	8.00%
Paraffins	SBP 5	64742-89-8	-0.4	Flammable Class IB		
Paraffins	SBP 80 / 110	64747-49-0	10.4	Flammable Class IB		
Paraffins	SBP 11	64742-88-7	100.4	Combustible Class II	1.00%	5.00%
Paraffins	Whitespirit	64742-82-1	102.2	Combustible Class II	0.60%	7.00%

Paraffins	D40	64742-48-9	109.4	Combustible Class II		
Paraffins	Methyl Cyclopentane	96-37-7	1.4	Flammable Class IB	1.20%	8.35%
Paraffins	Isododecane	13475-86-6	114.8	Combustible Class II		
Paraffins	Solvesso 100	64742-95-6	120.2	Combustible Class II		
Paraffins	Isopar H	90622-57-4	132.8	Combustible Class II	0.70%	5.40%
Paraffins	D60	64742-48-9	143.6	Combustible Class IIIA		
Paraffins	Isopar L	90622-58-5	145.4	Combustible Class IIIA	1.30%	8.80%
Paraffins	Solvesso 150	64742-94-5	147.2	Combustible Class IIIA		
Paraffins	D70	64742-47-8	154.4	Combustible Class IIIA	1.20%	6.00%
Paraffins	Isopar M	9622-58-5	168.8	Combustible Class IIIA	0.50%	4.90%
Paraffins	Isopar C	90622-56-3	19.4	Flammable Class IB	0.90%	6.30%
Paraffins	SBP 3	64742-89-8	19.4	Flammable Class IB		
Paraffins	Solvesso 200	64747-94-5	206.6	Combustible Class IIIB or NOS		
Paraffins	2-Methyl Pentane	107-83-5	-23.0	Flammable Class IA	1.00%	7.40%
Paraffins	Heptane	142-82-5	24.8	Flammable Class IB	1.05%	6.70%
Paraffins	Cypar 7	64742-89-8	24.8	Flammable Class IB		
Paraffins	Methyl Cyclohexane	108-87-2	25.7	Flammable Class IB	1.20%	6.70%
Paraffins	3-Methyl Hexane	589-34-4	25.7	Flammable Class IB	1.00%	7.00%
Paraffins	SBP 100/140	64742-49-0	33.8	Flammable Class IB	0.80%	8.00%
Paraffins	2-Methyl Heptane	592-27-8	42.8	Flammable Class IB	1.00%	5.80%
Paraffins	Cyclopentane	287-92-3	-43.6	Flammable Class IB	1.50%	8.70%
Paraffins	Isopar E	90622-56-3	44.6	Flammable Class IB	0.90%	6.20%
Paraffins	SBP 1	64742-89-8	-45.4	Flammable Class IB		
Paraffins	Pentane	109-66-0	-54.4	Flammable Class IA	1.50%	7.80%
Paraffins	Octane	111-65-9	55.4	Flammable Class IB	1.00%	6.50%
Paraffins	Ethyl Cyclohexane	1678-91-7	64.4	Flammable Class IB	0.90%	6.60%
Paraffins	Cypar 9	68553-14-0	75.2	Flammable Class IC		

(Continued)

Table A1-4 *(Continued)*

Type	Name	CAS #	Flash Point °F	NFPA Flammability Classification	LEL Vol %	UEL Vol %
Paraffins	D30	64742-48-9	82.4	Flammable Class IC		
Paraffins	Nonane	111-84-2	87.8	Flammable Class IC	0.80%	2.90%
Paraffins	DSP 145/160	8002-5-9	87.8	Flammable Class IC		
Paraffins	Hexane	110-54-3	-9.4	Flammable Class IB	1.10%	7.50%
Specials	Carbon Dioxide	124-38-9	None	NOS		
Specials	Tetrahydrofuran	109-99-9	1.4	Flammable Class IB	2.00%	11.80%
Specials	Dimethyl Formamide	68-12-2	136.0	Combustible Class II	2.20%	15.20%
Specials	Dimethyl Sulfoxide	67-68-5	185.0	ERR	2.60%	42.00%
Specials	Isophorone	78-59-1	194.0	ERR	0.80%	3.80%
Specials	N-Methyl Pyrrolidone	872-50-4	204.8	ERR	1.30%	9.50%
Specials	Gamma Butyrolactone	96-48-0	208.4	ERR	2.70%	15.60%
Specials	1,3 Dioxolane	646-6-0	21.2	Flammable Class IB	2.10%	20.50%
Specials	Dimethyl Acetamide	127-19-5	158.0	Combustible Class IIIA	1.80%	11.50%
Specials	Propylene Carbonate	108-32-7	275.0	ERR	4.70%	21.00%
Specials	Ethylene Carbonate	96-49-1	289.4	ERR	3.60%	16.10%
Specials	Soygold 1000	67784-80-9	300.0	ERR		
Specials	Acetonitrile	75-5-8	35.6	Flammable Class IB	3.00%	16.00%
Specials	1,4-Dioxane	123-91-1	53.6	Flammable Class IB	2.00%	22.00%
Specials	4-Bromo-3-Chloro-3,4, 4-Trifluoro-1-Butene	374-25-4	54.9	Flammable Class IB		
Specials	Dimethyl Carbonate	616-38-6	64.4	Flammable Class IB	4.22%	12.87%
Specials	Pyridine	110-86-1	68.0	Flammable Class IB	1.80%	12.40%
Specials	Cumene	98-82-8	96.8	Flammable Class IC	0.90%	6.50%
Specials	Water	7732-18-5	None	NOS		

Type	Name	Molar Vapor Concentration @ FP	Conc @ 20 °C > 25% of LEL ?	Evap. Rate n-BuAc = 1
Acids	Acetic Acid	4.62%	YES	0.97
Alcohols	Texanol	0.93%	NO	0.00
Alcohols	Methyl Isobutyl Carbinol	1.43%	YES	0.30
Alcohols	Amyl Alcohol	1.12%	NO	0.24
Alcohols	Diacetone Alcohol	0.69%	NO	0.08
Alcohols	1-Hexanol	1.54%	NO	0.05
Alcohols	Cyclohexanol	4.56%	YES	0.08
Alcohols	Furfuryl Alcohol	2.01%	NO	0.04
Alcohols	Tetrahydrofurfuryl Alcohol	2.70%	NO	0.00
Alcohols	Benzyl Alcohol	1.21%	NO	0.01
Alcohols	iso-Amyl Alcohol	1.68%	YES	0.20
Alcohols	Tertiary Butanol	1.59%	YES	0.03
Alcohols	Methanol	8.46%	YES	5.90
Alcohols	Ethanol	3.94%	YES	3.30
Alcohols	Isopropanol	3.01%	YES	2.83
Alcohols	N-Propanol	1.65%	YES	1.30
Alcohols	Secondary Butanol	2.26%	YES	1.30
Alcohols	Isobutanol	1.61%	YES	0.67
Alcohols	N-Butanol	1.04%	YES	2.80
Aromatics	Stoddard Solvent	1.70%	YES	0.15
Aromatics	Mineral Spirits	0.41%	NO	0.10
Aromatics	Benzene	2.07%	YES	0.16
Aromatics	Lactol Spirits	1.21%	YES	2.40
Aromatics	Toluene	1.27%	YES	1.98
Aromatics	VM&P Naptha	0.47%	YES	7.56

(Continued)

Table A1-4 *(Continued)*

Type	Name	Molar Vapor Concentration @ FP	Conc @ 20 °C > 25% of LEL ?	Evap. Rate n-BuAc = 1
Aromatics	o-Xylene	1.04%	YES	0.90
Aromatics	Turpentine, oil	1.15%	YES	0.86
Benzotrifluorides	Parachlorobenzotrifluoride	2.62%	YES	0.90
Benzotrifluorides	Benzotrifluoride	2.09%	YES	2.80
Brominateds	n-Propyl Bromide	17.36%	YES	4.50
CFCs	CFC-113	10.45%	YES	7.73
Chlorinateds	Perchloroethylene	6.47%	NO	2.60
Chlorinateds	Carbon Tetrachloride	1.79%	YES	7.50
Chlorinateds	Chloroform	2.70%	YES	10.20
Chlorinateds	1,1-Dichloroethane	5.16%	YES	0.00
Chlorinateds	Ethylene Dichloride	1.88%	YES	6.10
Chlorinateds	Methylene Chloride	12.70%	YES	14.50
Chlorinateds	1,1,1-Trichloroethane	4.88%	YES	6.00
Chlorinateds	Trans 1,2 Dichloroethylene	21.25%	YES	7.43
Chlorinateds	Trichloroethylene	13.35%	YES	6.40
Dibasic Acid Esters	DBE	2.93%	NO	0.00
Dibasic Acid Esters	DBE-4	121.62%	YES	0.00
Dibasic Acid Esters	DBE-6	1.79%	NO	0.00
Esters	Cyclohexyl Acetate	3.60%	YES	0.16
Esters	Ethyl-3-Ethoxypropionate	1.04%	NO	0.12
Esters	Methyl Acetate	6.32%	YES	5.50
Esters	Butyl Glycol Acetate	1.84%	NO	0.03
Esters	Dimethyl Succinate	1.43%	NO	0.00
Esters	Propylene Glycol Diacetate	1.43%	NO	0.04

Esters	2-Ethyl Hexyl Acetate	1.41%	NO	0.03
Esters	Methyl Formate	9.06%	YES	0.00
Esters	Ethylene Glycol Diacetate	3.44%	NO	0.02
Esters	Ethyl Acetate	2.90%	YES	4.17
Esters	Tertiary Butyl Acetate	2.05%	YES	2.80
Esters	Ethyl Formate	3.61%	YES	6.93
Esters	Isopropyl Acetate	2.66%	YES	2.88
Esters	n-Propyl Acetate	2.48%	YES	1.49
Esters	n-Propyl Propionate	1.28%	YES	1.20
Esters	n-Propyl Propionate	1.33%	YES	1.20
Esters	Isobutyl Acetate	2.07%	YES	1.70
Esters	Amyl Acetate	5.95%	YES	0.42
Esters	Amyl Acetate	6.30%	YES	0.42
Esters	n-Butyl Acetate	1.87%	YES	1.00
Esters	N-Butyl Propionate	0.92%	YES	0.45
Ethers	Methylal	1.00%	YES	0.88
Ethers	Methyl Butyl Ether	2.77%	YES	7.20
Ethers	Methyl t-Butyl Ether	2.56%	YES	19.95
Ethers	t-Butyl Ether	1.13%	YES	3.09
Ethers	Cyclopentyl Methyl Ether	1.25%	YES	3.50
Ethers	Amyl t-Butyl Ether	1.19%	YES	0.69
Glycol Ethers	Dipropylene Glycol Dimethyl Ether	1.17%	NO	0.13
Glycol Ethers	Dipropylene Glycol n-Propyl Ether	0.64%	NO	0.01
Glycol Ether Acetates	Ethylene glycol methyl ether acetate	1.72%	NO	0.40

(Continued)

Table A1-4 *(Continued)*

Type	Name	Molar Vapor Concentration @ FP	Conc @ 20 °C > 25% of LEL ?	Evap. Rate n-BuAc = 1
Glycol Ether Acetates	Propylene Glycol Methyl Ether Acetate	2.21%	YES	0.36
Glycol Ether Acetates	Propylene Glycol Ethyl Ether Acetate	1.61%	YES	0.21
Glycol Ether Acetates	Ethylene Glycol Ethyl Ether Acetate	1.67%	NO	0.23
Glycol Ether Acetates	Ethylene Glycol Butyl Ether Acetate	0.99%	NO	0.06
Glycol Ether Acetates	Dipropylene Glycol Methyl Ether Acetate	0.70%	YES	0.02
Glycol Ether Acetates	Diethylene Glycol Monoethyl Ether Acetate	1.69%	NO	0.08
Glycol Ether Acetates	Diethylene Glycol Monobutyl Ether Acetate	1.01%	YES	0.00
Glycol Ethers	Ethylene Glycol Monomethyl Ether	2.87%	YES	0.36
Glycol Ethers	Propylene Glycol Monoethyl Ether	2.85%	YES	0.41
Glycol Ethers	Propylene Glycol Monoethyl Ether	3.28%	YES	0.54
Glycol Ethers	Ethylene Glycol Monoethyl Ether	2.16%	NO	0.28
Glycol Ethers	Propylene Glycol Mono t-butyl Ether	0.62%	YES	0.25
Glycol Ethers	Propylene Glycol Monopropyl Ether	1.17%	YES	0.21
Glycol Ethers	Propylene Glycol n-Propyl Ether	1.03%	YES	0.21
Glycol Ethers	Ethylene Glycol Mono Isopropyl Ether	3.23%	YES	0.23

Glycol Ethers	Propylene Glycol Monobutyl Ether	1.43%	YES	0.09
Glycol Ethers	Ethylene Glycol Monobutyl Ether	1.19%	NO	0.07
Glycol Ethers	3-Methoxy-3-Methyl-1-Butanol	1.47%	NO	0.07
Glycol Ethers	Butyl Glycol Ether	1.71%	NO	0.07
Glycol Ethers	Dipropylene Glycol Monomethyl Ether	1.11%	NO	0.04
Glycol Ethers	Diethylene Glycol Monobutyl ether	0.20%	NO	0.00
Glycol Ethers	Propylene Glycol Methyl Propyl Ether	1.76%	YES	0.03
Glycol Ethers	Propylene Glycol Ethyl Propyl Ether	2.13%	YES	0.03
Glycol Ethers	Dipropylene Glycol Monoethyl Ether	2.34%	YES	0.00
Glycol Ethers	Diethylene Glycol Monomethyl Ether	2.01%	NO	0.01
Glycol Ethers	Diethylene Glycol Monomethyl Ether	2.47%	NO	0.02
Glycol Ethers	Dipropylene Glycol n-Butyl Ether	0.54%	NO	0.01
Glycol Ethers	Dipropylene Glycol Monobutyl Ether	0.86%	NO	0.01
Glycol Ethers	Dipropylene Glycol Monobutyl Ether	0.61%	YES	0.4
Glycol Ethers	Tripropylene Glycol Monomethyl Ether	0.84%	YES	0.00
Glycol Ethers	Tripropylene Glycol n-Propyl Ether	0.43%	YES	0.00

(Continued)

Table A1-4 *(Continued)*

Type	Name	Molar Vapor Concentration @ FP	Conc @ 20 °C > 25% of LEL ?	Evap. Rate n-BuAc = 1
Glycol Ethers	Triethylene glycol monomethyl ether	1.61%	YES	0.01
Glycol Ethers	Tripropylene Glycol n-Butyl Ether	86.98%	YES	0.01
Glycol Ethers	Propylene Glycol Monomethyl Ether	2.23%	YES	0.66
Glycol Ethers	Propylene Glycol Monomethyl Ether	3.00%	YES	0.62
Glycol Ethers	Ethylene Glycol Diethyl Ether	4.68%	YES	0.00
Glycols	Hexylene Glycol	1.26%	NO	0.00
Glycols	Propylene Glycol	3.61%	NO	0.01
Glycols	Ethylene Glycol	4.36%	NO	0.01
Glycols	Diethylene Glycol	0.87%	NO	0.00
Glycols	1,4-Butane Diol	2.55%	NO	0.00
Glycols	Dipropylene Glycol	3.10%	NO	0.00
Glycols	Triethylene Glycol	0.84%	NO	0.00
Glymes	Diethylene Glycol Dimethyl Ether (Diglyme)	1.67%	NO	0.36
Glymes	Diethylene Glycol Diethyl Ether (Ethyl Diglyme)	2.23%	NO	0.04
Glymes	Ethylene Glycol Dimethyl Ether (Monoglyme)	1.89%	YES	4.90
Glymes	Triethylene Glycol Dimethyl Ether (Triglyme)	1.47%	NO	0.00
Glymes	Diethylene Glycol Dibutyl Ether (Butyl Diglyme)	1.15%	YES	0.00
Glymes	Ethylene Glycol Diethyl Ether (Ethyl Glyme)	1.54%	YES	0.00

HCFCs	HCFC-225 ca / cb	5.69%	YES	6.73
HCFCs	HCFC-141b	11.34%	YES	13.32
HFCs	HFC-245 fa	17.07%	YES	104.00
HFCs	HFC 43-10 mee	3.48%	YES	29.00
HFCs	HFC-365mfc	6.24%	YES	
HFEs	HFE-7300	0.66%	YES	478.00
HFEs	HFE-7100	3.43%	YES	49.00
HFEs	HFE-8200	2.39%	YES	33.00
HFEs	HFE-7200	1.97%	YES	33.00
Iodateds	Perfluoro n-butyl Iodide	3.91%	YES	14.50
Ketones	Acetone	4.07%	YES	5.76
Ketones	Methyl Amyl Ketone	1.06%	YES	0.34
Ketones	Cyclohexanone	1.61%	YES	0.30
Ketones	Diisobutyl Ketone	1.01%	NO	0.20
Ketones	Dispropyl Ketone	0.89%	YES	0.00
Ketones	Methyl Ethyl Ketone	2.76%	YES	4.48
Ketones	Methyl Propyl Ketone	1.89%	YES	2.40
Ketones	Diethyl Ketone	2.48%	YES	2.30
Ketones	Methyl Isobutyl Ketone	1.63%	YES	1.21
Ketones	Acetyl Acetone	1.73%	YES	0.33
Ketones	Methyl Isoamyl Ketone	1.38%	YES	0.46
Lactates	Ethyl Lactate	1.35%	NO	0.18
Lactates	Methyl Lactate	3.42%	NO	0.26
Lactates	Isopropyl Lactate	1.93%	YES	0.18
Lactates	Propyl Lactate	2.84%	YES	0.11

(Continued)

Table A1-4 *(Continued)*

Type	Name	Molar Vapor Concentration @ FP	Conc @ 20 °C > 25% of LEL ?	Evap. Rate n-BuAc = 1
Lactates	Butyl Lactate	1.57%	NO	0.04
Lactates	2-Ethylhexyl Lactate	0.88%	YES	0.00
Limonene	Dipentene (Dl-Limonene)	1.19%	YES	0.20
Limonene	D-Limonene	0.98%	YES	0.20
OSs	OS-10 Hexamethyldisiloxane	1.31%	YES	3.80
OSs	OS-30 Decamethyltrisiloxane	61.36%	NO	0.15
OSs	OS-20 Octamethyltrisiloxane	0.58%	YES	1.00
Paraffins	Cyclohexane	1.47%	YES	4.84
Paraffins	SBP 5	0.70%	YES	3.90
Paraffins	SBP 80 / 110	1.64%	YES	4.50
Paraffins	SBP 11	0.55%	NO	0.29
Paraffins	Whitespirit	5.58%	YES	0.24
Paraffins	D40	0.81%	YES	0.12
Paraffins	Methyl Cyclopentane	1.99%	YES	5.93
Paraffins	Isododecane	0.23%	YES	0.14
Paraffins	Solvesso 100	1.78%	YES	0.16
Paraffins	Isopar H	0.95%	NO	0.09
Paraffins	D60	0.64%	YES	0.04
Paraffins	Isopar L	0.87%	NO	0.04
Paraffins	Solvesso 150	1.01%	YES	0.05
Paraffins	D70	0.78%	NO	0.02
Paraffins	Isopar M	0.42%	NO	0.01
Paraffins	Isopar C	1.12%	YES	5.45
Paraffins	SBP 3	1.08%	YES	2.70

Paraffins	Solvesso 200	0.69%	YES	0.00
Paraffins	2-Methyl Pentane	1.41%	YES	0.00
Paraffins	Heptane	1.31%	YES	3.18
Paraffins	Cypar 7	ERR	ERR	3.10
Paraffins	Methyl Cyclohexane	1.34%	YES	2.99
Paraffins	3-Methyl Hexane	1.97%	YES	4.75
Paraffins	SBP 100/140	1.22%	YES	1.90
Paraffins	2-Methyl Heptane	0.99%	YES	0.16
Paraffins	Cyclopentane	1.90%	YES	0.00
Paraffins	Isopar E	1.15%	YES	1.55
Paraffins	SBP 1	0.10%	YES	5.60
Paraffins	Pentane	1.36%	YES	10.46
Paraffins	Octane	0.88%	YES	1.23
Paraffins	Ethyl Cyclohexane	1.24%	YES	1.07
Paraffins	Cypar 9	ERR	ERR	0.68
Paraffins	D30	0.85%	YES	0.40
Paraffins	Nonane	0.82%	YES	0.41
Paraffins	DSP 145/160	0.76%	YES	0.60
Paraffins	Hexane	1.44%	YES	6.82
Specials	Carbon Dioxide	ERR	ERR	0.00
Specials	Tetrahydrofuran	2.96%	YES	6.30
Specials	Dimethyl Formamide	3.04%	NO	0.20
Specials	Dimethyl Sulfoxide	2.50%	NO	0.00
Specials	Isophorone	1.90%	NO	0.03
Specials	N-Methyl Pyrrolidone	2.65%	NO	0.06
Specials	Gamma Butyrolactone	3.14%	NO	0.03
Specials	1,3 Dioxolane	2.62%	YES	3.50

(Continued)

Table A1-4 *(Continued)*

Type	Name	Molar Vapor Concentration @ FP	Conc @ 20 °C > 25% of LEL ?	Evap. Rate n-BuAc = 1
Specials	Dimethyl Acetamide	2.20%	NO	0.14
Specials	Propylene Carbonate	3.78%	NO	0.00
Specials	Ethylene Carbonate	2.85%	NO	0.00
Specials	Soygold 1000	1.78%	YES	0.01
Specials	Acetonitrile	3.87%	YES	2.30
Specials	1,4-Dioxane	2.43%	YES	2.20
Specials	4-Bromo-3-Chloro-3,4,4-Trifluoro-1-Butene	6.48%	YES	244.00
Specials	Dimethyl Carbonate	3.31%	YES	3.22
Specials	Pyridine	2.13%	YES	0.00
Specials	Cumene	1.16%	YES	0.00
Specials	Water		NO	0.36

Table A1-5

Type	Name	CAS #	Disperse $MPa^{1/2}$	Polar $MPa^{1/2}$	H2-Bond $MPa^{1/2}$	Hildebrand SP $MPa^{1/2}$	Kb	Solubility of Water wt. %	Solubility in Water wt. %
Acids	Acetic Acid	64-19-7	14.5	8.0	13.5	21.37		100.00%	100.00%
Alcohols	1-Hexanol	111-27-3	15.9	5.8	12.5	21.04			
Alcohols	Amyl Alcohol	71-41-0	15.9	5.9	13.9	21.93		9.20%	1.70%
Alcohols	Benzyl Alcohol	100-51-6	18.4	6.3	13.7	23.79		8.37%	8.00%
Alcohols	Cyclohexanol	108-93-0	17.4	4.1	13.5	22.40			4.00%
Alcohols	Diacetone Alcohol	123-42-2	15.8	8.2	10.8	20.82			
Alcohols	Ethanol	64-17-5	15.8	8.8	19.4	26.52		100.00%	100.00%
Alcohols	Furfuryl Alcohol	98-0-0	17.4	7.6	15.1	24.26			
Alcohols	iso-Amyl Alcohol	123-51-3	15.8	5.2	13.3	21.30			2.00%
Alcohols	Isobutanol	78-83-1	15.1	5.7	15.9	22.66		16.40%	8.50%
Alcohols	Isopropanol	67-63-0	15.8	6.1	16.4	23.58		100.00%	100.00%
Alcohols	Methanol	67-56-1	14.7	12.3	22.3	29.41		100.00%	100.00%
Alcohols	Methyl Isobutyl Carbinol	108-11-2	15.4	3.3	12.3	19.98		6.30%	1.60%
Alcohols	N-Butanol	71-36-3	16.0	5.7	15.8	23.20		19.70%	7.80%
Alcohols	N-Propanol	71-23-8	16.0	6.8	17.4	24.60		100.00%	100.00%
Alcohols	Secondary Butanol	78-92-2	15.8	5.7	14.5	22.19			
Alcohols	Tertiary Butanol	75-65-0	15.2	5.1	14.7	21.75			
Alcohols	Tetrahydrofurfuryl Alcohol	97-99-4	17.8	8.2	12.9	23.46			
Alcohols	Texanol	25265-77-4	15.1	6.1	9.8	19.01		0.90%	
Aromatics	Benzene	71-43-2	18.4	0.0	2.0	18.51	105		0.18%
Aromatics	Lactol Spirits	64742-49-0	17.6	1.7	1.3	17.76	33		0.01%
Aromatics	Mineral Spirits	64741-41-9	16.1	1.6	1.6	16.26	29		
Aromatics	o-Xylene	95-47-6	18.5	3.6	2.4	19.00	95		

(Continued)

Table A1-5 *(Continued)*

Type	Name	CAS #	Disperse $MPa^{1/2}$	Polar $MPa^{1/2}$	H2-Bond $MPa^{1/2}$	Hildebrand SP $MPa^{1/2}$	Kb	Solubility of Water wt. %	Solubility in Water wt. %
Aromatics	Stoddard Solvent	8052-41-3	16.1	1.6	2.1	16.31	33		0.05%
Aromatics	Toluene	108-88-3	18.0	1.4	2.0	18.16	103	0.03%	0.52%
Aromatics	Turpentine, oil	8006-64-2	16.4	1.4	0.4	16.46	56		
Aromatics	VM&P Naptha	64742-89-8	15.0	2.7	2.7	15.48	34	0.05%	
Benzotrifluorides	Benzotrifluoride	98-08-8	18.6	8.1	1.5	20.34	49		
Benzotrifluorides	Parachlorobenzotrifluoride	98-56-6	18.5	5.2	1.7	19.29	64	0.02%	0.00%
Brominateds	n-Propyl Bromide	106-94-5	16.4	7.9	4.8	18.83	129	0.50%	0.25%
CFCs	CFC-113	76-13-1	14.7	1.6	0.0	14.79	31	0.01%	0.02%
Chlorinateds	1,1,1-Trichloroethane	71-55-6	16.8	4.3	2.0	17.46	124	0.33%	0.13%
Chlorinateds	1,1-Dichloroethane	75-34-3	16.5	7.8	3.0	18.50			
Chlorinateds	Carbon Tetrachloride	56-23-5	17.8	0.0	0.6	17.81		0.01%	0.08%
Chlorinateds	Chloroform	67-66-3	17.8	3.1	5.7	18.95		0.06%	0.82%
Chlorinateds	Ethylene Dichloride	107-06-2	18.0	7.4	4.1	19.89		0.16%	0.87%
Chlorinateds	Methylene Chloride	75-09-2	17.0	7.3	7.1	19.82	136	0.24%	1.85%
Chlorinateds	Perchloroethylene	127-18-4	18.3	5.7	0.0	19.17	90	0.00%	0.00%
Chlorinateds	Trans 1,2 Dichloroethylene	156-60-5	16.7	7.8	3.3	18.72	117		0.45%
Chlorinateds	Trichloroethylene	79-01-6	18.0	3.1	5.3	19.02	129	0.33%	0.13%
Dibasic Acid Esters	DBE	1119-40-0	17.0	4.7	9.8	20.16	51	3.10%	5.30%
Dibasic Acid Esters	DBE-4	106-65-0	16.4	8.4	9.2	20.54	35	3.80%	7.50%
Dibasic Acid Esters	DBE-6	627-93-0	16.4	8.7	7.9	20.15		2.40%	2.40%
Esters	2-Ethyl Hexyl Acetate	103-9-3	15.8	2.9	5.1	16.85		0.03%	0.55%
Esters	Amyl Acetate	628-63-7	15.8	3.3	6.1	17.26		0.01%	0.17%
Esters	Amyl Acetate	628-63-7	15.8	3.3	6.1	17.26		0.01%	0.17%

Esters	Butyl Glycol Acetate	112-07-2	15.3	7.5	6.8	18.35		1.83%	1.92%
Esters	Cyclohexyl Acetate	622-45-7	17.0	5.6	6.0	18.88		1.83%	1.92%
Esters	Dimethyl Succinate	106-65-0	16.4	8.4	9.2	20.54			0.85%
Esters	Ethyl Acetate	141-78-6	15.8	5.3	7.2	18.15		3.45%	6.79%
Esters	Ethyl Formate	109-94-4	15.5	8.4	8.4	19.53		17.00%	11.80%
Esters	Ethyl-3-Ethoxypropionate	763-69-9	16.2	3.3	8.8	18.73		2.20%	5.20%
Esters	Ethylene Glycol Diacetate	111-55-7	16.2	4.7	9.8	19.51		7.00%	16.40%
Esters	Isobutyl Acetate	110-19-0	15.1	3.7	6.3	16.77		1.25%	1.15%
Esters	Isopropyl Acetate	108-21-4	14.9	4.5	8.2	17.59		1.82%	1.72%
Esters	Methyl Acetate	79-20-9	15.5	7.2	7.6	18.70		8.20%	24.50%
Esters	Methyl Formate	107-31-3	15.3	8.4	10.2	20.22			3.00%
Esters	n-Butyl Acetate	123-86-4	15.8	3.7	6.3	17.41	28	1.26%	1.14%
Esters	N-Butyl Propionate	590-01-2	15.7	5.5	5.9	17.65		0.02%	0.20%
Esters	n-Propyl Acetate	109-60-4	15.3	4.3	7.6	17.62		1.20%	1.90%
Esters	n-Propyl Propionate	106-36-5	15.5	5.6	5.7	17.44			0.50%
Esters	n-Propyl Propionate	106-36-5	15.5	5.6	5.7	17.44			0.50%
Esters	Propylene Glycol Diacetate	623-84-7	16.1	5.5	9.1	19.30		4.10%	7.40%
Esters	Tertiary Butyl Acetate	540-88-5	15.0	3.7	6.0	16.57		0.71%	0.83%
Ethers	Cyclopentyl Methyl Ether	5614-37-9	16.7	4.3	4.3	17.77		1.50%	4.80%
Ethers	Methyl Butyl Ether	628-28-4	15.2	4.2	4.2	16.32			
Ethers	Methyl t-Butyl Ether	1634-04-4	14.6	2.7	6.8	16.33		4.20%	5.13%
Ethers	Methylal	109-87-5	15.0	1.8	8.6	17.38			33.00%
Ethers	t-Butyl Ether	6163-66-2	14.5	1.3	7.4	16.33			
Ethers	Amyl t-Butyl Ether	10100-95-5	15.1	2.0	5.4	16.16			
Glycol Ethers	Dipropylene Glycol Dimethyl Ether	111109-77-4	15.6	7.0	6.5	18.26		5.26%	35.00%
Glycol Ethers	Dipropylene Glycol n-Propyl Ether	29911-27-1	15.6	6.1	11.0	20.04		22.50%	17.50%

(Continued)

Table A1-5 *(Continued)*

Type	Name	CAS #	Disperse $MPa^{1/2}$	Polar $MPa^{1/2}$	H2-Bond $MPa^{1/2}$	Hildebrand SP $MPa^{1/2}$	Kb	Solubility of Water wt. %	Solubility in Water wt. %
Glycol Ether Acetates	Diethylene Glycol Monobutyl Ether Acetate	124-17-4	16.0	4.1	8.2	18.44			6.00%
Glycol Ether Acetates	Diethylene Glycol Monoethyl Ether Acetate	112-15-2	16.2	5.1	9.2	19.32			100.00%
Glycol Ether Acetates	Dipropylene Glycol Methyl Ether Acetate	88917-22-0	16.3	4.9	8.0	18.81		3.50%	16.00%
Glycol Ether Acetates	Ethylene Glycol Butyl Ether Acetate	112-07-2	15.3	7.5	6.8	18.35			1.50%
Glycol Ether Acetates	Ethylene Glycol Ethyl Ether Acetate	111-15-9	15.9	4.7	10.6	19.68			100.00%
Glycol Ether Acetates	Ethylene glycol methyl ether acetate	110-49-6	15.9	5.5	11.6	20.44			100.00%
Glycol Ether Acetates	Propylene Glycol Ethyl Ether Acetate	54839-24-6	15.6	6.3	7.7	18.50		2.38%	9.71%
Glycol Ether Acetates	Propylene Glycol Methyl Ether Acetate	108-65-6	15.6	5.6	9.8	19.26		4.76%	24.21%
Glycol Ethers	3-Methoxy-3-Methyl-1-Butanol	56539-66-3	16.0	6.3	12.9	21.50	200	100.00%	100.00%
Glycol Ethers	Butyl Glycol Ether	111-76-2	16.0	5.1	12.3	20.82		100.00%	100.00%
Glycol Ethers	Diethylene Glycol Monobutyl ether	112-34-5	16.0	7.0	10.6	20.43		100.00%	100.00%
Glycol Ethers	Diethylene Glycol Monomethyl Ether	111-77-3	16.2	7.8	12.6	21.96		100.00%	100.00%
Glycol Ethers	Diethylene Glycol Monomethyl Ether	111-77-3	16.2	7.8	12.6	21.96		100.00%	100.00%
Glycol Ethers	Dipropylene Glycol Monobutyl Ether	29911-28-2	15.7	6.5	10.0	19.72		12.00%	4.50%

Glycol Ethers	Dipropylene Glycol Monobutyl Ether	29911-28-2	15.7	6.5	10.0	19.72			5.00%
Glycol Ethers	Dipropylene Glycol Monoethyl Ether	30025-38-8	15.5	6.5	10.8	19.98			
Glycol Ethers	Dipropylene Glycol Monomethyl Ether	112-28-7	15.5	5.7	11.2	19.95			100.00%
Glycol Ethers	Dipropylene Glycol n-Butyl Ether	29911-28-2	15.7	6.5	10.0	19.72		14.00%	4.50%
Glycol Ethers	Ethylene Glycol Diethyl Ether	629-14-1	15.4	5.4	5.2	17.13			100.00%
Glycol Ethers	Ethylene Glycol Mono Isopropyl Ether	109-59-1	16.0	8.2	13.1	22.25			
Glycol Ethers	Ethylene Glycol Monobutyl Ether	111-76-2	16.0	5.1	12.3	20.82		100.00%	100.00%
Glycol Ethers	Ethylene Glycol Monoethyl Ether	110-80-5	15.9	7.2	14.0	22.38		100.00%	100.00%
Glycol Ethers	Ethylene Glycol Monomethyl Ether	109-86-4	16.0	8.2	15.0	23.41			100.00%
Glycol Ethers	Propylene Glycol Ethyl Propyl Ether	30025-38-8	16.0	7.2	11.0	20.71		100.00%	100.00%
Glycol Ethers	Propylene Glycol Methyl Propyl Ether	34590-94-8	15.5	5.7	11.2	19.95		100.00%	100.00%
Glycol Ethers	Propylene Glycol Mono t-butyl Ether	57018-52-7	15.3	6.1	10.8	19.70			17.00%
Glycol Ethers	Propylene Glycol Monobutyl Ether	1569-02-4	15.7	6.5	10.5	19.97		15.50%	5.50%
Glycol Ethers	Propylene Glycol Monoethyl Ether	1569-02-4	15.7	6.5	10.5	19.97		100.00%	100.00%
Glycol Ethers	Propylene Glycol Monoethyl Ether	1569-02-4	15.7	6.5	10.5	19.97		100.00%	100.00%
Glycol Ethers	Propylene Glycol Monomethyl Ether	107-98-2	15.6	6.3	11.6	20.44		100.00%	100.00%
Glycol Ethers	Propylene Glycol Monomethyl Ether	107-98-2	15.6	6.3	11.6	20.44		100.00%	100.00%

(Continued)

Table A1-5 *(Continued)*

Type	Name	CAS #	Disperse $MPa^{1/2}$	Polar $MPa^{1/2}$	H2-Bond $MPa^{1/2}$	Hildebrand SP $MPa^{1/2}$	Kb	Solubility of Water wt. %	Solubility in Water wt. %
Glycol Ethers	Propylene Glycol Monopropyl Ether	30136-13-1	15.8	7.0	9.2	19.58		100.00%	100.00%
Glycol Ethers	Propylene Glycol n-Propyl Ether	30136-13-1	15.8	7.0	9.2	19.58		100.00%	100.00%
Glycol Ethers	Triethylene glycol monomethyl ether	112-35-6	16.2	7.6	12.5	21.83		100.00%	100.00%
Glycol Ethers	Tripropylene Glycol Monomethyl Ether	25498-49-1	15.3	5.5	10.4	19.30			100.00%
Glycol Ethers	Tripropylene Glycol n-Butyl Ether	55934-93-5	15.8	5.7	8.5	18.79		8.00%	3.00%
Glycol Ethers	Tripropylene Glycol n-Propyl Ether	96077-04-2	15.6	5.3	6.0	17.53			4.50%
Glycols	1,4-Butane Diol	110-63-4	16.6	11.0	20.9	28.87		100.00%	100.00%
Glycols	Diethylene Glycol	111-46-6	16.6	12.0	19.0	27.94		100.00%	100.00%
Glycols	Dipropylene Glycol	2396-61-4	16.5	10.6	17.7	26.42			
Glycols	Ethylene Glycol	107-21-1	17.0	11.0	26.0	32.95		100.00%	100.00%
Glycols	Hexylene Glycol	107-41-5	15.7	8.4	17.8	25.18		100.00%	100.00%
Glycols	Propylene Glycol	57-55-6	16.8	10.4	21.3	29.05		100.00%	100.00%
Glycols	Triethylene Glycol	112-27-6	16.0	12.5	18.6	27.54		100.00%	100.00%
Glymes	Diethylene Glycol Dibutyl Ether (Butyl Diglym	112-73-2	15.8	4.7	4.4	17.06			
Glymes	Diethylene Glycol Diethyl Ether (Ethyl Diglyme	112-36-7	15.8	5.9	5.6	17.77		100.00%	100.00%
Glymes	Diethylene Glycol Dimethyl Ether (Diglyme)	111-96-6	15.7	6.1	6.5	18.05			
Glymes	Ethylene Glycol Diethyl Ether (Ethyl Glyme)	629-14-1	15.4	5.4	5.2	17.13			

Glymes	Ethylene Glycol Dimethyl Ether (Monoglyme)	110-71-4	15.4	6.0	6.0	17.58		100.00%	100.00%
Glymes	Triethylene Glycol Dimethyl Ether (Triglyme)	112-49-2	16.0	7.3	6.3	18.68		100.00%	100.00%
HCFCs	HCFC-141b	1717-00-6	15.7	4.0	1.0	16.23	56	0.18%	0.04%
HCFCs	HCFC-225 ca / cb	422-56-0/507-55-1	14.1	3.2	1.0	14.49	31	0.03%	0.03%
HFCs	HFC 43-10 mee	138495-42-8	14.2	1.1	0.0	14.24	9	0.05%	0.01%
HFCs	HFC-245 fa	460-73-1	15.4	3.0	0.0	15.69			0.16%
HFCs	HFC-365mfc	405-58-6	16.4	0.0	0.0	16.40	13	0.09%	0.09%
HFEs	HFE-7100	163702-07-6	13.7	2.2	1.0	13.91	5	0.01%	0.00%
HFEs	HFE-7200	163702-06-5	13.3	2.0	1.0	13.49	5	0.00%	0.00%
HFEs	HFE-7300	132182-92-4	11.5	4.1	1.0	12.25		0.01%	0.00%
HFEs	HFE-8200	163702-05-4/163702-06-	13.0	4.0	1.0	13.64		0.00%	0.01%
Iodateds	Perfluoro n-butyl Iodide	423-39-2	15.7	4.6	3.9	16.82			
Ketones	Acetone	67-64-1	15.5	10.4	7.0	19.94		100.00%	100.00%
Ketones	Acetyl Acetone	123-54-6	16.1	10.0	6.2	19.94		4.50%	16.60%
Ketones	Cyclohexanone	108-94-1	17.8	8.4	5.1	20.33	58.5	8.00%	2.30%
Ketones	Diethyl Ketone	96-22-0	15.8	7.6	4.7	18.15			
Ketones	Diisobutyl Ketone	108-83-8	16.0	3.7	4.1	16.93		0.70%	0.05%
Ketones	Dispropyl Ketone	123-19-3	15.8	7.6	4.9	18.20			
Ketones	Methyl Amyl Ketone	110-43-0	16.2	5.7	4.1	17.66		1.30%	0.50%
Ketones	Methyl Ethyl Ketone	78-93-3	16.0	9.0	5.1	19.05		12.00%	27.00%
Ketones	Methyl Isoamyl Ketone	110-12-3	16.0	5.7	4.1	17.47		1.20%	0.50%
Ketones	Methyl Isobutyl Ketone	108-10-1	15.3	6.1	4.1	16.97		1.90%	1.70%
Ketones	Methyl Propyl Ketone	107-87-9	16.0	7.6	4.7	18.33		4.20%	3.10%
Lactates	2-Ethylhexyl Lactate	186817-80-1	15.7	6.1	12.0	20.61	1000	3.00%	3.50%
Lactates	Butyl Lactate	138-22-7	15.8	6.5	10.2	19.90	1000	13.90%	4.50%

(Continued)

Table A1-5 *(Continued)*

Type	Name	CAS #	Disperse $MPa^{1/2}$	Polar $MPa^{1/2}$	H2-Bond $MPa^{1/2}$	Hildebrand SP $MPa^{1/2}$	Kb	Solubility of Water wt. %	Solubility in Water wt. %
Lactates	Ethyl Lactate	97-64-3	16.0	7.6	12.5	21.68	1000	100.00%	100.00%
Lactates	Isopropyl Lactate	63697-0-7	15.6	8.6	13.3	22.18	1000	100.00%	100.00%
Lactates	Methyl Lactate	27871-49-4	15.9	8.1	14.0	22.68	62	100.00%	100.00%
Lactates	Propyl Lactate	616-09-1	15.8	7.1	11.3	20.68		100.00%	100.00%
Limonene	D-Limonene	5989-27-5	17.2	1.8	4.3	17.82	67		0.00%
Limonene	Dipentene (Dl-Limonene)	138-86-3	17.2	1.8	4.3	17.82			
OSs	OS-10 Hexamethyldisiloxane	107-46-0	14.0	1.0	0.0	14.04	16.6		
OSs	OS-20 Octamethyltrisiloxane	107-51-7	11.7	1.0	0.0	11.74	15.1		
OSs	OS-30 Decamethyltrisiloxane	141-62-8	12.2	1.0	0.0	12.24	13.4		
Paraffins	2-Methyl Heptane	592-27-8	14.9	1.7	0.0	15.00			
Paraffins	2-Methyl Pentane	107-83-5	14.6	1.8	0.9	14.74			0.10%
Paraffins	3-Methyl Hexane	589-34-4	14.8	1.2	1.0	14.90			
Paraffins	Cyclohexane	110-82-7	16.8	0.0	0.2	16.80	58		
Paraffins	Cyclopentane	287-92-3	16.4	0.0	1.8	16.50			
Paraffins	Cypar 7	64742-89-8	0.0	0.0	0.0				
Paraffins	Cypar 9	68553-14-0	0.0	0.0	0.0				
Paraffins	D30	64742-48-9	15.5	0.0	0.0	15.50			10.00%
Paraffins	D40	64742-48-9	15.5	0.0	0.0	15.50			10.00%
Paraffins	D60	64742-48-9	15.5	0.0	0.0	15.50			
Paraffins	D70	64742-47-8	15.5	0.0	0.0	15.50	34		
Paraffins	DSP 145 / 160	8002-5-9	15.5	0.0	0.0	15.50		2.90%	
Paraffins	Ethyl Cyclohexane	1678-91-7	16.7	0.9	1.0	16.75	49		

Paraffins	Heptane	142-82-5	15.2	0.2	0.7	15.22		0.91%	0.04%
Paraffins	Hexane	110-54-3	14.9	0.0	0.0	14.90	27	1.11%	0.12%
Paraffins	Isododecane	13475-86-6	16.2	0.0	0.0	16.20			0.20%
Paraffins	Isopar C	90622-56-3	15.1	0.0	0.0	15.10	27	2.90%	
Paraffins	Isopar E	90622-56-3	15.1	0.0	0.0	15.10	29	2.90%	
Paraffins	Isopar H	90622-57-4	15.0	0.0	0.0	15.00	26	2.90%	
Paraffins	Isopar L	90622-58-5	14.9	0.0	0.0	14.90	27	2.90%	
Paraffins	Isopar M	9622-58-5	14.9	0.0	0.0	14.90	25		
Paraffins	Methyl Cyclohexane	108-87-2	16.0	0.0	1.0	16.03	27.5		
Paraffins	Methyl Cyclopentane	96-37-7	15.8	3.5	0.0	16.18			
Paraffins	Nonane	111-84-2	15.7	0.0	0.0	15.70		0.79%	
Paraffins	Octane	111-65-9	15.5	0.0	0.0	15.50			
Paraffins	Pentane	109-66-0	14.5	0.0	0.0	14.50	27	1.20%	0.38%
Paraffins	SBP 1	64742-89-8	0.0	0.0	0.0		33		
Paraffins	SBP 100/140	64742-49-0	15.5	0.0	0.0	15.50			
Paraffins	SBP 11	64742-88-7	0.0	0.0	0.0		38	0.05%	
Paraffins	SBP 3	64742-89-8	0.0	0.0	0.0		32		
Paraffins	SBP 5	64742-89-8	15.3	0.0	0.0	15.30	33		
Paraffins	SBP 80 / 110	64747-49-0	15.5	0.0	0.0	15.50	32		
Paraffins	Solvesso 100	64742-95-6	17.8	0.6	1.4	17.87		32	10.00%
Paraffins	Solvesso 150	64742-94-5	17.8	0.6	1.4	17.87			10.00%
Paraffins	Solvesso 200	64747-94-5	17.8	0.6	1.4	17.87			
Paraffins	Whitespirit	64742-82-1	16.2	0.0	0.0	16.20	36.5		
Specials	1,3 Dioxolane	646-6-0	18.1	6.6	9.3	21.39		100.00%	100.00%
Specials	1,4-Dioxane	123-91-1	17.5	1.8	9.0	19.76		100.00%	100.00%
Specials	4-Bromo-3-Chloro-3,4, 4-Trifluoro-1-Butene	374-25-4	15.6	2.4	2.3	15.95		0.00%	
Specials	Acetonitrile	75-5-8	15.3	18.0	6.1	24.40		100.00%	100.00%

(Continued)

Table A1-5 *(Continued)*

Type	Name	CAS #	Disperse $MPa^{1/2}$	Polar $MPa^{1/2}$	H2-Bond $MPa^{1/2}$	Hildebrand SP $MPa^{1/2}$	Kb	Solubility of Water wt. %	Solubility in Water wt. %
Specials	Carbon Dioxide	124-38-9	15.7	6.3	5.7	17.85			
Specials	Cumene	98-82-8	18.1	1.2	1.2	18.18	83		
Specials	Dimethyl Acetamide	127-19-5	16.8	11.5	9.4	22.42		100.00%	100.00%
Specials	Dimethyl Carbonate	616-38-6	15.5	8.6	9.7	20.21		41.12%	12.68%
Specials	Dimethyl Formamide	68-12-2	17.4	13.7	11.3	24.86			100.00%
Specials	Dimethyl Sulfoxide	67-68-5	18.4	16.4	10.2	26.68	164	100.00%	100.00%
Specials	Ethylene Carbonate	96-49-1	18.0	21.7	5.1	28.65			21.40%
Specials	Gamma Butyrolactone	96-48-0	18.0	16.6	7.4	25.58			
Specials	Isophorone	78-59-1	17.0	8.0	5.0	19.44		4.30%	1.20%
Specials	N-Methyl Pyrrolidone	872-50-4	18.0	12.3	7.2	22.96	350	100.00%	100.00%
Specials	Propylene Carbonate	108-32-7	20.0	18.0	4.1	27.22		100.00%	100.00%
Specials	Pyridine	110-86-1	19.0	8.8	5.9	21.75			
Specials	Soygold 1000	67784-80-9	16.4	3.3	4.7	17.38	61		
Specials	Tetrahydrofuran	109-99-9	16.8	5.7	8.0	19.46		100.00%	100.00%
Specials	Water	7732-18-5	18.1	17.1	16.9	30.09		100.00%	100.00%

Appendix A2

Calculations about Adsorption of Solvents

The purpose of this appendix is to provide a scientific and engineering basis for evaluating the performance of activated carbon adsorbents when removing cleaning solvent vapors from air. One may also project adsorption isotherms for solvents where there is no empirical data.

DESCRIPTION OF ADSORPTION EQUILIBRIUM

Adsorption is a dynamic process. This means that at any point in time solvent molecules are being attracted and attached to individual locations (sites) within the pores of pieces (or shapes) of activated adsorbent; and are simultaneously being released and diffusing into the air stream.

The reason why there is flow of solvent vapor (mass transfer) both to and from sites within activated carbon (or any other adsorbent) is that there are two competing and opposing driving forces.

- One driving force is the attraction of solvent *to* the activated site. The attraction is mutual; both the solvent molecules and the activated sites exert common (mostly polar) intermolecular forces upon one another.
- The other driving force is the concentration imbalance between the sites and the vapor phase. Because the attractive forces are so strong, the vapor phase becomes highly depleted (desirably so) of solvent molecules. But, that concentration difference is a driving force for solvent flow *away* from the activated site.

Each of these competing processes operates continually at some rate – related to the amount of solvent present in the incoming air (Chapter 4, Footnote 65). The balance of these two driving forces is a degree of fractional coverage of all the activated sites.

THE LANGMUIR EQUATION

This can be expressed mathematically with the following nomenclature:

$$r_{Adsorb} = k_{Adsorb} \times c \times (1 - f) \qquad \text{A2-1}$$

$$r_{Desorb} = k_{Desorb} \times c \qquad \text{A2-2}$$

r_{Adsorb} = Rate of adsorption of solvent on activated sites (Equation A2-1),
r_{Desorb} = Rate of desorption of solvent on activated sites (Equation A2-2),
k_{Adsorb} = Rate constant for the adsorption process,
k_{Desorb} = Rate constant for the desorption process,
c = Concentration of solvent in the vapor phase, ppm by volume,
f = Numerical fraction of sites to which solvent has been attached. The symbol f is the number of sites to which there are solvent molecules wedded, divided by the maximum number of sites within the activated adsorbent matrix.

So the rate of adsorption is proportional to the concentration of solvent in the vapor phase (c) times the fraction of activated sites to which solvent has *not* been attached (the number of available sites).

Similarly, the rate of desorption is proportional to the same concentration of solvent in the vapor phase. But, there is no dependence of the diffusion rate *away from* the activated sites on the number of unoccupied activated sites. Obviously, $k_{Adsorb} >> k_{Desorb}$ for solvents which are well adsorbed on activated adsorbents.

Equilibrium is a condition of stability – no net change over time. At equilibrium, the rates of adsorption and desorption are equal. Equilibrium is maintained until something changes – such as the concentration of solvent in the vapor, or the vapor temperature.

If one sets the two rates of mass transfer equal, and solves the combined equation for the equilibrium fraction (f) of activated sites to which adsorbed solvent is attached, one gets Equation A2-3, where K is the overall equilibrium constant; $K = \frac{k_{Adsorb}}{k_{Desorb}}$.

Equation A2-3 can be rearranged with a more explicit definition of the parameter f. Equation A2-4 shows f to be the ratio of activated sites with solvent molecules attached (s) to the total number of available activated sites (s_{max}).

$$f = \frac{K \times c}{1 + (K \times c)} \quad \text{A2-3}$$

$$f = \frac{s}{s_{max}} \quad \text{A2-4}$$

If one then combines Equations A2-3 and A2-4, and rearranges the outcome, one obtains Equation A2-5.

$$\frac{1}{s} = \frac{1}{s_{max}} + \frac{1}{K \times s_{max} \times c} \quad \text{A2-5}$$

This equation is linear, with the independent variable being $\frac{1}{c}$ and the dependent variable being $\frac{1}{s}$.

$$Intercept = \frac{1}{s_{max}} \quad \text{A2-6}$$

$$Slope = \frac{1}{K \times s_{max}} \quad \text{A2-7}$$

Given adsorption data of volumetric vapor composition (c) and adsorbed amount (s), one may plot them and calculate the intercept and slope of the linear plot, with the calculated outcomes having physical meaning.

$$s_{max} = \frac{1}{Intercept} = Capacity\ of\ Activated\ Sites \quad \text{A2-8}$$

$$K = \frac{1}{Slope \times s_{max}} = Equilibrium\ Constant \quad \text{A2-9}$$

Of course, from the intercept and slope one may estimate both the maximum number of available sites (the capacity of the activated adsorbent [s_{max}]), and the equilibrium constant (the net strength by which a given solvent is attached to an activated adsorbent [K]).

Equation A2-5 is called the Langmuir Equation (also known as the Langmuir Isotherm, or Langmuir Adsorption Equation), being published by Irving Langmuir[1] in 1916.

[1] Irving Langmuir (January 31, 1881 to August 16, 1957) was an American chemist and physicist. While at General Electric, from 1909–1950, Langmuir advanced several basic fields of physics and chemistry. He invented the gas-filled incandescent lamp, the hydrogen welding technique, and was awarded the 1932 Nobel Prize in Chemistry for his work in surface chemistry. He was the first industrial (rather than theoretical) chemist to become a Nobel laureate.

The Langmuir Equation is based on four assumptions, not all of which are strictly or always valid. They are:

1. The surface of the adsorbent is uniform; that is, all the adsorption sites are equivalent.
2. Adsorbed molecules do not interact.
3. All adsorption occurs through the same mechanism.
4. When all activated sites are filled with solvent molecules (i.e., $s = s_{max}$), only a monolayer of solvent molecules is formed; molecules of adsorbed solvent (adsorbate) do not deposit upon one other, only on the free surface of the adsorbent.

The second and fourth assumptions are seldom valid, as solvent molecules do adsorb to the monolayer of solvent attached to the activated adsorbent – forming layers. But, in the same way that the ideal gas law is useful, so also is the Langmuir Equation (isotherm). It is often the first choice for a model of the adsorption process because it has been shown to be valid for, and produces information about, the number of activated sites on (not within) a monolayer of an adsorbent.

THE FREUNDLICH EQUATION

Developed earlier, in 1906, by Herbert Max Finlay Freundlich[2], this equation was based on simple observations about adsorption of gases on solids. The observations were:

- At very low pressures, the amount of adsorption (S) on a solid was directly proportional to the pressure of the gas being adsorbed (P). This can be expressed by Equation A2-10.

$$s \ \alpha \ P^{m} \quad \text{A2-10}$$

- But at high pressures, the amount of adsorption (s) on a solid was independent of gas pressure (P). This could be expressed by Equation A2-11.

$$s \ \alpha \ P^{m=0} \quad \text{A2-11}$$

Freundlich reconciled his two observations with a third. It is that there is a smooth transition between the first two observations. This can be expressed by the somewhat more general Equation A2-12. The exponent is arbitrarily stated as $\frac{1}{n}$[3] and the proportionality constant is k_1.

$$s = k_1 \times P^{\frac{1}{n}} \quad \text{A2-12}$$

$$s = k_F \times c^{\frac{1}{n}} \quad \text{A2-13}$$

Equation A2-13 is produced by recognizing that in Equation A2-12 pressure (P) is a measure of concentration

[2] Freundlich, H.M.F., *Journal of Physical Electrochemical.* 1906;57:385.
[3] The exponent n is also known by some as α.

(c) of the solvent in the gas phase, the amount of adsorption (s) is the number of activated sites which are occupied, and that the proportionality constant (k_F) can be restated to allow these substitutions.

The Freundlich Equation, Equation A2-14, is Equation A2-13 after both sides have been converted to logarithms:

$$Log(s) = Log[k_F] + \frac{1}{n} \times Log(c) \qquad \text{A2-14}$$

$$k_F = 10^{+(Slope)} \qquad \text{A2-15}$$

$$n = \frac{1}{Intercept} \qquad \text{A2-16}$$

A plot of the logarithm of the solvent concentration in air (c) vs. the logarithm of the weight fraction of activated carbon (or other adsorbent) which is solvent (s) should be linear – with a slope and an intercept as shown in Equations A2-15 and A2-16.

Do not be confused; k_F is not K. K is the equilibrium constant in the Langmuir Equation. It is the ratio of two reaction rate constants. k_F is an empirical constant in a linear equation, determined by regression methods[4]. The subscript F in k_F is to recognize that the linear equation is the Freundlich Equation.

The Langmuir and Freundlich equations are different approaches to describing the same phenomenon – coverage of activated carbon pore surface with single molecules attached to those pores. Surface coverage is of a monolayer of solvent; there are no multiple layers of solvent in the phenomena described by the Langmuir and Freundlich equations; nor is volume of adsorbed solvent directly indicated.

THE BET EQUATION

Dr. Edward Teller should be known for more than being the "father of the hydrogen bomb." He was also one of a triumvirate of researchers who published a paper in 1938[5] which expanded capabilities to model adsorption of gases on solids.

Their approach was an extension of Irving Langmuir's work, which modeled adsorption of single layers of solvent onto solids. Instead, Stephen Brunauer, Paul Hugh Emmett, and Edward Teller (BET) considered multilayer adsorption.

Their BET analysis is based on three assumptions, which are:

1. Gas molecules physically and infinitely adsorb on a solid in layers with one activated site per solvent molecule. A new layer of gas molecules may be started without other layers being complete.
2. There is no interaction between each adsorption layer or between adjacent gas molecules in the same layer, except that the uppermost layer is in equilibrium with the gas phase.
3. The Langmuir theory can be applied to each layer.

Derivation of the BET equation is outside the scope of this book (Box A2.1). But it is shown as Equation A2-17.

$$v = \frac{\{v_{max} \times B \times c\}}{\{(1-c) \times (1 + [B-1] \times c)\}} \qquad \text{A2-17}$$

Box A2.1 Nomenclature in Adsorption Equations

The following nomenclature applies to the BET Equation (Equation A2-17), as well as the Langmuir Equation (Equation A2-5), and the Freundlich Equation (Equation A2-14).

Note the somewhat different bases and outcomes of the Langmuir/Freundlich equations which speak to the amount of activated sites in an adsorbent, whereas the BET Equation speaks to the volume of gas adsorbed on activated sites in an adsorbent.

B = An empirical constant only in the BET Equation, to be determined for each data set by regression analysis of data. It is dimensionless.

s_{max} = The maximum amount of sites upon which gas can be adsorbed as a *monolayer*[6] on the adsorbent (usually activated carbon) – used in the Langmuir Equation. It is dimensionless.

v_{max} = The maximum volume of gas in *all layers* of solvent molecules which can be adsorbed within the adsorbent – used in the BET Equation. Its units are of volume, usually in cc.

s = The total measured amount of sites upon which gas is adsorbed in a *monolayer* on the adsorbent[7]. This is certainly less than s_{max}. It too is dimensionless.

v = The total measured volume of gas in all layers of gas molecules within the adsorbent. Its units are of volume, usually in cc.

c = The measured concentration of solvent in the vapor phase, in volumes per volume[8].

[4] Yet, Irving Langmuir published a derivation of Freundlich's equation without using Freundlich's empirical observations in Langmuir, I., *Journal of the American Chemical Society*. 1916;38:2221.

[5] Brunauer, S., Emmett, P.H., and Teller, E., *Journal of the American Chemical Society*. 1938;30:309.

[6] This is not the total capacity of the activated carbon to hold solvent because the BET Equation allows for multilayer adsorption.

[7] As used in the Langmuir Equation, this parameter can take on values no greater than s_{max} (as solvent is present only in a monolayer).

[8] Vapor phase solvent concentration is often given in other units such as the ratio of equilibrium pressure to saturation pressure (via the Antoine Equation at the operating temperature). One unit which is not used in the BET Equation is ppm by pressure or volume – the BET Equation is based on fractional concentrations.

The BET Equation (Equation A2-17) can be rearranged in a linear form (Equation A2-17), as can the Langmuir and Freundlich equations. For the BET Equation, the independent variable is the volumetric concentration of solvent in the vapor phase (c). The dependent variable is the term $\left[\frac{c}{\{1-c\}} \times \frac{1}{s}\right]$, which contains only measured values.

$$\left\{\frac{c}{[v \times (1-c)]}\right\} = \frac{1}{[v_{max} \times B]} + c \times \left\{\frac{[(B-1)]}{v_{max} \times B}\right\} \qquad \text{A2-18}$$

In Equation A2-18 the intercept is $\frac{1}{[v_{max} \times B]}$, and the slope is $\left\{\frac{[B-1]}{[v_{max} \times B]}\right\}$.

$$v_{max} = \left\{\frac{1}{Slope + Intercept}\right\} \qquad \text{A2-19}$$

$$B = 1 + \left\{\frac{Slope}{Intercept}\right\} \qquad \text{A2-20}$$

Equations A2-19 and A2-20 allow calculation of the two adjustable parameters, v_{max} and B, from plots of a set of measured data.

ESTIMATE OF ADSORBENT SURFACE AREA

For many technologists the prime use of the BET Equation is to characterize an adsorbent for gases by estimating its total surface area available for adsorption (Box A2.2). This cannot be done directly with either the Langmuir or Freundlich equations because they are denominated in the amount of activated sites, and not in the volume of solvent molecules adsorbed[9]. The equation used is Equation A2-21.

$$A_{BET} = \frac{v_{max} \times A_x \times N}{V_m \times w} = \frac{[cc] \times \left[\frac{SM}{molecule}\right] \times \left[\frac{molecules}{mole}\right]}{\left[\frac{cc}{mole}\right] \times [gram]} = \frac{SM}{gram} \qquad \text{A2-21}$$

Box A2.2 Nomenclature Needed to Use the BET Equation

The nomenclature used in Equation 2-21 is that used in Equations A2-17 and A2-18, supplemented by some additional terms which appear similar but which have different meanings in an experiment to measure total adsorbent surface area (A_{BET}).

A_{BET} = The maximum (total) amount of adsorbent surface area, calculated using the BET theory and methodology (Equations A2-18 and A2-21)[10].

v_{max} = The measured maximum volume of gas in *all layers* of solvent molecules which can be adsorbed within the adsorbent – used in the BET Equation. Its units are of volume, usually in cc.

A_x = The known (from other experiments) cross-sectional area of the gas molecules being adsorbed. Its units are of area in square meters (SM) per molecule. For nitrogen, this value is 1.6×10^{-19} SM; for water, this value is 1.14×10^{-19} SM.

N = The number of molecules in a mole of gas. This is Avogadro's Number, 6.02×10^{-23}.

V_m = The calculated molar volume of the *gas*, at the conditions of the experiment. The units are of cc/g-mole. Recall that the molar volume of an ideal gas at 1 atmosphere pressure and 25°C is 22,414 cc (22.414 liters).

w = The measured mass of adsorbent used in the experiment, usually in units of grams.

Only v_{max} and w are measured in such an experiment; A_x, N, and V_m are presumed to be known constants or knowable. Nitrogen gas is commonly used, so the adsorption capacity of various activated adsorbents can be compared using the same adsorbate (gas). Effective activated adsorbents have one to several thousand SM of A_{BET} surface area.

DESCRIPTION OF ADSORPTION DATA

Measured adsorption data for four cleaning solvents (methylene chloride, trichloroethylene, perchloroethylene, and CFC-113) are plotted in Figure A2-1[11,12].

Which of these four solvents is most easily adsorbed?

---

[9] This point is often not made, or is ineffectively made, in the scientific literature. One reason for this outcome is inconsistent use of consistent nomenclature.

All three equations can be used to describe the "amount" of solvent vapor adsorbed on an amount of adsorbent (Figures A2-2 to A2-5). But different metrics are being used to define the word "amount." In the Langmuir and Freundlich equations, "amount" means of activated sites within a single layer of coverage; in the BET Equation, "amount" means of volume of gas molecules in all layers of coverage.

[10] The BET theory is not the only theory of adsorption used for this purpose.

[11] This plot is typically of most use in that a log-log scale allows a broad spectrum of airborne concentrations to be described.

[12] Data source is Kanegsberg, B. And Kanegsberg, E., *Handbook for Critical Cleaning*, Second Edition, CRC Publishing, April 2011, ISBN 1439828261; Table 22-1 is by J. McChesney.

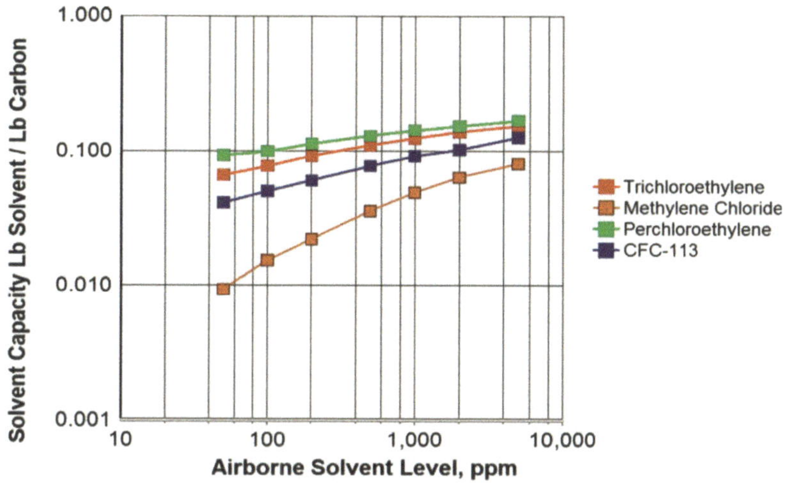

Figure A2-1 Measured Adsorption Isotherms for Four Cleaning Solvents

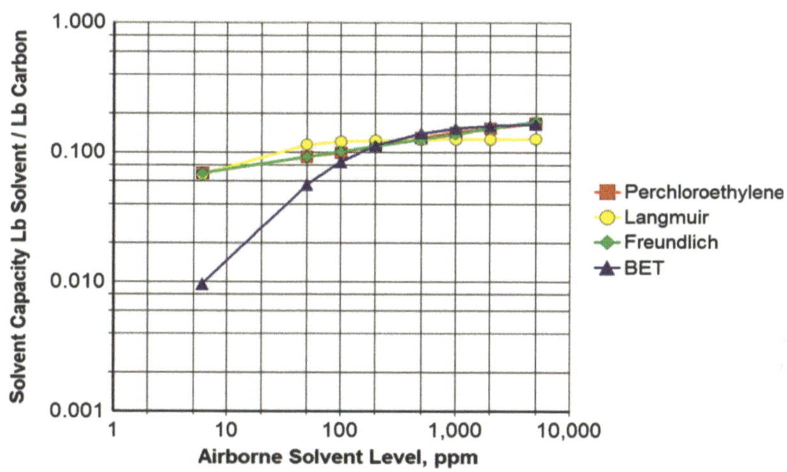

Figure A2-2 Measured and Calculated Adsorption Performance of Perchloroethylene on Activated Carbon

It is perchloroethylene. That solvent is adsorbed at any amount onto the adsorbent at the lowest vapor concentration. Recall, that perchloroethylene has no hydrogen-bonding intermolecular forces because it has no hydrogen atoms.

Actually, none of these four solvents is exceptionally well adsorbed. Exceptional adsorption means retention on the activated structure to at least the level of 20 weight percent, and perhaps as much as 30 weight percent.

Adsorption data of each of these four cleaning solvents can be described by all three of the adsorption isotherm equations above. This is shown in Figures A2-2 (perchloroethylene) through A2-5 (CFC-113).

The experimental data are described in each of these four plots as being the red square symbols[13]. This is significant because for each of the four solvents, one of the three isotherm equations nearly overlays the experimental data.

Note that the application of the BET Equation is unlike that of the Langmuir or Freundlich equations – at low airborne solvent concentrations. This is presumably

[13] The activated carbon adsorbent used to produce these data is produced from coal. The common commercial adsorbents are activated carbon produced from either coal or coconut shell.

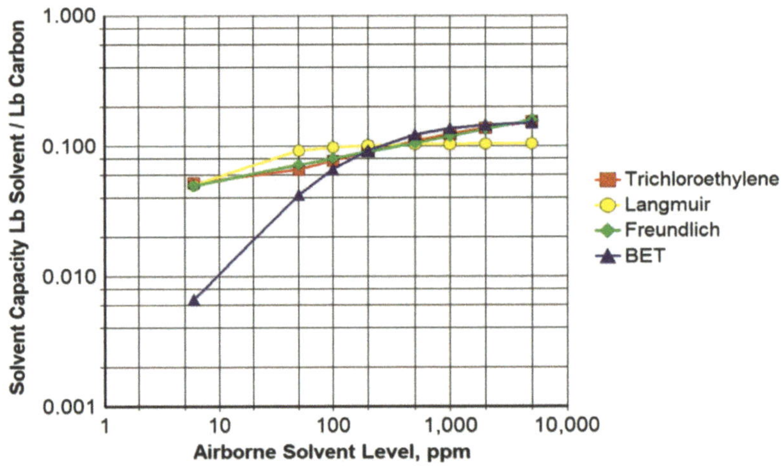

Figure A2-3 **Measured and Calculated Adsorption Performance of Trichloroethylene on Activated Carbon**

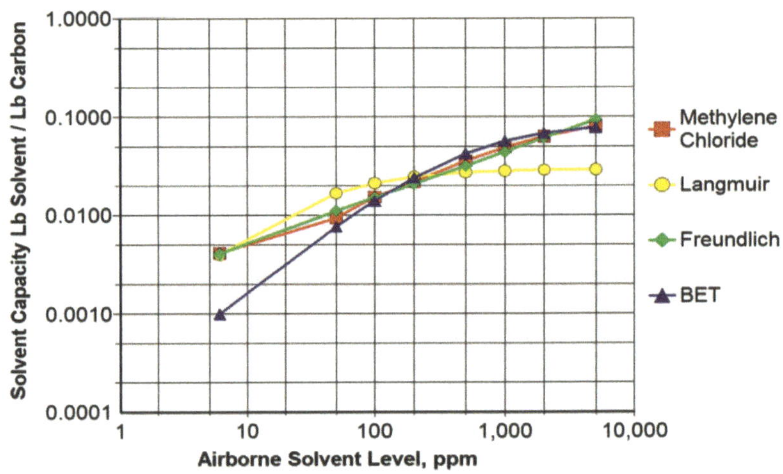

Figure A2-4 **Measured and Calculated Adsorption Performance of Methylene Chloride on Activated Carbon**

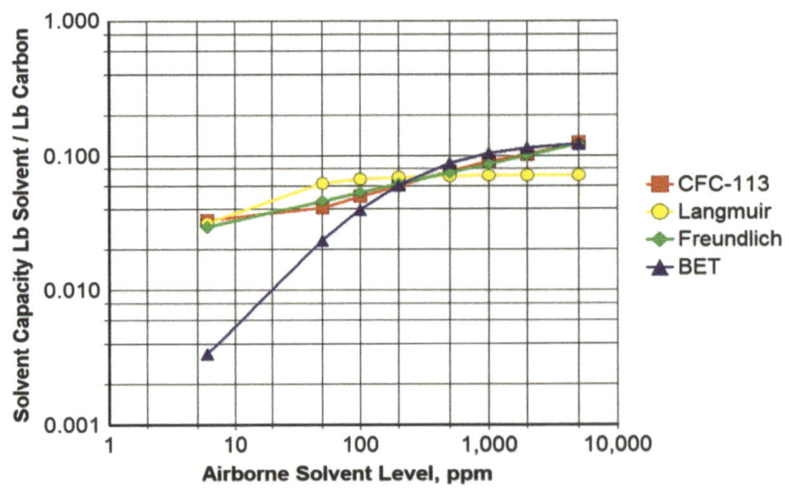

Figure A2-5 **Measured and Calculated Adsorption Performance of CFC-113 on Activated Carbon**

Table A2-1 Summary of Correlation Equations for Adsorption Isotherms

Isotherm Equation	Basic Equation	S_{max}, k_F, & V_{max}	K, n, & B
Langmuir	$\frac{1}{s} = \frac{1}{s_{max}} + \frac{1}{K \times s_{max} \times c}$	$s_{max} = \frac{1}{Intercept}$	$K = \frac{1}{Slope \times s_{max}}$
Freundlich	$Log(s) = Log[k_F] + \frac{1}{n} \times Log(c)$	$k_F = 10^{+(Slope)}$	$n = \frac{1}{Intercept}$
BET	$\left\{\frac{c}{[v \times (1-c)]}\right\} = \frac{1}{[v_{max} \times B]} + c \times \left\{\frac{[(B-1)]}{[v_{max} \times B]}\right\}$	$v_{max} = \left\{\frac{1}{[Slope + Intercept]}\right\}$	$B = 1 + \left\{\frac{Slope}{Intercept}\right\}$

Table A2-2 Comparison of Meaning of Three Adsorption Correlation Equations

Isotherm Equation	Application Geometry	Quantity Represented	Range of Use	Most Common Type of Use
Langmuir	Single layer of solvent molecules	Amount of activated sites	Low and moderate vapor concentrations	Represent (not predict) experimental data for design of adsorption equipment
Freundlich	Any configuration			
BET	Multiple layers of solvent molecules	Volume of adsorbed gas	Moderate and high vapor concentrations	Estimate adsorbent surface area

because the BET Equation is derived based on the assumption that the solvent vapor molecules are found in multiple layers within adsorbent pores, and this is unlikely to be so at low airborne concentrations of solvent.

COMPARISON OF CORRELATIONS

The Langmuir Equation, Freundlich Equation, or the BET Equation can be used to adequately represent experimental data. However, only the Langmuir or Freundlich equations are recognized as being reliable at very low solvent concentrations in the influent gas.

As with the Langmuir Equation, the Freundlich Equation and the BET Equation can be viewed as being linear.

- The Langmuir Equation is linear in reciprocal vapor concentration, and reciprocal amount of adsorption. The slope and intercept of linear plots have meaning.
- The Freundlich Equation is linear in the logarithm of vapor concentration, and the logarithm of amount of adsorption. The slope and intercept of linear plots do not have meaning.
- The BET Equation is linear in vapor concentration, and a function of both amount of adsorption and vapor concentration.

These details are listed and summarized in Table A2-1. The slope and intercept of linear plots do have meaning as summarized in Table A2-2.

PREDICTION OF ADSORPTION ISOTHERMS

The above three adsorption isotherm (equilibrium) equations describe measured adsorption data. They are methods of correlation of or interpolation among that data. Given two, or preferably more, measured values of adsorbed "amount" and measured vapor phase concentration one can reliably predict performance of a bed of activated adsorbent at nearly all other concentrations of vapor.

As such, these three equations are commonly used to design and manage adsorption facilities based on known data. That is their key value, and why they are enumerated in this appendix.

However, the Langmuir, Freundlich, and BET equations are not predictive. They do not allow prediction of the adsorbed loading which a concentration of a specific solvent vapor can be expected to obtain onto a specific activated adsorbent.

For predictive work, the Dubinin/Radushkevich (D/R) adsorption isotherm technology[14] (and augmentations upon it[15,16]) is considered by many to be the standard.

The D/R adsorption isotherm has a theoretical basis[17,18] which is outside the scope of this book. Its use requires three types of information: (1) a calculation of the thermodynamic amount of non-mechanical (i.e., chemical) work associated with a quantum of adsorption, without regard to the specific solvent or the specific adsorbent involved, but with regard to the relative volume of solvent available for adsorption; (2) independently determined empirical constants – known as affinity constants, noted by the symbol β – which allow extrapolation from one solvent to another; and (3) a characteristic energy of adsorption – noted by the symbol ΔE_0 – which is primarily based on the characteristic dimensions of the micropores.

$$W = W_0 \times \exp\left\{ - \left[\frac{A}{\beta \times \Delta E_0}\right]^2 \right\} \quad \text{A2-22}$$

The basic D/R equation is given by Equation A2-22, where:

W = The amount of solvent *vapor* adsorbed per unit mass of the adsorbent, in *molar* units (moles or millimoles) or adsorbate volume units[19].

W_0 = The maximum or limiting volume of adsorbed solvent vapor or the volume of the micropores (per unit *mass* of the adsorbent).

The quantity $\left\{\frac{W}{W_0}\right\}$ is useful to predict molar adsorption phenomena; what's wanted is the capability to predict mass adsorption phenomena. If it is modified to reflect that the adsorption is done from air, and that what is wanted is the mass amount of solvent adsorbed per mass amount of adsorbent, the desired quantity s can be computed as $s = \left\{\frac{W}{W_0}\right\} \times \left\{\frac{M_{air}}{M_{solvent}}\right\}$.

A = The thermodynamic change associated with converting a concentration of solvent in the bulk vapor state (not adsorbed) to the adsorbed state. This is called the Gibbs free energy of adsorption and is computed as $A = \left\{R \times T \times \ln\left[\frac{P}{P_{sat}}\right]\right\}$. R is the universal gas constant (0.00831 kJ/g-mole.K). T is the operating temperature in kelvin. P is the pressure of the solvent in any units; P_{sat} is the saturated vapor pressure of the solvent at the same temperature, via the Antoine Equation, in the same units. One may make the following substitution by incorporating the ideal gas law $\left(\left[\frac{P}{P_{sat}}\right] \approx \left[\frac{1}{c}\right]\right)$, where c is a volume fraction (not ppm).

ΔE_0 = The "characteristic energy" of adsorption. Apparently, this is neither a function of temperature or solvent, but is a function of the major dimension of the pore. This is because the intermolecular forces between solvent molecules and pore walls depend greatly upon the separation distance between them. Per the reference of Footnote 17 (page 373), ΔE_0 is inversely proportional to pore diameter, as $\Delta E_0 = \left\{\frac{29.6}{(D_{pore})}\right\}$ – with the units of ΔE_0 being kJ/g-mole, and the pore diameter given in nm. This relationship for ΔE_0 appears to be purely empirical.

β = An empirical constant, called an affinity constant, whose value is defined as unity for benzene. β values are in effect scaling divisors between benzene and some other solvent. Affinity in the D/R equation means similarity of the ability of the solvent to be polarized[20] (Box A2.3) to the same level as is benzene. The reference of Footnote 15 (pages 344 and 354) reports a correlation of β with molar polarizibility (P_e). $\beta = \left\{\frac{P_e}{P_{e\ reference}}\right\}^{0.9}$, with the reference P_e being that of benzene and being calculated as 26.274 cc/g-mole. Molar polarizibility is easily calculated with the Lorentz–Lorenz equation which involves the molar volume of the solvent (V_m) and its refractive index (n_D). The Lorentz–Lorenz equation (Box ?) is

$$P_e = V_m \times \left\langle \frac{\{[n_D]^2 - 1\}}{\{[n_D]^2 + 2\}} \right\rangle.$$

[14] Dubinin, M.M., Zaverina, E.D., and Radushkevich, L.V., "Sorption and Structure of Active Carbons. I. Adsorption of Organic Vapors," *Zh Fiz Khim*. 1947;21:1351–1362.

[15] Wood, G.O., "Affinity Coefficients of the Polanyi/Dubinin Adsorption Isotherm Equations: a Review with Compilations and Correlations," *Carbon*. 2001;39:343–356.

[16] Wood, G.O., "Review and Comparisons of D/R Models of Equilibrium Adsorption of Binary Mixtures of Organic Vapors on Activated Carbons," *Carbon*. 2002;40:231–239.

[17] Dubinin, M.M., "Generalization of the Theory of Volume Filling of Micropores to Nonhomogeneous Microporous Structures," *Carbon*. 1985;23(4):373–380.

[18] A very useful description of the theoretical basis is available by Professor Adam W. Marczewski at http://awmarcz.republika.pl/awm/ads/DR&DA.htm.

[19] Both D/R technology and BET technology treat adsorption as a process managed in the gas phase on a volumetric or molar basis, whereas Langmuir and Freundlich technology treat adsorption as a process managed on an area basis.

[20] As described in Ref. 2, Chapter 2.3 and Equation 2.3, there are three kinds of intermolecular forces: those dispersed (extended throughout the entire molecule), those involving hydrogen bonds (where hydrogen atoms bond by sharing electrons chiefly with oxygen or nitrogen atoms), and those involving polar interactions (aligning of a molecule's electron distribution about a center of charge).

The term polarizability is the relative tendency of a charge distribution, like the electron cloud of an atom or a molecule, to be distorted from its normal shape by proximity to the center of charge of another atom or molecule. In the case of adsorption, only molecules are of concern.

Box A2.3 Polarization of Solvent Molecules

Bonding of solvent gas molecules to an activated surface generally involves co-alignment of charge distributions between the solvent and the activated site. The phrase molar polarizibility refers to the polarization of one mole of solvent. It is identified by the symbol P_e

The Lorentz–Lorenz equation is named after the Danish mathematician and scientist Ludvig Lorenz, who published it in 1869, and the Dutch physicist Hendrik Lorentz, who discovered it independently in 1878.

The Lorentz–Lorenz equation as used by the author of the reference of Footnote 15 is implemented with the properties of the solvent in the liquid phase, not the vapor phase – which is the condition in which the solvent is applied to the adsorbent.

The refractive index (index of refraction or n_D) of a solvent is a measure of the speed of light in that solvent. It is expressed as a ratio of the speed of light in vacuum relative to that in the liquid solvent. Since motion of light isn't hindered by impact with other molecules in vacuum, n_D is always greater than unity.

Since V_m is molecular weight (M) divided by density (ρ), the correlation for β becomes:

$$\beta = \left\langle \frac{\left[\frac{M}{\rho}\right] \times \left\langle \frac{\{[n_D]^2 - 1\}}{\{[n_D]^2 + 2\}} \right\rangle}{26.274} \right\rangle^{0.9} .$$

To summarize, Equation A2-22 can be used to predict adsorption data for a given concentration of solvent in air, and an adsorbent whose pore diameter is known. The three parameters needed to implement Equation A2-22 for a given situation are:

$$A = \left\{R \times T \times \ln\left[\frac{1}{c}\right]\right\} \qquad \text{A2-23}$$

$$\Delta E_0 = \left\{\frac{29.6}{D_{pore}}\right\} \qquad \text{A2-24}$$

$$\beta = \left\langle \frac{\left[\frac{M}{\rho}\right] \times \left\langle \frac{\{[n_D]^2 - 1\}}{\{[n_D]^2 + 2\}} \right\rangle}{26.274} \right\rangle^{0.9} \qquad \text{A2-25}$$

And so Equation A2-22, can be restated as Equation A2-26, with the operating temperature (T) in °C and c as a volumetric fraction of solvent molecules in air. Equation A2-26 is most easily used when converted to a logarithmic form.

Equation A2-26 is useful with activated carbon structures which have a relatively homogeneous distribution of sizes of uniform pore diameters. Equation A2-22 has been modified for use with heterogeneous distributions of pore diameter (the reference of Footnote 17). That work is outside the scope of this book.

Consider the following example, with 100 ppm (by volume) trichloroethylene in air at 25°C being adsorbed into activated carbon (from coal) with a pore diameter of 1.34 nm. For this solvent, using physical property values in Appendix A1, M = 131.4 g/g-mole, n_D = 1.4770, and ρ = 1.4630 g/cc. The molecular weight of air is 28.97.

$$\ln[s] = \ln\left[\frac{28.97}{M}\right] - \left\{ \frac{[8.31 \times 10^{-3} \times (T + 273.16) \times (\ln\frac{1}{c})]}{\left[\left\langle \frac{\left[\frac{M}{\rho} \times \left\langle \frac{\{[n_D]^2-1\}}{\{[n_D]^2+2\}} \right\rangle\right]}{26.274} \right\rangle \times \left\{\frac{29.6}{D_{pore}}\right\}\right]} \right\}^2 \qquad \text{A2-26}$$

$$A = \left\{0.00831 \times (25 + 273.16) \times \ln\left[\frac{1}{0.0001}\right]\right\}$$
$$= -22.82 \qquad \text{A2-23}$$

$$\Delta E_0 = \left\{\frac{29.6}{1.34}\right\} = 22.09 \qquad \text{A2-24}$$

$$\beta = \left\langle \frac{\left[\frac{131.4}{1.4360}\right] \times \left\langle \frac{\{[1.4470]^2 - 1\}}{\{[1.4470]^2 + 2\}} \right\rangle}{26.274} \right\rangle^{0.9} = 0.9692 \qquad \text{A2-25}$$

$$\ln[s] = \ln\left[\frac{28.97}{131.4}\right] - \sqrt{\frac{26.81}{0.9692 \times 22.09}} = -2.678 \qquad \text{A2-26}$$

So, the concentration of trichloroethylene adsorbed on activated carbon with a pore diameter of 1.34 nm is exp[−2.678] = 6.87, or 6.87 percent.

The measured value is 9.1 percent.

Equation A2-26 is applied to the same concentrations of trichloroethylene in air as were plotted in Figure A2-3. The calculated results are plotted in Figure A2-6.

Similar plots of calculated adsorption of perchloroethylene, methylene chloride, CFC-113, and toluene (with 50% or less relative humidity) using the D/R technology vs. experimental data are given in Figures A2-7 through A2-10 respectively.

These plots should be compared to the correlations based on the Langmuir, Freundlich, and BET equations – Figures A2-2 through A2-5 respectively.

Figure A2-11 shows an adsorption isotherm for n-propyl bromide using the D/R technology. There is no available measured adsorption data for this solvent.

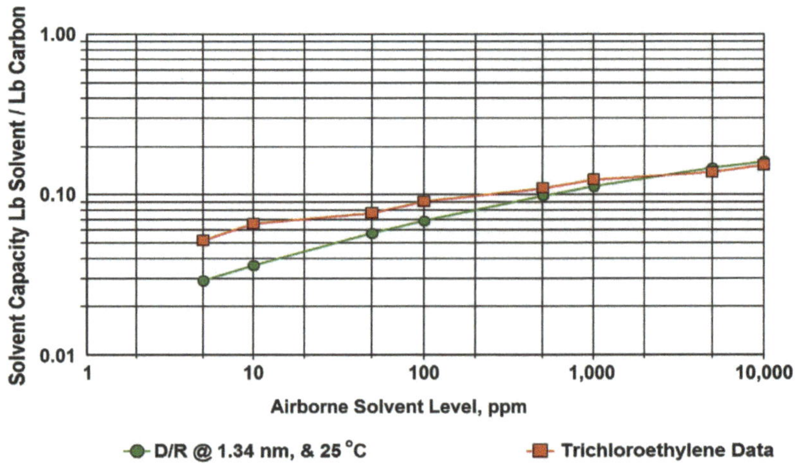

Figure A2-6 **Measured and Calculated Adsorption Data for Trichloroethylene**

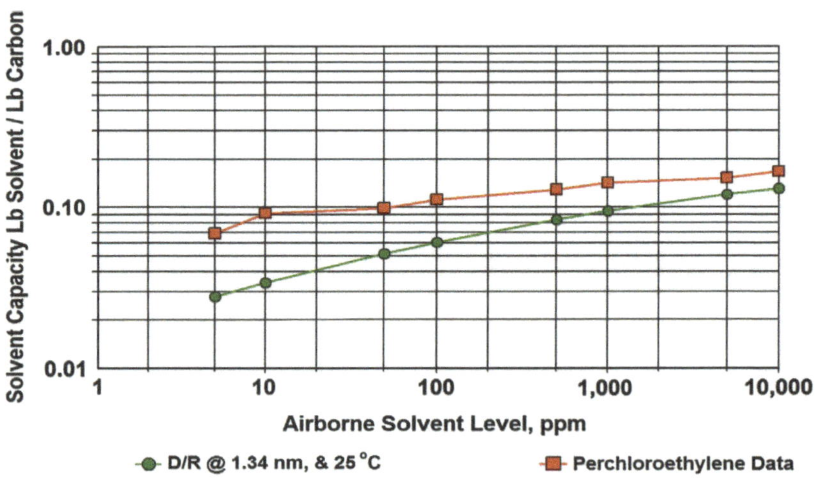

Figure A2-7 **Measured and Calculated Adsorption Data for Perchloroethylene**

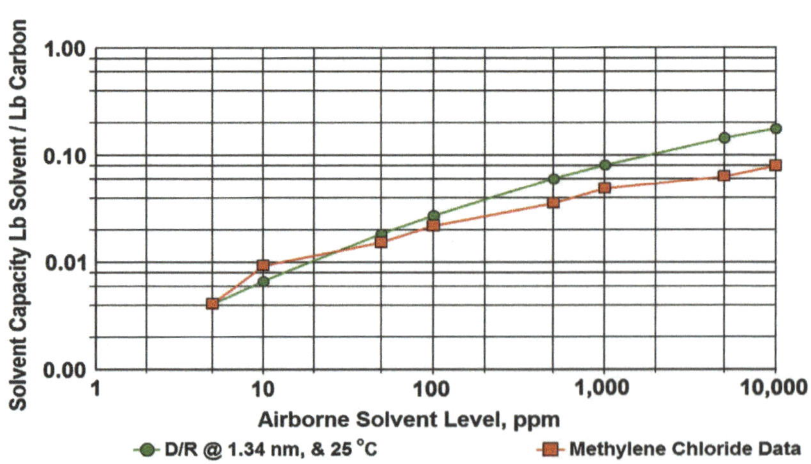

Figure A2-8 **Measured and Calculated Adsorption Data for Methylene Chloride**

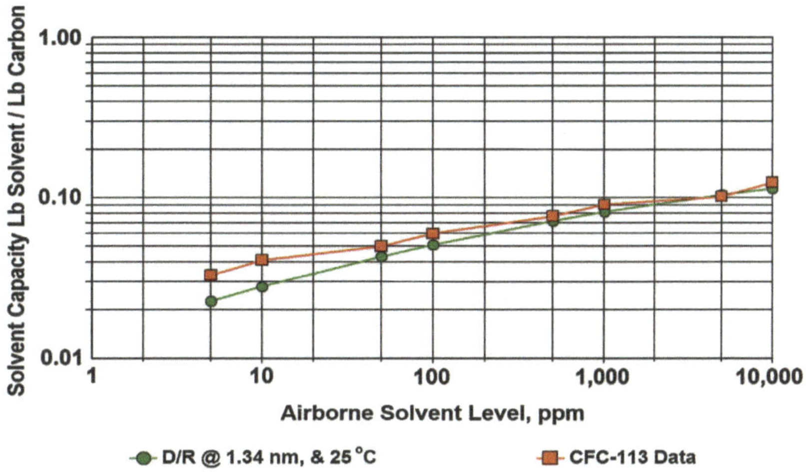

Figure A2-9 **Measured and Calculated Adsorption Data for CFC-113**

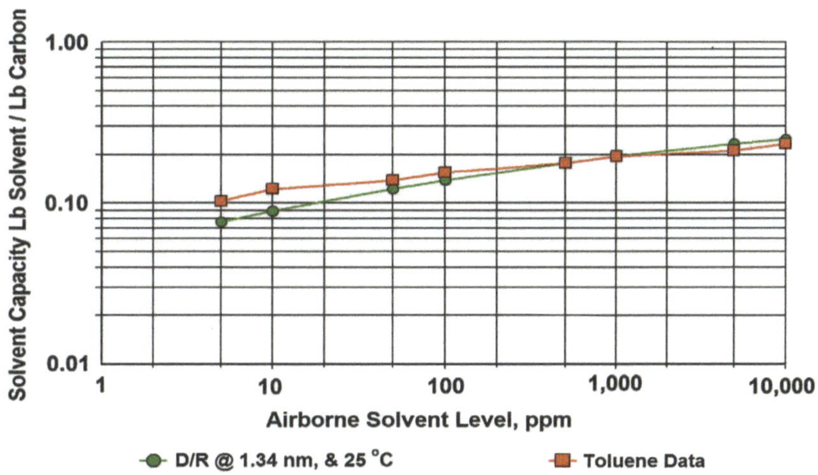

Figure A2-10 **Measured and Calculated Adsorption Data for Toluene**

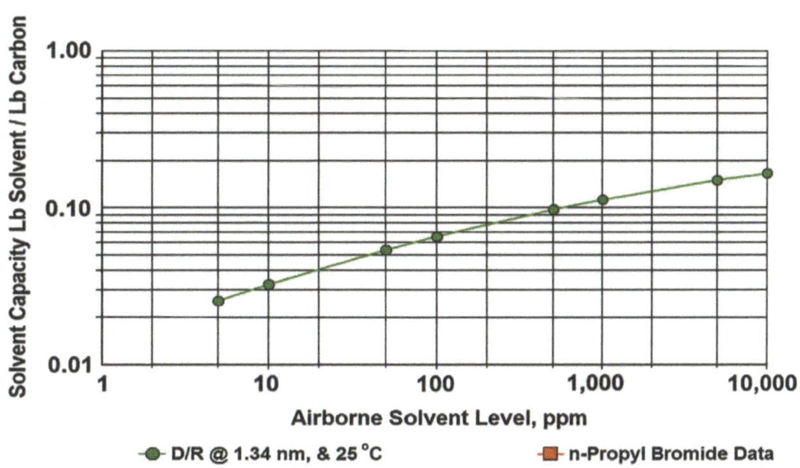

Figure A 2-11 **Measured and Calculated Adsorption Data for n-Propyl Bromide**

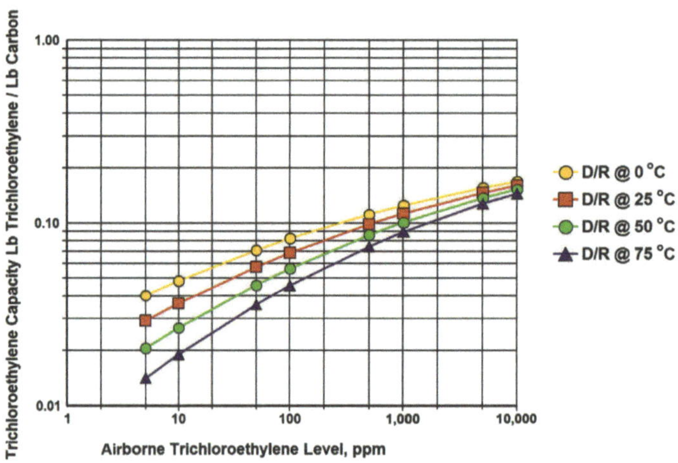

Figure A2-12 Calculated Variation in Adsorption Performance vs Operating Temperature

The Dubinin/Radushkevich methodology, initially disclosed in 1947, has been the basis for R&D by many scientists seeking controlled, understood, and improved adsorption of pollutant gases by activated surfaces[21]. The implementation described here allows exploration of the effect of operating variables on adsorption performance. This methodology includes: (1) properties associated with the adsorbent (pore diameter), (2) properties associated with the vapor being adsorbed (V_m and n_D), (3) properties associated with the environment [temperature], and (4) similarity to other solvents (β) which allows prediction about untested solvents.

THERMAL EFFECTS

Physical adsorption is an exothermic[22] process – heat is given off when solvent is adsorbed onto activated carbon. Heating can be offset to an extent by the inflow of cooler solvent-laden air.

Adsorption is somewhat temperature dependent. Improved capacity for solvent adsorption is achieved at lower temperatures. At higher temperatures, the motion of airborne molecules (adsorbates) becomes more energetic, and so molecules are more difficult to stably bond to activated sites. Usually, it is not possible to take advantage of this fact because it is practically and economically inconvenient to cool an air stream which exists only periodically (when the cleaning chamber of an enclosed machine is flushed, for example – Chapter 4.8.1.1), and which only contains hundreds of ppm of solvent. So, the temperature of small adsorption beds for treatment of solvent-laden air is not normally controlled.

However, scientists and developers who do testing of various adsorbents must measure and control the temperature at which their equilibrium measurements are made. Hence, an adsorption profile (amount adsorbed vs. inlet concentration) is measured at a fixed temperature (usually 25°C), and called an *isotherm*.

D/R methodology allows quantization of this effect through Equation A2-23[23] in which temperature is an explicit variable. For adsorption of trichloroethylene on activated carbon (from coal) with a pore diameter of 1.34 nm, the effect of operating temperature is clearly shown in Figure A2-12.

SMALL PORES ARE BEST

Certainly more significant is the effect of pore size on performance. Pore size is taken to be the average diameter in a homogeneous distribution of pore diameters. The calculated effect, described using D/R technology, is

[21] One may find adsorption isotherms for more than 300 chemicals plotted as above on the URL http://www.pacco-intl.com/isotherms.asp. They are calculated using a variant of the D/R technology, and as of publication are available at no charge.

[22] One explanation for why adsorption of gas molecules is exothermic is based on individual molecules having certain degrees of freedom in their allowable motion (energy states). Gases have the most degrees of freedom; liquids have less, and solids have still less. When a molecule loses degrees of freedom, it loses energy because opportunity for motion is limited. So the energy given off when solvent gases are converted to layers of solvent on activated sites makes the adsorption process exothermic.

[23] One may use D/R technology with both activated carbon and molecular sieves (Chapter 3, Footnote 94) based on carbon substrates. Per the reference of Footnote 15, the values of β are essentially the same.

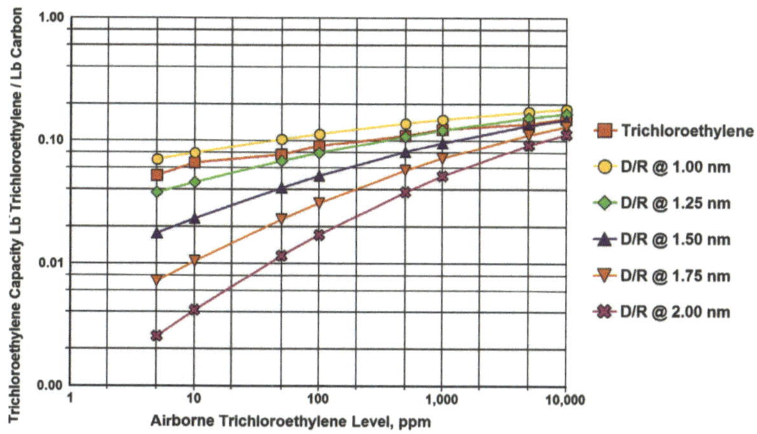

Figure A2-13 Calculated Variation in Adsorption Performance vs Adsorbent Pore Size

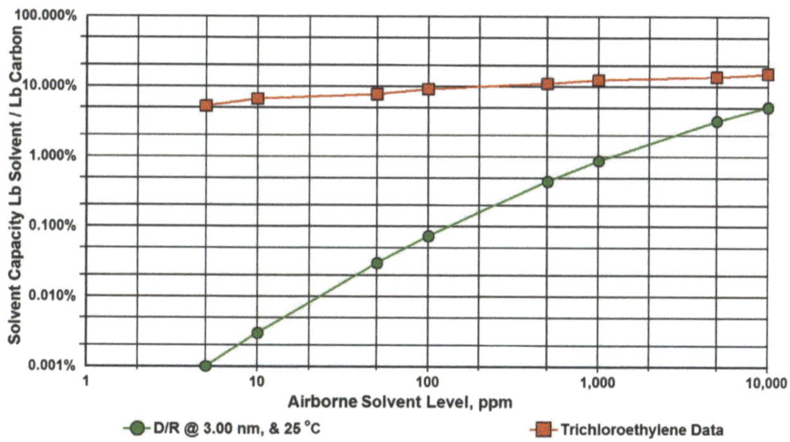

Figure A2-14 Calculated Variation in Adsorption Performance with a Huge Pore Size

particularly significant at low airborne concentrations of trichloroethylene, is shown in Figure A2-13.

The effect demonstrated in Figure A2-13 enables recognition of the nature of the short-range forces between solvent molecules and pore walls. At very low concentrations of solvent vapor, adsorptive capacity varies over roughly a decade when pore diameter is doubled from 1 to 2 nm.

At a pore diameter of 3 nm, calculated adsorptive capacity of trichloroethylene, whose kinetic diameter is 0.62 nm (Chapter 4, Footnote 6), changes greatly over a significant range of concentrations of solvent in air.

But in Figure A2-14 substantial calculated levels of adsorption do not really occur even at relatively high solvent concentrations. Even at several hundred ppm of solvent in the vapor, adsorption is essentially negligible when the pore diameter is relatively large.

Said another way, in order to be adsorbed, a solvent molecule has to be located close to a pore wall.

This is shown in Figure A2-15 for three molecules of trichloroethylene. This figure is proportionally correct.

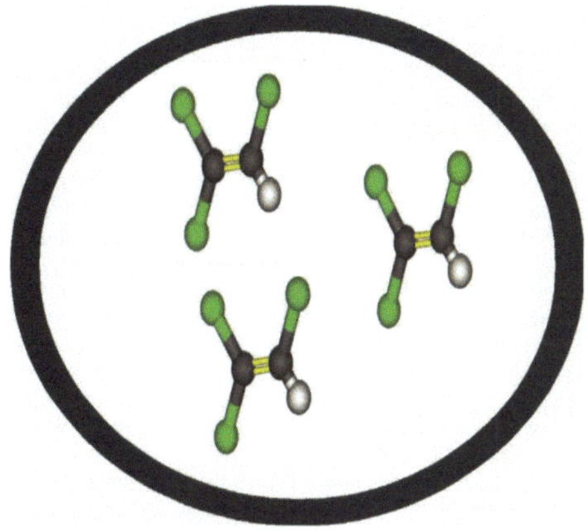

Figure A2-15 Hypothetical View of Trichloroethylene Molecules in a Pore of 1.34 nm

The circumference represents an idealized pore diameter of 1.34 nm The molecule's kinetic diameter is 0.62 nm. Pores of this size are produced in adsorbents so that trichloroethylene[24] solvent molecules fit easily within them, and are naturally located close to the pore walls.

CRUCIAL EFFECTS OF WATER (AND PORE SIZE)

Such is not the case with water. It is a much smaller molecule – its kinetic diameter is only 0.265 nm. Three water molecules fit nicely within pores of 1.34 nm or smaller in diameter.

But proximity to walls is another matter (Figure A2-16).

The circumference of the circle in Figure A2-16 represents the same pore diameter as in Figure A2-15. Similarly, there are three water molecules – proportionally sized to the pore diameter. With just a few veiy small[25] water molecules, there is a low probability that any one molecule will be located near the wall. So at low levels of water in the air fed to the adsorber bed, one would expect little adsorption of water.

When more water molecules are located within the pore diameter, as in Figure A2-17, useful adsorption performance is more likely. Predicted adsorption performance using the D/R technology is shown in Figure A2-18[26,27].

So, at higher concentrations of water in influent air (because of the condition shown in Figure A2-16), water is adsorbed from air on activated carbon. Either water or solvents can be adsorbed from a moving air stream. Actually, both can.

Two consequences flow from that observation – only one of which is an aid to effective cleanup of solvent fumes from emitted air.

The favorable consequence is that high concentrations of water (steam actually) in inlet air can be used to displace adsorbed solvents from activated carbon. Wet solvent removed from the adsorber bed can be collected and reused after drying of whatever small amount of water is soluble in the solvent (Chapter 4.17, right-hand columns of Table 4.8, and Figure 4.40).

And the adsorbed water can be evaporated from the activated carbon by hot gasses (usually nitrogen). So the favorable consequence is that water (low pressure steam, Chapter 4.16.2) is an enabling chemical which can render an adsorption bed saturated with solvent capable of being able to discharge that load of solvent and later repeatedly being able to adsorb more solvent fumes from moving air.

The unfavorable consequence is that activated sites (space) can be taken by water on an unsaturated adsorption bed. Solvent can't dislodge them. These sites can't be used to adsorb solvents – they represent a loss of bed capacity.

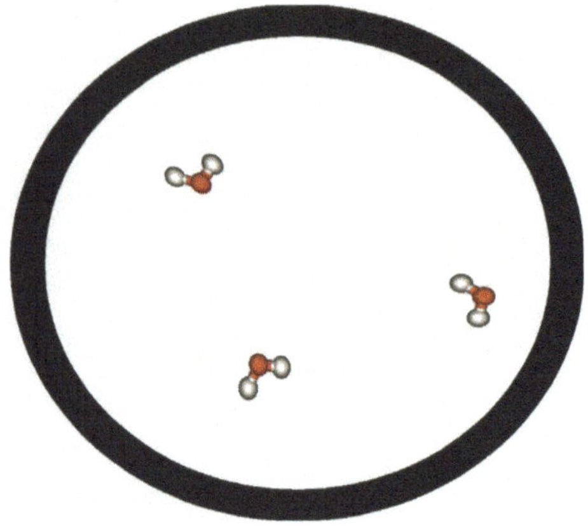

Figure A2-16 **Hypothetical View of Three Water Molecules in a Pore of 1.34 nm**

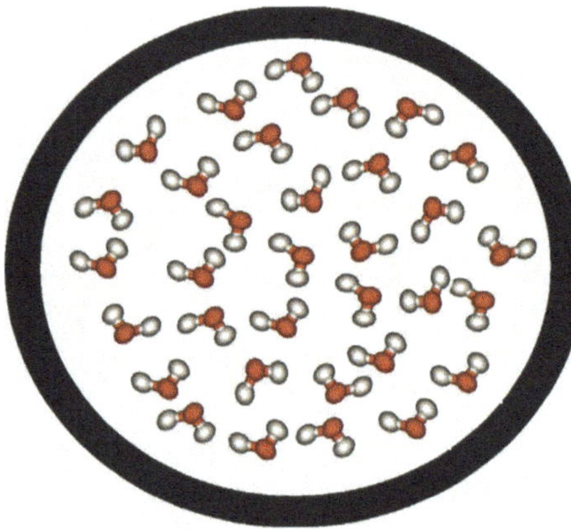

Figure A2-17 **Hypothetical View of Many Water Molecules in a Pore of 1.34 nm**

[24] A recent study reports that the preferred pore diameter for adsorption of trichloroethylene is about 0.8 to 1.1 nm. The study is reported in: Li, L., Quinlivan, P., and Knapp, D.D.R.U., "Effects of Activated Carbon Surface Chemistry and Pore Structure on the Adsorption of Organic Contaminants from Aqueous Solution," *Carbon*. 2002;40:2085–2100.

[25] The personal communication of Footnote 30 included the "rule of thumb" that solvents with a molecular weight less than 50 don't adsorb well. Viewing of Figure A2-16 suggests why this would be so.

[26] One hundred percent relative humidity at 25°C (dry bulb temperature) is equivalent to 18,700 ppm water in dry air.

[27] Use of the D/R technology to predict adsorption of water onto activated carbon is not as straightforward as with other solvents. This is because of: (1) the unusually small kinetic diameter of water relative to pore diameter (Figure A2-16), and (2) the unusually high level of polar bonding between water and activated surfaces. The issue is the value of the parameter β which should be used. An appropriate reference is that listed in Footnote 15. Author Wood notes that values of β calculated from experimental data range from 0.041 to 1.34 for various adsorbents, but recommends highest consideration with activated carbon derived from coal for independently determined values of 0.24, 0.25, and 0.29. The information plotted in Figure A2-18 is based on a value of β of 0.25.

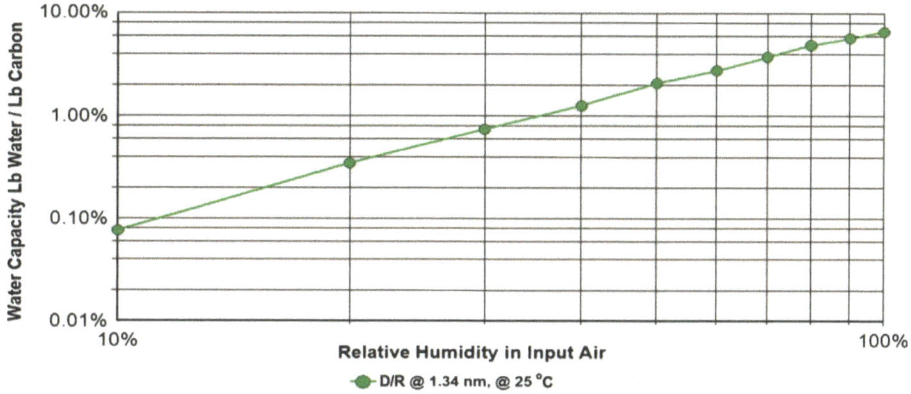

Figure A2-18 Adsorption of Water by Activated Carbon from Humid Air

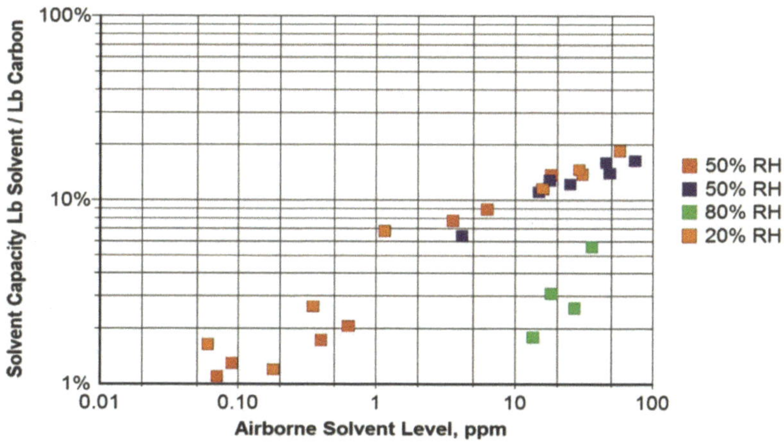

Figure A2-19 Measured Data of Adsorption of Water (Humidity) from Ambient Air

Performance data in Figure A2-19 show adsorbent capacity for toluene decline as a function of relative humidity (RH) in the inlet air stream. At 10 to 100 ppm toluene in the feed stream, the effect of RH is striking – adsorption at 80 % RH is substantially reduced from that at 20 (green symbols) and 50 % RH (orange and blue symbols respectively)[28–30].

Work sites conducting cleaning operations may not have steam available, and so ship their spent activated carbon back to its supplier[31] in a canister (Figure 4.44) along with money in return for fresh material (Chapter 4.16.6).

OXYGENATED SOLVENTS

It should not be inferred that effective adsorption of solvents on activated carbon is limited to halogenated solvents.

Other solvents described in this book are more, and less, effectively adsorbed.

This can be seen from the calculated[32] adsorption isotherms plotted in Figure A2-20 for several oxygenated solvents with different structures, and data of trichloroethylene.

[28] Noonan, D.C., James T. Curtis, J.T., *Groundwater Remediation and Petroleum: A Guide to Underground Storage Tanks*, CRC Press; 1st Edition, May 3, 1990, ISBN 087371217XS, page 41.
Personal communication from Kenneth Kikta, Sales Manager, Carbtrol Corporation, December 28, 2011.

[29] Predicted consequences for a binary mixture of solvents are the subject of the reference of Footnote 16, and are beyond the scope of this book.

[30] One must treat spent un-activated carbon adsorbent in the same manner that it was originally produced in order to reuse it.

[31] Sites that rely on thermal regeneration of activated carbon have to be of a certain size before it is economically viable to have regeneration facilities onsite.

[32] The D/R method is used as with most other estimations in this book where empirical data are not available. The activated carbon used for all solvents is the same – derived from coal carbon and with an internal pore diameter of 1.34 nm. Values for trichloroethylene are empirical measurements.

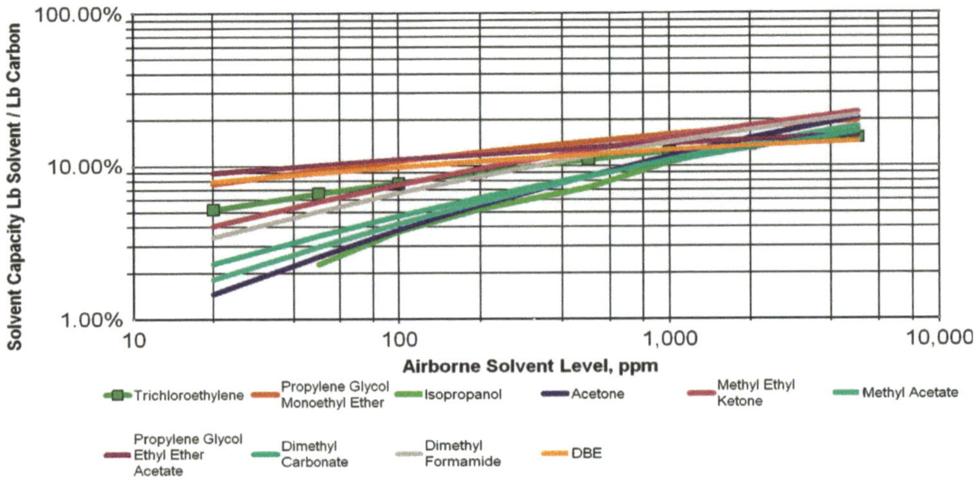

Figure A2-20 Calculated Adsorption Performance of Oxygenated Solvents

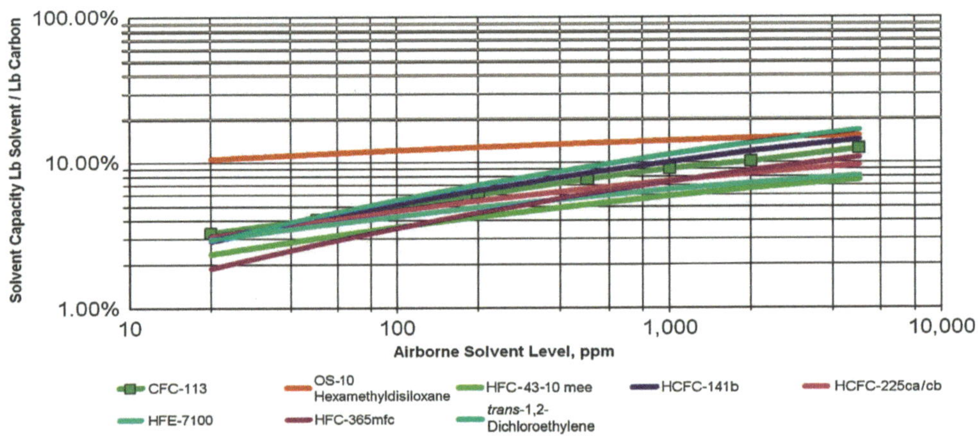

Figure A2-21 Calculated Adsorption Performance of Various Halogenated Solvents

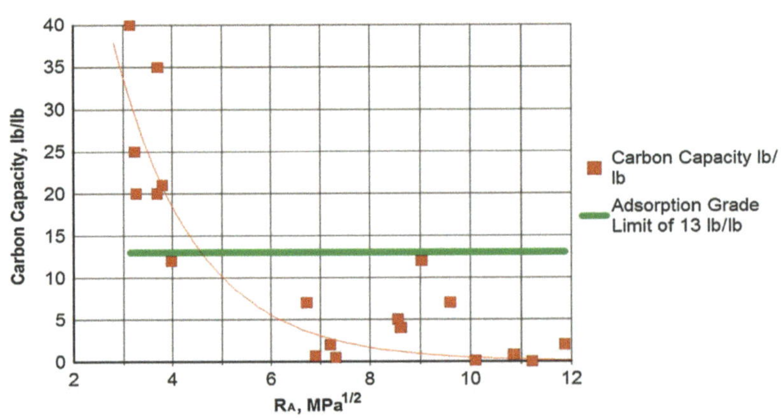

Figure A2-22 Definition of Selection of Limiting Adsorption Grade Value of 13 lb/lb

Table A2-3 Evaluation of HSP Values of an Activated Carbon (from Coal)

Solvent	Mol. Wt.	Liquid Density g/cc	Molar Volume cc/mol	Carbon Capacity lb/lb	Adsorbent Grade	δD	δP	δH	R_A	GOOD IN	GOOD OUT	BAD IN	BAD OUT	$R_O < R_A$	$R_O > R_A$	DF
					Grade Limit	ACTIVATED CARBON			R_O							
					13.00	19.14	3.65	2.60	3.84							
						HSP, $MPa^{1/2}$										
1,1,2,2-Tetrachloroethane	167.9	1.586	105.8	40	1	18.80	5.10	5.30	3.14	1	0	0	0	2.023	0.494	1.000
Tetrachloroethylene	165.8	1.613	102.8	35	1	18.30	5.70	0.00	3.71	1	0	0	0	1.138	0.879	1.000
Styrene	104.2	0.900	115.7	25	1	18.60	1.00	4.10	3.23	1	0	0	0	1.843	0.543	1.000
o-Xylene	106.2	0.877	121.1	21	1	17.80	1.00	3.10	3.80	1	0	0	0	1.043	0.959	1.000
Naphthalene	128.2	0.978	131.0	20	1	19.20	2.00	5.90	3.69	1	0	0	0	1.162	0.860	1.000
Toluene	92.1	0.864	106.6	20	1	18.00	1.40	2.00	3.26	1	0	0	0	1.793	0.558	1.000
Benzene	78.1	0.873	89.5	12	0	18.40	0.00	2.00	3.98	0	0	0	1	0.867	1.154	1.000
Methyl-t-Butyl Ether	88.2	0.736	119.8	12	0	14.80	4.30	5.00	9.02	0	0	0	1	0.006	178.237	1.000
1,1-Dichloroethane	99.0	1.168	84.7	7	0	16.50	7.80	3.01	6.72	0	0	0	1	0.056	17.865	1.000
Hexane	86.2	0.656	131.4	7	0	14.90	0.00	0.00	9.59	0	0	0	1	0.003	313.312	1.000
Ethyl Acrylate	100.1	0.917	109.2	5	0	15.50	7.10	5.50	8.56	0	0	0	1	0.009	111.743	1.000
Methyl Ethyl Ketone (MEK)	72.1	0.799	90.2	4	0	16.00	9.00	5.10	8.62	0	0	0	1	0.008	118.491	1.000
Methylene Dichloride	84.9	1.319	64.4	2	0	17.00	7.30	7.10	7.20	0	0	0	1	0.035	28.736	1.000
Acrylonitrile	53.1	0.802	66.2	2	0	16.00	12.80	6.80	11.86	0	0	0	1	0.000	3044.514	1.000
Acetone	58.1	0.787	73.8	0.8	0	15.50	10.40	7.00	10.85	0	0	0	1	0.001	1111.722	1.000
Vinyl Chloride	62.5	0.966	64.7	0.7	0	16.00	6.50	2.40	6.89	0	0	0	1	0.047	21.195	1.000
Ethyl Chloride	64.5	0.911	70.8	0.5	0	15.70	6.10	2.90	7.30	0	0	0	1	0.031	31.931	1.000
Bromotrifluoromethane (Freon 1381)	148.9	1.993	74.7	0.13	0	14.30	2.40	0.00	10.10	0	0	0	1	0.002	521.109	1.000
Methane (Liquid B.P.)	16.0	0.422	38.0	0.0003	0	14.00	0.00	0.00	11.21	0	0	0	1	0.001	1587.518	1.000

"DESIGNER" SOLVENTS

It is the same with the family of so-called "designer solvents" (Ref. 3 Chapters 20, 22, 24, and 25).

Calculated results, using the same basis, are plotted in Figure A2-21[33]. It is clear that the silicon-based solvent, OS-10, should be much better adsorbed than halogenated solvents.

HANSEN SOLUBILITY PARAMETERS

Adsorbents are chemicals. They are made of carbon which has been oxidized and thermally treated. As such, like other solid, semi-solid, and liquid chemicals, they display values of Hansen Solubility Parameters (HSPs). After all, it is the chiefly polar[34] intermolecular forces they radiate which enable reversible bonding of solvents to their surfaces – and one aim of HSPs is to quantify the effect of those forces.

One can characterize adsorbents as one does characterize cleaning solvents by their HSP values. The methodology of optimization is that of Ref. 2, Appendix C2 (PES polymer) and Ref. 2, Appendix C3 (bitumen) where a non-linear optimization routine is used to determine the HSP values for the adsorbent[35] which best place it in proper relation to a collection of 20 solvents.

The optimization shown in Table A2-3 is based on the standard that a solvent must be retained on the adsorbent to the level of at least 13 weight percent to be considered to have "GOOD" affinity for the adsorbent.

The effect of this standard on R_A is shown in Figure A2-22; one would expect at least acceptable adsorption performance if R_A is less than about 5 $MPa^{1/2}$.

The virtue of calculation of HSP for adsorbents is that comparison of them to the HSP values of solvents avoids the need for scouting tests to recover new solvents.

[33] Note that adsorption isotherms are not available at the reference of Footnote 23 for "designer" solvents.

[34] The nature of the attraction between solvents and adsorption sites depends upon the nature of the solvents being adsorbed. Solvents with polar centers (chlorine atoms in trichloroethylene, for example) are attracted by mutual alignment of electron structures. Solvents with oxygen atoms (alcohols or ethers, for example) are attracted through hydrogen bonding.

[35] The activated adsorbent upon which the solvents listed in Table A2-3 were adsorbed was "Virgin Vapor Phase Carbon" with a CCl_4 rating of 65 (quite high). All tests were conducted at 25°C with solvent vapor at 100 ppm concentration and relative humidity less than 50 percent. The data can be found at http://www.carbtrol.com/voc.pdf. The adsorbent was prepared from either coal or coconut shell substrate.

Index

Note: Page numbers with "f" denote figures; "t" tables; and "b" boxes.

Kolibri 360 GmbH
Ferdinand-Jühlke-Straße 7
99095 Erfurt, Deutschland
produktsicherheit@kolibri360.de